Biotechnology of Neglected and Underutilized Crops

Shri Mohan Jain · S. Dutta Gupta
Editors

Biotechnology of Neglected and Underutilized Crops

Springer

Editors
Shri Mohan Jain
Department of Agricultural Sciences
University of Helsinki
Helsinki
Finland

S. Dutta Gupta
Department of Agricultural
and Food Engineering
Indian Institute of Technology Kharagpur
Kharagpur
India

ISBN 978-94-017-8492-4 ISBN 978-94-007-5500-0 (eBook)
DOI 10.1007/978-94-007-5500-0
Springer Dordrecht Heidelberg New York London

Softcover re-print of the Hardcover 1st edition 2013

Printed on acid-free paper

Springer is part of Springer Science+Business Media (www.springer.com)

Preface

The availability of food supply worldwide is dependent on few crop species or 'major crops'. Till today, only 30 plant species are used to meet 95 % of the world's food requirements. They are widely and intensively cultivated crops; and developed by extensive selection from available large agro-biodiversity pool. There is a great need to expand the exploitation of the plant genetic diversity that would broaden the crop diversity for food supply in order to feed the ever growing human population and avoid dependence on few food crops, especially under the climatic change. The neglected crops could become an excellent source for useful gene source, e.g., nutrition, biofuel, abiotic and biotic stresses, medicinal compounds, and so on. Molecular marker analysis work would be of great help for analyzing genetic diversity, and exploit genetic resources for identification, isolation, conservation, and utilization. Moreover, the funding agencies have neglected underutilized crops and focused on major crops for feeding the population and earned economic gains. Several factors as physical appearance, taste, nutritional properties, cultivation methods, processing qualities, economic gains, and others are responsible for the promotion and acceptance of 'major crops' worldwide. The success of breeding programs of these crops is very much dependent on the availability of suitable gene pool and modern biotechnological tools. So far, all efforts have been made to conserve a narrow gene pool of 'major crops' and only a limited amount of underutilized and neglected crops. This situation raises questions how to maintain food security, poverty alleviation, economic growth, future supply of food, and risk of rural incomes and ecosystem conservation?

'What is a Neglected and Underutilised Crop?' This question is addressed at length in the book "Breeding of Neglected and Under-Utilized Crops, spices, and herbs" (Ochatt and Jain 2009). These crops are categorized as 'minor crops' and have lesser importance globally in terms of production and market value. Some crop species may be widely distributed worldwide but tend to have preference in the local ecology and local production and consumption system. They are traditionally grown in their centers of origin or in centers of diversity by farmers, and are still important for the subsistence of local communities and, more often than not, they also play a social role. They are locally well adapted to marginal lands and constitute an important part of the local diet, providing valuable nutritional elements (e.g. protein, vitamins, and minerals) and spices, often lacking in staple crops. Their role in traditional medicine is also well known. In addition, these crops are important sources of resistance genes for biotic and abiotic stress breeding that can be utilized also for the genetic improvement of commodity crops. As compared to the major crops, they require relatively low inputs and, therefore, contribute to sustainable agricultural production. These crops continue to be maintained by socio-cultural preferences and traditional uses. They remain inadequately characterized and underutilized for their potential to contribute to the income of rural poor and to the global food security and, until very recently, have been largely ignored by research and development, and conservation. This is due to the lack of government funding and interest of the private sector. These traditional crops are often low yielding and cannot compete with major crops, even though many of them have the potential to become economically viable. The lack of genetic improvement and often narrow genetic diversity for important agronomic traits hamper the development of these crops. Other constraints are a lack of knowledge on the taxonomy, reproductive biology, and genetics of agronomic and quality traits. International organizations have recognized the importance of underutilized and neglected crops toward food security and improving quality of rural poor lives by enhancing income.

This important reference is the first comprehensive resource worldwide that reflects research achievements in neglected and underutilized crop biotechnology, documenting research events during the past three decades, current status, and future outlook. This book has 16 chapters and divided into 4 parts. Part I has three chapters dealt with *Chenopodium* as a potential food source, thin cell layer technology in micropropagation of *Jatropha*, and *Panax vietnamensis*; Part II deals with three chapters on molecular biology and physiology of *Haberlea rhodopensis*, cell trait prediction in vitro and in vivo of legumes, and application of TILLING; Part III has five chapters on biotechnology of neglected oil crops, Quinoa, *Erucia sativa*, *Stylosanthes*, and *Miscanthus*; Part IV contains five chapters mainly on genetic transformation of Safflower, Jatropha, Bael, and Taro. This section also includes a chapter on genetic engineering of Mangroves.

This book is essential for researchers, policy makers, and commercial entrepreneurs concerned with neglected and underutilized crops. The book is invaluable for students and specialists. This book is written by an international team of experienced researchers from both academia and industry. We wish to express our deep sense of gratitude to all the contributors for agreeing to participate. We wish to thank Dr. A. K. Kukreja, CIMAP; Dr. M. K. Maity, IIT Kharagpur; and Dr. T. Gechev, University of Plovdiv for their helpful comments. All peer reviewed manuscripts were revised according to the reviewer suggestions. Finally, we appreciate Springer for giving us the opportunity to bring out this volume.

Finland S. Mohan Jain
India S. Dutta Gupta

Contents

Part III Section 3

Part IV Section 4

Part I
Section 1

Chapter 1
The Genus *Chenopodium*: A Potential Food Source

Eulogio de la Cruz Torres, Guadalupe Palomino Hasbach, Juan Manuel García Andrade, Cristina Mapes Sánchez, Josefina González Jiménez, Thelma Falcón Bárcenas and Octavio Vázquez Arriaga

Abstract The genus *Chenopodium* comprises important cultivated species such as quinua (*Chenopodium quinoa* Willd.), Chía roja, Huauzontle (*Chenopodium berlandieri* subsp. *nuttalliae*) and Cañahua (*Chenopodium pallidicaule*). These species had a relevant role in the development of pre-hispanic cultures from Meso and South America as source of food and in the religious context. Due to its high nutritive value (up to 19 % proteins) and to its tolerance to adverse factors such as drought, saline soils and frost, these species are considered as alternative crops for areas with extreme conditions, where also malnutrition prevails. The cytogenetic characterization by karyotyping and determination of DNA content by flow cytometry of seven cultivars of *Chenopodium* are reported. *Chenopodium quinoa* cultivar Barandales and *C. berlandieri* subsp. *nuttalliae* cultigens Huauzontle, Quelite and Chia roja showed $2n = 4x = 36$, $x = 9$. Statistically insignificant genome size differences for studied

E. de la Cruz Torres (✉) · J. M. García Andrade · J. González Jiménez · T. Falcón Bárcenas · O. Vázquez Arriaga
Departamento de Biología, Instituto Nacional de Investigaciones Nucleares, Carr. México-Toluca, S/N, La Marquesa C.P., 52750 Ocoyoacac, Estado de México, Mexico
e-mail: eulogio.delacruz@inin.gob.mx

J. García Andrade
e-mail: jmanuel.garcia@inin.gob.mx

J. González Jiménez
e-mail: josefina.gonzalez@inin.gob.mx

T. Falcón Bárcenas
e-mail: thelma.falcon@inin.gob.mx

O. Vázquez Arriaga
e-mail: octavio3502@yahoo.com.mx

G. Palomino Hasbach · C. Mapes Sánchez
Instituto de Biología-Jardín Botánico, Universidad Nacional Autónoma de México, 04510 México, D. F, Mexico
e-mail: palomino@ibiologia.unam.mx

C. Mapes Sánchez
e-mail: cmapes@ibunam2.ibiologia.unam.mx

S. M. Jain and S. Dutta Gupta (eds.), *Biotechnology of Neglected and Underutilized Crops*, DOI: 10.1007/978-94-007-5500-0_1,

varieties ranged from 2.96 pg/2C (1 Cx = 724 Mbp) in *C. quinoa* to 3.04 pg/2C (1 Cx = 743 Mbp) in Huauzontle. Also, in this chapter the molecular characterization by SSR of 38 accessions of cultigen Huauzontle, the study of the traditional growing system of cultigen Chía roja, the nutritional value of both cultigens and the floral development of quinua and Chía roja are described. Results of a mutation breeding program leading to a reduction in saponin content are also presented.

Abbreviations

NTS	Non-transcribed spacers
FISH	Fluorescent in situ hybridization
CMA	Chromomicine
ESTs	Expressed sequence tags
RAPD	Random amplified polymorphic DNA
SSR	Simple sequence repeats
SNP	Single nucleotide polymorphism
AFLP	Amplified fragment length polymorphism
NOR	Nucleolus organizing region

1.1 Introduction

Nowadays mankind is facing four main problems: (a) Population growth that will demand an increased supply of highly nutritive foods. (b) Climatic change. Models on climatic change foresee that in future drought conditions will be strengthened because an average increase in temperature of 3.5 °C is predicted for 2080. This increase will affect growing areas devoted to traditional crops. (c) Dependence on a reduced number of crops to supply food needs. It is estimated that there are between 300,000 and 500,000 plant species, from which 6,000 and 7,000 have been cultivated in some time and that other 25,000 have been used as medicine. As time has elapsed, most important crops have substituted many traditional cultivars and land races of less important crops and nowadays 120 plant species are cultivated at great scale, from which 30 are basic for human nutrition providing 95 % of the requirements of proteins and calories. From these species only 12 provide 75 % of the world food supplies and three of the most important crops rice, wheat and corn supply 50 % of the world food requirements (FAO 1996; Heywood 1999; Thies 2000). This reduction in biodiversity has contributed to worsen malnutrition problems making mankind more vulnerable to imminent problems such as climatic change. To solve this problems the searching for new crops among underutilized species is a feasible alternative. (d) Scarcity of fuels. Reduction of fossil fuels sources has prompted the development of bio-fuels that has exacerbated attempts to increase food production and food sovereignty. Underutilized species such as those from *Chenopodium* genus are alternatives to improve nutrition and food security.

1.2 Importance

Chenopodium genus is relevant because among them can be found cultivated species of high nutritive value and hardiness such as quinua (*Chenopodium quinoa* Willd.) and Huauzontle (*Chenopodium berlandieri* subsp. *nuttalliae*), being alternative crops for regions with marginal agricultural conditions (Heiser 1985).

Importance of species pertaining to this genus on food and agriculture of pre-Hispanic civilizations has been widely documented. Mendoza codex registers tributes for nearly 7,000 tons of seeds of *huautli,* Nahuatl name used by ancient Aztecs to refer both to seeds from *Amaranthus* and *Chenopodium* genus. This quantity was collected among towns ruled by the Aztec empire. West (1946) documents the growing of Chía roja (*Chenopodium berlanideri* subsp. *nuttalliae)* and amaranth (*Amaranthus hypochondriacus*) (alegría) in towns like Cherán and Pichátaro, in Michoacán México, mentioning elaboration of products such as *chapatas*, a sort of *tamales* (a lump of corn dough, prepared by mixing cornmeal with equal parts of Chía roja or Chía negra seeds and sugar, all of which is wrapped in corn husks '*totomoxtle*' and steam cooked) (López 2006), and the balls of popped amaranth seeds with honey known as *alegrías*. Mapes (1987) points out the role of Chía roja as complement of farmers diet, and also consumption of Chía negra and Chía blanca, pertaining these to *Amaranthus hypochondriacus*. Huauzontle contribution as source of protein has been exhibited by several studies (Carrillo and Engleman 1994; Martínez 2005) pointing out the high percentage of proteins near to 17 %.

Concerning to the improvement of species pertaining to *Chenopodium* genus, several problems are faced, being one of the most important the absence of uniform varieties or genotypes to start with, exception made of *Chenopodium quinoa* and *Chenopodium pallidicaule,* species that have been subjected to a process of characterization of germplasm and further breeding programs. Regarding to Huauzontle and Chía Roja, the cultivated materials correspond to land races that have not been subjected to a breeding program. The very small size of the flowers also makes difficult to proceed by methods such as hybridization, being selection and mutation breeding the desirable methods for improvement.

1.3 Genetic Resources of *Chenopodium* Genus and its Use in Meso and South America

In recent years, interest has grown in new crops of high nutritional value such as quinua (*Chenopodium quinoa*), staple crop of ancient South American cultures such as the Inca and *Chenopodium berlandieri* subesp. *nuttalliae* (Huauzontle and Chía roja) cultivars grown in Middle America since prehispanic times. These underutilized crops pertaining to *Chenopodium* genus have traits making them potential to become alternative crops for marginated zones (Müjica et al. 1999).

Chenopodium quinoa besides its high protein value exhibits tolerance to frost, drought and has adaptability to different latitudes and altitudes, but presence of saponins, products that confer bitter flavor, foam formation and that in high concentrations induce hemolysis, has been an inconvenient in the expansion of this crop (Peñafiel and Díaz 1988).

Chenopodium quinoa is important because the nutritive value of its seeds is comparable to the milk due to its high protein content and adequate essential amino acids balance, mainly lysine, methionine and tryptophan, besides a good quantity of vitamins and minerals (Miranda and Mújica 1985). *Chenopodium quinoa* is used for human consumption to prepare flour, soups, flakes, alcohol (Oelke et al. 2001) and the whole plant is used as forage in Bolivia, Peru, Ecuador and Colombia. There is no clear confirmation concerning to domestication of *C. quinoa,* however is supposed that this process occurred in different times and locations: in Peru (5,000 B.C), Chile (3,000 B.C) and Bolivia (750 B.C.) (Ruas et al. 1999). It has been documented that besides its value as food, quinua was used in religious and social life of prehispanic cultures (Wilson and Heiser 1979).

Some species pertaining to *Chenopodium* genus are capable to tolerate adverse conditions thus can thrive from extremely cold regions to desert and saline regions, to intermediate levels of temperature, humidity and type of soil due to its tolerance (Basset and Crompton 1982; Tanaka and Tanaka 1980). Quinua tolerate drought conditions provided that with only 200–300 mm of rainfall has acceptable yields of 1 ton/he in México (López 1988). In Mexico has tolerated −8 °C in vegetative growth (Mújica 1983). Quinua tolerate high salinity soils and pH up to 9.5. In laboratory trials, seeds have germinated in salt concentrations up to 3 % (Mújica 1983). Quinua can be adapted to several environments from sea level to 4,000 masl and due to its great plasticity can be found from Chile to Canada, even in countries of Africa and Europe (Gómez 1989).

Concerning to characterization, initially quinua classifications took in account color of plant and fruit and in some cases form and flavor of the seed. Humboldt et al. (1815) described two varieties: α and β that later Moquin-Tandon (1849) designated as *viridescens* and *rubescens* taking in account color (Hunziker 1952). In 1917 a Bolivian chemist González, made classification of quinua from the Bolivian high plateau, indicating presence of four species: C. *album, C. pallidus, C. ruber* and *C. niger* characterized by having sweet, sour, red and black seeds (Gandarillas 1982).

In 1943 Hunziker adds the variety *lutescens* and in 1944 Cárdenas (cited by Hunziker 1952) suggested to take into account panicle traits to separate three varieties (Amaranthiform, glomerulate and intermediate) (Hunziker 1952). Gandarillas (1982) suggests classifying in races, being intermediate between species and variety, considering traits such as height, color, panicle inflorescence size, color and flavor of the fruit. Taking into account these traits these races were established: Dulce, Achacahi, Puno, Copacaban, Real, Challapata, Potosí and Cochabamba among others (Gómez 1989).

Aiming to know variability of quinua germplasm, Rojas et al. (2000) evaluated 1,512 lines using 14 quantitative variables and 2 qualitative, using three multivariate methods, allowing to group accessions in seven clusters considering traits such as panicle form, length and diameter, number of branches, stem diameter and length. Nieto et al. (1999) evaluated 30 quinua lines dividing them in two groups, quinoas form the valley and quinoas from heights in 11 locations, utilizing only two traits, low and high saponin content, selecting during a 8 years period 9 lines (4 from the valley and 5 from heights) for a stability study.

1.4 Cytogenetics of *Chenopodium* Genus

Chromosome counting and karyotype: The basic number that predominates among chenopods is $x = 9$, being aneuploidy rare and polyploidy common. Diploid species predominate being almost 61 %.

Cytological studies have established that *Chenopodium quinoa* and C. *berlandieri* have a $2n = 4x = 36$ (Giusti 1970; Palomino et al. 1990; Bhargava et al. 2006). Tetraploid origin of these species has been corroborated by allele segregation studies of a gene and the morphologic results indicate a disomic heredity and in a lower proportion tetrasomic heredity according to its tetraploid origin. Walters (1987), through isozyme electrophoretic analysis concludes that C. *berlandieri* subsp. *nuttalliae* is an alotetraploid generated by hybridization of two diploid unidentified species from North America. *C quinoa* has exhibited variation in the number and position of chromosomes with satellite. Gandarillas and Luizaga (1967) did not find chromosomes with satellites in 12 varieties *of C. quinoa* from Bolivia and Peru, nor in the viridescens and melanospermum varieties neither on *C quinoa* subsp. *milleanum.* Bhargava et al. (2006) report a pair of chromosomes with satellites located on different pairs in a collection of *C. quinoa.* Palomino et al. (1990) observed two pairs of chromosomes with satellites on *C. berlandieri* subsp. *nuttalliae* cv. Huauzontle. Palomino et al. (2008) have reported $2n = 4x = 36$ with $x = 9$ in *C quinoa* cv. Barandales and 6 collections of *C. berlandieri* subsp. *nuttalliae*, cultivars Huauzontle, Quelite and Chía roja (Table 1.1).

Karyotype analysis revealed presence of four groups of nine metacentric chromosomes including two pairs of chromosomes with satellites in C. *berlandieri* subsp. *nuttalliae* cultivars Hauzontle, Quelite and Chía roja. First pair of satellites was located on the largest pair of chromosomes and the second in a different pair of chromosomes on the analyzed accessions. Variation among cultivars was evident on chromosome size, length of the genome (GL) and satellite location. Chía roja showed highest value for GL = 58.82 and largest chromosomes (2.04 μm). Huauzontle showed lowest value for genome length (GL = 45.02 μm) and shortest chromosomes (1.60 μm). Comparison of GL was statistically significant, allowing to distinguish three groups according to the use of these plants. First and second group was integrated by cultigens coming from locations where the consumed part of the plant is the seed and the third group was represented by cultigens used as vegetables (quelites) like Huauzontle (Table 1.1, Palomino et al. 2008).

Studies concerning DNA content refer to quantification of genome size in picograms (pg), its composition in mega base pairs of nulceotides (Mbp, a pg = 978 Mbp; Dolezel et al. 2003, 2000) and detection of polyploidy. These analyses are

Table 1.1 Provenance and karyotype analysis of *C. quinoa* and six collections of *C. berlandieri* subsp. *nuttalliae*

Taxa	Accession number	Site of collection	Karyotype formula (m)	Secondary constrictions (m)	Range of chromosome length (μm)	Genome length (μm) $\bar{x} \pm SE^2$	Genome length Tukey's grouping[1]	Index of asymmetry TF (%) $\bar{x} \pm SE^2$	Index of asymmetry Tukey's grouping
C. quinoa cv. Barandales	M-50	Toluca, México State, México	36	4	1.17–1.96	53.88 0.81	b	43.80 0.75	e
C. berlandieri ssp. *nuttalliae* cv. Huauzontle	8-01	Xonacatlan, México State, México	36	4	1.09–1.69	47.64 0.77	c	45.45 0.75	d
C. berlandieri ssp. *nuttalliae* cv. Huauzontle	7-01	Atlacomulco, México State, México	36	4	1.03–1.69	45.60 0.75	c	46.27 0.76	b
C. berlandieri ssp. *nuttalliae* cv. Huauzontle	2-01	Lerma, México State, México	36	4	0.94–1.60	45.02 0.73	c	46.82 0.75	a
C. berlandieri ssp. *nuttalliae* cv. Quelite	10-01	Coatepec, México State, México	36	4	1.11–1.59	47.81 0.77	c	45.91 0.75	c
C. berlandieri ssp. *nuttalliae* cv. Chia roja	1-99	Opopeo, Michoacan State, México	36	4	1.26–2.04	58.81 0.85	a	45.95 0.75	c
C. berlandieri ssp. *nuttalliae* cv. Chia roja	2-99	Opopeo, Michoacan State, México	36	4	1.06–1.69	45.32 0.75	c	43.58 0.74	e

[1] Same letters in columns indicate no statistical difference using Tukey test ($P < 0.05$), with an ANOVA $P < 0.0001$

[2] SE = Standard error

useful in taxonomy, biotechnology, and are important for planning programs of breeding and conservation (Dolezel et al. 2007). Concerning *Chenopodium* genus information related to DNA content is scarce.

Palomino et al. (2008) reported that DNA content of one accession of *C. quinoa* and 6 accessions of *C. berlandieri* subsp. *nuttalliae* cultigens ranged from 2.96 pg (1Cx = 724 Mbp) in Huauzontle to 3.04 pg (1 Cx = 744 Mbp) in *C. quinoa* cv. Barandales, exhibiting a difference of 2.7 % (Table 1.2, Fig. 1.1).

Average value of the seven studied cultigens was 2C DNA = 2.98 pg. Difference of 2.7 % was not statistically significant suggesting that any actual minor variation had originated during species differentiation. Changes in DNA content can be due to deletions, duplications or as response to environmental stress (Price 1976). Differences in DNA content often have their origin as contractions or expansions of repeated sequences arrays of DNA such as mobile genetic elements or transposons (Petrov 2001; Kidwell 2002). Bennett and Smith (1991) have reported similar values of DNA content for *C. quinoa* (2C = 2.66 pg). Obtained values indicate that chromosomes and genomes of these plants are very small and with stability in terms of DNA content, and that analysis of chromosomes support the tetraploid origin of *C. quinoa* and *C. berlandieri* subsp. *nuttalliae.* According to similarities between these species in relation to chromosome number 2n = 4 x = 36, x = 9, morphology of chromosomes, hybridization between *C. quinoa* and *C. berlandieri* subsp. *nuttalliae* could be developed as a strategy to reduce the saponin content and to improve adaptability of quinua to the Mexican High Plateau.

Table 1.2 Nuclear DNA content and genome size of *Chenopodium quinoa* and *Chenopodium berlandieri* subsp. *nuttalliae,* all with $2n = 4x = 36$

Taxa	Accession number	2C ADN (pg)[a] $\bar{x}$	± SE	Genome size 1C*x* Mpb
C. quinoa cv. Barandales	M50	2.96	0.01	724
C. berlandieri subsp. *nuttalliae* cv. Huauzontle	8–01	2.98	0.01	729
C. berlandieri subsp. *nuttalliae* cv. Huauzontle	7–01	3.04	0.02	744
C. berlandieri subsp. *nuttalliae* cv. Huauzontle	2–01	2.97	0.02	726
C. berlandieri subsp. *nuttalliae* cv. Quelite	10–01	2.99	0.01	731
C. berlandieri subsp. *nuttalliae* cv. Chia roja	1–99	2.96	0.01	724
C. berlandieri subsp. *nuttalliae* cv. Chia roja	2–99	2.97	0.02	726

[a] 1 picogram (pg) = 978 megabase pairs (Mbp) (Dolezel et al. 2003)
1C*x* = value represents DNA content of one monoploid genome with chromosome number *x* (Greilhuber et al. 2005)

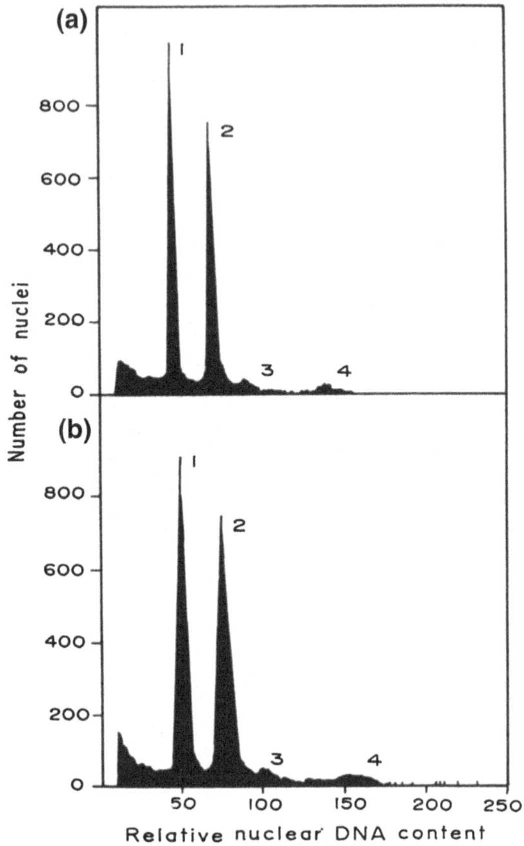

Fig. 1.1 Histograms of fluorescence intensity using flow cytometric analysis of nuclei isolated of Chenopodium taxa and *Lycopersicon esculentum* used as an internal standard. *Peaks* 1 and 3 represent G_1 and G_2 nuclei of **a** *Chenopodium quinoa* cv. Barandales and **b** *Chenopodium berlandieri* subsp. *nuttalliae* cv. Chía roja 1-99

1.5 Fluorescent in Situ Hybridization

Fluorescent in situ hybridization is based on the detection of determined sequences of DNA marked with fluorochromes, trough in situ hybridization (Schwarzacher and Heslop-Harrisson 2000). Pattern of distribution of sequences of ribosomal DNA (rDNA) in the chromosomes is performed with FISH using probes such as 18, 5.8, 28 and 5S. Characterization of these loci is performed through fluorescence; locus 5S stains in red, 45S stains green (Schwarzacher and Heslop-Harrisson 2000). FISH technique is used to examine changes and rearrangements at chromosomal level, especially those small and homogeneous, such as those observed in *Chenopodium* genus. This analysis can be used to study phylogeny and breeding, because through FISH are examined and can be observed the number of copies of loci 5S and 45S that contains the genome; analyzing homology of genomes form related species.

Sequences 45S and 5S are sequences repeated in tandems of genes of rRNA and form big blocks of DNAr positively identified by chromomicine (CMA). Genes 5S of rRNA also are organized in tandem, but do not form secondary constrictions in chromosomes in metaphase and are visible as heterochromatin blocks (Martins and Wasko 2004).

Kolano et al. (2001) applying FISH with probe for rDNA 45S and silver staining observed two hybridization sites (FISH) in a pair of chromosomes on the genome of *C. quinoa*. Maughan et al. (2006) have reported that *C. quinoa* has an interstitial hybridization site and another terminal in a pair of chromosomes with loci 5S rRNA and only a pair of NOR, suggesting a reduction on the loci number according to its alotetraploid origin. These authors observe that *C. berlandieri* subsp. *nuttalliae* have 3 loci 5S and one locus 45S of rRNA in variety Quelite and two loci 45S in the variety Huauzontle. This indicates that formation of *Chenopodium* section *Chenopodium* subsect. *Cellulata* comes from a monophyletic event, and one of the diploid ancestors could have at least two copies of locus 5S.

C. fremontii and *C. neomexicanum* are strong candidates to be considered as ancestors in *C. quinoa*, Quelite, Huauzontle and Chía roja, provided that exhibit a locus 5S as a donor ancestor and crosses of *C. neomexicanum* with *C. fremontii* produce viable hybrids, notwhistanding that *C. fremontii* does not pertain to subsection *Cellulata* (Sederberg 2008). These results also indicate duplication or deletion of loci 5S or 45S of rRNA, that has occurred during evolution of *Chenopodium* allotetraploid species from the new world. Reduction in the number of loci in polyploids has been observed in several species (Leitch et al. 1998).

1.6 Molecular Studies on *Chenopodium*

Molecular characterization is a valuable tool to perform studies concerning the assessment of genetic diversity, and the establishment of phylogenetic relationships and the improvement assisted with molecular markers. The first molecular studies on *Chenopodium* aimed to elucidate the genetic relationships between several species from this genus. Early studies used allozyme data to construct a phylogenetic tree of *Chenopodium* species which allowed differentiating among coastal and altiplano ecoptypes (Wilson 1988a). Wilson (1988b) and Fairbanks et al. (1990) reported studies using seed protein variation and morphological markers.

In relation to genetic characterization of *Chenopodium*, mainly on *C quinoa* several molecular markers have been utilized. Fairbanks et al. (1993) applied RAPDs to detect polymorphisms among different accessions of this species. RAPDs were also used by Bonifacio (1995), who reported these molecular markers as a tool to identify true hybrids from intergeneric crosses. Ruas et al. (1999) applied also RAPDs to assess the relationship among 19 species of *Chenopodium* finding that accessions formed groups according to its taxonomic classification. Recent studies have used markers different than RAPDs. Mason et al. (2005) developed microsatellite markers to be used in *C. quinoa*. They found that from 397 potential microsatellites, 208 were polymorphic in a group of 31 quinua accessions. Furthermore these markers were tested on three different *Chenopodium* species including *C. berlandieri* subsp. *nuttalliae,* where 99.5 %

of the SSR markers amplified in Huauzontle, illustrating the close relationship between this specie and quinua.

In *C quinoa*, besides SSR, also single nucleotide polymorphism markers (SNP) have been developed. Coles et al. (2004) obtained 424 expressed sequence tags (ESTs), having 349 of them homology with protein-encoding genes from other plants. 81 additional SNPs were obtained when quinoa was compared to *C. berlandieri* subsp. *nuttalliae.*

The first quinua genetic linkage map was developed by Maughan et al. (2004) using the aforementioned SSRs, based on the study of 82 F2 individuals from a cross between Ku-2 (Chilean lowland type) and 0654 (Altiplano type) and consisted of 230 AFLP, 19 SSR and 6 RAPD markers. The map spanned 1020 CM containing 35 linkage groups.

Stevens et al. (2006) developed a BAC library which was estimated to represent approximately nine times coverage of the haploid quinua genome. Maughan et al. (2006), sequenced the nucleolus organizing region (NOR) intergenic spacers (IGS and 5S rDNA non-transcribed spacers (NTS) from five quinua and one *C. berlandieri* accession. IGS sequences revealed length differences due to insertion/deletion (indels), differing in number of repeat copies and other rearrangements.

The molecular characterization by SSR markers of accession of *Chenopodium berlanideri* subsp. *nuttalliae* collected from the Mexican high plateau, mainly from the Toluca Valley is presented here.

1.6.1 Molecular Studies on Chenopodium berlandieri subsp. nuttalliae cv. Huauzontle

Huauzontle (*Chenopodium berlandieri* subsp. *nuttalliae*) is a pseudocereal native to Mexico and together with amaranth (*Amaranthus hypochondriacus* and *A. cruentus*) was known as *huautli* by ancient Aztecs, constituting in prehispanic times the fourth staple crop, after maize, beans and chili (Hunziker 1952). From amaranth the seeds are consumed in several ways (*atole*, tamales and tortillas) and also the young leaves and stems are eaten as vegetables (*quelites*), however Huauzontle was consumed mainly as quelite being the cultigen Chía roja the only chenopod whose seed was used for consumption in Middle America.

Though known and eaten both in rural and urban locations, Huauzontle production and consumption is relevant in rural areas in the central part of Mexico (México State, Puebla, Tlaxcala, Hidalgo) being a tradition mainly in villages inhabited by indigenous people from cultures such as Otomies, Matlatzincas, Mazahuas, Nahuatl and Totonacs. For people inhabiting rural areas where traditional farming systems are practiced and in many cases poverty and malnutrition prevails, the cultivation of this vegetable is important because during the growing season (from June to October) the consumption of Huauzontle is an important part of the diet of rural families eating tender leaves or panicles two or three days a week.

Studies concerning the nutritive value of Huauzontle panicle show that Huauzontle as *quelite* is a valuable source of carbohydrates, proteins (up to 29 %), lipids (3 %) and minerals (13 %) (Guzmán 2007), and the nutritive value of the seed will be presented further. Up till now Huauzontle cultivation is a tradition, being the farmers who conserve and exchange this valuable germplasm, so is not surprising that, even though México is the place of origin and domestication of this specie, there are no commercial cultivars officially released.

As occurs with most of neglected crops, the preservation of germplasm and the traditional knowledge about its cultivation and use corresponds to old people in the villages, and mostly the housewife is in charge of the care of the crop. Up till now little is known about the genetic diversity of this crop, so this study aims to evaluate the genetic diversity of some accessions collected from places pertaining to the Otomi ethnic group that, notwithstanding that its villages have been almost absorbed by the urbanization of the city of Toluca, in Mexico State, they preserve their traditions being the growing of Huauzontle one of the most rooted. Few molecular studies have been conducted on Huauzontle, because most of the molecular studies concerning the *Chenopodium* genus are devoted to *C. quinoa,* another outstanding pseudocereal that has received more attention from researchers as an important crop from South America.

Thirty eight accessions from cultigens Huauzontle were collected from twelve localities of México State in locations that ranged from 19°06′33″ to 20°17′31″ N and 99°27′22″ to 99°47′02″ W. Surveyed areas exhibited similar climatic conditions: temperate sub humid with summer rains, with a medium temperature ranging from 13.4 to 13.8 °C, annual average rainfall from 800 to 1,200 mm and the type of soil being vertisol. To collect the samples, healthy young leaves from 38 accessions were selected, following a directed sampling, aiming to obtain a representative sample of the existing genetic diversity.

DNA was extracted according to Dellaporta et al. (1983) method. Purity and concentration of DNA was determined by spectrophotometric techniques measuring absorbance at 230 and 260 nm. Integrity of DNA was verified by horizontal electrophoresis on 1.5 % agarose diluted with TBE 0.5X.

Horizontal electrophoresis was performed in a BIO-RAD equipment at 100 V during 50 min (Power source Sigma PS 250-1). Gels were stained with an ethidium bromide solution and were visualized on a transiluminator (BIO-RAD Universal Hood II). Twenty one SSR primers developed by Mason et al. (2005) were selected to carry out the study.

DNA was amplified with 15 μl a final reaction volume containing 1 X PCR Buffer (1.5 mM), 0.1 mM cresol red and 2 % (w/v) sucrose, 3 mM Mg Cl_2, DNTP (1 mM), 2 μM of each primer (forward and reverse), 0.3 U Taq polymerase (Biogenica) and 60 ng of DNA.

DNA was amplified on a thermocycler Perkin Elmer Cetus, according to the following thermal cycling profiles: 94 °C for 60 s, followed by 19 cycles of 94 °C for 60 s, 64 °C for 30 s (decreasing 0.5 °C every cycle), 72 °C for 60 s; 30 cycles of 94 °C for 60 s, 55 °C for 60 s, 72 °C for 60 s, followed by a final extension at 72 °C for 10 m.

PCR products were separated on 3 % agarose gels (Gibco BRL) 100 V for 50 min, using a power supply Sigma PS 250. All gels were run in 0.5 × TBE and was visualized using ethidium bromide staining with UV transillumination (BIO-RAD Universal Hood). Size of fragments was calculated from a standard curve based on known fragments from the 100 bp Biogenica ladder. A two entry matrix was established eliminating those diffuse fragments, and alleles were codified as present (1) and absent (0). A hierarchical conglomerate analysis utilizing Jacqard similitude index and UPGMA was used to generate the dendogram that is shown in Fig. 1.2.

Dendogram obtained (Fig. 1.2) allowed to classify accessions in five clusters. First cluster was integrated by 6 accesions (h_1 to h_6) collected in the localities of Lerma, Xonacatlán and Atlacomulco. Second cluster included six accessions (h_7 to h_{12}) collected in Villa Cuahutemoc, Xonacatlán and Toluca. Third cluster included 11 accessions originating from Toluca (h_{14} to h_{24}). Fourth cluster included six accessions, five from Toluca and one from Temascaltepec (h_{25}, h_{26}, h_{27}, h_{28}, h_{29}, h_{38}).

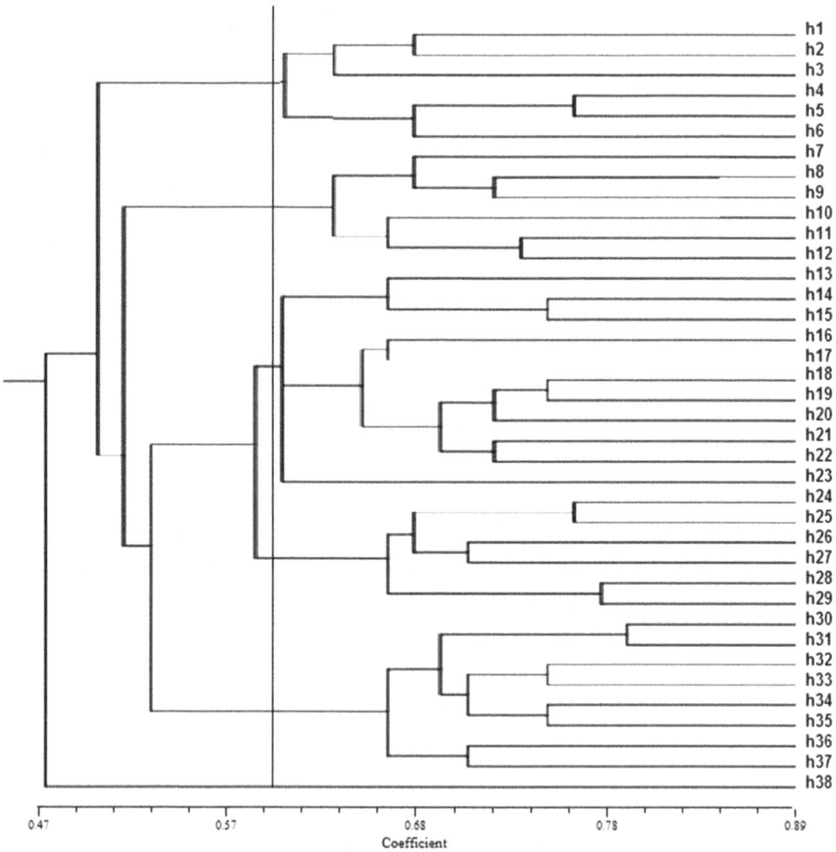

Fig. 1.2 Dendrogram obtained by hierarchical grouping UPGMA and Jaccard index of DNA amplified with SSR from 38 accessions of *Chenopodium berlandier*i ssp *nuttalliae*

Fifth cluster included eight accessions (h_{30}, h_{31}, h_{32} h_{33}, h_{34}, h_{36}, h_{35}, h_{37}) originating from Toluca Valley. Accession h_{13} from Almoloya de Juarez was different from all accessions integrating a separate cluster.

Results are in accordance with those obtained by Rana et al. (2010), in which cluster forming exhibited small but concrete differences in regard to morphological traits. All accessions showed a maximum homology of 0.887 among accessions h_{30} and h_{31}, corresponding to accession from the same locality (San Cristobal Huichochitlán), and a minimum homology of 0.309 among accessions h_{17} (Toluca) y h_3 (Atlacomulco). Generally speaking the degree of homology exhibited by accessions is in accordance with the autogamus nature of Huauzontle (Nelson 1968). These results indicate that even though most of the accession came from villages close among them, there is sufficient diversity to perform selection.

1.7 Chenopodium and Traditional Farming Systems

Traditional farming systems in México are based mainly on the use of local genotypes of staple crops such as corn and beans as well as land races of domesticated and semi domesticated plants all of which are grown in association to improve both income and diet of peasants inhabiting regions with marginal agricultural conditions. These traditional farming systems can be traced back to ancient cultures. *Chinampa* was developed by the Aztecs; the so called *solar* farming method by the Mayan civilization and the *milpa* has been in practice among Mesoamerican peoples since prehispanic times. All these systems promoted biodiversity by using several plant species which in turn satisfied food and housing requirements (Altieri 1999).

Nowadays there are regions where these traditional farming systems are still in practice, however, diverse factors such as mechanization, use of alien varieties and agrochemicals, changes in land usage and migration have all contributed to gradual decrease of these traditional methods, and consequently of the germplasm diversity (local land races, semidomesticated and wild edible plants). Among these valuable germplasm are pseudocereals pertaining to *Amaranthus* and *Chenopodium* genera, which like true cereals, are rich in mealy materials, able to be used in the elaboration of flour, bread and noodles. Pseudocereals were staple crops and were important not only from a socio-economic point of view but also in religious practices and for this reason that Spanish conquerors forbade *huautli* (nahuatl name referring both to *Chenopodium* and *Amaranthus* seeds) growing. This, added to introduction of Old World crops, contributed to their gradual decline (Iturbide 1994). However in some distant rural areas, pseudocereals in association with corn, beans and squash, are still grown. Tradition, remoteness, the need of food and income alternatives, as well as scarcity of economic resources has all contributed to preserve these prehispanic forms of culture. However as young people migrate to urban areas, the remaining old peasants find increasingly difficult to produce and sell these crops, some of which face the danger of extinction. Here we present the current status of *Chenopodium* pseudocereal cultivation on the Mexican region known as Purhépecha high plateau.

The studied areas comprised two communities located in the State of Michoacán, Opopeo within the municipality of Pátzcuaro and Santa María Huiramangaro, belonging to the municipality of Salvador Escalante (Fig. 1.3). Pátzcuaro is located at 19°31′ North and 101°36′ West, at an altitude of 2,140 masl. Climate is temperate with summer rainfalls. Annual rainfall is 983.3 mm and temperatures range from 9.2 to 23.2 °C. Its most important water reservoir is Pátzcuaro Lake. Mixed forests of pine, oak, cedar, spruce and juniper grow within this region. Ejido de Santa María Huiramangaro is located at 19°30′57″ North and 101°45′42″ West, at 2,263 masl. This village has 2,628 inhabitants, pertaining 586 to the economically active population (INEGI 2000; CIESEM 2001).

Pseudocereals are generally grown in association with corn, squash, and occasionally also with beans. In this multicrop system, corn and bean are sown in the same place, so that the former plants support the latter growth, which in turn helps to improve soil due to its nitrogen fixation capacity. Squash with its broad leaves covers the soil, protecting it from rainfall and sun direct actions thus diminishing erosion, evaporation and weed growth. Multi-crop systems bring about other benefits such as variety of harvested products per growing cycle, more labour occupation during the growing cycle and better land use.

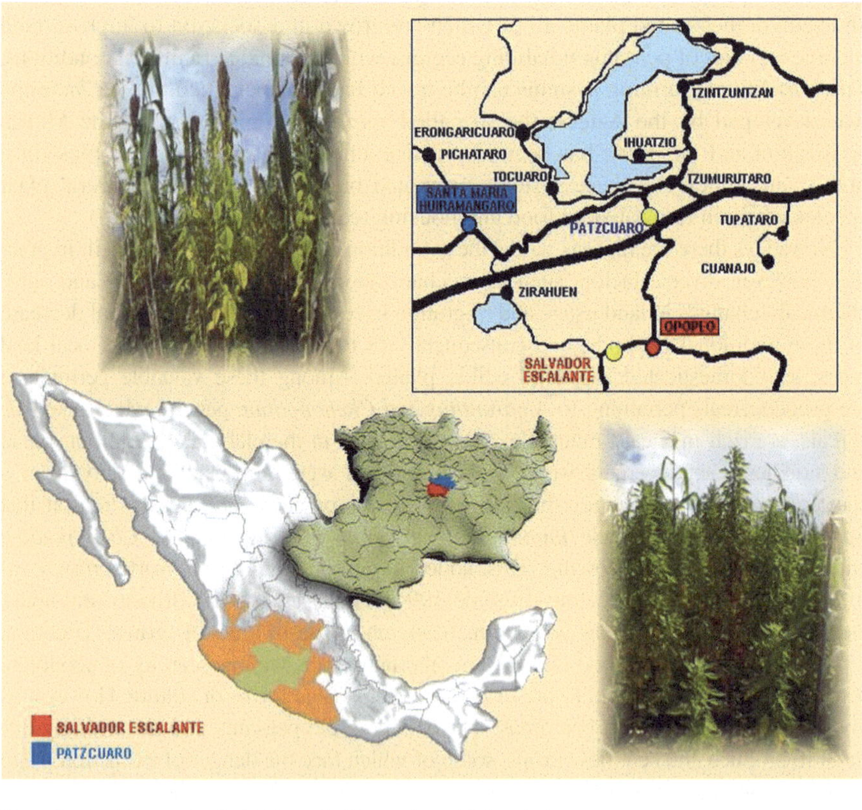

Fig. 1.3 Studied region in Michoacán State where Chía roja is grown

Chía roja (Fig. 1.4) is an erect herbaceous plant with a prominent, angular, red and branched stem with green, yellow and purple pigmented strips. Average number of branches ranges from 32 to 41. It has polymorphic leaves, lanceolate and rhombic with dented margins with 3–12 dents. Inferior leaves are mostly green, however in some cases red or purple leaves are also found. Panicles are terminal differentiated and amaranthiform with intermediate density but occasionally there are plants bearing a compact panicle. In some cases before maturity panicles exhibit red colour and upon maturity colour changes to purple (López 2006). One outstanding distinctive trait of this cultivar is a total absence of saponins, bitter compounds present in seeds of cultivars such as Huauzontle or other cultivated species of the same genus like quinua (De la Cruz et al. 2007).

Seeds of cultigen Chía roja are sharp cornered, slightly flat with smooth margins (Fig. 1.5). Diameter ranges from 1.5 to 1.7 mm, average thickness is around 0.95 mm, average weight ranges from 0.14 to 0.48 g per 100 seeds, and density is 0.66–0.76 g/ml. Colour is red purple exhibiting variability which according to the Royal Colour Chart corresponds to 59A and B, 185A and B, and 60 B (Royal Horticultural Society 2001). Seed yield is on average of 31 g per plant.

Pseudocereals are cultivated under a traditional multicrop system which depends entirely upon rainfall and are grown in association with corn, squash and occasionally beans. Basically two systems can be found: *milpa* and *mogote*.

Milpa system is practiced in plots located in plains or areas with moderate slope. Animal traction is used (horses or oxen). An individual guides the ploughing yoke, another follows sowing three corn seeds per step, another one sows pseudocereal and squash seeds mixed with either cow manure or chemical

Fig. 1.4 *Chenopodium berlandieri* subsp. *nuttalliae* cultivar Chía roja

Fig. 1.5 Seeds of *Chenpodium berlandieri* subsp. *nuttalliae* cv. Chía roja

fertilizer and finally someone covers the seeds using another plough. Cultural practices (weeding, thinning) are performed manually.

Mogote system of growing. Native plants are chopped and burned and pseudocereals grow among ashes and stones.

Mogote is a modality of the slash and burn growing system and is an ancient practice almost extinct nowadays (Fig. 1.6). Hills or high slope lands with naturally occurring broad leaved vegetation, shrubs of *Quercus spp.* and *Alnus jorullensis* are employed for this system. Land preparation begins during spring by cutting down and burning native plants. Upon establishment of the rainy season, corn and pseudocereal seeds are sown directly on burned vegetation ashes by means of a wooden

Fig. 1.6 *Mogote* system of growing. Native plants are chopped and burned and pseudocereals grow among ashes and stones

tool which digs holes allowing insertion of seeds, called *coa* (meaning digging stick). When solely pseudocereals are grown, seeds mixed with manure are spread by hand without covering. Weeding is performed manually and in multicrop system harvest is sequential: squash is harvested during August, fresh ears of corn (*elotes*) along September, pseudocereals (Chía roja, Chía blanca and Chía negra) are harvested in December, and dry corn in January (Pérez et al. 2005).

People from studied communities use Chía roja to make *chapatas*. *Chapatas* are a sort of *tamales* (a lump of corn dough, prepared by mixing nixtamalised cornmeal with equal parts of ground Chía roja seeds and sugar all of which is wrapped in corn husks (*totomoxtle*) and steam cooked (López 2006).

Yield estimation under the *milpa* system is 1 ton per hectare for corn and 700 kg per hectare for pseudocereals. Considering the ratio benefit/cost multicrop systems give a ratio of 2.5 in comparison to corn monocrops which gives a ratio of 1 (Pérez et al. 2005).

Thus multicrop systems bring about the possibility to diversify the impoverished, rural regions inhabitants diet, generating at the same products which through simple processing techniques (sweets, tamales etc.) bring an additional income to peasants who may obtain products even when climatic conditions are adverse.

1.8 Studies Concerning Nutritional Characteristics of *Chenopodium*

Regarding to nutritive value, samples of Chía roja collected from the studied region yielded protein and lipid contents ranging from 13.1 to 17.9 %; and 3.6–4.9 % respectively. Average proximal analysis and energy are presented in Table 1.3.

1.8.1 Composition of the Oil Fraction Concerning to Essential Fatty Acids

Concerning to the oil component of *Chenopodium berlandieri* subsp. *nuttalliae,* a study has been performed to quantify their percentage of fatty acids, and the results

Table 1.3 Proximal analysis of *Chenopodium berlandieri* subsp. *nuttalliae* cvs. Huauzontle and Chía roja (%)

	Chía Roja	Huauzontle
Moisture	8.24	8.015
Ashes	4.20	3.62
Etereo Extract	4.06	5.13
Proteins	15.78	17.8
Crude Fiber	10.44	4.02
Assimilable Carbohydrates	57.3	61.4
Total Carbohydrates	67.7	65.45
Energy (KJ)	1376.77	1519.8

Table 1.4 Identification of fatty acids in oil from seeds of *Chenopodium berlandieri* subsp *nuttalliae* cv Huauzontle

Fatty acids (percentage from the fatty fraction)			
Fatty acid		Huauzontle	Chía roja
$C_{16:0}$	Palmitic	10.4493	3.66
$C_{18:0}$	Estearic	0.2979	0.37
$C_{18:1}$ ω-9	Oleic	23.7925	19.54
$C_{18:2}$ ω-6	Linoleic	52.8209	53.55
$C_{18:3}$ ω-3	Linolenic	12.2794	20.70

Table 1.5 Comparison of the percentage of linoleic and linolenic acids in the fatty fraction of *Chenopodium berlandieri* subsp.*nuttalliae* cvs Huauzontle and Chía Roja, with other cereals

	Linoleic acid ω-6 (%)	Linolenic acid ω-3 (%)
Huauzontle	52.8209	12.2794
Chía roja	53.55	20.7
Corn	55.9	0.9
Nuts	56.7	12.3
Soybean	52.6	7.3

are presented in Table 1.4 (Falcón et al. 2007). It is important to note the high percentage of oleic, linoleic and linolenic acids which are essential for human nutrition. As can be seen in Table 1.5, the percentage of the fatty fraction of linoleic and linolenic acids is similar to that found in corn (55.9 %), nuts (56.7 %) and soybean (52.8209 %) according to Dhellot et al. (2006).

1.9 Morphological Description of Floral Development in *Chenopodium*

To advance in production of interspecific hybrids between Chía roja (*Chenopodium berlandieri* subsp. *nuttalliae*) and quinua (*Chenopodium quino*a Willd) is important to determine the different stages of development of gynoecium and pollen characterization in both species. The study was performed on flowering plants of both species detecting changes in form, size, color and receptivity of gynoecium and pollen grains. Sixty individuals from each species were selected. For gynoecium study 300 flowers per species were selected and for pollen grains 120 flowers were studied.

Flowers of both species are small (1,000–2,600 μm for quinua and 400–1800 μm for chía roja), lacking of petals with five sepals, being mostly hermaphrodites with androceus formed by 5 anthers and one gynoecium. However some pistilated flowers were observed mainly on Chía roja plants. Plants with glomerules containing only one hermaphrodite flower (apical) have been observed being the rest pistilated. Gandarillas (1967) mentions that the percentage of hermaphrodite and pistilated flowers is variable in a plant and so it could be authogamous or alogamous. Gynoecium is compossed of one superior ovary, a small style and a stigma with two stigmatic branches. Sometimes three branched stigmas can be found, that according to (Bertero and Diego 1996) could be related to androsterility.

1.9.1 Phases of Gynoecium Development

Phase 1

Gynoecium is very small, green, presenting an almost spherical form, and style begins to differentiate as a small protuberance, stigmatic branches are absent. In quinua flower size ranges from 800 and 1,000 μm, while in Chía roja is 400–450 μm (Fig. 1.7).

Phase 2

Gynoecium is formed, is green in color and its diameter is 350 μm in quinua and 200 μm in Chía. Only two stigmatic branches are observed. Fig. 1.8a, exhibits the three structures conforming gynoecium. Time required to pass from phase 1 to phase 2 is one to two days in Chía roja and from 4 to 6 days in quinua.

Phase 3

Gynoecium exhibits green color, its diameter is 400 μm for quinua and 350 μm for Chía. Time needed to reach this stage is 6 days in chía and 9–10 days in quinua. In this phase gynoeciums with two and three branches were observed (Fig. 1.8b).

Phase 4

In this phase gynoecium is 450 μm in quinua and 500 μm in Chía. In quinua some maturity events can be observed such as widening of the terminal end of stigma,

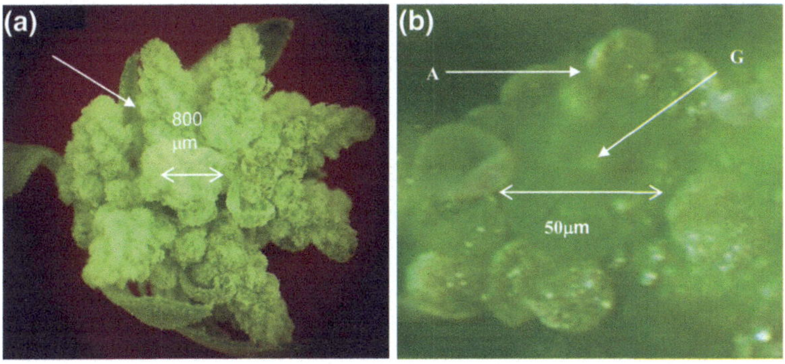

Fig. 1.7 Glomerules and gynoecium corresponding to stage 1. **a** Glomerules with small floral buttons 800 μm. **b** Reproductive organs: gynoecia (*G*), surrounded by 5 pairs of anthers (*A*)

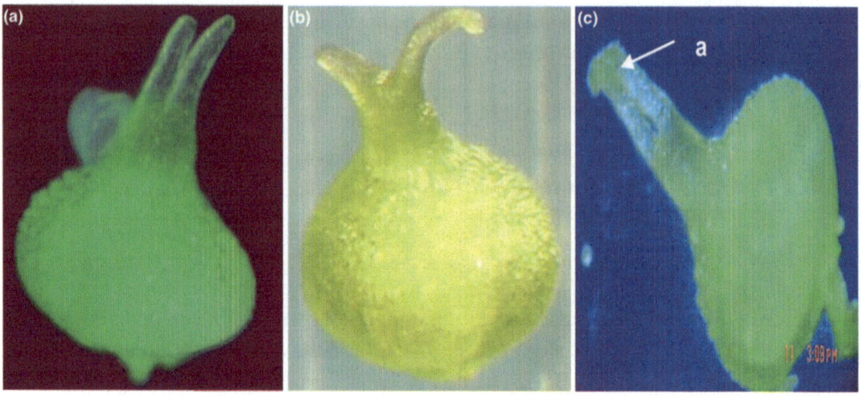

Fig. 1.8 **a** Gynoecium in phase 2, *two stigmatic branches* are shown. **b** Gynoecium in phase 3. Stigmatic branches are bigger and ovary has *rounded shape*. **c** Gynoecium in phase 4, *a* stigmatic papillae

indicating the beginning of development of stigmatic papillae (Fig. 1.8c). In Chía roja the structures are ready to perform fecundation (long styles up to 1,000 μm) with well developed stigmatic papillaes. Time required to reach this phase is 10 days for chía and 12–16 days for quinua.

Phase 5

Gynoecium in quinua reaches 500 μm, stigmatic papillae can be distinguished as small hair surrounding gyno ecium. This is the receptive phase in quinua and lasts 5–6 days. This phase requires nearly 21 days in quinua. In Chía roja, the development of the seed or the aging of gynoecium begins when there was no fecundation. If fecundation has occurred gynoecium increases its diameter to 600–650 μm, being the styles reduced and eventually disappear. In absence of fecundation gynoecium is reduced in diameter and styles tend to disappear (Fig. 1.9).

Phase 6

The development of gynoecium was observed in this stage in quinua. In quinua if fecundation has not occurred gynoecium reduces its size (Fig. 1.9).

Seed Formation

If fecundation occurs gynoecium grows reaching 700 μm diameter and 600 μm height. Style reaches 200 μm, stigmatic branches reach 700 μm as is shown in Fig. 1.10. Time required for this phase is 26–30 days.

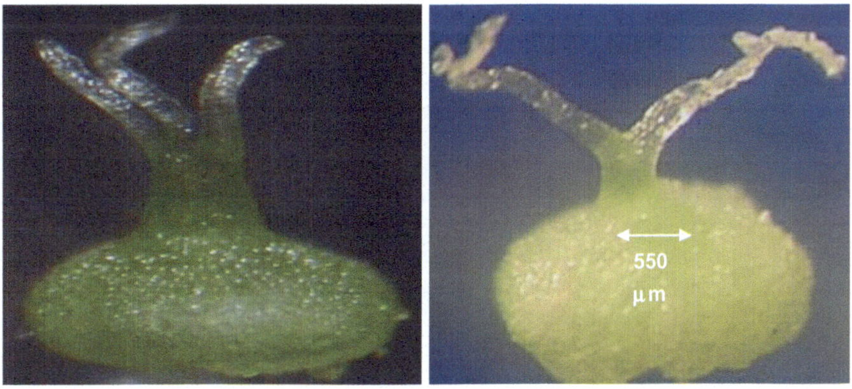

Fig. 1.9 Phase 5 receptive. Gynoecium with *three stigmatic* branches. Phase 6 aging gynoecium, its size is reduced and stigmatic branches are also reduced

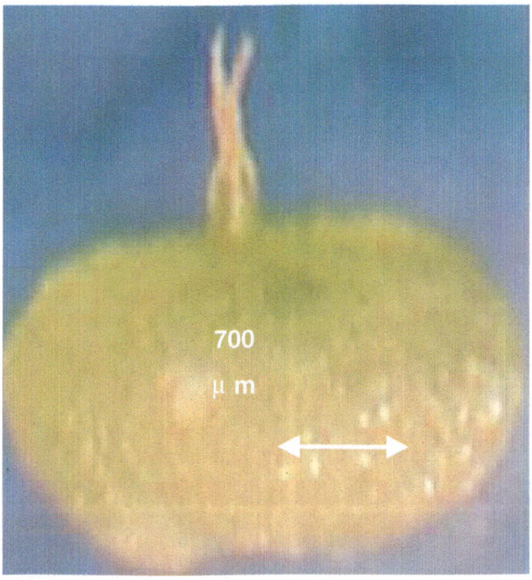

Fig. 1.10 Seed formation

1.9.2 Androecium and Pollen Characterization

Androecium is formed by five stamens with a narrow filament and an anther. Size of anthers is at the beginning 75–300 µm long and 50–175 µm wide for Chía roja and 75–350 µm long and 25–250 µm wide for quinua.

Phase 1: Pollen Mother Cells (1st Meiosis)

Anthers are green ovated 75 µm long for quinua and 50 µm for Chía roja, located at the base of gynoecium.

Phase 2: Tetrad (2nd Meiosis)

Anthers are green ovated 200 μm length for quinua and 150 μm for Chía. In acetocarmine preparation, tetrads are observed (80 %) and mother cells.

Phase 3 and 4 (Flowers with Microspores)

In the phases 3 and 4 anthers are ovated 300–200 μm in quinua and 175–200 μm in Chía roja color yellow-greenish. In acetocarmine preparations 95 % of pollen grains without exine layer are observed.

Flowers with Pollen

Flowers are 1,500 μm for chía and 2,000 μm for quinua in diameter (Fig. 1.11) during anthesis. Anthers are yellow, 300 μm long and 350 μm wide for quinua, being for Chía roja smaller. Anthers can exhibit the dehiscence line opened.

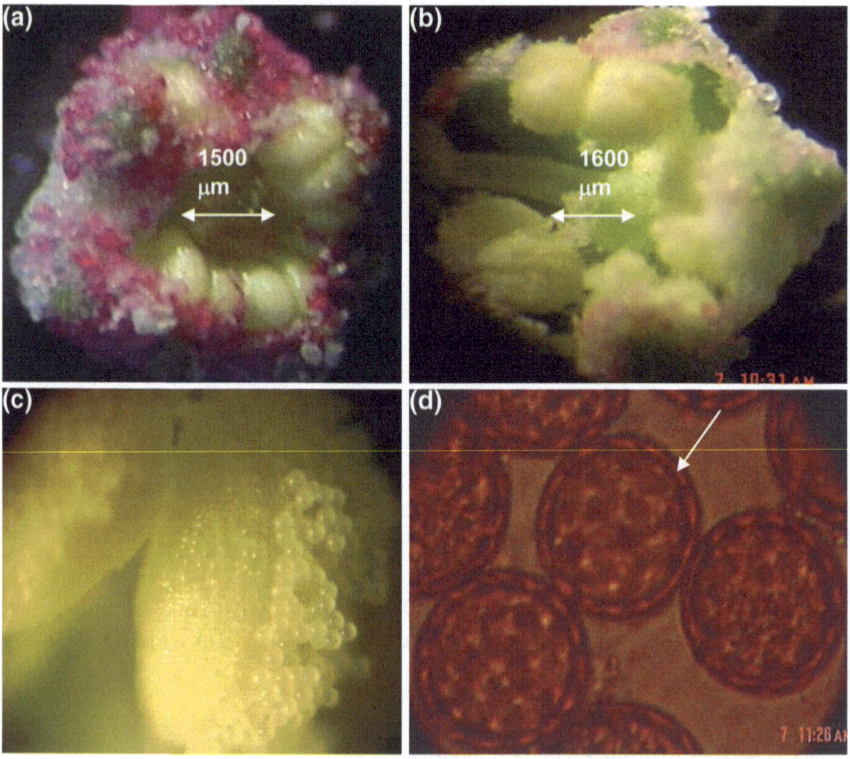

Fig. 1.11 **a** Flower exhibiting its five pairs of anthers with mature pollen grains. **b** Flower exhibiting its anthers with the dehiscence line opened. **c** Anther shedding mature pollen grains and **d** Pollen grains with exin layer

Fig. 1.12 Cross between *C. quinoa* (*left*) and *C. berlandieri* subsp. *nuttalliae* cv Chía roja (*right*) and F1 hybrid (*middle*)

Considering the different floral phases in quinua and Chía roja, interespecific crosses have been developed among *C. quinoa* and *C. berlandieri* subsp. *nuttalliae* cv. Chía Roja, obtaining viable seed formation as can be seen in Fig. 1.12. These results are important considering that quinua has good grain quality in size and color, being the presence of saponins an adverse factor. Absence of saponins in Chía roja is a desirable trait that can be transferred to quinua to obtain cultivars with high seed quality but low saponin content.

1.10 Mutation Breeding in *C. quinoa*

Research on pseudocereals, particularly on *Chenopodium quinoa* Willd. as an alternative crop for agricultural marginal conditions in México goes back to the early seventies in the Colegio de Postgraduados at Chapingo México, when the first trials on adaptability of cultivars and studies on physiology and genetic variance components were performed (Mújica 1983). In the late 1980s, the Instituto de Investigación y Capacitación Agropecuaria Acuícola y Forestal del Estado de México (ICAMEX) began an evaluation program searching for quinoa cultivars able to withstand adverse conditions such as low soil fertility, drought, and frosts, which prevail in many areas of the México State. As result of those evaluations, four cultivars emerged as a suitable alternative to be offered to peasants: Isluga, Barandales, Sierra Blanca and Lípez (Rodríguez 1992). However, their high saponin content was considered as a negative factor for their diffusion, and a coordinated research program among ICAMEX and the Instituto Nacional de Investigaciones Nucleares (ININ) was therefore launched in 1990, devoted to the application of radio-induced mutagenesis techniques to obtain low-saponin mutants of the cultivar Barandales.

The approach followed included first a radiosensitivity test concluding that the LD_{50} was 225 Gy (Hernández et al. 1994). Based on these results, seeds of Barandales were irradiated with 200 and 250 Gy. Selection of mutants was performed in the second generation (M_2), regarding traits such as earliness, size of panicle, productivity, and saponin content. From the evaluation of M_2, 88 putative mutants were selected and, from M_3 to M_7, a screening process led to the selection of 28 putative mutants.

The aim of this study was to generate low-saponin, high- yielding quinoa lines, also activities of exploration, recollection and characterization of native germplasm from *Chenopodium berlandieri* subsp. *nuttalliae* landrace Chía roja (the only chenopod whose seeds are consumed in México), were undertaken. This valuable germplasm could thus be incorporated into the breeding program of quinoa, as progenitor in future hybridizations aiming to introduce the low-saponin content trait from Chía roja to quinoa.

One of the reasons to incorporate quinoa as an alternative crop for peasants in regions with marginal agricultural and economical conditions is to diminish the degree of malnutrition. Therefore, the evaluation of saponin content is a relevant factor since low-saponin content cultivars will have a better acceptance. Thus, the saponin content was estimated in putative mutant lines, following the method established by CIRNMA (2001).

Figure 1.13 shows the percentage of saponins found in cv. Barandales considered as control exhibiting the highest saponin percentage (2.7 %), but also high protein and fat content, on the other hand mutant $20R_227$, exhibits minimum

Fig. 1.13 Seeds of quinua var. Barandales exhibitng high saponin content (*left*) Mutant line obtained trough radioinduced mutagenesis with no saponins (*Right*)

saponin content but also a reduction in proteins being the oil content not modified. So is important to take in account not only the saponin content but also the levels of other contents of nutritive compounds such as proteins, carbohydrates and oils, in any mutation breeding program.

1.11 Conclusion and Prospects

Middle and South America are two relevant centers of origin and domestication of many species with present and/or potential importance as food or as medicine, that includes corn, beans, chilli, avocado, vanilla, cocoa, husk tomatoes, tomatoes, chía, amaranth, and cultivated species of the *Chenopodium* genus such as *Chenopodium quinoa*, *Chenopodium berlandieri* subsp. *nuttalliae* and *C. pallidicaule* (Cañahua). Many of the aforementioned species are relevant because of their nutritive value or because their use in traditional healing practices. However most of these species remain underutilized because they are grown only at the local level. Even some of these species are endangered because the knowledge of its cultivation and use pertain to older generations in the Villages and the new generations leave rural communities looking for better opportunities, thus being the chain of transmission of traditional knowledge about these plants, its culture, and use, disrupted. On other hand, Latin America also share several of the main constraints to human development: (a) Poverty that reaches 39 % of population, being about 18 % in extreme poverty, according to the Banco Mundial (2006). (b) Undernourishment that affects 45 million inhabitants in Latin America according to FAO (2008) (period 2003–2005), with rural and marginated areas more affected. (c) Deterioration of natural resources in Latina America is increasing. According to FAO (1998) population pressure, accompanied by rising demand for food, fuel and construction materials, is placing progressively more intense pressure on the region's natural resources. It is estimated that soil erosion, acidification, loss of organic matter, compaction, nutrient impoverishment and salinization have reduced productivity on more than 3 million sq km of farmland, while almost 800,000 sq km of dry lands are threatened by desertification due to overgrazing, overexploitation of vegetation for domestic use, deforestation and inappropriate irrigation methods. One underappreciated factor potentially contributing to solving the problems is the improvement of underutilized and other important crops for sustainable agricultural development in rural communities. *Chenopodium quinua* is nowadays the most representative cultivated specie of *Chenopodium* genus, but cultigens from *Chenopodium berlandieri* subsp. *nuttalliae* and *C. pallidicaule* have great potential considering its present variability, and its use in traditional farming systems. Application of modern biotechnological techniques such as molecular markers, in vitro *culture*, dihaploid production, and mutation breeding could be very valuable to generate improved varieties that considering its high nutritive value and their adaptability to adverse conditions may contribute to solve malnutrition that prevails on marginal agricultural areas through a sustainable production system.

References

Altieri MA (1999) Applying agro ecology to enhance productivity of peasant farming systems in Latin America. Environ Dev Sustain 1:197–217

Basset J, Crompton W (1982) The genus *Chenopodium* in Canada. Can J Bot 60:602–610

Bennett MD, Smith J (1991) Philosophical transactions of the royal society of London B. Ann Bot 334:309–345

Bertero D, Diego M (1996) Changes on apical morphology during floral initiation and reproductive development in quinoa (*Chenopodium quinoa* hill). Ann Bot 79:317–324

Bhargava A, Shukla S, Ohri D (2006) Karyotypic studies on some cultivated and wild species of *Chenopodium* (Chenopodiaceae). Genet Res Crop Evol 53:1309–1320

Bonifacio A (1995) Interspecific and intergeneric hybridization in chenopod species. Brigham Young University, Provo

Carrillo AO, Engleman EM (1994) Anatomía de la semilla de *chenopodium berlandieri* ssp. *nuttalliae* (chenopodiaceae) "Huauzontle". Bol Soc Bot Méx 54:17–34

CIESEM (2001) Municipio en Cifras: secretaria de educación en el estado, centro de información económica y social del estado de Michoacán. Municipio Pátzcuaro. pp 112

CIRNMA (2001) Calidad del grano: centro de investigaciones en recursos naturales y medio ambiente. Puno

Coles ND, Coleman CE, Christensen SA, Jellen EN, Stevens MR, Bonifacio A, Rojas-Beltran JA, Fairbanks DJ, Maughan PJ (2004) Development and use of an expressed sequenced tag library in quinoa (*Chenopodium quinoa* Willd.) for the discovery of single nucleotide polymorphisms. Plant Sci 168(2):439–447

De la Cruz TE, Rubluo IA, Palomino HG, García AJM, Laguna CA (2007) Characterization of *chenopodium* germplasm, selection of putative mutants and its cytogenetic study. In: Ochatt S, Mohan-Jain S (eds) Breeding of neglected and under-utilized crops, spices and herbs. Science Publishers, Enfield, pp 124–136

Dellaporta SL, Wood J, Hicks JE (1983) A plant DNA minipreparation: version II. Pl Mol Biol Rep 1(4):19–21

Dhellot JR, Matouba E, Maloumbi MG, Nzikou JM, Dzondo MG, Linder M, Parmentier M, Desobry S (2006) Extraction and nutritional properties of *Solanum nigrum* L seed oil. Afr J Biotechnol 5(10):987–991

Dolezel J, Lysak MA, Kubalakova M, Simkova H, Macas J (2000) Sorting of plant chromosomes in methods in cell biology: flow cytometry. In: Darzynkiewicz Z (ed) 3rd edn. Academic Press, San Diego

Dolezel J, Bartos J, Voglmayr H, Greilhuber J (2003) Nuclear DNA content and genome size of trout and human. Cytometry A 51A:127–128

Dolezel J, Greilhuber J, Suda J (2007) Flow cytometry with plants: an overview. In: Flow cytometry with plant cells. In: Dolezel J, Greilhuber J, Suda y J (eds) Willey-Vch Verlag. CmbH and Co. KGaA, Weinheim, Federal Republic of Germany. pp 41–65

Fairbanks DJ, Burgener KW, Robison LR, Andersen WR, Ballón E (1990) Electrophoretic characterization of quinoa seed proteins. Plant Breed 104:190–195

Fairbanks DJ, Waldrigues DF, Ruas CF, Ruas RM, Maughan PJ, Robison LR, Andersen WR, Riede CR, Panley CS, Caetano LG, Arantes OMN, Fungaro MH, Vidotto MC, Jankevicius SE (1993) Efficient characterization of biological diversity using field DNA extraction and RAPD markers. Brazil J Genet 16:11–33

Falcón BT, Vázquez AO, Luna GMA (2007) Acidos grasos en las semillas de *Chenopodium berlandieri* ssp *nuttalliae* raza local Huauzontle, Congreso Técnico Científico ININ-SUTIN

FAO (Food and Agriculture Organization) (1996) The state of the world's plant genetic resources: preparatory materials for the 4th international technical conference on plant genetic resources. Leipzig, Germany, p 511

FAO (2008) The state of food insecurity in the world. Rome Italy. 56p

FAO (1998) Land conservation in Latin America. http://www.fao.org/ag/magazine/spot4.htm. Consulted january, 2013

Gandarillas H (1967) Observaciones sobre la biología reproductiva de la quinoa (*Chenopodium quinoa* Wild.). Saya. Sociedad de Ingenieros Agrónomos de Bolivia. Abril-Noviembre. La Paz Bolivia, pp 4

Gandarillas H (1982) Quinoa production. IBTA-CIID (trans: Sierra-Blanca Assoc) Denver, pp 57

Gandarillas H, Luizaga J (1967) Número de cromosomas de *Chenopodium quinoa* Willd. en radículas y raicillas. Turrialba 17:275–279

Giusti L (1970) El género *Chenopodium* en argentina. l. Número de Cromosomas Darwiniana 16:98–105

Gómez MME (1989) ¿Qué grado de alogamia tiene la Quinua (*Chenopodium quinoa* Willd.). Tesis Profesional, Depto. Fitotecnia, Universidad Autónoma Chapingo, México

Greilhuber J, Dolezel J, Lysak MA, Bennett MD (2005) The origin, evolution and proposed stabilization of the terms 'Genome Size' and 'C-Value' to describe nuclear DNA contents. Ann Botany 95:255–260.

Guzmán ME (2007) Evaluación nutritiva y toxicologica de la inflorescencia del Huauzontle (*Chenopodium nuttalliae*) cruda y cocida. Tesis profesional. Facultad de Química. UNAM, p 76

Heiser CH (1985) Chenopods: from weeds to the halls of Montezuma, Of plants and people. University of Oklahoma Press, Oklahoma, pp 82–99

Hernández AM, de la Cruz TE, Rodríguez DR (1994) Evaluación de características agronómicas de quinua (*Chenopodium quinoa* Willd) generación M_4. Memorias del VI Congreso Técnico–Científico ININ–SUTIN, México

Heywood V (1999) Trends in agricultural biodiversity. In: Janik J (ed) Perspectives on new crops and new uses. ASHS Press, Alexandria, pp 2–14

Humboldt AL, Bonpland A, Kunth CS (1815) Nova genera et species plantarum, 6e Part. Botanique (2) Paris, p 184

Hunziker TA (1952) Los pseudocereales de la agricultura indígena de América. ACME AGENCY Soc. Resp. Ltda, Suipacha

INEGI (2000) XII Censo de Población y Vivienda. Instituto Nacional de Estadística Geografía e Informática, México

Iturbide GA (1994) Grain amaranths. In: Hernándo BJE, León J (eds) Neglected crops: 1492 from a different perspective. FAO, Rome, pp 93–101

Kidwell MG (2002) Transponsable elements and evolution of genome size in eukaryotes. Genetica 115:49–63

Kolano B, Pando LG, Maluszynska J (2001) Molecular cytogenetics studies in *Chenopodium quinoa* and *Amaranthus caudatus*. Acta Societalis Botanicorum Pol 70:85–90

Leitch AR, Lim KY, Leitch IJ, O'Neill M, Chye M, Low F (1998) Molecular cytogenetics studies in rubber, hevea brasilensis mull arg (Euphorbiaceae). Genome 41:464–467

López MA (2006) Estudio de los recursos genéticos de los pseudocereales Chía roja (*Chenopodium berlandieri* subsp. *Nuttalliae*) y Amaranto (*Amaranthus spp.)* en los Ejidos de Santa María Huiramangaro y Opopeo: michoacán. Universidad Michoacana de San Nicolás de Hidalgo, México, p 154

López GA (1988) Potencial genético de la quinua (*Chenopodium quinoa* Willd) centro de genética, Resúmenes del XII congreso de fitogenética, pp 16–17. UA Ch, Chapingo

Mapes SC (1987) El maíz entre los Purhepechas de la cuenca del Lago de Pátzcuaro Michoacán. Méx Am Indígena 27(2):346–373

Martínez GNC (2005) Caracterización física y bromatológica de germoplasma de pseudocereales: *Chenopodium berlandieri* ssp. *nuttalliae* y *Amaranthus spp*: Tesis de Licenciatura. Químico en Alimentos. Universidad Autónoma del Estado de México, México, p 78

Martins C, Wasko AP (2004) Organization and evolution of 5S ribosomal DNA in the FISH genome. In: Williams CR (ed) Focus on genome research. New Science Publishers, Hauppauge, pp 289–318

Mason SL, Stevens MR, Jellen EN, Bonifacio A, Fairbanks DJ, Coleman CE, MacCarty RR, Rasmussen AG, Maughan PJ (2005) Development and use of microsatellite

markers for germplasm characterization in quinoa (*Chenopodium quinoa* Willd). Crop Sci 45:1618–1630
Maughan PJ, Bonifacio A, Jellen EN, Stevens MR, Coleman CE, Ricks M, Mason S, Jarvis D, Gardunia B, Fairbanks D (2004) A genetic linkage map of quinoa (*Chenopodium quinoa* Willd), based on AFLP, RAPD, and SSR markers. Theor Appl Genet 109:1188–1195
Maughan PJ, Kolano BA, Maluszynska J, Coles ND, Bonifacio A, Rojas J, Coleman CE, Stevens MR, Fairbanks DJ, Parkins SE, Jellen EN (2006) Molecular and cytological characterization of ribosomal RNA genes in *Chenopodium quinoa* and *Chenopodium berlandieri*. Genome 49:825–839
Miranda CS, Mujíca SA (1985) Avances en la investigación genética en quinua (*Chenopodium quinoa* Willd) colegio de postgraduados. Centro de Genética, Chapingo, p 43
Moquin-Tandon CHBA (1849) Amarantaceae salsolaceae. In: De Candolle AP (ed) Prodromus systematis naturalis regni vegetabilis, vol 13(2), pp 211–230
Mújica SMA (1983) Selecciones de variedades de quinua: colegio de postgraduados, centro de genética. Tesis de Maestría, Portugal 110
Mújica SMA, Izquierdo J, Marathee JP, Morón C, Jacobsen S (1999) Reunión Técnica y Taller de Formulación de Proyecto Regional sobre Producción y Nutrición Humana en Base a Cultivos Andinos. Arequipa, Perú
Mundial B (2006) La pobreza y creciemiento: circulos virtuosos y circulos viciosos. Washington, p 216
Nelson DC (1968) Taxonomy and origins of *Chenopodium quinoa* and *Chenopodium nuttalliae*. Ph D thesis, University of Indiana, Bloomington
Nieto CC, Vimos NC, Caicedo VC, Monteros JC, Rivera MM (1999) Obtención de dos variedades de quinua de bajo contenido de saponina, para La Sierra Ecuatoriana. Producción y Procesamiento de Quinua en Ecuador
Oelke EA, Putnam DH, Oplinger ES (2001) Quinoa: alternative field crops manual. University of Minnesota, Minnesota
Palomino GH, Segura MD, Bye RB, Mercado MR (1990) Cytogenetic distinction between *Teloxys* and *Chenopodium* (Chenopodiaceae). Southwestern Nat 35(3):351–353
Palomino GL, Trejo HL, De la Cruz TE (2008) Nuclear genome size and chromosome analysis in *Chenopodium quinoa* and *Chenopodium berlandieri* subsp. *nuttalliae*. Euphytica 164:221–230
Peñafiel CCE, Díaz VL (1988) Determinación espectrofotométrica de ácido oleanólico y saponinas de quinua (*Chenopodium quinoa* Willd) Variedad Kancolla. Arch Latinoam Nutr 38:113–125
Pérez AE, de la Cruz TE, Mapes SC, García AJM (2005) Las comunidades campesinas: un importante reservorio de recursos para la humanidad. Leisa Rev Agroecología 20(4):20–23
Petrov DA (2001) Evolution of genome size: new approaches to an old problem. Trends Genet 17:23–28
Price HJ (1976) Evolution of DNA content in higher plants. Bot Rev 42:27–52
Rana TS, Narzary D, Ohri D (2010) Genetic diversity and relationships among some wild and cultivated species of *Chenopodium* L. (Amaranthaceae) using RAPD and DAMD methods. Curr Sci 98(6):840–846
Rodríguez DR (1992) Tecnología para la producción de quinoa (*Chenopodium quinoa* Willd) en el Estado de México. Resúmenes del Primer encuentro de Ciencia y Tecnología del Sector Agropecuario y Forestal del Estado de México
Rojas W, Barriga P, Figueroa H (2000) Multivariate analysis of the genetic diversity of Bolivian quinua germplasm. Plant Genet Resour Newsl 122:16–23
Royal Horticultural Society (2001) RHS colour chart. The Royal Horticultural Society, London
Ruas PM, Bonifacio A, Ruas FC, Fairbanks JD, Andersen RW (1999) Genetic relationship among 19 accessions of six species of *Chenopodium* L. by random amplified polymorphic DNA fragments (RAPD). Euphytica 105:25–32
Schwarzacher T, Heslop-Harrison SJ (2000) Reprobing of preparations In: Kingston F (ed.) Practical in situ hybridization, BIOS Scientific Publishers Limited. Oxford, UK, p 110
Sederberg MC (2008) Physical mapping of ribosomal RNA genes in new world members of the genus *Chenopodium* using fluorescence in situ hybridization. Master of Science Thesis. Department of Plant and Wildlife Sciences. Brigham Young University, USA, p 53

Stevens MR, Coleman CE, Parkinson SE, Maughan PJ, Zhang HB, Balzotti MR, Kooyman DL, Arumuganathan K, Bonifacio A, Fairbanks DJ, Jellen EN, Stevens JJ (2006) Construction of a quinoa (*Chenopodium quinoa* Willd.) BAC library and its use in identifying genes encoding seed storage proteins. Theor Appl Genet 112(8):1593–1600
Tanaka R, Tanaka A (1980) Karyomorphological studies in halophitic plants. l: some taxa of Chenopodium. Cytologia 45:257–269
Thies E (2000) Promising and underutilized species crops and breeds. Deutsche Gesellschaft fur Technishe Zusammenarbeit (GTZ) GmbH. Eschborn, Germany, p 14
Walters TW (1987) Electrophoretic evidence for the evolutionary relationship of the tetraploid *Chenopodium berlandieri* to its putative diploid progenitors. Selbyana 10:36–55
West CR (1946) Cultural geography of the modern Tarascan area. Smithsonian Institution, Institute of Social Anthropology. Greenwood Press, Publishers Westport. Publication No 7, pp 42–43
Wilson HD (1988a) Quinoa biosystematics I: domesticated populations. Econ Bot 42:461–477
Wilson HD (1988b) Quinoa biosystematics II: free-living populations. Econ Bot 42:478–494
Wilson H, Heiser BC (1979) The origin and evolutionary relationships of huauzonthe (*Chenopodium nuttalliae*) domesticated chenopod of Mexico. Am J Bot 66:198–206

Chapter 2
Thin Cell Layer Technology in Micropropagation of *Jatropha curcas* L.

Duong Tan Nhut, Do Dang Giap, Bui Van The Vinh, Nguyen Thi Kim Loan, Nguyen Phuc Huy, Nguyen Thanh Hai, Hoang Xuan Chien, Tran Trong Tuan and Thai Xuan Du

Abstract Physic nut plant (*Jatropha curcas* L.) is considered as one of potential sources for a non-edible biofuel-producing energy crop throughout the world. The conventional propagation methods for this species present many problems including poor seed viability, low germination, unstable yield and high oil content. In addition to the problems associated with establishing seedlings, *Jatropha curcas* was found to be largely recalcitrant to in vitro regeneration and unresponsive to many plant growth regulators. Recently, a rapid and effective system of somatic embryogenesis and organogenesis from leaf transverse thin cell layers (tTCLs) of *Jatropha curcas* L. was established. By inducing several thin cell layers precisely with various plant growth regulators, numerous small shoot clumps formed directly. These small shoot clumps were then transferred to a new medium for shoot elongation. Besides, tTCL promoted indirect organogenesis, through callus formation, on media containing kinetin and IBA. Nodular callus structures started to differentiate into shoot buds when hard compact ones were transferred to fresh medium containing kinetin only. Embryogenic calli were induced and proliferated on MS medium supplemented with kinetin and 2,4-D, including vigorous somatic embryos. These embryos developed to plants with normal phenotype and rooted easily in growth regulator-free half-strength MS medium. Regenerated plantlets from organogenesis and somatic embryogenesis were acclimatized under the controlled greenhouse conditions with a high survival rate.

Abbreviations:

tTCL	Thin cell layer
2,4-D	2,4,-dichloropheoxyacetic acid
IBA	Indolebutyric acid

D. T. Nhut (✉) · D. D. Giap · B. Van The Vinh · N. T. K. Loan · N. P. Huy · N. T. Hai · H. X. Chien · T. T. Tuan · T. X. Du
Tay Nguyen Institute of Biology, 116 Xo Viet Nghe Tinh, Dalat, Lam Dong Province, Vietnam
e-mail: duongtannhut@gmail.com

S. M. Jain and S. Dutta Gupta (eds.), *Biotechnology of Neglected and Underutilized Crops*, DOI: 10.1007/978-94-007-5500-0_2,

TDZ	Thidiazuron
BA	6-benzylaminopurine
2-iP	6-γ,γ-dimethylallylaminopurine
GA	Gibberrelic acid
PGRs	Plant growth regulators

2.1 Introduction

Jatropha curcas L. (physic nut) is a bush plant belonging to the family Euphorbiaceae, which comprises approximately 8,000 species. It is a drought resistant species which is widely cultivated in the tropics as a living fence. Oil from seeds of *J. curcas* is considered one potential source of a non-edible biofuel producing energy crop throughout the world. Many parts of the plants are also used in traditional medicine, such as leaves (cough-suppressant and antiseptic after birth), stem sap (anti-hemorrhagic), and oil (purgative, skin disease treatment, pain killer for rheumatism). The oil is not edible due to the presence of toxic substances including a lectin (curcin), phorbol esters, saponins, protease inhibitors and phytates (Sujatha et al. 2005), and it is conventionally used for making soaps, candles, paints and lubricants (Sujatha and Mukta 1996).

The demand for large amount of *J. curcas* planting materials for industrial and medicinal purposes is difficult to be achieved using the conventional propagation methods namely propagation by seeds and by cuttings. The direct seedling method presents numerous problems including poor seed viability, low germination, unstable yield and high oil content while plants propagated by cuttings show a lower longevity and possess a lower drought and disease resistance than those propagated by seeds (Heller 1996). Thus, alternative technologies are needed for the production of uniform planting material in large numbers.

Currently, the sustainable and reliable propagation of *J. curcas* can be achieved via in vitro techniques. Micropropagation has been reported in *J. curcas* through both direct and callus-mediated shoot regeneration and somatic embryogenesis using different explant sources including epicotyls, hypocotyls, peduncles, axillary buds, nodal segments, immature embryo, leaf and shoot tips (Sujatha et al. 2005; Wei et al. 2004; Jha et al. 2007; Deore and Johnson 2008; Varshney and Johnson 2010). Although extensive work has been carried out using the above explants, there is no report on the morphogenesis of leaf transverse thin cell layers (tTCLs). Thin cell layers include a small number of cells from different tissue types: epidermal, cortical, cambium, perivascular and medullar tissue, parenchyma cells (Van Tran Thanh 1980). Specific cell or tissue layers in TCL systems depended on controlled growth conditions including PGRs and facilitated the in vitro induction of specific morphogenic responses (Teixeira da Silva 2003). Many previous researchers reported that tTCL explants from the surface of floral branches of tobacco could be induced to form callus, vegetative buds, flowers or roots by adjusting the pH and the ratio of auxin to cytokinin in the culture medium (Nhut et al. 2003). Hence, application of tTCL in regeneration and micropropagation of *J. curcas* is considered as a breakthrough for overcoming current difficulties.

Fig. 2.1 Organogenesis and somatic embryogenesis from *J. curcas* leaf tTCLs. Direct organogenesis from tTCLs: shoot clumps **a**$_1$; globular structures **a**$_2$; shoot buds **a**$_3$; elongated shoots **a**$_4$. Indirect organogenesis: compact, *green calli* **b**$_1$; globular structures **b**$_2$; shoot buds **b**$_3$; shoots **b**$_4$. Indirect somatic embryogenesis: soft, friable, *yellowish calli* **c**$_1$; globular embryo **c**$_2$; *heart-shape embryo* **c**$_3$; *Torpedo-shape embryo* **c**$_4$. 21-day-old plantlets from somatic embryogenesis **d** and complete hardened plants growing under controlled greenhouse conditions **e**

2.2 Direct Shoot Organogenesis From Leaf tTCLs

Direct multiple shoot formation from leaf explants is critical in order to produce a large number of uniform plants from the elite lines of *Jatropha curcas* plants. The current work showed that on MS (Murashige and Skoog 1962) medium supplemented with kinetin and BA small shoot clumps and globular structures were formed at the edge of explants without the formation of callus tissue after 20 days

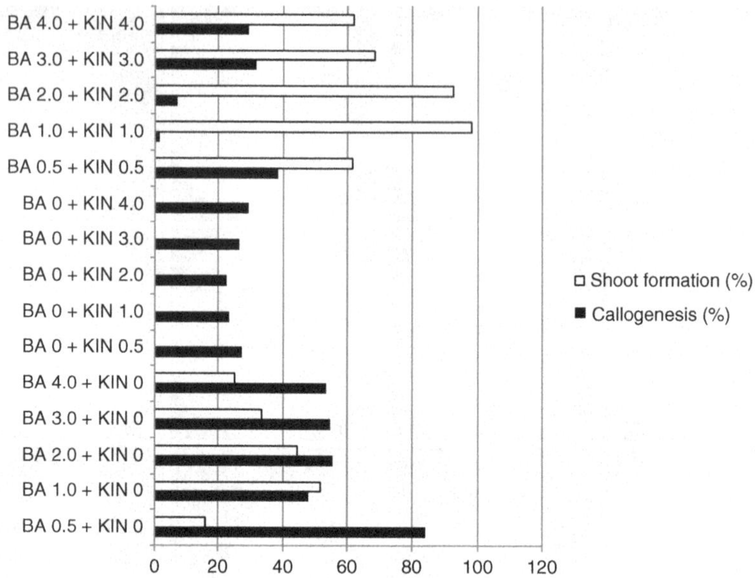

Fig. 2.2 Effects of BA and kinetin on direct formation of shoot clumps from leaf tTCL explants of *J. curcas*

of culture (Fig. 2.1a_1, a_2). Subsequently, these structures developed to form shoot buds (Fig. 2.1a_3). In order to obtain plantlets, clumps of regenerated shoots were transferred to MS medium augmented with 0.5 mg l^{-1} GA_3 and 20 % coconut water for shoot elongation (Fig. 2.1a_4). The highest rate of shoot clump regeneration from tTCLs was observed in response to 1 mg l^{-1} BA combined with 1 mg l^{-1} kinetin (98.27 %, Fig. 2.2).

In most of the Euphorbiaceae members, the presence of cytokinin alone gave optimal shoot proliferation such as 2-iP (0.3 mg l^{-1}) in *Euphorbia lathyrus* (Lee et al. 1982), BA in *E. peplus* (2.0 mg l^{-1}) and *E. tannensis* (2.0 mg l^{-1}) (Tideman and Hawker 1982). The present study showed that BA alone in media was not suitable for direct shoot regeneration from tTCLs of *J. curcas*. Sujatha and Mukta (1996) obtained shoots along with calli from *J. curcas* hypocotyl explants on MS medium supplemented with 9.12 μM zeatin while kinetin and BA incorporated separately in MS medium at various concentrations failed to support morphogenesis and induced necrosis. In another study, Sujatha et al. (2005) evaluated the response of *J. curcas* leaf explants to combinations of BA and IBA. They noted that adventitious buds originated directly from the region close to the induced greenish white callus. The highest frequency of shoot bud regeneration was recorded using 22.2 μM BA combined with 4.9 μM IBA (79 %). Deore and Johnson (2008) found that *J. curcas* leaf discs increased in size and adventitious shoot buds originating from the cut ends in contact to the medium surface using 2.27 μM TDZ, 2.22 and 0.49 μM IBA (53.5 %).

2.3 Indirect Shoot Organogenesis and Somatic Embryogenesis from Leaf tTCLs

Indirect morphogenesis is the development of shoots or somatic embryos subsequent to callus formation (Sharp et al. 1986). In vitro regeneration of plants via somatic embryogenesis has some distinct features, such as single-cell origin, low frequency of malformation, and the production of a high number of regenerants. tTCL explants were cultured on MS medium supplemented with BA (0.5–4.0 mg l^{-1}) alone or in combination with kinetin (0.5–4.0 mg l^{-1}) for inducing indirect shoot regeneration. Small shoot clumps were transferred to MS medium containing 0.5 mg l^{-1} GA_3 and 20 % young coconut water for shoot elongation. In order to obtain callus, MS media containing kinetin (0.5–2.0 mg l^{-1}) in combination with IBA (0.1–1.0 mg l^{-1}) or 2,4-D (1.0–2.0 mg l^{-1}) were used. The soft, friable, yellow calli exhibited somatic embryogenesis while the hard, compact, green calli gave indirect shoot bud organogenesis.

2.3.1 Callus Induction

There was no visible sign of callus proliferation in plant growth regulator-free media. For callus induction from leaf tTCLs of *J. curcas*, kinetin in combination with 2,4-D and IBA were necessary. Treatments containing kinetin and 2,4-D stimulated higher rates of callus induction than those containing kinetin and IBA

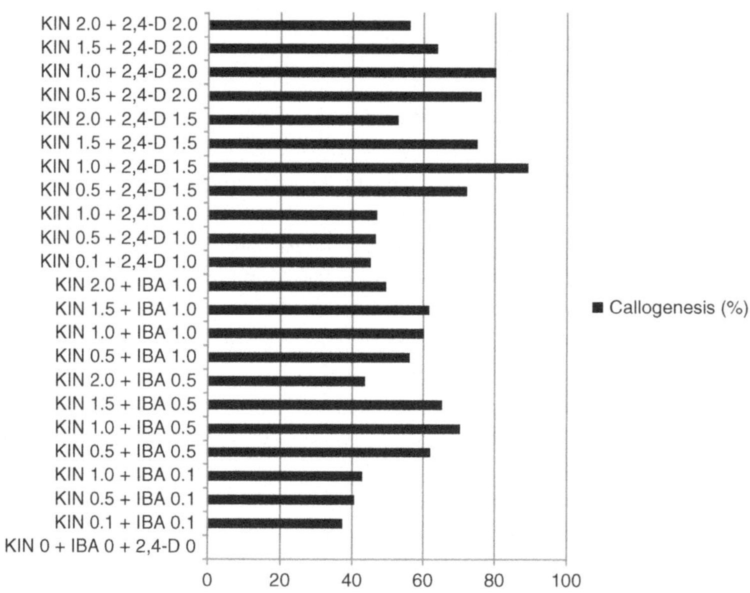

Fig. 2.3 Effects of kinetin in combination with IBA or 2,4-D on callus induction from leaf tTCL explants of *J. curcas*

(Fig. 2.3). Two types of callus were observed depending on the hormonal treatments. Media containing kinetin in combination with IBA stimulated the formation of green compact calli (Fig. 2.1b_1) while soft, friable, yellow calli were formed on media containing kinetin and 2,4-D (Fig. 2.1c_1). Green compact calli and yellow friable calli resulted in shoot bud induction and somatic embryogenesis, respectively (as described in Sects. 2.3.2 and 2.3.3).

Transverse thin cell layers from leaf explants cultured on PGR-free MS medium enlarged significantly after 7–8 days of culture. These explants, however, turned brown and became necrotic after 4 weeks of culture. These results are consistent with the report on the morphogenetic capacity of TCL explants of sugar beet, which was strongly dependent upon the presence of PGRs in the medium (Detrez et al. 1988).

Similarly, Varshney and Johnson (2010) also obtained these two types of callus using different growth regulators (2,4-D, IAA, IBA, NAA, Picloram, BA and Kinetin) to induce callus formation from immature embryo cultures of *J. curcas*. In certain cases, the callus became brownish after 4 weeks in culture when the media contained 1.5–2 mg l^{-1} kinetin and 1 mg l^{-1} IBA. Brown callus exhibited decreased regenerability, poor growth and even necrosis. The changing of callus from green to brown and dark brown was recorded by Yang et al. (2009). They found that during browning process of callus derived from *J. curcas* hypocotyls, chlorophylls and carotenoids concentrations decreased steadily. The darkening of callus might be caused by the release of phenolic compounds by the explants (Monacelli et al. 1995).

2.3.2 Indirect Shoot Organogenesis

When hard compact calli were subcultured to fresh medium containing 1.0 mg l^{-1} kinetin, growth in callus resumed and within 4 weeks nodulation was observed (Fig. 2.1b_2). Kinetin (1 mg l^{-1}) was found to be most effective for shoot bud induction wherein the nodular structures started to differentiate into an organized structure of shoot buds (Fig. 2.1b_3). These buds developed into healthy shoots after 4 weeks of culture (Fig. 2.1b_4). Nema et al. (2007) also recorded in *Tylophora indica* a similar pattern of shoot bud differentiation. They used 5 μM BA to induced shoot bud in *Tylophora indica*. The higher concentration of BA (10 μM) decreased the number of shoot buds.

To induce rooting, shoots (3–4 cm long) from either direct or indirect organogenesis were separated and transferred to half-strength MS medium supplemented with various concentrations of NAA and IBA. The most effective medium for root induction in *J. curcas* is medium contained 0.5 mg l^{-1} NAA (Fig. 2.4). Media supplemented with different concentrations of IBA were found to be unsuitable for rooting because of the callogenesis at the basal cut end of the shoots. The rooting of shoots through basal callogenesis was not recommended because in most cases, proper acclimatization of the plantlets could not be achieved. Similar findings were reported by Sujatha and Mukta (1996) and Kalimuthu et al. (2007).

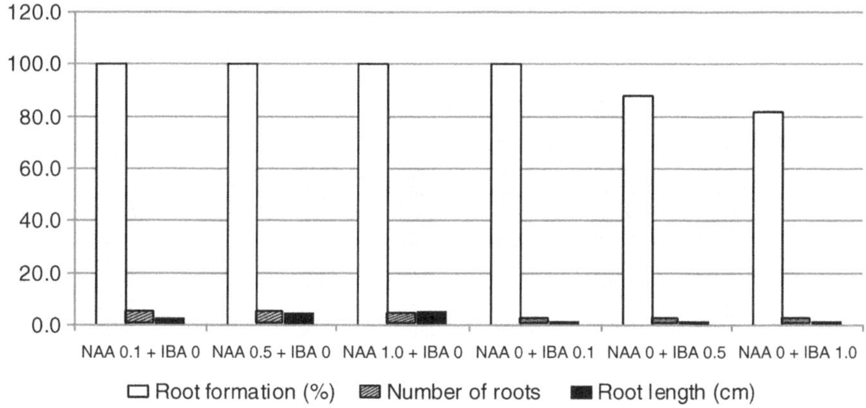

Fig. 2.4 Effects of NAA and IBA on the root formation of *J. curcas* in vitro shoots

2.3.3 Indirect Somatic Embryogenesis

Plant growth regulators are required for induction of somatic embryogenesis; and the most commonly used plant growth regulators for this purpose are 2,4-D, dicamba and picloram (Roostika and Mariska 2003). Kalimuthu et al. (2007) found that somatic embryos were induced directly from green cotyledon explants on MS medium supplemented with 2.0 mg l^{-1} BAP. The results on direct somatic embryogenesis showed that BAP was essential for inducing somatic embryogenesis from cotyledonary explants of *J. curcas*. However, Ramasamy et al. (2005) reported that auxin in combination with cytokinin not only influenced the frequency but also had a significant impact on the maturation of somatic embryos of *Solanum surattense*. Embryogenic callus (soft, friable, yellowish callus) was induced from leaf tTCLs of *J. curcas* cultured on MS medium supplemented with kinetin (0.5–2 mg l^{-1}) and 2,4-D (1–1.5 mg l^{-1}) (Figs. 2.1c_1, 2.3). Indirect somatic embryogenesis occurred when these calli were transferred to media with lower concentrations of 2,4-D in combination with 1.0 mg l^{-1} kinetin (Fig. 2.5).

These results differ from those of Jha et al. (2007) whom obtained embryogenic calli from leaf explants on MS basal medium supplemented with only 2.0 mg l^{-1} kinetin. These embryogenic calli were transferred to fresh media containing kinetin and IBA at various concentrations. They concluded that the most effective combination of kinetin and IBA for somatic embryo induction in *J. curcas* were 0.5 and 0.2 mg l^{-1}, respectively. This may be due to the size and source of explants. Jha et al. (2007) used 2.0 × 1.0 cm leaf explants of 7-month-old ex vitro plants compared to 0.5 × 10 mm leaf tTCLs explants excised from 45-day-old in vitro plants. The capacity of a TCL to regenerate depends upon a number of factors, including the physiological states and origin of TCL as well as the stress factors applied to the TCL (Teixeira da Silva 2003).

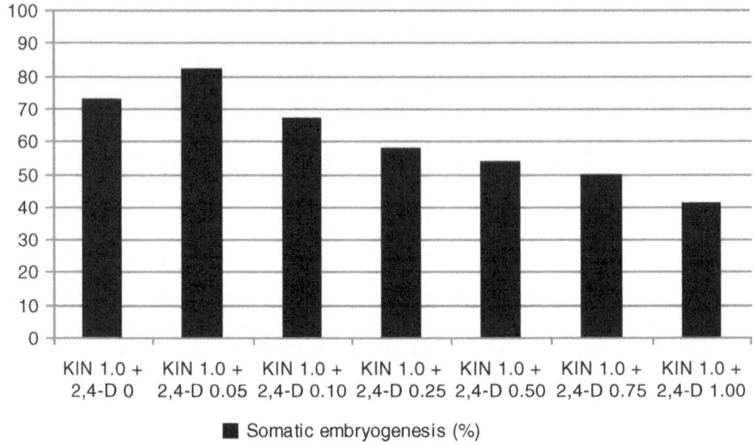

Fig. 2.5 Induction of somatic embryogenesis in calli on media containing lower concentrations of 2,4-D together with 1.0 mg l^{-1} kinetin

Within 14 days after transferring the calli to MS medium containing 1 mg l^{-1} kinetin and 0.05 mg l^{-1} 2,4-D, small globular glossy somatic embryos appeared from the upper surface of the callus (Fig. 1c$_2$). In concurrence with Jha et al. (2007) observations, embryos at different stages of development were isolated from the peripheral callus mass (Fig. 1c$_2$, c$_3$, c$_4$). These structures further developed into dark green colored embryos. Soon after mature somatic embryos were obtained, they were transferred to plant growth regulator-free half-strength MS medium to obtain fully developed cotyledonary leaves and vigorous plants. Root and shoot initiation took place simultaneously in these bipolar structures.

Plantlets with fully expanded leaves and well developed roots regenerated by either organogenesis or somatic embryogenesis (Fig. 2.1d) were acclimatized under the greenhouse conditions at high survival rates reaching 70 and 90 %, respectively (Fig. 2.1e).

2.4 Conclusion and Prospects

Leaf tTCLs culture of *J. curcas* was capable of undergoing different morphogenetic processes, including direct organogenesis, indirect organogenesis and indirect somatic embryogenesis. The factors determining regenerative competence and the redirection of plant growth and development remain largely undefined; however, it is apparent from recent studies that the interactions of environmental factors and thin cell layers can determine the developmental and morphogenic pathway of *J. curcas*.

In this respect the *J. curcas* regeneration system though tTCLs offers an interesting experimental system for investigation of the factors that favor somatic embryogenesis since explants exposed to the same culture medium underwent different morphological processes. It is likely that different levels of endogenous hormones predetermined the differential morphogenic potential of individual explants. An efficient procedure for micropropagation via organogenesis and somatic embryogenesis from tTCLs provides a suitable system for further plant selection and breeding investigations and for the mass production of high-quality *J. curcas* for the marketplace.

References

Deore AC, Johnson TS (2008) High-frequency plant regeneration from leaf-disc cultures of *Jatropha curcas* L.: an important biodiesel plant. Plant Biotechnol Rep 2:7–11

Detrez C, Tetu T, Sangwan RS, Sangwan-Norreel BS (1988) Direct organogenesis from petiole and thin cell layer explants in sugar beet cultured in vitro. J Exp Bot 39:917–926

Heller J (1996) Physic nut *Jatropha curcas* L. Plant Genetics and Crop Plant Research Institute, Rome

Jha TB, Mukherjee P, Datta MM (2007) Somatic embryogenesis in *Jatropha curcas* L., an important biofuel plant. Plant Biotech Rep 1:135–140

Kalimuthu K, Paulsamy S, Senthilkumar R, Sathya M (2007) In vitro propagation of the biodiesel plant *Jatropha curcas* L. Plant Tiss Cult Biotech 17:137–147

Lee CW, Yeckes J, Thomas JC (1982) Tissue culture propagation of *Euphorbia lathyris* and Asclepias erosa. HortScience 17:533

Monacelli B, Pasqua G, Cuteri A, Varusio A, Botta B, Monache GD (1995) Histological study of callus formation and optimization of cell growth in *Taxus baccata*. Cytobios 81:159–170

Murashige T, Skoog F (1962) A revised medium for rapid growth and bioassays with tobacco tissue cultures. Physiol Plant 15:473–479

Nema RK, Ramawat KG, Gupta GD, Tanwar YS, Mathur M (2007) Rapid micropropagation of *Tylophora indica*. Pharm Mag 3:52–55

Nhut DT, Aswath CR, Silva Teixeira da JA, Bui VL, Thorpe T, Van Tran Thanh K (2003) Tobacco thin cell layer morphogenesis. In: Thorpe T (ed) Thin cell layer cultures system: regeneration and transformation applications. Kluwer Academic Publishers, Dordrecht, pp 65–134

Ramasamy N, Ugandhar T, Praveen M, Venkataiah P, Rambabu M, Upender M, Subhash K (2005) Somatic embryogenesis and plantlet regeneration from cotyledons and leaf explants of *Solanum surattense*. Indian J Biotech 4:414–418

Roostika IT, Mariska I (2003) In vitro culture of Pineapple by organogenesis and somatic embryogenesis: its utilization and prospect. Buletin Agro Bio 6:34–40

Sharp WR, Sondhal MR, Caldar RS, Maraffa SB (1986) The physiology of in vitro asexual embryogenesis. Hortic Rev 2:268–310

Sujatha M, Makkar HPS, Becker K (2005) Shoot bud proliferation from axillary nodes and leaf sections of non-toxic *Jatropha curcas* L. Plant Grow Reg 47:83–90

Sujatha M, Mukta N (1996) Morphogenesis and plant regeneration from tissue cultures of *Jatropha curcas*. Plant Cell Tiss Org Cult 44:135–141

Teixeira da Silva JA (2003) Thin cell layer technology in ornamental plant micropropagation and biotechnology. Afr J Biotecnol 2:683–691

Tideman J, Hawker JS (1982) In vitro propagation of latex producing plants. Ann Bot 49:273–279

Van Tran Thanh K (1980) Control of morphogenesis by inherent and exogenously applied factors in thin cell layers. Intl Rev Cytol 32:291–311

Varshney A, Johnson TS (2010) Efficient plant regeneration from immature embryo cultures of *Jatropha curcas*, a biodiesel plant. Plant Biotechnol Rep 4:139–148
Wei Q, Lu WD, Loao Y, Pan SL, Xu Y, Tang L, Chen F (2004) Plant regeneration from epicotyl explants of *Jatropha curcas*. J Plant Physiol Mol Biol 30:475–478
Yang H, Xiulian G, Ran L, Bei N, Vijaya P, Pei H, Feng C, Ying X, Fang C (2009) Changes in morphology and biochemical indices in browning callus derived from *Jatropha curcas* hypocotyls. Plant Cell Tiss Org Cult 98:11–17

Chapter 3
New Achievement in *Panax vietnamensis* Research

Duong Tan Nhut, Nguyen Thanh Hai, Nguyen Phuc Huy, Hoang Xuan Chien and Nguyen Ba Nam

Abstract *Panax vietnamensis* Ha et Grushv., an endemic *Panax* genus of Vietnam, is a well known Vietnamese ginseng (Ngoc Linh ginseng) rich in pharmaceutical compounds, most importantly saponin. Its cultivation takes a long time, generally 5–7 years, and needs extensive efforts to quality control in the face of environmental stresses including soil, shade, climate, pathogens and pests. In vitro techniques have been widely explored for rapid and efficient production of ginseng biomass and ginsenosides. The establishment of cell and adventitious root cultures of *P. vietnamensis* opens the way to commercial applications. Various physiological and biochemical parameters affecting the biomass production and ginsenoside accumulation have been investigated. These parameters are effect of various phytohormones, sucrose and activated charcoal (AC) on shoot regeneration and proliferation from callus, and adventitious and secondary root formation. The saponin analysis of calli and roots showed the presence of ginsenoside-Rg_1, majonoside-R_2, and ginsenoside-Rb_1. These results indicated that *P. vietnamensis* biomass has a great potential to produce saponin as a new source for the pharmaceutical and cosmetic industry.

Abbreviations

AC	Activated charcoal
MS	Murashige and Skoog
NAA	α-napthaleneacetic acid
2, 4-D	2, 4-dichlorophenoxyacetic acid
TDZ	Thidiazuron
IBA	Indole-3-butyric acid

D. T. Nhut (✉) · N. T. Hai · N. P. Huy · H. X. Chien · N. B. Nam
Tay Nguyen Institute of Biology, 116 Xo Viet Nghe Tinh, Dalat, Lam Dong, Vietnam
e-mail: duongtannhut@gmail.com

S. M. Jain and S. Dutta Gupta (eds.), *Biotechnology of Neglected and Underutilized Crops*, DOI: 10.1007/978-94-007-5500-0_3,

3.1 Introduction

The word "ginseng" derives from the Chinese term, based on the roots of several distinct species of plants, mainly Korean or Asian ginseng (*Panax ginseng*), Siberian ginseng (*Eleutherococcus senticosus*), and American ginseng (*Panax quinquefolius*). All these species belong to the Araliaceae family. They are well known for various effects on human health. Recently, researchers in Asia and the West have been paying great attention to ginseng research. So far, more than 200 ginsenosides and non-saponin constituents have been isolated (Zang 2006). Ginsenosides show significant biological activities in physiological and pathological aspects. The demand for ginseng has dramatically increased worldwide, and has become very expensive due to inability to increase production. Moreover, traditional methods take longer time, 5–7 years, and challenge the production cycles. Plant biotechnology is an efficient alternative approach for the mass production of ginseng and its active components. Several commercial products including health foods, drinks and cosmetics are produced from powder and extracts of ginseng cell cultures. The ginseng culture has continued to attract considerable research and development efforts in the recent years and scientists seek to understand and optimize the culture conditions. In Vietnam, *P. vietnamensis* Ha et Grushv. was found on highland of Central Vietnam in 1973, and was regarded as a precious medicinal plant of local residents. *P. vietnamensis* is routinely used by the Sedang ethnic group, unknown to others, and use as a life-saving plant drug to treat many serious diseases, and to enhance body stamina during long journeys in high mountains. Nowadays *P. vietnamensis* supply is very limited because it grows slowly mainly in Ngoc Linh mountain area. Due to excessively harvesting, the ginseng is among 250 endangered species at high risk of extinction (Vietnam's Red Data book). This ginseng species is used specifically for certain physical actions: anti-stress, anti-depression, in vitro and in vivo antioxidation, etc. Saponin triterpenoic compounds are the main effective group of compounds in the species. Chemical studies of the constituents of the plant have identified 23 saponins including 14 new compounds. Some of them are common to *P. ginseng*, such as protopanaxadiol and protopanaxatriol saponins, however, high concentration in this species. In addition, high concentration of ocotillol saponins, i.e., majonoside-R_2 (5.3 % of the dried rhizome) was determined. Moreover, *P. vietnamensis* possessed the highest dammaran-frame saponin (12–15 %) and saponin content among *Panax* genus. With these special features, this ginseng is one of the most valuable species in Vietnam and worldwide (Dong et al. 2007).

The initial work on propagation and culture medium improvement of this ginseng species had limited success Dung et al. (1995). Nhut et al. (2009a, 2009b) studied secondary root formation of *P. vietnamensis*, and the best results were obtained on SH medium supplemented with 3.0 mg/l 2,4-D. Jacques et al. (2007) investigated optimum conditions to increase ginseng biomass in bioreactor. Recently, Duong et al. (2008) analyzed to quantify ginsenoside-Rg_1, -Rg_2, -Rd in cell extract from ginseng biomass by HPLC. Apparently, biomass collection and saponin analysis are essential to validate in vitro growth. In the current research, the work is being carried out on the effects of medium compositions, culture conditions, explant size for in vitro multiplication of *P. vietnamensis* from callus induction to shoot and root regeneration and quantification of saponin from biomass collected from in vitro cultures.

3.2 Micropropagation and Morphogenetic Programs

Since the traditional propagation of *P. vietnamensis* through seed and stem cuttings is recalcitrant, *in vitro* techniques especially micropropagation would be an ideal approach for mass production; ginseng large-scale cell suspension culture for the production of pharmaceutical active compounds and fresh raw materials. Leaves and petioles explants of *P. vietnamensis*, grown at Tay Nguyen Institute of Biology, used to induce calli; washed with 5 % sodium hypochlorite solution (Nhut et al. 2010), and followed by 2 h continuous washing with water. Explants were treated with 70 % alcohol while shaking for 30 s, and continuously rinsed 4–5 times with sterile distilled water; treated with 0.1 % $HgCl_2$ containing a few drops of Tweens-20, for 5 min, and washed in distilled water to remove $HgCl_2$. The sterilized leaves were chopped in small pieces with a scalpel, size 1.0 × 10 cm. The petioles were vertically cleft, and cut in 1.0 cm long pieces.

The culture media used were MS basal medium (Murashige and Skoog 1962), modified MS ½ (originated essential minerals and half of microminerals) and modified MS ½ (half essential and microminerals) media supplemented with 30 g/l sucrose, 8.0 g/l agar and pH 5.7. The suitable culture conditions for callus induction and development, shoot regeneration and proliferation are: 25 ± 2 °C, 16 h photoperiod, light intensity of 40 μ $mol.s^{-1}.m^{-2}$ fluorescent light, and relative humidity 75–80 %. Root regeneration and proliferation carried out in the darkness.

3.2.1 Callus Induction

It is important to standardize the protocol for ginseng shoot regeneration through callus. The regenerated plants are genetically identical to the original material, pathogen free and produce a large number of plantlets in a very short duration. However, the response of explants for callus induction and regeneration differ in different media. Therefore, it is required to establish an effective protocol for rapid plant regeneration from leaf and petiole explants.

3.2.1.1 Effect of Auxin Types and Concentration on Callus Induction from Leaf and Petiole

Research on *Panax* genus showed that callus induction stage usually required a combination of cytokinins and auxins. When Korean ginseng seeds are cultured,

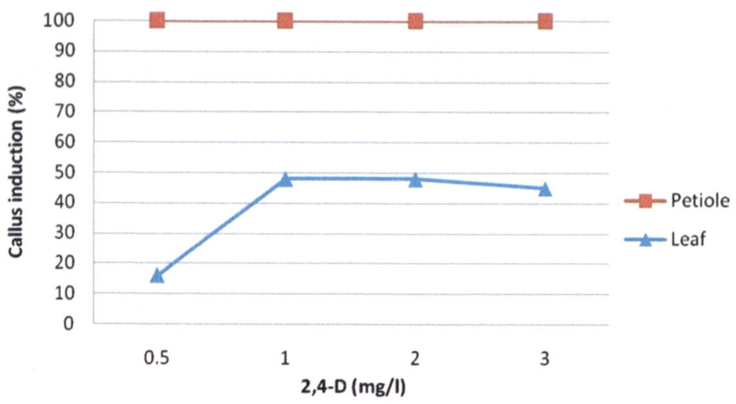

Fig. 3.1 Effect of 2,4-D on callus induction from leaf and petiole

MS induction medium supplemented with 1 mg/l 2,4-D and 0.01 mg/l kinetin was used (Arya et al. 1993) while leaf and other explants were cultured MS medium containing 1 mg/l 2,4-D and 0.1 mg/l kinetin (Lim et al. 1997). Nhut et al. (2011) used 0.2 mg/l TDZ in combination with 1.0 mg/l 2,4-D to investigate the effects on callus induction from *P. vietnamensis* leaf explants.

The effect of 2,4-D on callus induction from leaf and petiole explants after 8 weeks of culture is shown in Fig. 3.1. Among the 3 auxins (2,4-D, IBA, NAA) tested in the induction medium, 2,4-D exhibited the ability to stimulate callus formation from leaf and petiole. On the induction culture medium containing 1.0 mg/l 2,4-D, cultured explants produced highest callus formation, 90 % from leaf and 100 % from petiole explants with a high number of rigid structure and bright yellow calli. 3.0 mg/l 2,4-D induced crystal callus formation on the same explants (Nhut et al. 2009a, 2009b). Radhakrishnan et al. (2001) suggested that the cells can only utilize limited amount of auxin and over-use of auxins at any level inhibits cell growth. Therefore, above 3.0 mg/l 2,4-D is not suitable for callus induction from *P. vietnamensis* leaves.

3.2.1.2 Effect of Lighting Condition on Callus Induction from Leaf and Petiole

Normally, induced callus cultures are kept under light or in the darkness depending on the type of explants, however, leaf-derived callus is maintained in the darkness. In some cases, culture explants produce better calli under the light. The previous reports indicated that the number and quality of calli, grown in the darkness, are lower than in the light due to crystalline formation especially in the medium supplemented with 3.0 mg/l 2,4-D (Fig. 3.2a, b). These observations validate the results on Korean ginseng (Lim et al. 1997). Therefore, the lighting period 16 h/day is ideal to stimulate callus formation from *P. vietnamensis* leaves.

Fig. 3.2 Callus formation and multiplication of *P. vietnamensis*. **a** Callus formation from leaves on medium containing 3.0 mg/l 2,4-D under 16 h photoperiod. **b** Callus formation from leaves on medium containing 3.0 mg/l 2,4-D in the darkness. **c** Callus multiplication on medium containing 1.0 mg/l 2,4-D under 16 h photoperiod. **d** Callus multiplication on medium containing 1.0 mg/l 2,4-D in the darkness

3.2.1.3 Effect of Auxin Types and Concentration on Callus Multiplication

Our results showed that after multiplication stage, calli cultured on the medium supplemented with 0.5 mg/l IBA and the medium containing 1.0 mg/l 2,4-D had the highest dry weight and highest increase in callus dry weight, respectively (Fig. 3.2c, d).

Medina et al. (1998) demonstrated that carbohydrate is responsible for most of callus dry weight. The main carbon source in the culture medium is sucrose, and sugar utilization by callus depends on the medium and explant source - cotyledon, petiole, hypocotyl and leaf. The utilization of both auxin and cytokinin may help calli improving sugar and other nutrients absorption from the culture medium, and results in callus development and biomass dry weight increase. According to our results, IBA is more effectively used together with TDZ, rather with NAA and 2,4-D. Dry weight ratio of calli in 0.5 mg/l IBA-containing medium is highest among 3 treatments using 3 auxins. Although dry weight of calli in IBA-containing medium is highest, 2,4-D increased the highest dry weight ratio and relative high callus fresh weight. Alternatively, calli had the best conditions for growth and were highly regenerative in 2,4-D-containing medium.

3.2.1.4 Effect of Explant Size on Callus Development

Explant size is one of the most critical factors in *in vitro* multiplication. The initial difference in explant size can lead to a significant difference in cell density during the multiplication process, which can lead to many other changes in the culture medium (Akalezi et al. 1999). When investigating the effects of initial callus size on callus development, we learned that the smallest size (0.5 × 0.5 cm) gave the highest increase in fresh and dry weight. Some products released during the callus growth in the culture medium, which have toxic feedback to calli growth. Garcia and John (1983) observed that tobacco calli grown in the presence of 0.5–25 mg 2,4-D, higher internal ethane and ethylene were produced in the medium, which later decelerated callus multiplication. Also at same 2,4-D concentration, bigger callus size could produce more ethylene and ethane would be toxic to further callus development of *P. vietnamensis*. Therefore, small callus size is ideal to produce less ethylene and ethane and has higher rate of development.

3.2.2 Shoot Formation

3.2.2.1 Shoot Regeneration from Callus

The auxin and cytokinin ratio plays an important role in shoot regeneration. Cytokinins usually promote shoot formation, which can be stimulated with a low concentration of auxins. The studies on the shoot regeneration ability in *P. vietnamensis* callus with the addition of BA and NAA in the culture medium indicated that 1.0 mg/l BA and 1.0 mg/l NAA gave the highest shoot number (6.3 shoots/explant) with an average weight 0.185 g (Fig. 3.3a).

3.2.2.2 Effect of BA on Shoot Development

With the addition of 1.0 mg/l BA and 0.5 mg/l NAA in the culture medium, the best response was on shoot regeneration from *P. vietnamensis* callus, 0.87 g fresh weight of new shoots and height 6.16 cm.

3.2.2.3 Effect of Sucrose Concentration on Shoot Development

Normally, shoot regeneration requires 30–120 g/l sucrose in the culture medium. *P. vietnamensis* shoot development showed that adding sucrose into culture

Fig. 3.3 Shoot regeneration from *P. vietnamensis* callus. **a** Shoot regeneration from callus. **b**, **c** Root formation on medium containing AC

medium had positive effects on shoot growth. The increase of sucrose concentration in the culture medium is not only stimulated shoot development but also effectively increase their fresh weight; 50 g/l was ideal to produce best results on weight, height and leaf number (Fig. 3.4).

3.2.2.4 Effect of AC on Shoot Development

Activated charcoal is not a plant growth regulator, however, added to the medium to maintain pH and absorb chemicals preventing the development of tissues. Moreover, AC is beneficial to shoot development and increase in shoot

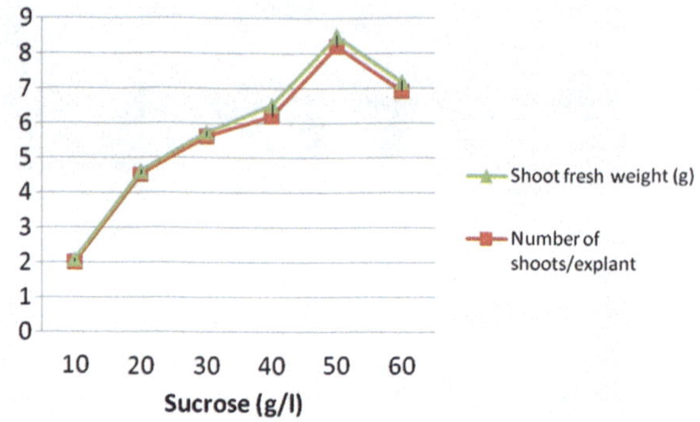

Fig. 3.4 Effect of sucrose concentration on shoot regeneration from callus

fresh weight. Our results indicated that an increase in AC concentration could lead to a considerable change in either shoot weight or height, but not the number of leaves. The highest shoot weight was 1.01 g/shoot with the addition of 2.0 g/l AC in the culture medium, an increase by 1.9 fold with control (Fig. 3.3b, c). Thus, 2.0 g/l AC is the optimal concentration for *P. vietnamensis* shoot development.

The calli of *P. vietnamensis* was successfully induced from rhizome explants on MS medium containing 2,4-D and TDZ under 16 h photoperiod. The highest callus biomass yield was obtained on MS medium supplemented with 1.0 mg/l 2,4-D and 0.2 mg/l TDZ. The highest number of regenerated shoots from callus was in MS medium amended with 1.0 mg/l BA, 1.0 mg/l NAA and 50 g/l sucrose. In plantlet formation phase, 1/2 MS medium containing 1.0 mg/l BA, 0.5 mg/l NAA. 50 g/l sucrose and 2.0 g/l AC. After 6 months, under the lighting conditions 70 % red LEDs and 30 % blue LEDs is the most suitable medium and condition for plantlet formation. After 6 months growth in the nursery, in vitro derived plantlet growth rate was faster than seed-derived plant growth rate and the survival rate was 87 %.

3.3 Adventitious Root Formation

In general, ginseng adventitious roots are an excellent starting material for the production of bioactive secondary metabolites without forging genes. However, adventitious root regeneration requires an understanding of basic research on plant physiology and biochemistry. So far, few reports are available on *P. vietnamensis* adventitious root and secondary root cultures. The more knowledge of the metabolism and mineral nutrient uptake in root cultures, the more strategies for large-scale production need to be accomplished. Therefore, several studies were carried out to investigate to optimize environmental factors, which have crucial role in growth and development of ginseng explant in vitro.

3.3.1 Adventitious Root Formation from Callus

While investigating the effects of three auxins (IAA, IBA, NAA) we learned that IAA was not suitable for *P. vietnamensis* root formation. NAA and IBA could well stimulate rooting process. The two best results were obtained with the addition of 3.0 mg/l NAA and 5.0 mg/l IBA which gave 100 % root formation, 8.7 roots/explant, fresh weight root to explant ratio by an average of 21.88 %, and a longest root average of 13 mm; and 100 % root formation, 4.8 roots/explant, fresh weight root to explant ratio by an average of 15.81 %, and a longest root average of 18 mm, respectively (Fig. 3.5a). These phenomena could be explained by higher activity of synthetic auxins (IBA, NAA, 2,4-D) than the natural one (IAA). IAA was not able to stimulate root formation due to its low biological activity and its sensitivity to enzyme activity. George and Sherington (1984) observed that IAA, IBA and NAA used for root formation, among them, IBA was most effective. Moreover, auxin not only stimulated root development but also increased explant fresh weight; NAA and IBA were more effective than IAA (Kull and Arditti 2002).

3.3.2 Adventitious Root Multiplication

In order to choose the most suitable auxin for adventitious root regeneration and multiplication in *P. vietnamensis*, we continued to multiply adventitious roots using two auxin types, IBA and NAA. Our experiments indicated that NAA was suitable for adventitious root multiplication of *P. vietnamensis*. NAA at concentration 5.0 mg/l was optimal for root multiplication with the highest root formation rate (60%), the highest secondary root formation (9 roots/explant) and the highest weight increase (average fresh weight: 390 ± 20 mg, increase 3.5 folds

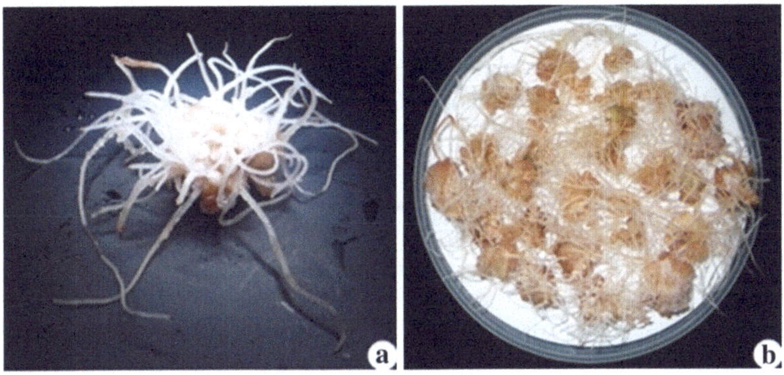

Fig. 3.5 Effects of auxins on root formation and multiplication from *P. vietnamensis* callus. **a** Adventitious root formation from callus. **b** Root multiplication on medium supplemented with NAA

over the original one). In addition, five among six treatments supplemented with NAA showed root formation while IBA showed four among six treatments. As a result, 3.0 mg/l NAA is most suitable for root formation from callus and 5.0 mg/l NAA is most suitable for adventitious root multiplication at *P. vietnamensis*.

3.3.3 Saponin of In Vitro Cultured P. vietnamensis Biomass

TLC analysis was used to determine Rf values of the compounds by their positions and colors showing on the plate. The results indicated that MR_2 G-Rg_1, and G-Rb_1 were present in calli when compared with explant color and position on the plate with the standard compounds. Especially, the color chart from root weight showed the presence of three standard ginsenosides.

Ginsenoside types of callus and in vitro cultured biomass of ginseng genus depended on explant sources and supplemented auxin (Bonfill et al. 2002; Furuya et al. 1986). The ratio of group Rb/Rg in 2 year-old Korean ginseng rootstalk was 0.49 in the presence of 2,4-D after 5 weeks of culturing (Bonfill et al. 2002). William (2000) reported Rb group amount in 2 year-old Korean ginseng rootstalk was lower than Rg (0.6 vs. 1.0 %), indicating that saponin accumulation in Korean Ginseng callus is similar with natural explants. When trying to produce saponin from adventitious roots of Korean ginseng, Langhansova et al. (2005) found that total weight of ginsenosides of cultured roots in a bioreactor was about 14.48 mg/g biomass, while natural ginseng root contained 33.12 mg/g. In the total biomass, 5.02 mg/g G-Rb and 9.46 mg/g G-Rg compounds are present when compared with 15.06 mg/g G-Rb and 18.06 mg/g G-Rg in roots of Korean ginseng (Langhansova et al. 2005). These results showed that even in in vitro conditions, collected biomass is able to synthesize compounds, which are similar to the original explants.

Although there is no evidence of ginsenoside Rb and Rg group presence, our results showed that G-Rg_1, a representative of 20(*S*)-protopanaxatriol group in callus, is present in very low amount in the original *P. vietnamensis* leaves. Analyzing saponin components in *P. vietnamensis* leaves showed 20(*S*)-protopanaxadiol derivatives, but not G-Rg_1, hold a high ratio among saponin in stems and leaves (Dong et al. 2007); shows inability of G-Rb_1 detection in leaf and shoot-derived callus. Moreover, majonoside-R_2 was present in callus, absent in saponin components from leaves. Therefore, auxin and cytokinin had some influence on the multiplication process, and callus cells can synthesize majonoside-R_2 themselves. This saponin is critical for medical uses of this ginseng.

Metabolite extracts from *P. vietnamensis* tissue grown in nature (1), and biomass of the petiole lTCL-derived calli (2) were run on TLC plate along with authentic standards of MR_2, G-Rb_1 and G-Rg_1 (Table 3.1, Fig. 3.6). The results showed that biomass of calli included MR_2, G-Rb_1 and G-Rg_1. Furthermore, metabolite extracted from biomass of calli also had the other bands corresponding to those present in extract from plant grown in nature, suggesting that calli had

Table 3.1 Result of the saponin content in petiole lTCL-derived callus sample of *P. vietnamensis*

Weight of sample (mg)	Standard	Peak area	Content of saponin in sample (μg)	(%)
565.3	G-Rg_1	924368 ± 23	0.320324 ± 0.00001	0.061
	MR_2	483832 ± 28	2.213238 ± 0.00013	0.424
	G-Rb_1	378858 ± 25	0.454827 ± 0.00003	0.087

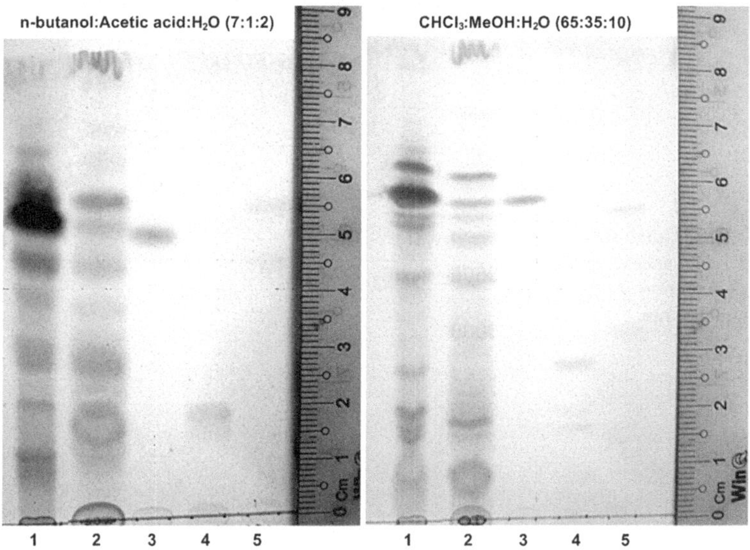

Fig. 3.6 Fractions eluted from petiole lTCL-derived callus sample and reference sample. **1** Reference sample. **2** Biomass of calli. **3** MR_2. **4** G-Rb_1. **5** G-Rg_1

similar chemical profile with those in natural environment. Saponins from *in vitro* *P. vietnamensis* calli were also analyzed using high-performance liquid chromatography with photodiode array detector at 190 nm (for MR_2), and 203 nm (for G-Rb_1, and G-Rg_1) (Figs. 3.7, 3.8). With authentic saponin standards, HPLC analysis revealed that all three important saponins of *P. vietnamensis* were present in the petiole lTCL-derived calli at high abundance. They included MR_2 (0.424 %), G-Rg_1 (0.061 %), and G-Rb_1 (0.087 %).

In *P. vietnamensis* adventitious roots, all three main saponin groups were present: G-Rb_1 representative of 20(*S*)-protopanaxadiol group, G-Rb_1 representative of 20(*S*)-protopanaxatriol group, and MR_2 representative of ocotillol group. Although these compounds were not quantified yet, according to the strength of visualized colors, we could infer that MR_2 has the highest amount, followed with G-Rg_1 and finally with G-Rb_1. These results are consistent with saponin

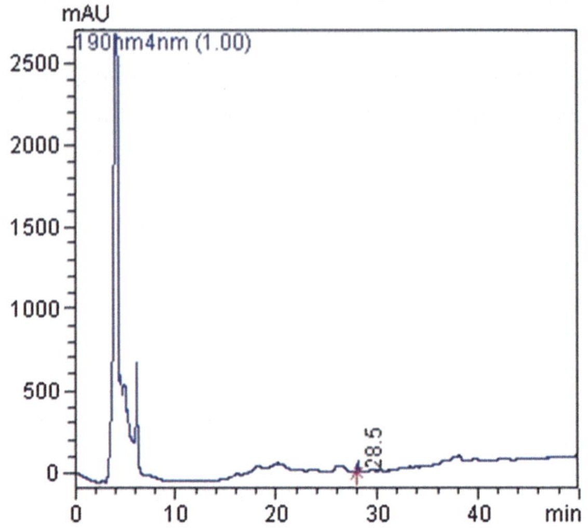

Fig. 3.7 HPLC analysis of *in vitro P. vietnamensis* callus with PDA detection at UV wavelength 190 nm

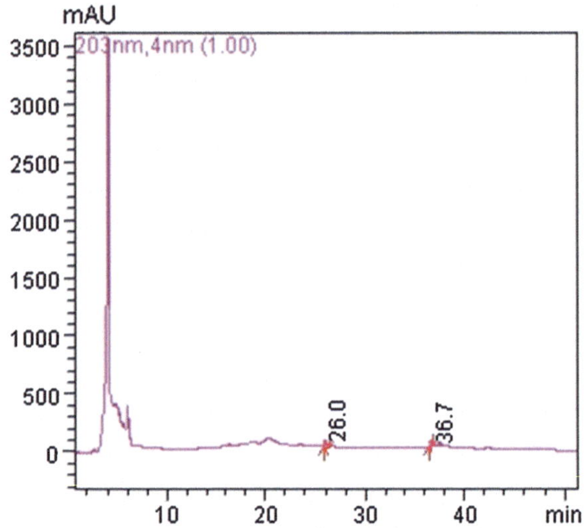

Fig. 3.8 HPLC analysis of *in vitro P. vietnamensis* callus with PDA detection at UV wavelength 203 nm

components presenting in natural *P. vietnamensis* roots, with 50 % MR_2 in total saponin in root and rootstalk.

Besides chemical bands, which had the same positions with standard ones, there were different bands with other colors (green, or yellow, etc.) on the TLC plate. These alien bands indicated that auxin and cytokinin had stimulated the synthesis of non-saponin compounds in callus development. Types and compositions of these compounds are yet to be identified. In addition, there were some other bands, which almost had the same positions with two standard samples, Korean ginseng and *P. vietnamensis*. There may be some other saponins, besides the standard ones, with low concentration, in ginseng callus cultures.

These results suggested that this ITCL morphogenesis protocol is feasible for *in vitro* culture of *P. vietnamensis*. There is no difference in saponin profile obtained from calli compared to field- grown plants.

3.4 Conclusion and Prospects

In recent years, *P. vietnamensis* has been extensively studied on pharmacological aspects, micropropagation and biomass production. The results suggested that *P. vietnamensis* regeneration and adventitious root protocols should strictly follow the steps regarding many biological and physical factors in the culture condition. *P. vietnamensis* callus was successfully induced from leaves and petioles cultured in MS medium supplemented with 1.0 mg/l 2,4-D, 0.2 mg/l TDZ under 16 h photoperiod. In callus multiplication stage, MS medium containing 1.0 mg/l 2,4-D, 0.3 mg/l TDZ was suitable for callus induction. Callus with small size (~0.5 × 0.5 cm) showed high growth rate. Callus regeneration was optimal on MS medium amended with 1.0 mg/l BA and 1.0 mg/l NAA. Subsequently, ½ MS

Fig. 3.9 Regeneration protocol of *P. vietnamensis*. **a** Callus induction. **b** Shoot and plantlet formation from callus. **c** Plantlet acclimatization. **d** Six-month-ginseng plant derived from *in vitro* plantlets

medium containing 1.0 mg/l BA, 0.5 mg/l NAA, 50 g/l sucrose and 2.0 g/l AC was most suitable for shoot development. In case of callus-derived root formation, the callus culture on MS ½ medium added with 3.0 mg/l NAA produced the highest root formation rate, root number and root fresh weight. MS ½ medium containing 5.0 mg/l NAA stimulated root multiplication and gave the highest secondary root formation. Saponin analysis showed that G-Rg_1, MR_2 and G-Rb_1 present in callus and root biomass.

Recently Nhut et al. (2011) demonstrated high efficiency of plant regeneration with applied biotechnology in *P. vietnamensis* (Fig. 3.9). Biomass production through adventitious root is applicable to commercial propagation of *P. vietnamensis*. However, many problems still exist to scale-up and to improve saponin quantity in biomass. The main problems are: difficulties in controlling biochemical pathway, preparation of explant and determination of the optimum culture conditions. They all depend on explant type, and different factors affected in culture process.

The efficient production of *P. vietnamensis* adventitious roots and biomass has great future prospect to enhance productive of pharmacy and cosmetic ginsenosides.

Acknowledgments The authors would like to thank the National Foundation for Science and Technology Development (Ministry of Science and Technology, Vietnam), the Committee for Local Science and Technology (Ministry of Science and Technology, Vietnam) and the Department of Application and Development of Technology (Vietnam Academy of Science and Technology) for the financial support.

References

Akalezi CO, Liu S, Li QS, Yu JT, Zhong JJ (1999) Combined effects of initial sucrose concentration and inoculum size on cell growth and ginseng saponin production by suspension cultures of *Panax ginseng*. Process Biochem 34:639–642

Arya S, Arya IDI, Eriksson T (1993) Rapid multiplication of adventitious somatic embryos of *Panax ginseng*. Plant Cell Tiss Org Cul 34:157–162

Bonfill M, Rosa MC, Javier P, Teresa PM, Morales C (2002) Influence of auxins on organogenesis and ginsenoside production in *Panax ginseng* calluses. Plant Cell Tiss Org Cul 68:73–78

Dong NT, Luan TC, Huong NTT (2007) Ngoc Linh Ginseng and some medicinal plants belong to Ginseng family. Science and Technology Publishing House, Vietnam

Dung NN (1995) Propagation of *Panax vietnamensis* by biological technology. Agricultural Publishing Houses, Vietnam

Duong VB, Dao Don DV, Long NV, Luong HV, Lau TV (2008) Study on quantitative analysis of some main ginsenosides of Ngoc Linh Ginseng by HPLC method. Med J 390:41–43

Furuya T, Yoshikawa T, Ushiyama K, Oda H (1986) Formation of plantlets from callus cultures of ginseng (*Panax ginseng*). Experientia 42:193–194

Garcia FG, John WE (1983) Ethylene and ethane production in 2,4-D treated and salt treated tobacco tissue cultures. Ann Bot 51:287–295

George EF, Sheringtoln PD (1984) Plant propagation by tissue culture. Exegetics Ltd., Eversley

Jacques P, Kevers C, Gaspar T, Dommes J, Thonart P (2007) Condition *Panax vietnamensis* cell mass production in bioreactors. Acta Bot Gallica 154(1):21–26

Kull T, Arditti J (2002) Orchid biology reviews and perspective, vol 8. Kluwer Academic Publishers, The Netherland, pp 443–487

Langhansová L, Maršík P, Vaněk T (2005) Production of saponins from *Panax ginseng* suspension and adventitious root cultures. Biol Plantarium 49(3):463–465

Lim HT, Lee HS (1997) Regeneration of *Panax ginseng* C. A. Meyer by organogenesis DNA analysis of regenerants. Plant Cell Tiss Org Cul 49:179–187

Medina M, Villalobos N, Cruz PJDL, Dorado A, Guerra H (1998) Effect of culture medium and light conditions on the morphological characteristics and carbohydrate contents of *Medicago strasseri* calli. Acta Physiol Plant 20(4):383–392

Murashige T, Skoog F (1962) A revised medium for rapid growth and bioassays with tobacco tissue cultures. Physiol Plant 15:473–497

Nhut DT, Luan VQ, Binh NV, Phong PT, Huy BN, Ha DTN, Tam PQ, Nam NB, Hien VT, Vinh BT, Hang LTM, Ngoc DTM, Thao LB, Luan TC (2009a) The effects of some factors on in vitro biomass production of vietnamese ginseng (*Panax vietnamensis* ha et grushv.) and preliminary analysis of saponin content. J Biotechnol Viet 7(3):365–370

Nhut DT, Tam PQ, Luan VQ, Binh NV, Ha DTN, Chien HX, Nhan NCT Nam NB, Truc NB, Thuy LNM, Nguyen PLH, Hien VT, Hang LTM, Hang NTT, Hai NT (2009b) Adventitious and secondary root formation of *Panax vietnamensis* cultured in vitro. In: Proceedings of biotechnology conference of southern provinces of Vietnam. Science and Technology Publishers, pp. 252–258

Nhut DT, Hang LTM, Vinh BT, Tam PQ, Nam NB, Nhan NCT, Chien HX, Thuy LNMT, Hien VT, Binh NV, Luan VQ, Luan TC, Duc DT (2010) Identification of saponin content and residue of some phytohormones in callus, shoot and root of *Panax vietnamensis* cultured in vitro. J Biotechnol Viet 8(2):189–202

Nhut DT, Huy NP, Luan VQ, Binh NV, Nam NB, Thuy LNM, Ha DTN, Chien HX, Huong TT, Cuong HV, Cuong LK, Hien VT (2011) Shoot regeneration and micropropagation of *Panax vietnamensis* Ha et Grushv. from *ex vitro* leaf-derived callus. Afr J Biotechnol 10(84):19499–19504

Radhakrishnan T, Murthy TGK, Chandran K, Banyopadhyay A (2001) Somatic embryogenesis in *Arachis hypogaea* revisited. Aust J Bot 49:753–759

William EC (2000) Ginseng. Harwood Academic Publishers, The Gordon and Breach Publishing Group

Zhang JT (2006) The chemistry, metabolism and biological activities of ginseng. Chemical Industry Press, Beijing, pp 3–17

Part II
Section 2

Chapter 4
Molecular Biology and Physiology of the Resurrection Glacial Relic *Haberlea Rhodopensis*

Maria Benina, Veselin Petrov, Valentina Toneva, Atanaska Teneva and Tsanko Gechev

Abstract *Haberlea rhodopensis* is a glacial relic with impressive tolerance to desiccation and freezing stress. It is mainly known as an ornamental plant and its other potential uses were largely neglected. Transcriptome analyses by next generation sequencing and cDNA–AFLP technologies identified transcription factors, stress-related and novel genes that could contribute to drought tolerance. These recent molecular studies have raised possibilities of gene discovery for crop improvement and established *H. rhodopensis* as one of the models to study desiccation tolerance. Furthermore, the abundant secondary metabolites of *H. rhodopensis* are rich sources of compounds with medicinal properties. Extracts from *Haberlea* possess radioprotective, anticlastogenic and antioxidant activities and can stimulate regeneration of human fibroblasts *in vitro*. All this has rejuvenated the interest in *H. rhodopensis* and indicated potential applications of this species in biology and medicine.

Abbreviations

AFLP	Amplified fragment length polymorphism
ELIPs	Early-light inducible proteins/genes
ESC	Expressed sequence contigs
GABA	γ-aminobutyric acid
LEA	Late embryogenesis-abundant proteins
NGS	Next generation sequencing
ROS	Reactive oxygen species
RWC	Relative water content

Maria Benina and Veselin Petrov authors contributed equally to this work.

M. Benina · V. Petrov · V. Toneva · T. Gechev (✉)
Department of Plant Physiology and Molecular Biology, University of Plovdiv, 24 Tsar Assen Str, Plovdiv 4000, Bulgaria
e-mail: tsangech@uni-plovdiv.bg

A. Teneva
University of Forestry, 10 St.Kl.Ochridsky Blvd, Sofia 1756, Bulgaria

S. M. Jain and S. Dutta Gupta (eds.), *Biotechnology of Neglected and Underutilized Crops*, DOI: 10.1007/978-94-007-5500-0_4,

4.1 Introduction

Haberlea rhodopensis, an endemic herbaceous species for the Balkan Peninsula, has long been known for its beautiful flowers and used in gardening (Fig. 4.1). This plant was first described in the Rhodope mountain in 1834 by the renowned explorer of the Balkans, the Hungarian Imre Frivaldszky, and became known to science as a new species in 1835. Frivaldszky named it after his teacher—the Hungarian botanist Haberle (Ganchev 1950). It is a perennial, rock-dwelling, poikilohydric plant from the family Gesneriaceae. *H. rhodopensis* is a glacial (tertiary) relic that inhabits shaded mountain zones (within the range of 250–1,400 m above the sea level) with relatively high humidity and forms ground dense tufts of thick leaves with one to five purple flower-stalks in spring time (Fig. 4.1, Georgieva et al. 2008).

As a perennial plant, *H. rhodopensis* must survive the harsh winters and it has therefore developed freezing tolerance. However, it is best known for its remarkable tolerance to drought. *H. rhodopensis* is a world-record holder in desiccation tolerance, able to survive the loss of more than 95 % of its cellular water content for up to 31 months and to resume its normal growth and physiological functioning within hours after re-watering (Djilianov et al. 2005). It is one of the first plants recognized as a genuine resurrection species (Ganchev 1950).

Fig. 4.1 *Haberlea rhodopensis* in its natural habitat, Rhodope Mountains, Bulgaria

Resurrection plants have the unique ability to survive desiccation of their vegetative tissues, a feature normally restricted to pollen grains and seeds (Gaff 1971). For the purpose, they use different protection and repair mechanisms resulting in a variety of physiological, biochemical, and genetic adaptations. For example, leaf folding is one of the most obvious morphological changes accompanying drought stress and plays an important role in reducing the surface directly exposed to light, thus limiting the irradiation-induced oxidative stress (Scott 2000; Farrant et al. 2003; Vander Willigen et al. 2004; Farrant et al. 2007; Vicré et al. 2004a). Mechanical stress occurring during water loss is overcome by protection mechanisms that allow avoidance of plasmalemma rupture and cell wall collapse. The inherent cell wall characteristics of resurrection plants provide their cells with remarkable elasticity, which ensures that stable, but reversible conformational changes can be achieved during re/dehydration (Vicré et al. 2004a). In some species, like *Craterostigma plantagineum* and *Selaginella lepidophylla,* the cell wall was shown to possess the ability to fold in order to preserve the structural integrity (Vicré et al. 2004b; Thomson and Platt 1997). Membrane fluidity is another important parameter for cell viability during desiccation. Generally, it is achieved by depositing larger amounts of polyunsaturated acids in the lipid bilayers and is supposed to contribute to cell and tissue flexibility during periods of water deficit (Hoekstra 2005; Moon et al. 1995). A range of protective compounds are synthesized by resurrection plants in response to drought. These include different types of sugars, usually accumulated in very high amounts during the periods of water deficit, secondary metabolites, as well as proteins like LEA and heat shock proteins (Le and McQueen-Mason 2006). In *C. plantagineum*, a 22 kDa ELIP-like desiccation-induced protein (dsp 22), which accumulates in the thylakoid membranes, was also shown to contribute to the protection against photoinhibition caused by dehydration (Alamillo and Bartels 2001). Another common marker of desiccation tolerance is the very well developed antioxidant system reinforced by a highly efficient network of free-radical scavenging enzymes that limit the oxidative damage. Unlike in other plants, most resurrection species maintain their ROS-scavenging activity even during drying (Sherwin and Farrant 1998). For example, the leaves of *Haberlea* manage to keep significant superoxide dismutase and peroxidase activities even in an air-dry state (Yahubyan et al. 2009). When the antioxidant pathways are compromised under high light exposure, the plant cannot recover from the desiccation anymore. This demonstrates the essential role of antioxidant machinery in recovery mechanisms (Kranner et al. 2002; Bartels and Hussain 2011).

H. rhodopensis, as other homoiochlorophyllous desiccation-tolerant plants, switches off the photosynthesis but keeps its photosynthetic machinery intact so that it can restore its functionality quickly after rehydration. These plants are therefore exposed to a risk of photooxidative damage, a threat countered by the above mentioned leaf folding and by molecular mechanisms such as induction of ELIPs and antioxidant enzymes. In contrast, the poikilochlorophyllous resurrection plats, most of which are monocotyledonous, degrade their chlorophyll and lose their thylakoid membranes during desiccation. They benefit from minimized risk of ROS accumulation due to the lack of excitable chlorophyll, but also need much more time to resynthesize de novo their photosynthetic machinery after rehydration (Bartels and Hussain 2011).

Due to its tolerance to desiccation and freezing, *H. rhodopensis* emerges as one of the most interesting models to study the effects of adverse abiotic factors of the environment. The aim of the present chapter is to summarize and extend the available up-to-date physiological and molecular data on the responses of *H. rhodopensis* to dehydration and to outline possible mechanisms that contribute to the desiccation tolerance of this unique species.

4.2 Molecular Responses of *H. rhodopensis* to Drought Stress and Desiccation

H. rhodopensis can lose almost all of its water, stay prolonged time in a desiccated state and completely recover after rewatering (Fig. 4.2). During desiccation, the relative water content (RWC) of the leaves could reach a value as low as of 4 %. Shrinkage and rolling of the leaves is observed at that time. However, the RWC and the leaf morphology quickly recover to those of unstressed controls several days after rehydration (the kinetics of recovery depends on the period of desiccation). No significant membrane damage, oxidative stress, and cell death are observed during desiccation, but the loss of water is accompanied by reduction of chlorophyll pigments Gechev et al. 2013. Interestingly, the measurements of the variable chlorophyll fluorescence (Fv/Fm) indicate non-functional photosystems only during the most extreme desiccation and not during the first stages of drought stress.

The work of Georgieva et al. (2007) shows that the initial decline of net CO_2 assimilation rate is due to a decrease in stomatal conductance, while at the later stages of dehydration the decline in CO_2 assimilation is caused both by stomatal closure and a decrease in the photochemical activity of photosystem II. *H. rhodopensis* is much more sensitive to dehydration combined with high irradiance (Georgieva et al. 2008).

Fig. 4.2 *Haberlea rhodopensis* during normal vegetative growth (fully hydrated, **a**), one month after desiccation (**b**) and four days after desiccated plants were watered (rehydration, **c**)

In addition to desiccation, our own experiments show that desiccated plants can withstand freezing stress of −20 °C for 48 h. Apparently, genes and metabolites induced by drought can serve protective roles (cross-protection) against low temperature/freezing stress. This observation is in accordance with the fact that in other plants many genes are induced by both drought and cold stress and supports the notion of existing crosstalk between the drought and the cold stress signaling pathways (Shinozaki et al. 2003; Shinozaki and Yamaguchi-Shinozaki 2000).

A combination of transcriptome and metabolome analyses was carried out recently in order to gain knowledge on the molecular mechanisms behind drought stress tolerance. The genome of *H. rhodopensis* is about 1.372 (approximately 10 times the size of *Arabidopsis*) but the number of genes is not known (Zonneveld et al. 2005). The transcriptome analysis was carried out by next generation sequencing (NGS) at four time points: control conditions, moderate drought (50 % RWC), severe desiccation (5 % RWC), and rehydration Gechev et al. 2013. The NGS revealed 96353 expressed sequence contigs (ESC), many of which related to unknown transcripts with no similarity to sequences from other genomes Gechev et al. 2013. Comparative analysis using BLAST showed that most of the transcript fragments were homologous to sequences from *Vitis vinifera*, *Populus trichocarpa*, and *Ricinus communis* (Fig. 4.3). Among the most responsive genes whose regulation changes during different water regimes, are genes related to the photosynthesis (chlorophyll a/b binding protein, light harvesting complex lhc1, rubisco small subunit). Many of them are induced during rehydration, which confirms the functional plant recovery on molecular level. Two catalases are highly

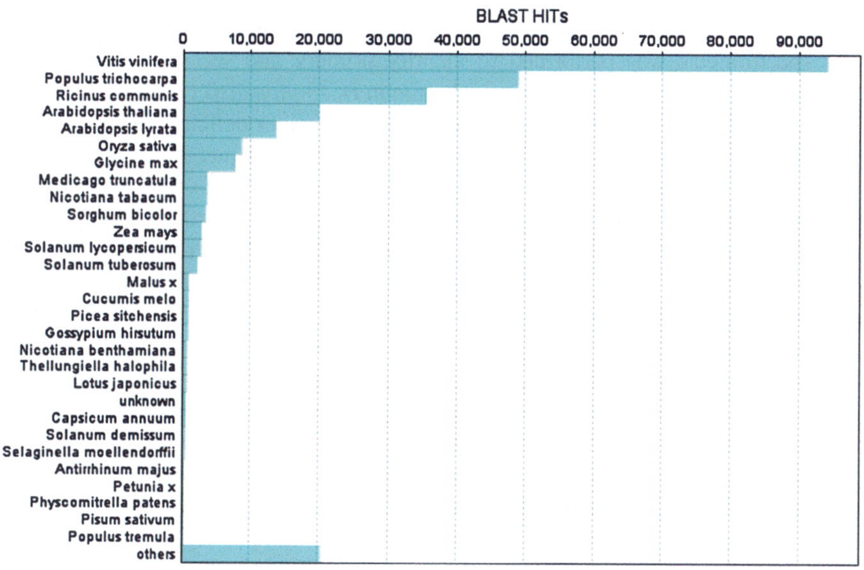

Fig. 4.3 Species distribution of *H. rhodopensis* expressed sequence contigs identified by next generation sequencing

expressed in all time points and several more are highly induced during drought and desiccation, in parallel with some LEA genes. This could provide a plausible explanation for the lack of electrolyte leakage and oxidative stress in the desiccated samples. In addition, big changes in expression patterns of genes related to the sugar metabolism were detected, supporting the key role of carbohydrate homeostasis and turnover during desiccation. Highly expressed in stressed samples was a stachyose synthase-related gene. This corresponds to previous results showing considerable induction of the β-amylase gene even in mild drought stressed samples. A possible interpretation is that starch breakdown processes occur at the onset of drying, which provide a substrate for stachyose synthesis, probably used as an energy source during rehydration reactions. Sugar conversions are very important not only for the synthesis of osmoprotectants but also for the cell wall remodeling, for the mechanical stabilization and flexibility during severe drought stress.

The transcriptomics analysis also showed an extensive antioxidant gene network. It seems that *H. rhodopensis* contains much more superoxide dismutases, catalases, monodehydroascorbate reductases and glutathione reductases than *Vitis vinifera* and *Arabidopsis thaliana* Gechev et al. 2013.

Recently, AFLP analysis was used as an alternative approach to identify genes highly regulated in *Haberlea rhodopensis* during drought (Georgieva et al. 2011). Genes upregulated at the earlier drought time point included protein kinases, transcription factors, succinate dehydrogenase, and a xyloglucan endotransglycosylase, while early downregulated genes included an elongation factor for RNA polymerase II, Ca^{2+}-ATPase, a subunit of the mitochondrial ATP synthase, and others (Georgieva et al. 2011).

In parallel, metabolome analysis by gas- and liquid chromatography coupled with mass spectrometry was performed to identify primary and secondary metabolites during dehydration and subsequent rehydration in *H. rhodopensis* Gechev et al. 2013.

Sucrose, maltose, and verbascose accumulated in water deficient plants, in contrast with the reduced levels of glucose and fructose. Sugar metabolism appears to be drastically regulated in other resurrection plants as well and some of these sugars have prominent osmoprotectant functions (Hoekstra et al. 2001). Comparative analysis of *H. rhodopensis* with the stress-sensitive *A. thaliana* reveals higher levels of sucrose and raffinose in the stress-tolerant *Haberlea* (Benina, unpublished results), suggesting that the maintenance of constantly high levels of sugars can be used as an important factor for establishing a resurrection phenotype. While some amino-acids like glutamate, glutamine, proline, pyroglutamate, aspartate, threonine and asparagine decreased at water limitation conditions, others like β-alanine accumulated. Under drought and desiccation, the levels of two stress-related metabolites: spermidine and γ-aminobutyric acid (GABA), are significantly higher. It is known that spermidine plays an important osmoprotective role also in other resurrection species (Alcazár et al. 2011), while GABA is a signaling molecule related to stress-responses and growth control (Renault et al. 2010, 2011). The comparative metabolite analysis of unstressed *H. rhodopensis* and *A. thaliana* revealed much higher initial levels of sugars and GABA in desiccation tolerant *Haberlea* than in the desiccation sensitive *A. thaliana*. Finally, the metabolite profiling detected many unknown compounds, especially in the desiccated samples. A large part of them is probably related to cell protection mechanisms against these severe conditions or play an important role as an energy source upon rehydration.

4.3 Recent Advances of *Haberlea Rhodopensis* Biotechnology, Future Prospects and Practical Implications

H. rhodopensis seems to be valuable not only as a decorative plant but also as a model to study drought and chilling/freezing stress. It can be a possible source of genes that can be used for engineering abiotic stress tolerance, and as a source of metabolites with medicinal properties. However, *H. rhodopensis* is an endangered species and as such, its collection from nature is prohibited. In the last few years, a successful tissue culture of *H. rhodopensis* was established and our team has created an in vitro bank of *H. rhodopensis* plants from different localities (Djilianov et al. 2005; Gechev et al. 2013). These developments will facilitate the research on *Haberlea* and serve as a platform for achieving the above mentioned goals.

H. rhodopensis is particularly rich in polyphenols; these compounds were recognized as protectors against several adverse abiotic stress factors and in particular as important molecules that secure the transition from desiccation to recovery. A dense luminal substance, most probably a phenolic compound, was proposed to protect and stabilize thylakoid membranes in *Haberlea* at low light intensities, while its absence correlates with poor chloroplast recovery after rehydration (Georgieva et al. 2010). Ebrahimi et al. explored the *Haberlea* fraction enriched in polyphenols and produced a phytochemical profile which includes 3 new flavones C-glycosides (Ebrahimi et al. 2011). Dell'acqua and Schweikert (2012) isolated a fraction rich in the caffeoyl phenylethanoid glycoside (Fig. 4.4) and other polyphenols.

Moreover, they showed that myconoside can strongly stimulate antioxidant skin defenses and extracellular matrix protein synthesis. Hydrogen peroxide-treated normal human dermal fibroblasts showed increased collagen VI (+822 %),

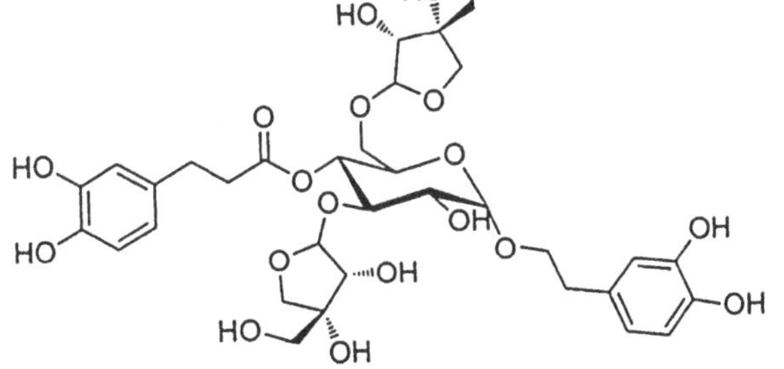

Fig. 4.4 Structure of myconoside, an abundant glycoside in *Haberlea rhodopensis*

collagen XVI (+928 %) and elastin (+144 %) mRNA synthesis, measured by RT-qPCR. This effect was superior to those obtained with benchmarks retinoic acid and retinol. Indeed, *H. rhodopensis* extracts stimulated the synthesis of elastin in a dose-dependent manner much better than retinol (Dell'acqua and Schweikert 2012). We also found high amounts of myconoside in unstressed tissues of *H. rhodopensis*, as well as many unidentified compounds presumably unique to this species Gechev et al. 2013. A recent study described the radioprotective, anti-clastogenic and antioxidant effects of total *H. rhodopensis* extract on rabbit blood samples exposed to gamma radiation in vitro (Popov et al. 2011). These results suggest that the strong antioxidant potential of this resurrection plant should be the aim of further extensive experiments which could provide strategies and solutions in combating various free-radical mediated human diseases.

Conclusion

For many years, *Haberlea rhodopensis* was regarded as an exotic plant, glacial relic and resurrection species suitable for gardening. The recent molecular studies on *H. rhodopensis* identified novel genes and metabolites with potential roles in drought and desiccation tolerance. In addition, *Haberlea's* extracts rich in antioxidants, polyphenols, and other specific secondary metabolites were shown to possess antioxidative, regenerative, radioprotective, and other medicinal properties. All this has established *H. rhodopensis* as a model plant to study molecular mechanisms of drought/desiccation tolerance and indicated potential biotechnological applications of this species in medicine.

Acknowledgments This work was supported by EC FP7 (project Biosupport 245588), University of Plovdiv Research Fund (project BF006), NSF of Bulgaria, project DO2-1068, and the Swiss Enlargement Contribution in the framework of the Bulgarian-Swiss Research Programme, SNSF project No. IZEBZ0_143003/1.

References

Alamillo J, Bartels D (2001) Effects of desiccation on photosynthesis pigments and the ELIP like dsp 22 protein complexes in the resurrection plant *Craterostigma plantagineum*. Plant Sci 160:1161–1170

Alcázar R, Bitrián M, Bartels D et al (2011) Polyamine metabolic canalization in response to drought stress in *Arabidopsis* and the resurrection plant *Craterostigma plantagineum*. Plant Sig Behav 6:243–250

Bartels D, Hussain S (2011) Resurrection plants: physiology and molecular biology. In: Lüttge U, Beck E, Bartels D (eds), Plant desiccation tolerance ecological studies, vol 215. Springer, Berlin, pp 339–364

Dell'acqua G, Schweikert K (2012) Skin benefits of a myconoside-rich extract from resurrection plant *Haberlea rhodopensis*. Int J Cosmet Sci 34:132–139

Djilianov D, Genova G, Parvanova D et al (2005) *In vitro* culture of the resurrection plant *Haberlea rhodopensis*. Plant Cell Tiss Org 80:115–118

Ebrahimi S, Gafner F, Dell'Acqua G et al (2011) Flavone 8-C-glycosides from *Haberlea rhodopensis* Friv (Gesneriaceae). Helv Chim Acta 94:38–45

Farrant J, Brandt W, Lindsey G (2007) An overview of mechanisms of desiccation tolerance in selected angiosperm resurrection plants. Plant Stress 1:72–84

Farrant J, Vander Willigen C, Loffell DA et al (2003) An investigation into the role of light during desiccation of three angiosperm resurrection plants. Plant Cell Environ 26:1275–1286

Gaff DF (1971) Desiccation-tolerant flowering plants in southern Africa. Science 174:1033–1034

Ganchev I (1950) Anabiotic dry tenacity and other biological particularities of *Haberlea rhodopensis*. Rep Inst Bot Bulg Acad Sci 1:191–214

Gechev TS, Benina M, Obata T, Tohge T, Sujeeth N, Minkov I, Hille J, Temanni MR, Marriott AS, Bergström E, Thomas-Oates J, Antonio C, Mueller-Roeber B, Schippers JH, Fernie AR, Toneva V (2013) Molecular mechanisms of desiccation tolerance in the resurrection glacial relic *Haberlea rhodopensis*. Cell Mol Life Sci 70:689–709

Georgieva K, Sigeti Z, Sarvari E et al (2007) Photosynthetic activity of homoiochlorophyllous desiccation tolerant plant *Haberlea rhodopensis* during dehydration and rehydration. Planta 225:955–964

Georgieva K, Lenk S, Buschmann C (2008) Responses of the resurrection plant *Haberlea rhodopensis* to high irradiance. Photosynthetica 46:208–215

Georgieva K, Sarvari E, Keresztes A (2010) Protection of thylakoid against combined light and drought by a lumenal substance in the resurrection plant *Haberlea rhodopensis*. Ann Bot 105:117–126

Georgieva T, Christov N, Djilianov D (2011) Identification of desiccation-regulated genes by cDNA-AFLP in *Haberlea rhodopensis*: a resurrection plant. Acta Physiol Plant 34:1055–1066

Hoekstra F (2005) Differential longevities in desiccated anhydrobiotic plant systems. Integr Comp Biol 45:725–733

Hoekstra F, Golovina E, Buitink J (2001) Mechanisms of plant desiccation tolerance. Trends Plant Sci 6:431–438

Kranner I, Beckett R, Wornik S et al (2002) Revival of a resurrection plant correlates with its antioxidant status. Plant J 31:13–24

Le T, McQueen-Mason S (2006) Dessiccation-tolérant plants in dry environments. Rev Environ Sci Biotechnol 5:269–279

Moon B, Higashi S, Gombos Z, Murata N (1995) Unsaturation of the membrane lipids of chloroplasts stabilizes the photosynthetic machinery against low-temperature photoinhibition in transgenic tobacco plants. Proc Natl Acad Sci USA 92:6219–6223

Popov B, Georgieva S, Gedjeva V, Petrov V (2011) Radioprotective anticlastogenic and antioxidant effects of total extract of *Haberlea rhodopensis* on rabbit blood samples exposed to gamma radiation in vitro. Revue de Médecine Vétérinaire 162:34–39

Renault H, Roussel V, El Amrani A et al (2010) The *Arabidopsis pop2-1* mutant reveals the involvement of GABA transaminase in salt stress tolerance. BMC Plant Biol 10:20

Renault H, El Amrani A, Palanivelu R et al (2011) GABA accumulation causes cell elongation defects and a decrease in expression of genes encoding secreted and cell wall-related proteins in *Arabidopsis thaliana*. Plant Cell Physiol 52:894–908

Scott P (2000) Resurrection plants and the secret of eternal leaf. Ann Bot 85:159–166

Sherwin H, Farrant J (1998) Protection mechanisms against excess light in the resurrection plants *Craterostigma wilmsii* and *Xerophyta viscose*. Plant Growth Regul 24:203–210

Shinozaki K, Yamaguchi-Shinozaki K (2000) Molecular responses to dehydration and low temperature: differences and cross-talk between two stress signaling pathways. Curr Opin Plant Biol 3:217–223

Shinozaki K, Yamaguchi-Shinozaki K, Seki M (2003) Regulatory network of gene expression in the drought and cold stress responses. Curr Opin Plant Biol 6:410–417

Thomson W, Platt K (1997) Conservation of cell order in desiccated mesophyll of *Selaginella lepidophylla* ([hook and grev] spring). Ann Bot 79:439–447

Vander Willigen C, Pammenter N, Mundree S, Farrant J (2004) Mechanical stabilization of desiccated vegetative tissues of the resurrection grass *Eragrostis nindensis*: does a TIP 3; 1 and/or compartmentalization of subcellular components and metabolites play a role? J Exp Bot 55:651–661

Vicré M, Farrant J, Driouich A (2004a) Insights into the cellular mechanisms of desiccation tolerance among angiosperm resurrection plant species. Plant Cell Environ 27:1329–1340

Vicré M, Lerouxel O, Farrant J et al (2004b) Composition and desiccation-induced alterations of the cell wall in the resurrection plant *Craterostigma wilmsii*. Physiol Plant 120:229–239

Yahubyan G, Gozmanova M, Denev I et al (2009) Prompt response of superoxide dismutase and peroxidase to dehydration and rehydration of the resurrection plant *Haberlea rhodopensis*. Plant Growth Regul 57:49–56

Zonneveld B, Leitch I, Bennett M (2005) First nuclear DNA amounts in more than 300 angiosperms. Ann Bot 96:229–244

Chapter 5
Cell Morphometry as Predictor of Protein Legume In Vitro Growth

S. J. Ochatt

Abstract A detailed characterization of crops in terms of their adequacy as food or feed is a prerequisite to their exploitation for either of these purposes, but also in terms of their taxonomical distinction when this is incomplete or not up-to-date, as is frequently the case with neglected crops. One additional, and important, feature of cell parameters is their impact of the competence for regeneration in vitro and for growth and development in vivo. The summation of such cell traits is the cell morphology while the numerical measurement of them is the cell morphometry. In this chapter, the role of such cell traits on all of these aspects is addressed and their influence on the further agro-economic relevance of the crops considered and their breeding is discussed.

5.1 Introduction

In the present socio-economical context, the use of plant protein as feed represents a really strategic choice. Protein legumes in general and pea (*Pisum sativum* L.) and some underutilised species of grasspea (*Lathyrus spp.*) and Bambara groundnut (*Vigna subterranea*) in particular are such sources of proteins. Indeed, pea seeds are rich in lysine, their starch content is close to that of cereals and nitrogen content is intermediate between soybean and cereals (Cousin 1992). As a result, pea is routinely added to industrial feed formulations for piglets, ruminants and poultry (Duc et al. 2001). However, the digestible efficiency of such diet differs among them because of the differences existing between their respective digestions (UNIP 2011, http://www.prolea.com/unip/, accessed December 2011). Thus, when producing feed for ruminants or monogastrics (Mariscal-Landin et al. 2002), it is relevant to distinguish between flours and grains that will be better adapted for either group of animals, particularly in terms of particle size (UNIP 2011) and cell wall contents (Chanliaud et al. 2002).

S. J. Ochatt (✉)
Lab Physiologie Cellulaire Morphogenèse & Validation (PCMV), INRA, UMR 1347 Agroécologie,Pôle GEAPSI, 17 rue Sully, B.P. 86510, 21065 Dijon cedex, France
e-mail: Sergio.ochatt@dijon.inra.fr

S. M. Jain and S. Dutta Gupta (eds.), *Biotechnology of Neglected and Underutilized Crops*, DOI: 10.1007/978-94-007-5500-0_5,

Spring genotypes of peas are of significant importance in industry but their large-scale development is countered by several factors, including a low and irregular yield as compared to wheat (Carrouée et al. 2003), their use essentially limited to pork feed (Mariscal-Landin et al. 2002), a strong competition by soybean and cereals (UNIP), and other problems which are specific to pea (including pests and diseases) (Cousin 1992; UNIP 2011). Against this background, winter pea appears as an interesting and realistic alternative, with the potential to widen the area of protein pea cultivation, with a higher productivity, and new market openings as feed (UNIP 2011). The same can be said about various grasspea genotypes, which exhibit a rather high protein content and are, in addition, quite plastic when faced with several stress factors, both abiotic and biotic.

The research for winter pea ideotypes to feed monogastric and ruminants requires a prospective analysis of the use of such genotypes as based on an evaluation prior to their actual cultivation, the identification of specific breeding criteria of interest from the agronomic, zootechnical and economical viewpoints, and the development of methodologies suitable for an assessment of the economic impact of such novelties (Mariscal-Landin et al. 2002). Likewise, underutilised grain legumes require this and even more studies on their traits of interest before farmers and the feed industry may be incline to widen their cultivation. Among these traits, particle size and cell wall contents in the feed are, as evoked above, crucial in terms of such impact, since they will modulate the actual digestibility and, *in fine*, weight gain for animals fed with formulations enriched with winter pea (Duc et al. 2001). Recent studies have shown that the primary cell walls and their biomechanical properties can play a paramount role in the digestibility of feed (Chanliaud et al. 2002). In this context, published information on the final size of seeds (Ambrose et al. 1987; Lemontey et al. 2000) and on the size and characteristics of cotyledonary cells in protein legumes in general (Brillouet and Carré 1983) is scanty.

Similar considerations can be made with respect to the neglected and underutilised grain legumes, such as Bambara groundnut (*Vigna subterranea*) used mostly as food in sub-Saharan Africa (Koné et al. 2007; Massawe et al. 2007) but also the grass peas (*Lathyrus sativus* and related *Lathyrus* species) which are used as forage, food and feed (Ochatt et al. 2007; Vaz Patto et al. 2006).

5.2 Cell Morphometry Parameters and Biotechnology Approaches

When embarking on the description of a plant and its organs and tissues, shape is one of the main parameters taken into consideration, and the summation of the different characteristics observed in an individual constitute its morphology, while their measurement is its corresponding morphometry. The latter permits to stress useful biodiversity aspects which might otherwise remain unobserved (Roy and Foote 1997; McLellan and Endler 1998) and are, nevertheless, frequently essential to characterize in detail the biodiversity observed, as for example when describing genetic novelties such as those obtained via chemical, physical or insertional mutagenesis. In this respect, common volumetric measurements are routinely based on the use of equipments such as the Coulter densitometer, without considering that this apparatus assimilates all objects to a sphere and there is therefore the risk that two small cells stuck to each other might be counted as a single, larger one. Conversely, when measuring microscopically, say, the cell surface, the operator will see in two dimensions but is unable to assess cell depth, thereby causing bias in the data. These considerations are particularly relevant for the detailed analyses of those plant cells and tissues that are responsible for the agro-economic importance of a crop and its products. Parallel to these aspects, characterising the morphometry of cells has a significant impact on their competence to respond to biotechnological approaches, particularly in terms of regeneration capacity (Ochatt et al. 2000, 2011) and embryogenic ability (Ochatt et al. 2008; Ochatt and Moessner 2010). Cell morphometry is also paramount for the establishment of yield, as maximum cell volume is responsible *in fine* of the final size of organs (including seed), and their protein storage capacity is under genetic and environmental determinism (Lemontey et al. 2000). It is evident that this intrinsic limitation of organ size has a strong incidence not only on the yield of a culture but also on the protein content of its seeds. Furthermore, the cell wall properties not only determine the ability of cells to expand and grow but also their division competence (Ochatt 2011; Ochatt et al. 2008, 2010), which is another determinant of the total cell number in an organ and, ultimately, also influences its final size.

The phenomena of cell volume limitation must be linked to specific cytological and biochemical events in order to "date" the duration of the cell division and storage protein accumulation phases in the developing seed (Gallardo et al. 2006, 2007; Ochatt 2011). In addition to this, Yeung (1995) indicated that the formation of embryogenic cells may be correlated with characteristic morphological changes in most embryogenic systems, a result later confirmed with pea, *Medicago truncatula* and *Lathyrus* cell suspensions (with or without embryogenic competence) by Ochatt et al. (2008).

Toonen et al. (1994) classified the single-cell fraction of established embryogenic cultures into five morphological groups by video cell tracking the fate of embryogenic carrot cells. Shortly after that, Zeng et al. (2007) also characterised the morphology of cotton embryogenic cells. More recently, in our work with protein legume cell cultures, we also showed that cell morphology can be used as an early marker of embryogenic competence (Ochatt et al. 2008), the same as their relative nuclear relative DNA content is a good predictor of their regeneration competence in general (Elmaghrabi and Ochatt 2006; Ochatt 2008, 2011; Ochatt et al 2011.).

In this respect, Ochatt et al. (2008) showed that, for a number of genotypes in three different legume species (pea, *Medicago truncatula* and *Lathyrus sativus*) but also for the model species *Arabidopsis thaliana* (where a wild type and one insertion mutant derived from it were tested) the acquisition of competence of cells to undergo somatic embryogenesis is strongly correlated with their cell wall thickness, their internal osmolarity (hence also the osmolarity of the external medium in which they grew) and their surface.

5.3 Biodiversity: When Shape Counts

Comparing the different aspects of diversity is an essential feature underlying the patterns of biodiversity observed in nature (Roy and Foote 1997; McLellan and Endler 1998), as it provides with useful measurements of traits that are not always and not necessarily picked up by functional, taxonomic or phylogenetic metrics.

In our team, we have thus studied 35 genotypes of pea (Roche et al. 2004; Fig. 5.1), where we determined the morphometry (number, surface, volume) and the wall thickness of cotyledonary cells isolated enzymatically (with Pectinase) from mature seeds without affecting their shape, and we were able to group the genotypes in different classes according to these parameters. Interestingly, we verified some differences in the ranking of genotypes analysed either in terms of their belonging to a particular class or within a same class of size. These differences mostly concerned surface versus volume of cells and may logically be ascribed to the cell shape. Thus, we undertook a new series of studies this time to

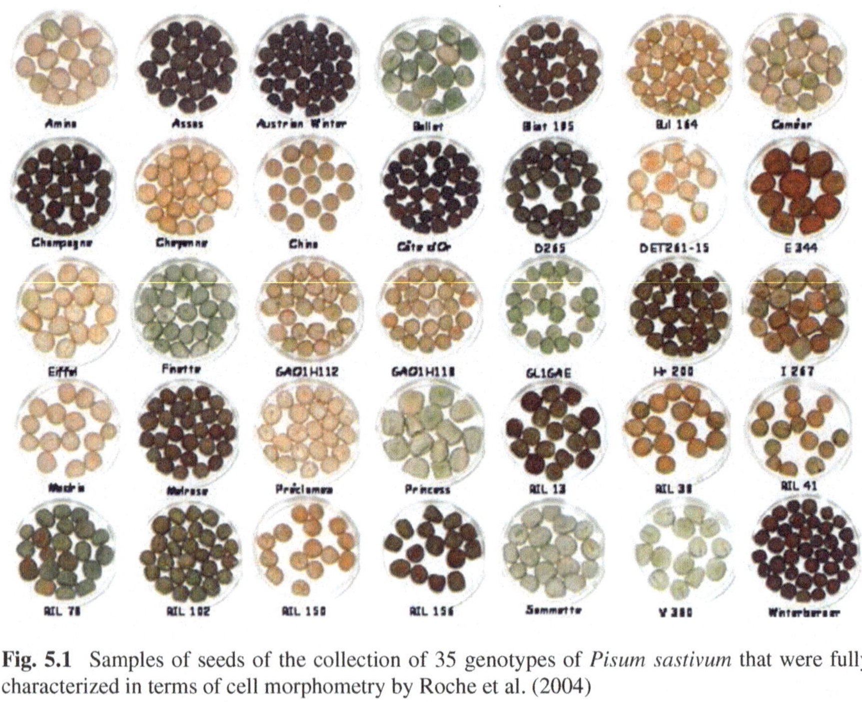

Fig. 5.1 Samples of seeds of the collection of 35 genotypes of *Pisum sastivum* that were fully characterized in terms of cell morphometry by Roche et al. (2004)

devise a shape coefficient that would be reliable and robust enough to avoid such discrepancy (Ochatt and Moessner 2010).

Shape is probably the first and most widespread problem encountered for the microscopic and volumetric analysis of intact plant cells (McLellan and Endler 1998; Chanliaud et al. 2002) because, different from animal ones, plant cells are rarely spherical. As a result of this, sometimes errors are introduced in the measurements, particularly when fluorochrome signals have to be used, as for flow cytometry (Ochatt 2008), image analysis (McLellan and Endler 1998), and cell surface (Ochatt et al. 2008; Atif et al. 2011) or volume measurements (Roche et al. 2004; Atif et al. 2011).

These various analytical problems with plant cells come not only from their shape but also from their rigid cell wall (Kerstens and Verbelen 2003), and the most obvious solution imagined by researchers to resolve them was to eliminate the latter and produce protoplasts, i.e. wall-less cells which, being spherical are not expected to disturb a light trajectory as irregularly-shaped objects tend to do (Ochatt 2008). However, this requires a far more difficult, long and costly preparation than that of cells and encompasses the use of enzyme solutions for tissue digestion whose composition is strongly dependent on the plant genotype, source tissue and its developmental stage (Ochatt and Power 1992). Much more importantly, they also require the use of osmotica to prevent isolated protoplasts from bursting and such osmotic agents will also alter the original shape of cells and thereby biasing the results of analyses.

Biodiversity can be characterized by taxonomic patterns, but these may sometimes mask other interesting spatial patterns of morphological, functional or phylogenetic diversity (Roy and Foote 1997). In this respect, scores of articles on the nature of biodiversity start by recognizing that the richness of a species is only one of several ways of measuring biodiversity even if they thereafter proceed treating it as if it were the whole measure of biodiversity (Roy and Foote 1997; McLellan and Endler 1998). This is probably due to the relatively limited success of different methods to measure and describe the shape of biological objects to date, particularly plant cells (McLellan and Endler 1998; Lemontey et al. 2000), and also to the limited reliability in the classing of different species, or genotypes within a species, when different methods are employed to characterize their morphometry.

The results reported by Roche et al. (2004) and later on by Ochatt and Moessner (2010) showed the great relevance of shape as an additional parameter for cell characterization, particularly for the establishment of ideotypes better adapted as feed for either monogastrics or ruminants. Figure 5.1 shows the range of seed phenotypes observed for those genotypes which was, however, insufficient to characterize them fully in terms of their digestibility as food or feed for ruminants or mongogastrics.

Thus, in their work with 35 different genotypes of pea, Roche et al. (2004) reported a fine characterization in terms of cotyledonary cell morphometry (surface and volume) and wall thickness (Fig 5.2) for a large range of winter genotypes compared with several spring varieties in protein pea.

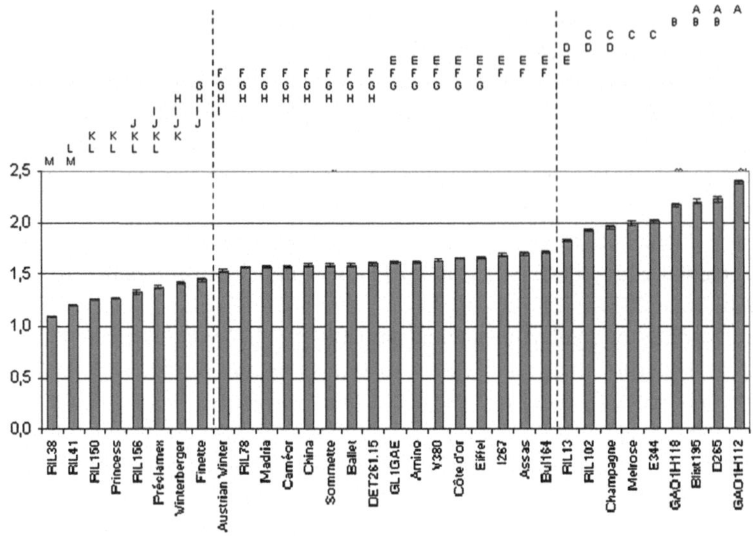

Fig. 5.2 Statistical grouping and mean ±SD of the cell wall thickness for 33 pea genotypes analysed

It was then clear that, apart from the obvious colour and, for some genotypes, also size of seeds, other differences required a more detailed, cytological, analysis. Firstly, a Coulter densitometer was used to assess the cell volume of this range of genotypes but it soon appeared that if cell surface instead of volume was considered, the genotypes studied could not be statistically grouped in the same size class. It was therefore decided to measure them in a way that may take into account possible differences in cell shape. For this though, a microscopic observation (i.e. 2-D) of cotyledonary cells of all genotypes was a prerequisite. The range of shape variability found was surprisingly large, as illustrated in Fig 5.3 for 20 of the 35 genotypes studied. It appeared that, unless a coefficient suitable to take into account the differences in cell shape was developed, it would not be possible to characterize the genotypes any further, nor to assign them to a given class of feed and/or food. Once devised, this coefficient would, in turn, also help in the taxonomical distinction between closely related genotypes of a number of other legume species including neglected and underutilised ones, while simultaneously providing a unique insight as to their potential subsequent use for human or animal nutrition and to correlate it with the digestibility expected from flours or seed of a given genotype.

The shape coefficient devised by Ochatt and Moessner (2010), also of use for other studies where the typical non-sphericity of plant cells might bias analyses, such as in flow cytometry (Ochatt 2008) or quantitative microfluorometric studies (Ochatt et al. 2008), is an additional tool to characterize genotypes in terms of their adaptability for the production of feed flours of different digestibility.

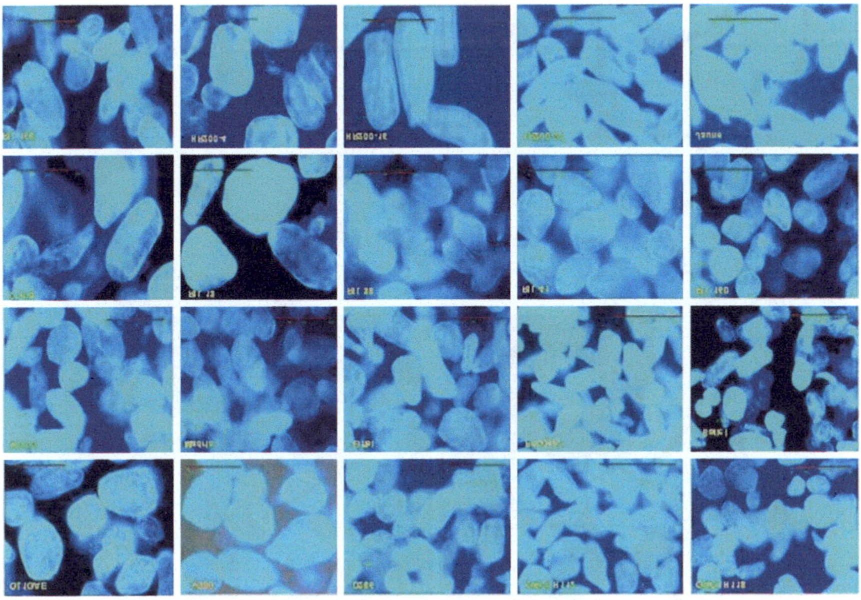

Fig. 5.3 Representative microscopic fields showing the differences in shape of cotyledonary cells of 20 spring and winter type genotypes of pea. *Scale bars* = 200 μm

This shape coefficient (SC) was calculated as:

$$SC = \frac{\sqrt{a^2 - b^2}}{a}$$

where, a represents half the length of the cell at its longest and b half the length of the cell at its shortest (Fig. 5.4).

These values were measured on cotyledonary cells that had been isolated using an enzyme solution containing 1 % (w/v) Pectinase (Sigma Aldrich) at 37 °C, with gentle shaking, for 1 h. Then, cells were stained with Calcofluor White (Fig. 5.5; Rattee and Breur 1974), which labels cellulose by covalent binding at the β1–4 bonds of the molecule, and were observed under UV light (Ochatt et al. 2008; Ochatt and Moessner 2010).

5.4 Predicting Embryogenic Ability in Vitro: Why Bother About Thickness

In plant cell biology, the role of the cell wall has always attracted much attention (Burk and Ye 2002; Kerstens and Verbelen 2003); whereas the implications of cell wall composition is a matter of intense research (Fagard et al. 2000).

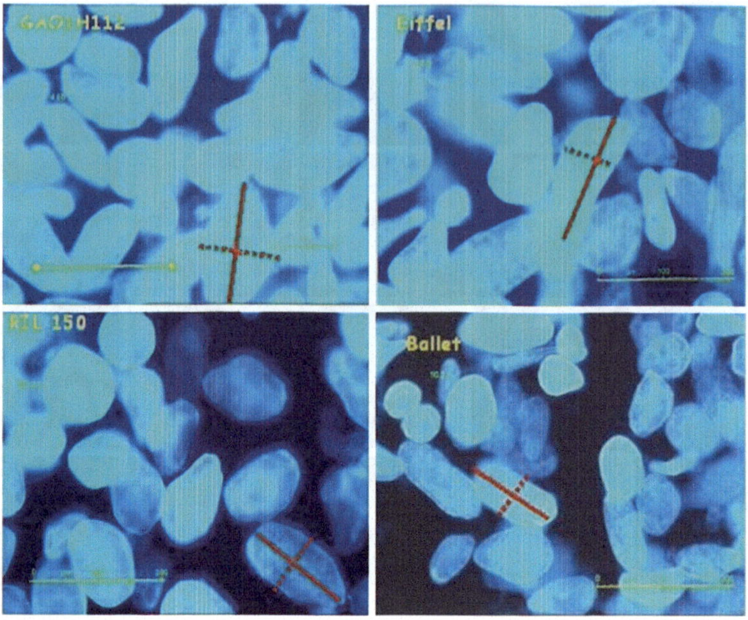

Fig. 5.4 Representative samples of cotyledonary cells of genotypes GAO1H112, Eiffel, RIL150 and ballet stained with calcofluor white and observed under UV light with an inverted microscope. Images were captured with Archimedes pro and treated with histolab (Microvision, France) for the calculation of shape coefficient. *Full lines* = 2a, *dotted lines* = 2b, *Scale bars* = 200 μm

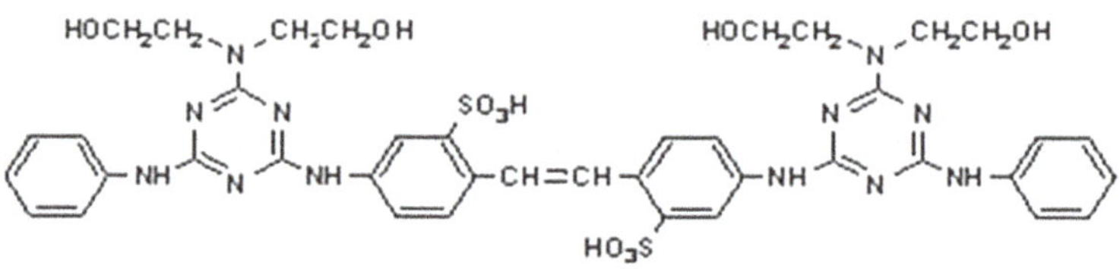

Fig. 5.5 Calcofluor white

Yet, knowledge on the mechanisms through which the form of plants and plant organs is established at the cell and tissue levels is still largely insufficient (Frugis et al. 1999). In vitro plant regeneration of grain legumes has been reported by several groups, but they are still regarded as recalcitrant (Ochatt et al. 2000a). Therefore, tools permitting an early distinction of regeneration-competent cultures (Ammirato 1983) from those that are not would be of interest, whatever the regeneration pathway. For instance, we have previously shown that in pea (Ochatt et al. 2000a, b) and *Medicago truncatula* (Elmaghrabi and Ochatt 2006), hypernodulating mutants were highly embryogenic whereas the wild type was not, a feature that could be reverted by adding exogenous lipo-chito oligosaccharins to the medium (Ochatt et al. 2005). We also found that grass pea was consistently recalcitrant to embryogenesis in our hands (Ochatt et al. 2002, 2007).

We studied the kinetics of cell surface, cell wall thickness and also the kinetics of the medium and intracellular osmolarity at the onset of embryogenesis from cell suspensions and during embryo growth, on various species genotypes (Ochatt et al. 2008). Thus, we worked with embryogenic and non-embryogenic cell suspensions of several pea, grass pea, *M. truncatula* and Arabidopsis genotypes, and identified novel cyto-physical parameters which served as early indicators of the competence of cells to undergo somatic embryogenesis. This approach may also be useful to study the kinetics of seed development and storage protein synthesis and, *in fine*, will contribute to a faster and more efficient characterization of genotypes and ideotypes in terms of their adaptation as either food or feed.

5.5 Conclusion and Prospects

The data in this chapter shows the importance of morphometric measurements as an additional parameter for cell characterization, particularly for the identification of grain legume ideotypes better adapted as feed for either monogastrics or ruminants. More generally, the shape coefficient we devised should also find use in other studies (e.g. taxonomy) and with other species, where the non-sphericity typical of plant cells might bias analyses, such as in flow cytometry or in quantitative microfluorometric studies, and also for the validation of gene function of genotypes that were modified genetically (by insertional or chemical mutagenesis) for genes involved in embryo development and seed filling, where the reliable characterization of their cotyledonary cells will be required.

The summation of the parameters evoked here serves as a reliable indicator of the technological grinding ability of different genotypes and, eventually, as predictors of the adequability of one given genotype for production of feed for different types of animals (ruminants versus monogastrics). Importantly, though, before any conclusion as to the nutritional quality of one genotype can be drawn, it's content in antinutritional factors, digestible energy and available amino acid content should not be disregarded (Mariscal-Landin et al. 2002). The results obtained during these experiments might be useful for the identification and characterization of ideotypes adapted as feed, and it would be interesting to further analyse the various wall fractions in cotyledonary cells of grain crops in order to assess their impact on the subsequent grinding and digestibility of seeds. From a zootechnical viewpoint, such a study would validate the interest of this kind of cell morphometric characterization of genotypes. Finally, in socio-economic terms, the availability of novel grain legume genotypes, better adapted to either ruminant or monogastric animals would allow for a better competition of this group of crops versus other protein sources in the elaboration of feed and thereby foster an extension of their production zones. In this context, neglected and underutilised crops should greatly benefit from the setup of these approaches as an additional tool for their taxonomical distinction but also for their further agro-economic exploitation and, in general, for their breeding.

References

Ambrose MJ, Wang TL, Cook SK, Hedley CL (1987) An analysis of seed development in *Pisum sativum*. J Exp Bot 38:1909–1920

Ammirato PV (1983) Embryogenesis. In: Evans DA, Sharp WR, Ammirato PV, Yamada Y (eds) Handbook of plant cell culture: techniques for propagation and breeding, vol 1. MacMillan, New York, pp 82–123

Atif RM, Sanchez M, Thompson RD, Lesignor C, Verdier J, Ochatt S (2011) Characterization of the role of selected transcription factors in seed filling in *Medicago truncatula*. In: Model legume congress, Saint Maxime, pp 53–P51

Brillouet J-M, Carré B (1983) Composition of cell walls from cotyledons of *Pisum sativum*, *Vicia faba* and *Glycine max*. Phytochem 22:841

Burk DH, Ye Z-H (2002) Alteration of oriented deposition of cellulose microfibrils by mutation of a katanin-like microtubule severing protein. Plant Cell 14:2145–2160

Carrouée B, Crépon K, Peyronnet C (2003) Les protéagineux: intérêt dans les systèmes de production fouragers français et européens. Fourrages 174:163–182

Chanliaud E, Burrows KM, Jeronimidis G, Gidley MJ (2002) Mechanical properties of primary cell wall analogues. Planta 215:989–996

Cousin R (1992) Les protéagineux (Proteaginous). In: Gallais A, Bannerot H (eds) Amélioration des espèces végétales cultivées (Improvement of the cultivated vegetable species). INRA Editions, Paris, pp 173–226

Duc G, Bertrand D, Bourdon D, Carré B, Faurie F, Gallant D, Hess V, Huart, M, Marget P, Melcion J-P, Moessner A, Mousset-Declas C, Sève B, Thibault J-F, Tisserand J-L (2001) Cotyledon cell-wall composition and distribution in *Pisum sativum L.*: some variability and its relationship with the seed grinding response and the nutritional value. IVth European Conf on Grain Legumes, Cracow, Poland, pp 84–85

Elmaghrabi A, Ochatt S (2006) Isoenzymes and flow cytometry for the assessment of true-to-typeness of calluses and cell suspensions of barrel medic prior to regeneration. Plant Cell Tiss Org Cult 85:31–43

Fagard M, Höfte H, Vernhettes S (2000) Cell wall mutants. Plant Physiol Biochem 38:15–25

Frugis G, Giannino D, Mele G, Nicolodi C, Innocenti AM, Chiapetta A, Bitonti MB, Dewitte W, Van Onckelen H, Mariotti D (1999) Are homeobox knotted-like genes and cytokinins the leaf architects? Plant Physiol 119:371–373

Gallardo K, Kurt C, Thompson R, Ochatt S (2006) In vitro culture of immature *Medicago truncatula* grains under conditions permitting embryo development comparable to that observed in vivo. Plant Sci 170:1052–1058

Gallardo K, Firnhaber C, Zuber H, Hericher D, Belghazi M, Henry C, Kuster H, Thompson R (2007) A combined proteome and transcriptome analysis of developing *Medicago truncatula* seeds: Evidence for metabolic specialization of maternal and filial tissues. Mol Cell Prot 6:2165–2179

Kerstens S, Verbelen J-P (2003) Cellulose orientation at the surface of *Arabidopsis* seedling. Implication for biomechanics in plant development. J Struct Biol 144:262–270

Koné M, Patat-Ochatt EM, Conreux C, Sangwan RS, Ochatt SJ (2007) In vitro morphogenesis from cotyledon and epicotyl explants and flow cytometry distinction between landraces of Bambara groundnut [*Vigna subterranea* (L.) Verdc], an under-utilised grain legume. Plant Cell Tiss Org Cult 88:61–75

Lemontey C, Mousset-Déclas C, Munier-Jolain N, Boutin JP (2000) Maternal genotype influences pea seed size by controlling both mitotic activity during early embryogenesis and final endoreduplication level/cotyledon cell size in mature seeds. J Exp Bot 51:167–175

Mariscal-Landin G, Lebreton Y, Sève B (2002) Effect of tannins in sorghum on amino acid ileal digestibility and on trypsin (E.C.2.4.21.4) and chymotrypsin (E.C.2.4.21.1) activity of growing pigs. Anim Feed Sci Technol 97:183–198

Massawe FJ, Mwale SS, Azam-Ali SN, Roberts JA (2007) Towards genetic improvement of Bambara groundnut [*Vigna subterranea* (L) Verdc.]. In: Ochatt S, Mohan Jain S (eds) Breeding of neglected and under-utilized crops, spices and herbs, Science Publication, Enfield, pp 61–80

McLellan T, Endler JA (1998) The relative success of some methods for measuring and describing the shape of some complex objects. Syst Biol 47:264–281

Ochatt SJ (2008) Flow cytometry in plant breeding. CytometryA 73:581–598

Ochatt SJ (2011) Immature seeds and embryos of *Medicago truncatula* cultured in vitro. In: Thorpe TA, Yeung EC (eds) Plant embryo culture: methods and protocols, methods in molecular biology, vol 710. Springer, Berlin, pp 39–52

Ochatt SJ, Power JB (1992) Plant regeneration from cultured protoplasts of higher plants. In: Moo-Young M, Warren GS, Fowler MW (eds) Comprehensive biotechnology, second supplement. Pergamon Press, New York, pp 99–127

Ochatt SJ, Moessner A (2010) Rounding up plant cells. Int J Plant Biol 1–e8:40–42

Ochatt SJ, Mousset-Déclas C, Rancillac M (2000a) Fertile pea plants regenerate from protoplasts when calluses have not undergone endoreduplication. Plant Sci 156:177–183

Ochatt SJ, Pontécaille C, Rancillac M (2000b) The growth regulators used for bud regeneration and shoot rooting affect the competence for flowering and seed set in regenerated plants of protein pea. In Vitro Cell Dev Biol Plant 36:188–193

Ochatt SJ, Munneaux E, Machado C, Jacas L, Pontécaille C (2002) The hyperhydricity of in vitro regenerants of grass pea (*Lathyrus sativus* L.) is linked with an abnormal DNA content. J Plant Physiol 159:1021–1028

Ochatt SJ, Delaitre C, Lionneton E, Huchette O, Patat-Ochatt EM, Kahane R (2005) One team, PCMV and one approach, in vitro biotechnology, for one aim, the breeding of quality plants with a wide array of species. In: Dris R (ed) Crops: growth quality and biotechnology. WFL Publisher, Helsinki, pp 1038–1067

Ochatt SJ, Abirached-Darmency M, Marget P, Aubert G (2007) The Lathyrus paradox: "poor men's diet" or a remarkable genetic resource for protein legume breeding? In: Ochatt SJ, Mohan Jain S (eds) Underutilised and neglected crops, herbs and spices. Science Press, Enfield, pp 41–60

Ochatt S, Muilu R, Ribalta F (2008) Cell morphometry and osmolarity as early indicators of the onset of embryogenesis from cell suspension cultures of grain legumes and model systems. Plant Biosyst 142:480–486

Ochatt SJ, Atif RM, Patat-Ochatt EM, Jacas L, Conreux C (2010) Competence versus recalcitrance for in vitro regeneration. Not Bot Hort Agrobot Cluj 38(2):102–108

Ochatt SJ, Patat-Ochatt EM, Moessner A (2011) Ploidy level determination within the context of in vitro breeding. Plant Cell Tiss Organ Cult 104:329–341

Rattee ID, Breur MM (1974) Calcofluor white. In: The physical chemistry of dye absorption, Academic, New York, pp 180–182

Roche JB, Ochatt S, Moessner A, Etévé G, Duc G (2004) Genotypic and environmental effects on the relationship between seed size, cotyledon cell volume, cell shape and cell wall thickness in protein pea. In: Vth European conference on grain legumes, Dijon, p 268

Roy K, Foote M (1997) Morphological approaches to measuring biodiversity. Tree 12:277–281

Toonen MAJ, Hendriks T, Schmidt EDL, Verhoeven HA, Van Krammen A, De Vries SC (1994) Description of somatic-embryo forming single cells in carrot suspension cultures employing video cell tracking. Planta 194:565–572

UNIP (2011) http://www.prolea.com/unip/

Vaz Patto MC, Skiba B, Pang ECK, Ochatt SJ, Lambein F, Rubiales D (2006) Lathyrus improvement for resistance against biotic and abiotic stresses: from classical breeding to marker assisted selection. Euphytica 147:133–147

Yeung EC (1995) Structural and developmental patterns in somatic embryogenesis. In: Thorpe TA (ed) In vitro embryogenesis in plants. Kluwer, Dordrecht, pp 205–248
Zeng F, Zhang X, Jin S, Cheng L, Lianf S, Hu L, Guo X, Nie Y, Cao J (2007) Chromatin reorganization and endogenous auxin/cytokinin dynamic activity during somatic embryogenesis of cultured cotton cell. Plant Cell Tiss Org Cult 90:63–70

Chapter 6
Application of TILLING for Orphan Crop Improvement

Korinna Esfeld, Cristobal Uauy and Zerihun Tadele

Abstract People in developing countries mostly depend for their diet on special staple crops, so called orphan crops. These crops play a key role in food security since they are grown by many resource-poor farmers and consumed locally. Despite their huge importance in the economy and livelihood of the developing world, orphan crops have received little attention in terms of scientific improvement. Although conventional breeding is widely implemented to improve crop plants, alternative methods such as marker-assisted breeding and reverse genetics approaches have proved to be efficient in developing crop cultivars. In this review, we present detailed description of a non-transgenic and reverse genetics technique called TILLING (Targeting Induced Local Lesion IN Genomes). The method was originally optimized in the model plant *Arabidposis thaliana* and subsequently applied to crops such as maize, wheat, and rice. We also present detailed procedures for several TILLING strategies and discuss their benefits and drawbacks. The application of the technique for orphan crop improvement is also discussed based on several TILLING platforms currently carried-out on these understudied crops of the world.

Abbreviations

AFLPs	Amplified fragment length polymorphisms
CAPS	Cleaved amplified polymorphic sequence
CJE	Celery juice extract
CODDLE	Codons optimized to detect deleterious lesions

K. Esfeld · Z. Tadele (✉)
Tef Improvement Project, Institute of Plant Sciences, University of Bern,
Altenbergrain 21, 3013 Bern, Switzerland
e-mail: Zerihun.tadele@ips.unibe.ch

C. Uauy
John Innes Centre, Norwich Research Park, Norwich NR4 7UH, UK

C. Uauy
National Institute of Agricultural Botany, Cambridge CB3 0LE, UK

S. M. Jain and S. Dutta Gupta (eds.), *Biotechnology of Neglected and Underutilized Crops*, DOI: 10.1007/978-94-007-5500-0_6,

dCAPS	Derived cleaved amplified polymorphic sequences
EMS	Ethyl methanesulphonate
HF	Hydrogen fluoride
IRD	Infra-red dye
MAB	Marker-assisted breeding
MAS	Marker-assisted selection
MNU	N-methyl-N-nitrosourea
ODAP	β-N-Oxalyl-L-α, β-diaminopropanoic acid
PARSESNP	Project aligned related sequences and evaluate SNPs
SIFT	Sorting intolerant from tolerant
SNPs	Single nucleotide polymorphism
SSR	Simple sequence repeats
TILLING	Targeting induced local lesion IN genomes

6.1 Introduction

Most people in developing countries depend on special crops as their food source and/or income generation. These crops are mostly known as orphan crops but also as understudied-, lost-, disadvantaged- crops, or NUS (neglected and underutilized crops). According to the Global Facilitation Unit for Underutilized Species (GFU), several criteria must be met for a species to be considered as an orphan crop, including proven food or energy value and that the plant has been widely cultivated in the past, or is currently cultivated in a limited geographical region (http://www.underutilized-species.org/species/about_species.asp; accessed August 2011). Orphan crops are annually grown on more than 250 million ha in developing countries (Naylor et al. 2004). In general, these crops play a key role in the livelihood of resource-poor farmers and consumers in developing countries where they perform better than major crops under extreme soil and climate conditions prevalent in the region.

6.1.1 Types and Diversity of Orphan Crops

Orphan crops belong to the major groups of crops including cereals [e.g., finger millet (*Eleusine coracana*), tef (*Eragrostis tef*), fonio (*Digitaria* spp)], legumes [e.g., cowpea (*Vigna unguiculata*), bambara groundnut (*Vigna subterranea*), grass pea (*Lathyrus sativus*)], and root crops [e.g., cassava (*Manihot esculenta*), yam (*Dioscorea* spp), enset (*Ensete ventricosum*)]. Although orphan crops are many in numbers, brief descriptions are provided below for the most important ones in terms of area they are grown or population they feed.

Finger millet is the most important millet cultivated in Africa and Asia predominantly as a staple food grain (http://test1.icrisat.org/SmallMillets/SmallMillets.htm;

accessed July 2011). The plant is tolerant to drought and the seeds contain methionine (NRC 1996), a valuable amino acid lacking in the diets of millions of the poor people who live on starchy staples such as cassava. Finger millet is also a popular food among diabetic patients because of its low glycemic index and slow digestion (Chandrashekar 2010).

Tef is a cereal crop mainly grown in the Horn of Africa, and its annual cultivation in Ethiopia alone accounts for over 2.8 million ha of land (CSA 2011). The crop is tolerant to abiotic stresses especially to poorly drained soils where other crops such as maize and wheat do not withstand. In addition, tef is considered a healthy food, since the seeds do not contain gluten, the cause for celiac disease (Spaenij-Dekking et al. 2005; Hopman et al. 2008). Despite many benefits associated with the cultivation of tef, the grain yield obtained from the crop is extremely low. The main cause for this extremely low productivity is the susceptibility of the tef plant to lodging (Assefa et al. 2009). Tef plants possess tall and tender stems which are susceptible to damage by wind and rain, therefore lodging (the permanent displacement of the stem from the up-right position) contributes to significant loss in production.

Fonio (acha) is an indigenous West African crop mainly grown by small farmers for home consumption. The crop is not only tolerant to drought, but also a very fast maturing crop. It is also considered as highly nutritious, since it is rich in methionine and cysteine, the two amino acids essential for human health but deficient in major cereals such as wheat, rice, and maize (IPGRI 2004).

Bambara groundnut is an annual legume crop grown for human consumption and is the third most important grain legume in Africa after cowpea and groundnut (Asiwe 2009). The seeds of bambara groundnut are known as a complete food, because they contain adequate quantities of protein (19 %), carbohydrate (63 %), and fat (6.5 %; NRC 2006).

Grass pea is another legume commonly grown for human consumption in Asia and Africa. The plant is extremely tolerant to drought and is considered as an insurance crop, since it produces reliable yields when all other crops fail due to moisture scarcity. Like other legumes, grass pea is a source of protein, particularly for resource poor farmers and consumers. However, the seeds of grass pea contain a neuron-toxic substance called ODAP (β-N-Oxalyl-L-α, β-diaminopropanoic acid; Yan et al. 2006). ODAP is the cause of a disease called neurolathyrism, a neurodegenerative disease that results in paralysis of the lower body. Serious neurolathyrism epidemics were reported during famines when grass pea was the only food source (Getahun et al. 2003).

Cassava (manioc) is a woody shrub, native in South America and intensively cultivated in the tropic and sub-tropic areas of the world (Nassar 2009). It is tolerant to drought and also performs better than other crops on soils poor in nutrients. The major problems related to cassava are low protein content and presence of poisonous compounds called cyanogenic glycosides (CG) in the edible part which release cyanide (Ceballos et al. 2004). Konzo is the paralytic disease associated with consumption of insufficiently processed cassava.

Yam combines different species under the genus *Dioscorea*. It is grown on about 5 million hectares of land worldwide (IITA, http://www.iita.org/; accessed July 2011) and is regarded as a staple food in West Africa. The edible roots look like sweet potato (*Ipomoea batatas*), although these species are not taxonomically related.

Enset is commonly known as 'false banana' for its close resemblance to the domesticated banana plant. Unlike banana, where the fruit is consumed, in enset the pseudostem and the underground corm are the edible parts. Enset is the major food for over 10 million people in densely populated regions of Ethiopia. The plant is considered as extremely drought tolerant and adapts to different soil types (Brandt et al. 1997). Although food products from enset are rich in starch, they are deficient in other essential nutrients; hence enset-based diets need supplementation with proteins.

6.1.2 Role of Orphan Crops in Developing Countries

Feeding an increasing population is a major global challenge, especially in the face of climate change. Understudied- or orphan crops play a key role in food security, nutrition, and income generation to resource-poor farmers and consumers in developing countries and are compatible with agro-ecology and socio-economic conditions. Adaptation or tolerance of orphan crops to abiotic stresses such as moisture scarcity also increases their importance in food security. For example, orphan crops such as finger millet and bambara groundnut are extremely tolerant to prolonged periods of drought or to moisture scarcity (Padulosi et al. 2009; Stadler et al. 2009).

According to Naylor et al. (2004) food security can only be achieved if emphasis is given to the improvement of crops widely cultivated by the poor. Among these, sorghum, millet, and cassava are considered critical crops in African agriculture (Ejeta 2010). In Ethiopia, tef is the main staple cereal grown on the largest area allocated for cereal crops (CSA 2011). In Uganda, sweet potato is a dominant staple crop which ranks third next to cassava and banana in providing the diet for poor rural households (Naylor et al. 2004). Furthermore, cowpea is a major source of dietary protein for the resource poor people in the Sub-Saharan Africa, South Asia, and parts of Latin America (Boukar and Fatokun 2009). Minor millets are also grown primarily by poor farmers in Africa and South Asia (Naylor et al. 2004). In India, for instance, minor millets are annually cultivated on about three million hectares of land, equivalent to 12 % of the whole area under coarse cereals (Paludosi et al. 2009).

6.1.3 Major Constraints of Orphan Crops Cultivation

Despite many beneficial aspects, orphan crops are also bound by a number of challenges especially related to poor productivity. As indicated above, the major bottlenecks affecting these crops are traits such as low yield (e.g., tef and millet), poor nutritional value (e.g., cassava and enset), and the production of toxic substances (e.g., cassava and grass pea).

In addition, crop productivity is affected by a variety of abiotic and biotic stresses. Major abiotic stresses are drought, soil salinity, and acidity. There is some evidence that in recent decades agricultural land has been lost to desertification, salinization, soil erosion, and other consequences of unsustainable land use (Godfray et al. 2010). Furthermore, due to the presence of high density and diversity of pests, diseases, and weeds in tropical regions, productivity of crops significantly decreases. Other factors that affect food production are the use of agricultural land for urbanization and biofuel production, and environmental challenges such as global warming (Godfray et al. 2010).

6.1.4 Need for Orphan Crop Improvement

Crop production could be increased by either expanding the arable area or through intensification, i.e., using improved seeds, fertilizer, fungicides, herbicides, irrigation, etc. According to the Food and Agriculture Organization (FAO), agricultural intensification represents about 80 % of future increases in crop production in developing countries (FAO 2002). Based on this goal, crop breeders and scientists are focusing towards achieving improved cultivars that produce higher yields and at the same time tolerate the sub-optimal soil and climatic conditions.

Among traits that contributed to higher crop productivity in the last century, those which alter the architecture of plants rank first. Architectural changes include alteration in branching pattern and reduction in plant height. The major achievement of the Green Revolution in the 1960s was due to the development and introduction of semi-dwarf crop varieties of wheat and rice along with optimum level of inputs for crop production. These broadly adapted semi-dwarf cultivars were responding to fertilizer application and this led to tremendous increases in productivity. Currently, a number of genes affecting plant height are known in major cereal crops including wheat, rice, and maize (for review, Wang and Li 2006). According to the International Food Policy Research Institute (IFPRI), the Green Revolution represented the successful adaptation and transfer of scientific revolution in agriculture (IFPRI 2002). However, this agricultural revolution which boosted crop production in Asia and Latin America did not occur in Africa. This is mainly due to the fact that the Green Revolution focused on rice and wheat, but not on African crops such as sorghum and millet (Ejeta 2010). Therefore, new strategies and approaches have to be implemented to boost crop productivity in African orphan crops.

6.2 Major Crop Improvement Techniques

Improvement of existing crop varieties and cultivation needs integrative research strategies. However, due to the lack of attention and funds, orphan crop researchers had little exposure to advanced scientific technologies. Crop improvement techniques are broadly grouped into (1) *conventional* approaches that include various types of selection methods, introgression (hybridization), and mutation breeding, and (2) *biotechnological or molecular* approaches that include transgenic and non-transgenic methods such as marker-assisted breeding (MAB) and TILLING (Targeting Induced Local Lesions IN Genomes).

6.2.1 Selection

Selection is an ancient breeding method that is still implemented at large-scale to improve crop plants particularly orphan crops. The technique relies mainly on the selection of plants according to their phenotype and performance. Diverse types of selection techniques have been developed for a variety of crops depending on the pollination behaviour and other factors. For example, mass selection is applied to a certain level in self-fertilizing plants and is an effective method for improving landraces, especially for transferring highly heritable traits (Ghosh 2011). Mass selection refers to the technique whereby individual plants are selected based on their phenotypic performance, and bulk seeds from selection are used to produce the next generation.

6.2.2 Introgression

Introgression or artificial hybridization refers to crossing closely related species in order to create genetic variation which can be utilized for improving traits of choice. According to Baenzinger et al. (2006) the success in hybridization depends mainly on the selection of parents. Hybridization can be broadly grouped into intra-specific (crossing within the species) or inter-specific (crossing between species). Successes in intra-specific crosses resulted in semi-dwarf cultivars of wheat and rice which boosted the productivity of both crops during and after the Green Revolution (Borlaug 2007). The major breakthrough from the inter-specific crossing was the development of an artificial cereal called Triticale. Triticale is a cross between wheat and rye and proved to be tolerant to abiotic stresses such as soil acidity (Kim et al. 2001).

6.2.3 Mutation Breeding

Mutation breeding relies on the implementation of either physical or chemical agents in order to create variability in the population of interest. Mutagenesis created by these agents is widely used for crop improvement, and foods derived from the process are readily accepted by the public for consumption (Jain 2007). During the last seventy years, more than two thousand varieties of crop plants derived from mutation breeding have been released either as direct mutants or as parents to derive new varieties. Most mutation breeding programs aimed at altering traits such as plant height and disease resistance in well-adapted plant varieties. The technique of mutation breeding has been widely implemented in crops such as rice, barley, wheat, chickpea, and soybean (Ahloowalia et al. 2004). In rice alone, over 400 improved varieties have been derived from this technique (Ahloowalia et al. 2004). Among these, mutation in the semi-dwarf cultivar 'Calrose 76' significantly contributed to rice improvement. In this particular cultivar the Sd 1 (semi-dwarf 1) gene is down-regulated using gamma ray mutagenesis. Two high-yielding and semi-dwarf cultivars of barely, namely 'Golden Promise' and 'Diamant' which were also obtained through induced mutation played key role in the development of the European brewing industry. Orphan crops have also benefited from mutation breeding. For example, the release of early maturing and virus resistant mung bean, and high-yielding and blight resistant chick pea were the major contributions of mutation breeding (Ahloowalia et al. 2004; Jain 2007). In addition, the FAO-IAEA project "Genetic improvement of underutilized and neglected crops in low income food deficit countries" (LIFDCs) implemented irradiation and related techniques in order to improve orphan crops (Jain 2009).

6.2.4 In Vitro Culture

Plant tissue culture is the aseptic culture of cells, tissues, organs, and their components under defined physical and chemical conditions *in vitro* (Thorpe 2007). Developing an efficient regeneration system requires optimization for various types of explants and media components. Hormones and growth regulators play key role in determining the conversion of somatic cells to embryogenic tissues (Jimenez 2005). The tissue culture techniques have been successfully implemented in diverse types of plants including cereals (Maqbool et al. 2001; Vasil 2005), legumes (Lakshmanan and Taji 2000), vegetables (Bal and Abak 2007; Ochoa-Alejo and Ramirez-Malagon 2001), oil plants (Wahid et al. 2005), trees (Giri et al. 2004), and forestry (Golle et al. 2009). Tissue culture also enables to rescue and utilize desirable properties of endangered plant species in crop improvement (Bapat et al. 2008). Among diverse tissue culture techniques, the doubled haploids are becoming a popular method in crop improvement (Germana 2011).

6.2.5 Marker-Assisted Breeding

Marker-assisted breeding (MAB or MAS for marker-assisted selection) is the utilization of molecular markers located near genes, which can be traced to breed for traits that are difficult to observe. Tester and Langridge (2010) indicated the benefits of applying new technologies and molecular markers in crop improvement.

These molecular markers are utilized to effectively assemble favorable alleles in phenotypic selection (Charcosset and Moreau 2004). Commonly applied markers are Amplified Fragment Length Polymorphisms (AFLPs), and microsatellites (or Simple Sequence Repeats; SSRs), while markers based on single nucleotide polymorphism (SNPs) are rapidly emerging. According to Collard and Mackill (2008) the following factors should be considered before selecting the type of DNA marker to be used in MAS: reliability, quantity and quality of DNA required, technical procedure for marker assay, level of polymorphism, and cost.

6.2.6 Transgenic Approach

Transgenic technology is also considered as another approach to improve the productivity of crops. To date, only a few crops have benefited from transgenic improvements, but due to high adoption rate, the global area under transgenic crops has increased tremendously in the last decade (James 2010). Although the technology has shown significant impact in increasing crop productivity, due to extensive regulatory procedures and negative public perception, its expansion to other crops and geographical regions remains restricted (Tester and Langridge 2010). Some of the recent investigations on transgenics dealt with solving the major concerns related to the technology to promote the acceptance by the public. Among the concerns, the presence of antibiotic resistance and non-plant promoters were the major ones. However, Ayalew and Stewart (2005) enabled to substitute antibiotic resistant markers with those without any adverse effects. In addition, Jacobsen and Schouten (2007) introduced the modified cisgenesis method whereby plant specific promoters are used instead of foreign promoters from bacteria or other organisms. Advances in increasing the efficiency of gene targeting as demonstrated by Shukla et al. (2009) and Townsend et al. (2009) using zinc-finger proteins will also promote specific or targeted gene transfer, and avoids unwanted or unnecessary pieces of DNA movement to the final crop of interest.

6.2.7 Orphan Crop Improvement

Orphan crop improvement largely depends on conventional techniques such as selection and hybridization as molecular techniques have not yet been implemented. Only a limited number of orphan crops breeding programs use modern techniques such as marker-assisted breeding (for details, Tadele 2009). In addition, genomic information such as whole-genome sequences is not yet available for the majority of orphan crops. This situation is rapidly changing with the increased interest in orphan crops. One example is the draft cassava genome sequence which was recently released (Cassava Genome Project 2010, http:://www.phytozome.net/cassava). In addition, comparative mapping studies have revealed that the genomes of plant species within families are highly conserved. Therefore, orthologous genes from orphan crops can be identified and isolated based on information from sequenced major crops (Moore et al. 1995; Schmidt 2000; Choi et al. 2004; Devos 2005; Jung et al. 2009).

In summary, research on orphan crops needs to focus not only on conventional approaches but also on modern and high-throughput techniques. One of the most promising modern techniques is called TILLING (Targeting Induced Local Lesion IN Genomes) which will be described in detail in the following sections.

6.3 TILLING: An Efficient and Rapid Method of Mutation Discovery

TILLING is a non-transgenic and relatively low-cost reverse genetics method which uses traditional mutagenesis followed by high-throughput mutation detection. In contrast to the forward genetics approaches where mutants are first identified based on phenotypes, in reverse genetics the candidate genes responsible for the expected phenotype are first discovered. TILLING identifies single base pair changes, or small deletions, in specific targeted genes (McCallum et al. 2000; Till et al. 2003; Henikoff et al. 2004; Comai and Henikoff 2006).

6.3.1 History of TILLING and its Applications

TILLING was developed in the year 2000 in the model plant *Arabidopsis thaliana* (McCallum et al. 2000). Nowadays, the technique is successfully adapted to numerous animal and plant species, since it can be applied to all organisms independent of the genome size, reproductive system, generation time, and ploidy level (McCallum et al. 2000; Till et al. 2003; Henikoff et al. 2004; Comai and Henikoff 2006). While TILLING identifies induced mutations in mutagenized populations, the modified form known as EcoTILLING detects naturally occurring SNPs, especially in landraces and wild accessions (Comai et al. 2004; Haughn and Gilchrist 2006). The latter has additional applications in genetic mapping, breeding and genotyping, and also provides information concerning gene structure, linkage disequilibrium, population structure, or adaptation (Haughn and Gilchrist 2006).

TILLING has proved to have additional benefits in addition to identifying polymorphism in genomes. The first part of TILLING requires the development of large number of mutagenized populations. Mutagenesis has been widely applied by breeders for many decades as a conventional improvement technique, and it has enabled the release of many crop cultivars (Ahloowalia et al. 2004; Wang et al. 2006). However, through TILLING, this random mutagenesis is better exploited by screening for mutations in defined genes controlling the trait of interest (Wang et al. 2006). Since TILLING directly introduces genetic variation on improved

Table 6.1 TILLING and EcoTILLING projects in crop plants including orphan crops

Crop	Scientific name	Ploidy level	Mutagen	Concentration	Mutation frequency	Mutation detection	Reference
Cereals							
Barley	*Hordeum vulgare*	Diploid	EMS	0.3, 0.4 %	1/1,000 kb	CEL I transgenomic WAVE-HS	Caldwell et al. 2004
			NaN_3	1.5 mM	1/2,500 kb	CEL I ABI PRISM® 377 DNA sequencer	Lababidi et al. 2009
			NaN_3	1, 5, 10 mM	1/374 kb	CEL I LI-COR	Talamè et al. 2008
Bread wheat	*Triticum aestivum*	Hexaploid	EMS	0.9–1 %	1/38 kb	CEL I non-denaturing polyacrylamide gels	Uauy et al. 2009
			EMS	0.8, 1, 1.2 %	1/24 kb	CEL I LI-COR	Slade et al. 2005
			EMS	0.5, 0.6, 0.7 %	1/23.3–1/37.5 kb	CEL I agarose	Dong et al. 2009a
			EMS	0.5, 0.6, 0.7 %	>1/12 kb	HRM & sequence analysis (Mutation Surveyor®)	Dong et al. 2009b
					Proof-of-concept	Illumina sequencing	Tsai et al. 2011
Durum wheat	*Triticum durum*	Tetraploid	EMS	0.7–0.8 %	1/51 kb	CEL I non-denaturing polyacrylamide gels	Uauy et al. 2009
			EMS	0.8, 1 %	1/40 kb	CEL I LI-COR	Slade et al. 2005
					Proof-of-concept	Illumina sequencing	Tsai et al. 2011
Maize	*Zea mays*	Polyploid	EMS (pollen)	1 %	1/485 kb	CEL I LI-COR	Till et al. 2004a
			EMS (pollen)	0.06 %	1/1,075, 1/476 kb	CEL I LI-COR	Weil and Monde 2007
Oat	*Avena sativa*	Hexaploid	EMS	0.9 %	1/20–1/40 kb	MALDI-TOF	Chawade et al. 2010
Rice	*Oryza sativa*	Diploid			Proof-of-concept	Cel I agarose	Raghavan et al. 2007
			MNU	1 mM	1/135 kb	CEL I capillary gel electrophoresis	Suzuki et al. 2008
			Gamma-ray	500 Gy	1/6,190 kb	BPE agarose gel	Sato et al. 2006
			Az-MNU	1–15 mM	1/265 kb	CEL I LI-COR	Till et al. 2007a
			EMS	1.5 %	1/294 kb		
					Proof-of-concept	Illumina sequencing	Tsai et al. 2011
Sorghum	*Sorghum bicolor*	Diploid	EMS	0.1–0.6 %	1/526 kb	CEL I LI-COR	Xin et al. 2008

(continued)

Table 6.1 (continued)

Crop	Scientific name	Ploidy level	Mutagen	Concentration	Mutation frequency	Mutation detection	Reference
Tef	*Eragrostis tef*	Tetraploid	EMS	0.2 %	~1/500 kb	CEL I LI-COR	Esfeld et al. prep.
Legumes							
Common bean	*Phaseolus vulgaris*	Diploid	EMS	0.4–0.6 %	–	Phenotypic screening	Porch et al. 2009
Mung bean	*Vigna radiata*	Diploid	–	EcoTILLING	–	CEL I LI-COR	Barkley et al. 2008
Pea	*Pisum sativum*	Diploid	EMS	0.05 %	1/669 kb	ENDO 1 LI-COR	Triques et al. 2007
Soybean	*Glycine max*	Polyploid	NMU	2.5 mM	1/140 kb	CEL I LI-COR	Cooper et al. 2008
			EMS	0.5, 0.6 %	1/140; 1/250; 1/550 kb		
Vegetables							
Cabbage	*Brassica oleracea*	Genetically diploid	EMS	0.4 %	1/447 kb	CEL I LI-COR	Himelblau et al. 2009
Mustard	*Brassica rapa*	Diploid	EMS	0.1, 0.6, 0.8, 1 %	1/60 kb	CEL I capillary ABI3730 sequencer	Stephenson et al. 2010
Rape seed	*Brassica napus*	Amphidiploid	EMS	0.3, 0.6 %	1/41.5 1/130.8 kb	CEL I LI-COR	Wang et al. 2008
Potato	*Solanum tuberosum*	Diploid-tetraploid	Gamma-ray	See Al-Safadi and Arabi 2007	Pilot study	CEL I LI-COR	Elias et al. 2009
Tomato	*Solanum lycopersicum*	Diploid	EMS	0.5 %	Proof-of-concept	KeyPoint® GS FLX sequencing	Rigola etal. 2009
			EMS	0.7, 1 %	1/322–1/574 kb	ENDO 1 LI-COR	Minoia et al. 2010
Fruits							
Banana and plantain	*Musa spp*	Diploid, triploid	–	EcoTILLING	–	CEL I LI-COR	Till et al. 2010
Melon	*Cucumis melo*	Diploid	–	EcoTILLING	–	ENDO 1 LI-COR	Nieto et al. 2007

or elite germplasm, it avoids the need for introgression of a mutant allele in a non-adapted background into current high-yielding varieties and avoids the problem of linkage drag. Therefore, the introduction of agriculturally undesirable traits is avoided (Slade and Knauf 2005; Uauy et al. 2009; Sestili et al. 2010).

So far, TILLING and EcoTILLING have been implemented in the following crops: maize (Till et al. 2004a); wheat (Slade et al. 2005; Dong et al. 2009a, b; Uauy et al. 2009; Tsai et al. 2011); rice (Sato et al. 2006; Raghavan et al. 2007; Till et al. 2007a; Suzuki et al. 2008); barley (Caldwell et al. 2004; Talamè et al. 2008; Lababidi et al. 2009); sorghum (Xin et al. 2008); soybean (Cooper et al. 2008); pea (Triques et al. 2007); potato (Elias et al. 2009); and orphan crops such as banana (Till et al. 2010); and tef (Tadele 2009; Esfeld et al. prep., Table 6.1).

6.3.2 The TILLING Method

The TILLING procedure includes the following steps: mutagenesis, development of a non-chimeric population, DNA isolation, and mutation detection (Comai and Henikoff 2006; Tadele et al. 2009; Wang et al. 2010). EcoTILLING follows a similar set of procedures except for using natural populations as the experimental samples instead of induced mutagenized populations. The standard TILLING procedures are presented in the following sections and are summarized in Fig. 6.1 (see also Wang et al. 2010).

6.3.2.1 Mutagenesis: The Critical Step in Generating Experimental Material

As already mentioned, conventional mutagenesis has a long history in crop breeding and thus the broad experience simplifies its application (Henikoff et al. 2004; Comai and Henikoff 2006; Tadele et al. 2009). Mutagenesis can be applied to all species including orphan crops, even if they lack advanced genetic tools (Cooper et al. 2008). Since no exogenous DNA is introduced into the plant, mutagenesis is considered as non-transgenic. The products are, therefore, exempted from regulatory restrictions or procedures imposed on transgenic products (Slade and Knauf 2005; Till et al. 2007b; Cooper et al. 2008).

Selection of the Mutagen

Broadly, mutagens are grouped under chemical and physical agents. Commonly used chemical mutagens include ethyl methanesulfonate (EMS), sodium azide (NaN_3), N-methyl-N-nitrosourea (MNU), methyl methanesulfonate (MMS), hydrogen fluoride (HF), and hydroxylamine (NH_2OH). Physical mutagens such as gamma-ray, x-ray, and fast neutrons are also widely used. In general, mutagens are

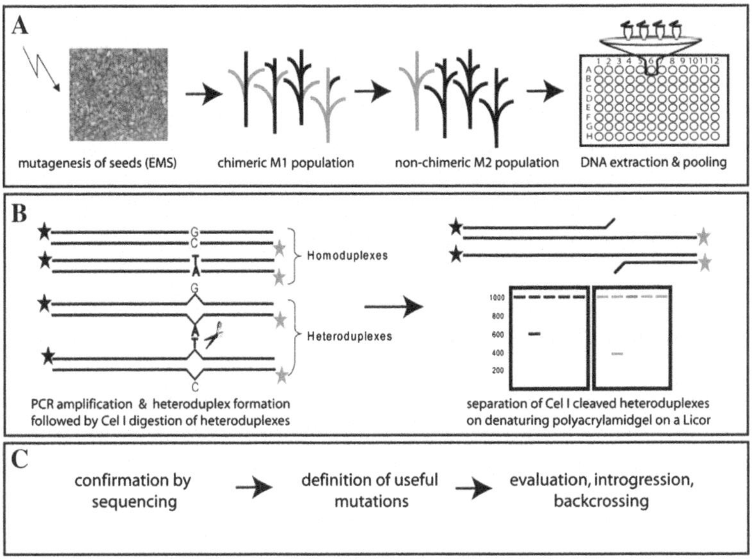

Fig. 6.1 Overview of a standard TILLING protocol. **a** The initial step is establishing a mutant population and assembling the DNA of individual M_2 plants into pools and plates for efficient screening. **b** The mutation detection step involves cleaving of PCR products followed by resolution on gel based systems, such as LI-COR. **c** The last step involves the confirmation of the putative SNP through direct sequencing, characterizing SNPs based on their predicted effect on protein function (using PARSESNP and SIFT), and the subsequent phenotypic characterization and deployment into breeding programs

known to create diverse types of mutations in the genome which range from point mutations to double-strand breaks. Since TILLING detects nucleotide polymorphisms, mutagens which create point mutations are preferentially selected to generate populations for screening. A mutagen proved to induce point mutations and widely used in developing TILLING populations is EMS (Caldwell et al. 2004; Slade et al. 2005; Xin et al. 2008). EMS normally creates G:C to A:T transitions in the genome, due to the alkylation of G nucleotide residues which then pairs with T instead of C (Comai and Henikoff 2006; Parry et al. 2009). In *Arabidopsis*, wheat, and maize about 99 % of EMS induced point mutations are transitions, however, in rice and barley these figures range between 70–84 % (see Cooper et al. 2008). Some studies showed that in rice and barley mutagens such as sodium azide and MNU were more efficient in introducing mutations (Cooper et al. 2008; Suzuki et al. 2008; Talamè et al. 2008).

Mutagenesis is the critical step, since the balance between optimum mutation density and a feasible germination rate contribute to the success of the TILLING population (Haughn and Gilchrist 2006; Parry et al. 2009). Before embarking on large-scale mutagenesis, pilot studies should be made to determine the right mutagen, optimum concentration, and handling procedures (Tadele et al. 2009). While working with mutagens, it is also important to consider the safety precaution measures, since mutagens are carcinogenic and extremely toxic to living organisms, including humans.

In general, the optimum concentration of the mutagen is lower in diploid species than in polyploids, as the former have a lower tolerance towards mutagens than the latter. The high tolerance of polyploids to mutagen is mainly due to complementation of essential genes by homoeologous copies. As a consequence, to obtain the same number of mutants, larger sized populations need to be screened in diploid species than in polyploid species (Parry et al. 2009).

Developing Mutagenized Population

Once the type and optimum concentration of the mutagen is identified, large-scale mutagenesis is made using the explant of choice. In most crop plants, the seeds are used for mutagenesis except for maize, where the pollen grains are mutagenized using EMS followed by introgression to the un-mutagenized female parent (Comai and Henikoff 2006; Till et al. 2007b). The first generation of mutagenized seeds (defined as M_1 population) is typically chimeric; i.e., different cells make different genotypes, due to the multicellular stage of embryos in seeds (Tadele et al. 2009). Hence, M_1 plants are self-pollinated to establish the subsequent M_2 populations. At this point, the mutations do not segregate within cells of an individual plant, and therefore the induced changes are considered stable and heritable (Till et al. 2007b; Cooper et al. 2008). Tissue samples are collected from individual M_2 plants for DNA isolation, and M_3 seeds are harvested and stored for subsequent studies (Henikoff et al. 2004).

6.3.2.2 DNA Sampling and Pooling

Tissue from individual M_2 lines is used for genomic DNA isolation using either high-throughput 96-well plate procedures or small scale methods. After DNA isolation, the quality and quantity of the DNA needs to be checked. The easiest and best way to do this is by loading an aliquot of DNA on electrophoresis gels, as the quality of DNA can be easily observed on an agarose gel. The DNA concentration can also be estimated by comparing it to DNA markers of known concentration. Once the concentration of each sample is estimated, the DNA samples are normalized to the same concentration level to ensure equal representation of each sample within a DNA pool and also later as template in the PCR reaction. After achieving identical DNA concentration for all samples, pooling takes place as this is a pre-requisite to detect mutations. DNA samples are pooled from 2-fold up to 8-fold in a one- or two-dimensional range to reduce screening costs and time, (Till et al. 2007b; Tadele et al. 2009). One can either pool one wild-type (wt) and one mutant DNA, however, most often individual mutant plants are pooled together. Since only one plant of the pool is likely to carry a mutation in the target sequence it ensures that wt and mutant sequences are represented in a pool, while simultaneously screening in a time and cost efficient manner (Haughn and Gilchrist 2006). The dimension refers to the way of pooling as each sample is either represented once on a plate or at two different unique positions. Two-dimensional pooling has the additional advantage of reducing false positives, since candidate mutants are visualized at two independent positions (Rigola et al. 2009; Tadele et al. 2009). Furthermore, individual mutants can be directly targeted in a one-step screening, as the pools do not have to be deconvoluted and re-analysed to identify the actual mutant individual within the pool. On the other hand, two-dimensional pooling decreases the high-throughput, since more pools have to be analysed. Still, the advantage of direct single candidate detection and avoidance of false positives often justifies the additional effort of that particular pooling scheme.

For EcoTILLING, the pooling strategy is different. Here, DNAs from two genotypes are combined in a ratio of 1:1, and in general a single genotype is used as the reference for all lines. This different pooling strategy reflects the fact that more SNPs are expected between distantly related genotypes (Raghavan et al. 2007). Therefore, pooling of individuals with too many nucleotide differences, as expected if genotypes were pooled, would reduce the detection efficiency. In addition, the use of a common reference genotype for all the collection allows the immediate identification of haplotypes within the diverse germplasm and a targeted re-sequencing of only the unique haplotypes.

6.3.2.3 PCR Amplification

TILLING is used to screen for mutations in specific genes that are expected or known to correspond to the trait of interest. With an established TILLING population, virtually any gene can be targeted, however, some prerequisites have to be met.

Primer Design

Designing the right primer pair is crucial and sometimes a challenge in TILLING projects. Primers need to be specific to amplify only the gene of interest and the appropriate gene region. Multiple amplifications reduce the detection efficiency. Primer specificity is especially important for members of multi-gene families, or in polyploid species where multiple homoeologues are present (Slade and Knauf 2005; Tadele et al. 2009; Uauy et al. 2009). This is commonly achieved by using copy-specific primers. In polyploid species, specificity can be achieved by designing primers in more divergent regions, particularly in the introns or in the 5′ and 3′ UTR region (Slade et al. 2005; Uauy et al. 2009; Sestili et al. 2010). In addition, in cases where only a few differences between homologues or regions exist, appropriate primers can be selected based on the following two criteria: (1) aligning the first nucleotide from the 3′ end of the primer to the genome-specific SNP, and (2) introducing a mismatch in the primer at the third or fourth position from the 3′ end (Uauy et al. 2009).

The problem of multiple target PCR amplification was also solved using a High Resolution Melting (HRM) analysis of mixed PCR amplicons containing homoeologous gene fragments and sequence analysis using Mutation Surveyor® software that simultaneously detects mutations in homoeologous genes (Dong et al. 2009b; Ishikawa et al. 2010). Although not widely implemented at present, the third option of avoiding unspecific PCR amplification is made by pre-treating the DNA with a restriction enzyme that specifically cuts only a single copy of the genome (Cooper et al. 2008).

Although prior information of the genome sequence is not required to perform TILLING, the presence of full-length genomic sequence for the gene of interest improves the chance of success. The knowledge or presence of the genome sequence accelerates the development of suitable targets, especially the designing of effective primers (Henikoff et al. 2004; Parry et al. 2009).

Depending on the detection method, the length of the amplified products can range from 0.3 to a maximum of ~3 kb (Slade and Knauf 2005; Raghavan et al. 2007). To ensure that the appropriate gene region is amplified by PCR, the CODDLE (Codons Optimized to Detect Deleterious LEsions; http://www.proweb.org/coddle/; accessed August 2011) software can be used. CODDLE defines the best region to be screened for mutations by detecting functional domains or domains which are likely to be the most sensitive to amino-acid substitutions. The use of CODDLE increases the probability of detecting deleterious mutations in the gene of interest, obtaining regions with high frequency of stop codons, and those which are evolutionary conserved. Hence it is useful for providing an allelic series of mutations (Colbert et al. 2001).

PCR Amplification and Hetero-duplex Formation

To assure the specificity and to maximise the yield of PCR products for the subsequent steps, most TILLING projects use touch-down PCR protocols. Non-specific amplification can lead to additional heteroduplex formation, reducing the detection frequency and resulting in cleaved products even in control samples (Raghavan et al. 2007). Copy-specific primers deduced from unique gene regions overcome this problem and allow a normal PCR protocol to be used, resulting in an appropriate amount of the specific PCR product for mutation detection. Proof-reading polymerases can be used (although the costs are high), but are not necessary in our experience.

The PCR amplification is followed by the heteroduplex formation step where the PCR amplified products are first denatured and then allowed to slowly cool, which facilitates the formation of heteroduplex molecules (see Fig. 6.1; Till et al. 2003; Gilchrist and Haughn 2005). This is usually introduced as an additional step at the end of the PCR program.

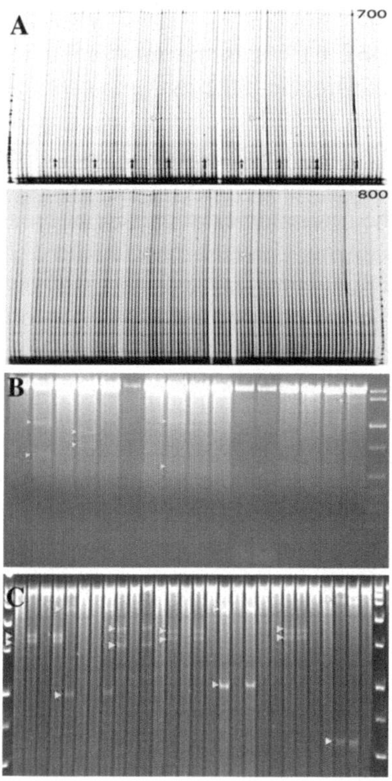

Fig. 6.2 Three options of gel-based mutation detection methods in TILLING: (**a**) LI-COR DNA analyzer, (**b**) agarose gel, and (**c**) non-denaturing PAGE gel. In the LI-COR method, products are PCR amplified using infra-red dye labelled primers, and digested using endonucleases. They are then loaded on a denaturing polyacrylamide gel for mutation detection and visualized as complementary bands in the gel images produced for each of the dyes (as shown above for channels 700 and 800). The agarose (**b**) and non-denaturing polyacrylamide gels (**c**), however, do not require labelled primers, since mutations can be visualized on gels stained with ethidium bromide. *Yellow arrows* indicate the presence of complementary bands corresponding to the two digested products originating from a mutant individual within the pooled DNA samples

6.3.2.4 Mutation Detection

Mutations can be detected using different techniques (Fig. 6.2). Initially, the detection of single base pair differences was done using denaturing HPLC (DHPLC; McCallum et al. 2000), whereas nowadays single-strand cleaving endonucleases are used to detect mismatches in heteroduplexes (Henikoff et al. 2004; Gilchrist and Haughn 2005). Several single-strand specific nucleases, members of the S1 nuclease family (e.g. CEL I or mung bean nuclease), recognize and cleave the mismatches formed in heteroduplexes (Till et al. 2004b; Comai and Henikoff 2006).

The comparison among different nucleases in terms of efficiency has been studied (Till et al. 2004b). CEL I is the most commonly used and preferred enzyme for mutation detection in TILLING projects (Till et al. 2004b, 2006), despite a slight preference of CEL I in cleaving specific types of mismatches (see Triques et al. 2007). CEL I cleaves to the 3′ side of mismatches in heteroduplexes while leaving homoduplexes intact (Henikoff et al. 2004). Thereby, CEL I cleaves each strand of the heteroduplex at the site of the mismatch that produces two complementary fragments (Haughn and Gilchrist 2006). Although CEL I can be purchased through commercial suppliers (Surveyor Mutation Detection Kit; Transgenomic®), the majority of TILLING platforms isolate the enzyme from celery stalks (celery juice extract; CJE) through an easy, cheap, and very robust extraction method established by Till et al. (2006). Recently, other endonucleases such as *Brassica* petiole extract (BPE; Sato et al. 2006) and ENDO1, which is extracted from *A. thaliana* have been proven to be as efficient, or even more efficient than CEL I in cleaving heteroduplexes (Triques et al. 2007).

Mutation detection after cleavage of heteroduplexes can be done using different methods. The most commonly used method is through a denaturing polyacrylamide gel run on a LI-COR DNA analyser (referred as LI-COR, Fig. 6.2a). For this approach, PCR products are amplified using infra-red dye (IRD) labelled primers. Both, the forward- and reverse- primers are labelled at the 5′ end with a specific dye that can be detected in one of the two channels of the LI-COR. Since the amplification efficiency of labelled primers is lower than that of unlabelled primers, the addition of both labelled and unlabelled primers in every PCR reaction is required to obtain a high-yielding PCR product. After PCR amplification and endonuclease digestion, products are purified using either Sephadex® purification (Till et al. 2006) or ethanol precipitation.

The purified samples are loaded on 5–6 % denaturing polyacrylamide gels. The common loading is done by pipetting the samples first in a comb loading tray (Gel Company Inc, San Francisco, USA) before they are loaded to a membrane comb (Gel Company Inc) by immersing the comb into the samples in a tray. A labelled fragment of defined size could be applied at specified distance in order to facilitate the counting of the lanes. Before loading the samples onto the gel, a pre-run of the LI-COR is made for about 20 min. After the successful pre-run, samples are loaded onto the membrane comb and then the comb is inserted on the top of the gel. The run-time for the gel is dependent on the product size. Since forward- and reverse primers are labelled with different infra-red dyes, cleaved products are visible in both channels of the LI-COR (Fig. 6.1). The sum of the two cleaved products gives the size of the original PCR product. The GelBuddy program (http://www.proweb.org/gelbuddy; accessed April 2011) is used to automate band calling in polyacrylamide gels (Till et al. 2006).

Modified mutation detection methods have been developed by various TILLING platforms. These methods include detection by agarose (Fig. 6.2b) and non-denaturing polyacrylamide gels (Fig. 6.2c; Sato et al. 2006; Raghavan et al. 2007; Dong et al. 2009a; Uauy et al. 2009). These alternative methods have applications especially in developing countries, where resources are limited and frequent power failures result in complete loss of data when using a LI-COR system. The information or data from the electrophoresis systems can easily be recovered, even after multiple power interruptions. In addition, the LI-COR system is not affordable in many individual labs as it uses expensive primers and equipment.

Alternative detection methods which do not require labelled primers can be easily adopted in developing countries laboratories as well as for orphan crop improvement. Raghavan et al. (2007) and Dong et al. (2009a) used conventional agarose gels for mutation detection, a method appropriate for low-budget and ill-equipped laboratories. Another modification by Uauy et al. (2009) used non-denaturing polyacrylamide gels stained with ethidium bromide to detect mutations. Advantages of these alternative TILLING methods include the use of only unlabelled primers which are easier to obtain and handle. In addition, the size of the PCR product can be increased to 2–3 kb (instead of a maximum of 1.5 kb in the case of the LI-COR). Furthermore, no prior purification of the digested samples is necessary, which further reduces the cost and time (Raghavan et al. 2007; Dong et al. 2009a). The lower costs and simplicity of the technique allows further applications in germplasm characterization and mapping studies (Raghavan et al. 2007). Despite all these benefits in terms of ease of use and relative simplicity, alternative methods are not as sensitive as the LI-COR and do not provide an exact location or site of the mutation within the target amplicon.

In addition to the above mentioned methods, Next Generation Sequencing (NGS) platforms have recently been implemented to detect mutations in TILLING populations (Rigola et al. 2009; Tsai et al. 2011). Rigola et al. (2009) used the 454-FLX platform (Roche Applied Science) to discover mutants in the tomato eIF4E gene, by screening more than 3,000 families in a single sequencing run. Tsai et al. (2011) applied the Illumina GA sequencing platform to identify mutations in populations of rice and wheat. Conventional endonuclease based TILLING projects are limited to eight samples per pool because of the detection efficiency. As a consequence, analyses with more than eight samples per pool might result in data noises (Rigola et al. 2009). Some of the benefits of sequencing based TILLING approaches are: (1) the sequence of each mutation and its impact on protein sequence is directly determined, (2) it does not rely on either labelled primers or endonucleases, (3) it is based on an objective statistical method and not on visual inspections, (4) it is flexible with respect to changing numbers of samples and amplicons, and (5) since it is based on highly redundant sequencing, the likelihood of identifying false positives is reduced (Rigola et al. 2009; Tsai et al. 2011). Despite these advantages, TILLING by sequencing requires access to expensive and technically challenging NGS equipment and experienced bioinformaticians for data analysis. Due to these reasons, TILLING by sequencing is currently implemented mainly by service providers. An overview summarizing the advantages and disadvantages of various TILLING strategies is shown in Table 6.2.

Table 6.2 Comparison of different TILLING detection methods

	LI-COR DNA analyser	Non-denaturing PAGE	Agarose gels	NGS
Primers	Labelled	Unlabelled	Unlabelled	Unlabelled
DNA pooling (number of individuals)	Up to 8	Up to 8	Up to 8	96 or more individuals
Amplicon size	Up to 1.5 kb	Up to 1.8 kb	Up to 3 kb	Flexible
Purification	Sephadex or ethanol precipitation	Not required	Not required	Requires NGS library preparation
Time	1–2 days	1 day	1 day	Several weeks
Throughput	Single amplicon	Single amplicon	Single amplicon	Multiple amplicons[a]
Resolution	Within 30–50 bp (directional)	Within 100–200 bp (non-directional)	Within 100–200 bp (non-directional)	Single base resolution
Bioinformatic analysis of results	Web-based programs available	Not required	Not required	Extensive bioinformatic pipeline required

[a]Number of amplicons depends on the size of each amplicon, pooling depth, and NGS throughput

6.3.2.5 Confirmation by Sequencing

Mutations detected by gel-based TILLING methods need to be confirmed by sequencing. If the LI-COR is used, the detected mutation pinpoints directly to the location of the polymorphism and makes confirmation by sequencing quite efficient. The labelled primers on the LI-COR provide a directionality (5′ or 3′ end), which allows the sequencing reaction to target the specific site (Gilchrist and Haughn 2005; Till et al. 2007b). On the other hand, the alternative screening methods, which use unlabelled primers, do not provide an exact position of the mutation. Since the size of the fragments on the gels indicates only a putative but not exact location, sequencing should be carried out in order to pinpoint the mutation. In addition, the longer fragments (~2–3 kb) that can be amplified, imply that sometimes an additional sequencing primer within the amplicon might be needed to reveal the mutation.

Mutations are randomly induced and can target every gene (Colbert et al. 2001; Parry et al. 2009). The advantage of the chemical mutagenesis in TILLING is that it creates an allelic series of mutations. These include (1) *nonsense mutations*: a single base pair change converts an amino acid codon into a stop codon, (2) *missense mutations*: a single base pair change alters the amino acid encoded by a particular codon, (3) *silent mutations*: a single base pair change does not alter the amino acid encoded by a particular codon, and (4) *splice junction mutations*: a single base pair change alters the canonical GT/AG splice sites encoded within the introns and leads to mis-spliced messenger RNA, and in most cases, to a truncated protein (McCallum et al. 2000). The allelic series of induced mutations can potentially confer various phenotypes ranging from subtle to strong. Mutations in the coding region of the gene might alter plant metabolism or the effective level of a gene product that might be useful for breeding. TILLING is the only reverse genetics technique that can deliver such genetic diversity (Haughn and Gilchrist 2006) and several studies estimate the numbers or percentages of missense-, nonsense-, and silent mutations. The numbers vary slightly, but in general around 50 % of EMS-induced changes in the target are missense mutations, and ~5 % are truncations (either nonsense or splice junction mutations), whereas the remaining are either silent or located within non-coding regions (Haughn and Gilchrist 2006). Similar findings were reported for soybean (Cooper et al. 2008), *Arabidopsis* (Greene et al. 2003; Till et al. 2003; Till et al. 2004a), *Brassica rapa* (Stephenson et al. 2010), and durum- and bread-wheat (Slade et al. 2005; Dong et al. 2009a; Uauy et al. 2009).

To investigate whether the detected mutations have an influence on the protein, web-based software, namely PARSESNP (http://www.proweb.org/parsesnp/; accessed August 2011) and SIFT (http://sift.jcvi.org/; accessed August 2011) can be used. PARSESNP (Project Aligned Related Sequences and Evaluate SNPs) reveals the changes in the nucleotide and amino-acid sequences and documents any restriction endonuclease site that has been altered (Taylor et al. 2003; Gilchrist and Haughn 2005). SIFT (Sorting Intolerant From Tolerant) uses alignments between closely related sequences to predict whether the amino acid change is expected to have deleterious effects on the protein (Ng and Henikoff 2003; Henikoff et al. 2004).

6.3.2.6 Estimation of Mutation Frequency

The mutation frequency of a population can be calculated based on the confirmed mutations identified across several TILLING targets. This value is calculated as the total number of base pairs screened divided by the total number of mutations scored; whereby the total number of base pairs refers to amplicon size x screened individuals. The amplicon size needs to be adjusted by excluding the regions close to either primer (roughly 5–10 % of the total size), due to technical limitations in mutation detection.

The highest mutation density was obtained from the two polyploid wheat species, namely bread wheat and durum wheat. The mutation frequencies reported were one mutation per 25 kb in hexaploid bread wheat and one mutation per 40 kb in tetraploid durum wheat (Slade et al. 2005). The mutation density reported for other major crops were 1/374 to 1/1,000 kb for barley (Caldwell et al. 2004; Talamè et al. 2008), 1/300 to 1/600 kb for rice (Till et al. 2007a), and 1/526 kb for sorghum (Xin et al. 2008). According to Weil (2009), a mutation rate of 1/500 kb or less is considered feasible in TILLING. Higher mutation frequencies imply that the number of individuals to be screened in order to obtain a certain number of mutations is too large to make TILLING a practical endeavour. Table 6.1 shows mutation frequencies for different crop species. In general, mutation frequencies are determined by the type and concentration of the mutagen, genetic backgrounds as well as the gene region screened (Uauy et al. 2009). Hence, specific criteria should be applied to calculate the mutation frequency, so that the values obtained are comparable.

6.4 Beyond TILLING: What Follows Mutation Detection

Most TILLING and EcoTILLING projects on different crops so far are implemented as a proof-of-concept to test the applicability of the technique. Hence, the projects ceased once the intended goals were achieved. Several studies, however, implemented TILLING directly for crop improvement. Among these, granule-bound starch synthase genes responsible for producing waxy wheat were investigated in TILLING populations. Here, two mutation detection systems were used, namely LI-COR and thin (≤4 mm) agarose gels (Slade et al. 2005; Dong et al. 2009a). Although both detection methods enabled the development of complete waxy wheat, the latter method took only 18 months after crossing two truncation mutants obtained from the TILLING population (Dong et al. 2009a). Several other groups have targeted the starch branching enzyme gene (*SBE-IIa*) in wheat, as reduced transcript levels of this gene by RNA interference (RNAi) led to transgenic wheat with high amylose (Regina et al. 2006). The initial proof-of-concept by RNAi provides strong evidence that a TILLING based approach will work. Currently, many of these lines are being backcrossed and combined for evaluation (Uauy et al. 2009). The use of TILLING for crop improvement is also being pursued by companies such as Arcadia Biosciences (http://www.arcadiabio.com/), who are developing tomato lines that suffer less postharvest loss. The number of examples continue to increase with rapid advances in the knowledge of gene function (which should provide new targets for TILLING) and as several mutants make their way through breeding programs.

To apply TILLING in crop improvement, mutants discovered from screening need to be investigated further. Phenotypic screening will usually be the first option to link mutations in the candidate gene with the trait of interest. To save time and cost, only the most promising candidates should be screened, i.e. those that encode for nonsense-, splice- junctions- or radical missense- mutations as determined by PARSESNP and SIFT. Evidence of a causal effect between the target gene and the trait of interest is further advanced, if independent mutant alleles result in the same phenotype (Haughn and Gilchrist 2006). However, phenotypic screening is a challenge in polyploid species, since genetic buffering makes it less likely that recessive mutations reveal the expected phenotype. Therefore, it may be necessary to identify mutations in each homoeologous copy of the targeted gene and combine them by crossing (Parry et al. 2009). In addition to the use of the mutant alleles for breeding, backcrossing to the original line, or introgression to other elite cultivars is also necessary. The multiple backcrosses to the recurrent parent remove unlinked mutations that could lead to undesirable phenotypes

(Slade and Knauf 2005; Sestili et al. 2010). Several methods exist to prove if the desired mutation is still in the subsequent generation. These include, (1) CAPS (Cleaved Amplified Polymorphic Sequence) or dCAPS (Derived Cleaved Amplified Polymorphic Sequences) markers that take advantage of restriction sites, (2) sequencing of the mutation region, (3) using TILLING with the parents as reference preferentially using the alternative TILLING strategy on agarose gels, and (4) designing SNP-specific assays such as KASPar (Allen et al. 2011). Marker-assisted backcrossing (MABC) can help to introduce the desired mutation in widely-adapted and high-yielding cultivars. In summary, the use of MABC to remove background mutations is needed to breed a stable line with good agronomic performance.

6.5 Application of TILLING and EcoTILLING to Orphan Crops

So far, TILLING and EcoTILLING are not widely implemented in orphan crops. However, there are some initiatives to apply these methods and we present some of these projects below.

6.5.1 Cassava (Manihot esculenta)

Conventional cassava breeding is challenging, due to limited genetic variation in elite germplasm. Therefore, radiation mutation of selected elite lines as well as wide crosses involving wild *Manihot* species and castor bean are done to broaden the genetic base of the germplasm pool, and expand the industrial uses of cassava. In collaboration with the International Atomic Energy Agency (IAEA) three different methods of irradiation have been used for inducing mutations. About 4,000 seeds from six different cassava clones were irradiated with gamma rays or with fast neutrons, which led to the harvest of about 5,000 M_2 seeds (Ceballos et al. 2010). The International Institute of Tropical Agriculture (IITA; http://liferay.iita.org/web/mab/home/tilling;jsessionid=8166693C8CBE40D02B12544406EB36AF; accessed August 2011) is conducting a pilot study on EMS-induced mutagenesis to focus on a single trait. However, once the protocol is established, it shall be adapted to discover mutations for other important traits, and to other crops of the IITA. The aims of the project are (1) identification of mutants with desired traits that will be incorporated into the existing cassava improvement program, (2) marker development for MAS, and (3) to design large-scale reverse-genetics experiments to tag genes controlling desirable traits in cassava and other IITA mandate crops. The expected outputs are to (1) establish a catalogue of characterized mutant population, (2) identify DNA sequences of candidate genes as basis for marker development, and (3) to develop standard operating procedures (SOPs) for TILLING using IRD-labelled primers on a gene analyzer instrument.

6.5.2 Banana and Plantain (Musa spp)

Banana provides staple food for approximately 400 million people and is important for both export and local markets. The EcoTILLING project on banana and plantain utilized 80 accessions to discover and characterize nucleotide polymorphisms. From these accessions alone, over 800 novel alleles were obtained by tilling 14 gene targets (Till et al. 2010). Thus, EcoTILLING contributes to understanding genetic diversity, and improving *Musa* for food production. Till et al. (2010) used the technique for the simultaneous discovery of heterozygous and homozygous polymorphisms in diploid accessions, which permit fast evaluation of nucleotide diversity in accessions of the same genome type. The high base pair resolution allows the grouping of accessions based on shared banding patterns, and it is a fast method to compare and barcode a large number of accessions at relatively low cost. It can also be used to tag accessions in gene banks and for classification of newly acquired samples, and even phylogenetic approaches are possible. Additionally, EcoTILLING can be applied for functional genomic studies and mutation breeding (Nieto et al. 2007) and natural allele mining can serve as primary selection for material to be used in breeding programs. In addition to the intensive use of EcoTILLING, an EMS-mutagenized TILLING population of the triploid 'Grande Naine' was developed and large-scale efforts are currently being made (Till et al. 2010).

6.5.3 Tef (Eragrostis tef)

The Tef Improvement Project (TIP) focuses on tackling lodging, the major yield limiting factor in tef cultivation. Developing lodging resistant semi-dwarf tef cultivars is the main goal of the TIP next to developing drought tolerant cultivars, since moisture scarcity is becoming widespread in Ethiopia. So far around 4,500 M_2 individuals were screened for several genes of interest (Esfeld et al. in prep). One of the major obstacles related to tef TILLING is the presence of two homoeologous gene copies due to the allotetraploid nature of tef. However, this particular problem is overcome by designing copy-specific primers for each gene of interest and amplifying one copy at a time (see Sect. 6.3.2.3.1. for further detail).

6.5.4 Pearl Millet (Pennisetum glaucum)

Pearl millet is one of the drought tolerant crops widely grown in Asia and Africa. The International Crop Research Institute for Semi-Arid Tropics (ICRISAT, http://www.icrisat.org/bt-gene-discovery.htm; accessed August 2011) produced a

TILLING population using the inbred line "P1449-2-P1". A total of 31,000 seeds were mutagenized in three different batches, using 5.0, 7.5, 9.0, and 10.0 mM EMS. DNA from at least 7,458 M_2 TILLING lines is available, and this number is currently being increased to 10,000 M_2 lines, as this was calculated to be the ideal population size for mining allelic variants in candidate genes. Eight-fold pooling of normalized genomic DNA from the 7,458 M_2 lines has been completed. The pools of DNA will be made available for the international pearl millet community (http://www.icrisat.org/bt-gene-discovery.htm), however, the latest progress is not communicated on the web site.

6.5.5 Chickpea (*Cicer arietinum*)

Chickpea is an important food legume, extensively cultivated in Asia and Africa. Next to pearl millet, chickpea was chosen for TILLING at the ICRISAT (http://www.icrisat.org/bt-gene-discovery.htm; accessed August 2011). Over 20,000 seeds were treated with the EMS and roughly 9,000 germinated, of which only 6,000 set seeds.

6.5.6 Mung Bean (*Vigna radiata*)

Mung bean is an important economic crop and dietary staple in many developing countries. Barkley et al. (2008) performed an EcoTILLING study to discover and verify DNA polymorphisms in the mung bean collection USDA-ARS PGRCU. This collection has been evaluated for various morphological characters and some of the core accessions were evaluated with SSR markers. They showed both limited genetic and morphological diversity. Thus, EcoTILLING was used to further assess *V. radiata* for molecular diversity, and to evaluate the EcoTILLING method for rapid detection of SNPs in plant germplasm. The intron region was used to detect the polymorphism, as this region is ideal when assessing polymorphism in a species with high genetic identity. Thus, the overall aim of the study was to reveal variation among *V. radiata* var. *radiata* accessions, and its progenitor *V. radiata* var. *sublobata*, mine for SNPs within the *V. radiata* var. *radiata* accessions, and determine the number of different haplotypes in the collection. A total of 157 DNA polymorphisms were detected from ten primer sets, which resulted in 45 haplotypes. Numerous DNA polymorphisms between the two varieties were found, however, the core radiata collection is genetically similar. The results suggest, that either *V. radiata* var. *radiata* generally has a narrow genetic base or that the collection of the USDA-ARS PGRCU has to be expanded. Subsequent studies might use the obtained data to develop SNP markers that can be advantageous in MAS.

6.6 Conclusions

TILLING and EcoTILLING are high-throughput mutation detection systems that exploit non-transgenic allelic variation. The technique has proven highly efficient in obtaining desirable mutant lines and is also extremely versatile. This includes the possibility to develop many in-house protocols and reagents such as CJE, and the ability to implement detection methods using both basic technology platforms, such as agarose gels, as well as NGS equipment. The flexibility of the system has allowed the technology to become widely adopted in many crop plants, including maize, wheat, and rice since its first development in *Arabidopsis* a decade ago. Consequently, there is also growing interest in applying the technique in orphan crops that lack well developed genetic tools.

There are still several major challenges to overcome. The first is, the expansion of these approaches to orphan crops and the training of local expertise to implement these technologies. The development of TILLING requires support from local agencies, so that the populations generated, and downstream techniques and supplies, can be used efficiently by the research community. The second is, the development of NGS approaches and the associated technologies, such as RNAseq and genome capture which will greatly increase the possibility to rapidly access mutations in orphan crops. The possibility to rapidly sequence gene space by RNAseq and the ability to capture these sequences from mutant populations (via genome capture) now opens the possibility to re-sequence complete TILLING populations. Although these efforts require large financial and technical investments at the moment, we expect that in the near future the *in silico* cataloguing of mutations for a complete TILLING population will be a routine practice. Once mutations of interest have been obtained, the major challenge of shifting these mutations into breeding lines or varieties with advanced agronomic performance for resource poor farmers remains. It is only at this point, that TILLING will be delivering on the promise of improving food security.

Acknowledgments K. E. and Z. T. express their gratitude to Syngenta Foundation for Sustainable Agriculture and University of Bern for supporting the Tef Improvement Project. The TILLING research in the Uauy lab is funded by the UK Biotechnology and Biological Sciences Research Council (BBSRC) (BB/I000712/1).

References

Ahloowalia BS, Maluszynski M, Nichterlein K (2004) Global impact of mutation-derived varieties. Euphytica 135:187–204

Allen AM, Barker GLA, Berry ST, Coghill JA, Gwilliam R, Kirby S, Robinson P, Brenchley RC, D'Amore R, McKenzie N, Waite D, Hall A, Bevan M, Hall N, Edwards KJ (2011) Transcript-specific, single-nucleotide polymorphism discovery and linkage analysis in hexaploid bread wheat (*Triticum aestivum* L.). Plant Biotechnol J. doi:10.1111/j.1467-7652.2011.00628.x

Al-Safadi B, Arabi MIE (2007) In vitro induction, isolation and selection of potato mutants tolerant to salinity. Adv Hort Sci 20:127–132

Asiwe, JAN (2009) Field evaluation of bambara groundnut germplasm for important traits and resistance to root knot nematode infestation. In: Tadele Z (ed) New approaches to plant breeding of orphan crops in Africa: Proceedings of an International Conference, Bern, Switzerland. Stämpli AG, Bern, 19–21 Sept 2007

Assefa K, Belay G, Tefera H, Yu J-K, Sorrells ME (2009) Breeding tef: conventional and molecular approaches. In: Tadele Z (ed) New approaches to plant breeding of orphan crops in Africa: Proceedings of an International Conference, Bern, Switzerland. Stämpli AG, Bern, 19–21 Sept 2007

Ayalew M, Stewart N (2005) Overexpression of *Arabidopsis thaliana* ABC transporter confers kanamycin resistance to transgenic plants. Nat Biotechnol 23:1177–1180

Baenziger PS, Russell WK, Graef GL, Campbell BT (2006) Improving lives: 50 years of crop breeding, genetics and cytology. Crop Sci 46:2230–2244

Bal U, Abak K (2007) Haploidy in tomato (*Lycopersicon esculentum* Mill.): a critical review. Euphytica 158:1–9

Bapat VA, Yadav SR, Dixit GB (2008) Rescue of endangered plants through biotechnological applications. Natl Acad Sci Lett 31:201–210

Barkley NA, Wang ML, Gillaspie AG, Dean RE, Pederson GA, Jenkins TM (2008) Discovering and verifying DNA polymorphisms in a mung bean [*V. radiata* (L.) R. Wilczek] collection by EcoTILLING and sequencing. BMC research notes: 1:28. doi:10.1186/1756-0500-1-28

Borlaug NE (2007) Sixty-two years of fighting hunger: personal recollections. Euphytica 157:287–297

Boukar O, Fatokun C (2009) Strategies in cowpea breeding. In: Tadele Z (ed) New approaches to plant breeding of orphan crops in Africa. In: Proceedings of an International Conference, Bern, Switzerland. Stämpli AG, Bern, 19–21 Sept 2007

Brandt SA, Spring A, Hiebsch C, McCabe JT, Tabogie E, Diro M, Wolde-Michael G, Yntiso G, Shigeta M, Tesfaye S (1997) The tree against hunger: enset-based agricultural systems in Ethiopia. American Association for the advancement of science

Caldwell DG, McCallum N, Shaw P, Muehlbauer GJ, Marshall DF, Waugh R (2004) A structured mutant population for forward and reverse genetics in barley (*Hordeum vulgare* L.). Plant J 40:143–150

Ceballos H, Iglesias CA, Pérez JC, Dixon AG (2004) Cassava breeding: opportunities and challenges. PlantMol Biol 56:503–516

Ceballos H, Okogbenin E, Pérez JC, López-Valle LAB, Debouck D (2010) Cassava. In: Bradshaw JE (ed) Hand book of plant breeding: root and tuber crops. Springer, New York

Chandrashekar A (2010) Finger millet: *Eleusine coracana*. Adv Food Nutr Res 59:215–262

Charcosset A, Moreau L (2004) Use of molecular markers for the development of new cultivars and the evaluation of genetic diversity. Euphytica 137:81–94

Chawade A, Sikora P, Bräutigam M, Larsson M, Vivekanand V, Nakash MA, Chen T, Olsson O (2010) Development and characterization of an oat TILLING-population and identification of mutations in lignin and beta-glucan biosynthesis genes. BMC Plant Biol 10:86

Choi HK, Mun JH, Kim DJ, Zhu H, Baek JM, Mudge J, Roe B, Ellis N, Doyle J, Kiss GB, Young ND, Cook DR (2004) Estimating genome conservation between crop and model legume species. Proc Natl Acad Sci USA 101:15289–15294

Colbert T, Till BJ, Tompa R, Reynolds S, Steine MN, Yeung AT, McCallum CM, Comai L, Henikoff S (2001) High-throughput screening for induced point mutations. Plant Physiol 126:480–484

Collard BC, Mackill DJ (2008) Marker-assisted selection: an approach for precision plant breeding in the twenty-first century. Philos Trans R Soc Lond B Biol Sci 363:557–572

Comai L, Henikoff S (2006) TILLING: practical single-nucleotide mutation discovery. Plant J 45:684–694

Comai L, Young K, Till BJ, Reynolds SH, Greene EA, Codomo CA, Enns LC, Johnson JE, Burtner C, Odden AR, Henikoff S (2004) Efficient discovery of DNA polymorphisms in natural populations by Ecotilling. Plant J 37:778–786

Cooper JL, Till BJ, Laport RG, Darlow MC, Kleffner JM, Jamai A, El-Mellouki T, Liu S, Ritchie R, Nielsen N, Bilyeu KD, Meksem K, Comai L, Henikoff S (2008) TILLING to detect induced mutations in soybean. BMC Plant Biol 8:9

CSA (Central Statistical Agency; 2011) Agricultural sample survey 2010/11. Statistical Bulletin 505, Addis Ababa, Ethiopia

Devos KM (2005) Updating the "Crop Circle". Curr Opin Plant Biol 8:155–162

Dong C, Dalton-Morgan J, Vincent K, Sharp P (2009a) A modified TILLING method for wheat breeding. Plant Genome 2:39–47

Dong C, Vincent K, Sharp P (2009b) Simultaneous mutation detection of three homoeologous genes in wheat by High Resolution Melting analysis and Mutation Surveyor. BMC Plant Biol 9:143

Ejeta G (2010) African Green Revolution needn't be a mirage. Science 327:831–832

Elias R, Till BJ, Mba C, Al-Safadi B (2009) Optimizing TILLING and EcoTILLING techniques for potato (*Solanum tuberosum* L.). BMC research notes 2:141. doi:10.1186/1756-0500-2-141

FAO (Food and Agricultural Organization; 2002) World agriculture: towards 2015/2030. FAO, Rome

Germana MA (2011) Gametic embryogenesis and haploid technology as valuable support to plant breeding. Plant Cell Rep 30:839–857

Getahun H, Lambein F, Vanhoorne M, Stuyft PV (2003) Food-aid cereals to reduce neurolathyrism related to grasspea preparations during famine. Lancet 362:1808–1810

Ghosh A (2011) Modern methods for selection of plants with better characteristics. http://www.biotecharticles.com/Agriculture-Article/Modern-Methods-For-Selection-of-Plants-With-Better-Characterisitcs-860.html. Accessed 15 May 2012

Gilchrist EJ, Haughn GW (2005) TILLING without a plough: a new method with applications for reverse genetics. Curr Opin Plant Biol 8:211–215

Giri CC, Shyamkumar B, Anjaneyulu C (2004) Progress in tissue culture, genetic transformation and applications of biotechnology to trees: an overview. Trees-Struct Funct 18:115–135

Godfray HCF, Beddington JR, Crute IR, Haddad L, Lawrence D, Muir JF, Pretty J, Robinson S, Thomas SM, Toulmin C (2010) Food security: the challenge of feeding 9 billion people. Science 327:812–817

Golle DP, Reiniger LRS, Curti AR, Bevilacqua CB (2009) Forestry improvement: emphasis on biotechnology application. Cienc Rural 39:1606–1613

Greene EA, Codomo CA, Taylor NE, Henikoff JG, Till BJ, Reynolds SH, Enns LC, Burtner C, Johnson JE, Odden AR, Comai L, Henikoff S (2003) Spectrum of chemically induced mutations from a large-scale reverse-genetic screen in Arabidopsis. Genetics 164:731–740

Haughn GW, Gilchrist EJ (2006) TILLING in the Botanical Garden: a reverse genetic technique feasible for all plant species. Floriculture Ornamental Plant Biotechnol 1:476–482

Henikoff S, Till BJ, Comai L (2004) TILLING. Traditional mutagenesis meets functional genomics. Plant Physiol 135:630–636

Himelblau E, Gilchrist EJ, Buono K, Bizzell C, Mentzer L, Vogelzang R, Osborn T, Amasino RM, Parkin IAP, Haughn GW (2009) Forward and reverse genetics of rapid-cycling *Brassica oleracea*. Thero Appl Genet 118:953–961

Hopman E, Dekking L, Blokland ML, Wuisman M, Zuijderduin W, Koning F, Schweizer J (2008) Tef in the diet of celiac patients in The Netherlands. Scand J Gatroentero 43:277–282

IFPRI (International Food Policy Research Institute (2002) Green Revolution: curse or blessing?. IFPRI, Washington

IPGRI (International Plant Genetic Resources Institute, 2004) Promoting fonio production in West and Central Africa through germplasm management and improvement of post-harvest technology. Final report. (http://www.underutilized-species.org/%20Documents/PUBLICATIONS/fonio_rapport_final_GTZ_project1.pdf).Accessed Aug 2009)

Ishikawa T, Kamei Y, Otozai S, Kim J, Sato A, Kuwahara Y, Tanaka M, Deguchi T, Inohara H, Tsujimura T, Todo T (2010) High-resolution melting curve analysis for rapid detection of mutations in a Medaka TILLING library. BMC Mol Biol 11:70

Jacobsen E, Schouten HJ (2007) Cisgenesis strongly improves introgression breeding and induced translocation breeding of plants. Trends Biotechnol 25:219–223

Jain SM (2007) Biotechnology and mutagenesis in genetic improvement of cassava (*Manihot esculenta*). Gene Conserve 6:329–343

Jain SM (2009) Mutation induced genetic improvement of neglected crops. In: Tadele Z (ed) New approaches to plant breeding of orphan crops in Africa. In: Proceedings of an International Conference, Bern, Switzerland. Stämpli AG, Bern, 19–21 Sept 2007

James C (2010) Global status of commercialized biotech/GM Crops: 2010. International Service for the Acquisition of Agri-biotech Applications (ISAAA) Briefs No. 42. Ithaca, NY. ISBN: 978-1-892456-49-4. http://www.isaaa.org/. Accessed Aug 2011

Jimenez VM (2005) Involvement of plant hormones and plant growth regulators on in vitro somatic embryogenesis. Plant Growth Regul 47:91–110

Jung S, Jiwan D, Cho I, Lee T, Abbott A, Sosinski B, Main D (2009) Synteny of *Prunus* and other model plant species. BMC Genomics 10:76. doi:10.1186/1471-2164-10-76

Kim BY, Baier AC, Somers DJ, Gustafson JP (2001) Aluminum tolerance in triticale, wheat, and rye. Euphytica 120:329–337

Lababidi S, Mejlhede N, Rasmussen SK, Backes G, Al-Said W, Baum M, Jahoor A (2009) Identification of barley mutants in the cultivar 'Lux' at the *Dhn* loci through TILLING. Plant Breeding 128:332–336

Lakshmanan P, Taji A (2000) Somatic embryogenesis in leguminous plants. Plant Biol 2:136–148

Maqbool SB, Devi P, Sticklen MB (2001) Biotechnology: Genetic improvement of sorghum (*Sorghum bicolor* (L.) Moench). In Vitro Cell Dev Biol Plant 37:504–515

McCallum CM, Comai L, Greene EA, Henikoff S (2000) Targeted screening for induced mutations. Nat Biotechnol 18:455–457

Minoia S, Petrozza A, D'Onofrio O, Piron F, Mosca G, Sozio G, Cellini F, Bendahmane A, Carriero F (2010) A new mutant genetic resource for tomato crop improvement by TILLING technology. BMC Res Notes 3:69

Moore G, Devos KM, Wang Z, Gale MD (1995) Grasses, line up and form a circle. Curr Biol 5:737–739

Nassar NMA (2009) Wild species to improve orphan crops: the case of cassava. In: Tadele Z (ed) New approaches to plant breeding of orphan crops in Africa. In: Proceedings of an International Conference, Bern, Switzerland. Stämpli AG, Bern, 19–21 Sept 2007

Naylor RL, Falcon WP, Goodman RM, Jahn MM, Sengooba T, Tefera H, Nelson RJ (2004) Biotechnology in the developing world: a case for increased investments in orphan crops. Food Policy 29:15–44

Ng PC, Henikoff S (2003) SIFT: Predicting amino acid changes that affect protein function. Nucleic Acids Res 31:3812–3814

Nieto C, Piron F, Dalmais M, Marco CF, Moriones E, Gómez-Guillamón ML, Truniger V, Gómez P, Garcia-Mas J, Aranda MA, Bendahmane A (2007) EcoTILLING for the identification of allelic variants of melon eIF4E, a factor that controls virus susceptibility. BMC Plant Biol 7:34

NRC (National Research Council, 1996). Finger millet. In: Lost Crops of Africa. Volume I: Grains. National Academy of Press, Washington, DC

NRC (National Research Council, 2006) Bambara bean. In: Lost crops of Africa. Volume II: Vegetables. National Academy of Press, Washington, DC

Ochoa-Alejo N, Ramirez-Malagon R (2001) In vitro chili pepper biotechnology. In Vitro Cell Dev Biol Plant 37:701–729

Padulosi S, Mal B, Ravi SB, Gowda J, Gowda KTK, Shanthakumar G, Yenagi N, Dutta M (2009) Food security and climate change: role of plant genetic resources of minor millets. Indian J Plant Genet Resour 22:1–16

Parry MA, Madgwick PJ, Bayon C, Tearall K, Hernandez-Lopez A, Baudo M, Rakszegi M, Hamada W, Al-Yassin A, Ouabbou H, Labhilili M, Phillips AL (2009) Mutation discovery for crop improvement. J Exp Bot 66:2817–2825

Porch TG, Blair MW, Lariguet P, Galeano C, Pankhurst CE, Broughton WJ (2009) Generation of a mutant population for TILLING common bean genotype BAT 93. J Amer Soc Horti Sci 134:348–355

Raghavan C, Naredo MEB, Wang HH, Atienza G, Liu B, Qiu FL, McNally KL, Leung H (2007) Rapid method for detecting SNPs on agarose gels and its application in candidate gene mapping. Mol Breed 19:87–101
Regina A, Bird A, Topping D, Bowden S, Freeman J, Barsby T, Kosar-Hashemi B, Li Z, Rahman S, Morell M (2006) High-amylose wheat generated by RNA interference improves indices of large-bowel health in rats. Proc Natl Acad Sci USA 103:3546–3551
Rigola D, van Oeveren J, Janssen A, Bonné A, Schneiders H, van der Poel HJ, van Orsouw NJ, Hogers RC, de Both MT, van Eijk MJ (2009) High-throughput detection of induced mutations and natural variation using KeyPoint technology. PLoS ONE 4:e4761
Sato Y, Shirasawa K, Takahashi Y, Nishimura M, Nishio T (2006) Mutant selection from progeny of gamma-ray-irradiated rice by DNA heteroduplex cleavage using *Brassica* Petiole extract. Breeding Sci 56:179–183
Schmidt R (2000) Synteny: recent advances and future prospects. Curr Opin Plant Biol 3:97–102
Sestili F, Botticella E, Bedo Z, Phillips A, Lafiandra D (2010) Production of novel allelic variation for genes involved in starch biosynthesis through mutagenesis. Mol Breeding 25:145–154
Shukla VK, Doyon Y, Miller JC, DeKelver RC, Moehle EA, Worden SE, Mitchell JC, Arnold NL, Gopalan S, Meng X, Choi VM, Rock JM, Wu YY, Katibah GE, Zhifang G, McCaskill D, Simpson MA, Blakeslee B, Greenwalt SA, Butler HJ, Hinkley SJ, Zhang L, Rebar EJ, Gregory PD, Urnov FD (2009) Precise genome modification in the crop species *Zea mays* using zinc-finger nucleases. Nature 459:437–441
Slade AJ, Knauf VC (2005) TILLING moves beyond functional genomics into crop improvement. Transgenic Res 14:109–115
Slade AJ, Fuerstenberg SI, Loeffler D, Steine MN, Facciotti D (2005) A reverse genetic, non-transgenic approach to wheat crop improvement by TILLING. Nat Biotechnol 23:75–81
Spaenij-Dekking L, Kooy-Winkelaar Y, Koning F (2005) The Ethiopian cereal tef in celiac disease. N Engl J Med 353:1748–1749
Stadler F, Mohler V, Kilian A, Bonin A, Thümmler F, Mayes S, Ros B, Wenzel G (2009) Initial results from applying diversity arrays technology (DArT) and massively parallel signature sequencing (MPSS) to bambara groundnut. In: Tadele Z (ed) New approaches to plant breeding of orphan crops in Africa. In: Proceedings of an International Conference, Bern, Switzerland. Stämpli AG, Bern, 19–21 Sept 2007
Stephenson P, Baker D, Girin T, Perez A, Amoah S, King GJ, Østergaard L (2010) A rich TILLING resource for studying gene function in *Brassica rapa*. BMC Plant Biol 10:62
Suzuki T, Eiguchi M, Kumamaru T, Satoh H, Matsusaka H, Moriguchi K, Nagato Y, Kurata N (2008) MNU-induced mutant pools and high performance TILLING enable finding of any gene mutation in rice. Mol Genet Genomics 279:213–223
Tadele Z (ed) (2009) New approaches to plant breeding of orphan crops in Africa: Proceedings of an International Conference, Bern, Switzerland. Stämpli AG, Bern, 19–21 Sept 2007
Tadele Z, Mba C, Till BJ (2009) TILLING for mutations in model plants and crops. In Mohan Jain S, Brar DS (eds) Molecular techniques in crop improvement. Springer, Netherlands
Talamè V, Bovina R, Sanguineti MC, Tuberosa R, Lundqvist U, Salvi S (2008) TILLMore, a resource for the discovery of chemically induced mutants in barley. Plant Biotechnol J 6:477–485
Taylor NE, Greene EA (2003) PARSESNP: A tool for the analysis of nucleotide polymorphisms. Nucleic Acids Res 31:3808–3811
Tester M, Langridge P (2010) Breeding technologies to increase crop production in a changing world. Science 327:818–822
Thorpe TA (2007) History of plant tissue culture. Mol Biotechnol 37:169–180
Till BJ, Reynolds SH, Greene EA, Codomo CA, Enns LC, Johnson JE, Burtner C, Odden AR, Young K, Taylor NE, Henikoff JG, Comai L, Henikoff S (2003) Large-scale discovery of induced point mutations with high-throughput TILLING. Genome Res 12:524–530
Till BJ, Reynolds SH, Weil C, Springer N, Burtner C, Young K, Bowers E, Codomo CA, Enns LC, Odden AR, Greene EA, Comai L, Henikoff S (2004a) Discovery of induced point mutations in maize genes by TILLING. BMC Plant Biol 4:12

Till BJ, Burtner C, Comai L, Henikoff S (2004b) Mismatch cleavage by single-strand specific nucleases. Nucleic Acids Res 32:2632–2641

Till BJ, Zerr T, Comai L, Henikoff S (2006) A protocol for TILLING and Ecotilling in plants and animals. Nat Protoc 1:2465–2477

Till BJ, Cooper J, Tai TH, Colowit P, Greene EA, Henikoff S, Comai L (2007a) Discovery of chemically induced mutations in rice by TILLING. BMC Plant Biol 7:19

Till BJ, Comai L, Henikoff S (2007b) TILLING and EcoTILLING for crop improvement. In Varshney RK, Tuberosa R (eds) Genomics-Assisted Crop Improvement, vol 1: genomics approaches and platforms. Springer, Netherlands

Till BJ, Jankowicz-Cieslak J, Sági L, Huynh OA, Utsushi H, Swennen R, Terauchi R, Mba C (2010) Discovery of nucleotide polymorphisms in the *Musa* gene pool by Ecotilling. Theor Appl Genet 121:1381–1389

Townsend JA, Wright DA, Winfrey RJ, Fu F, Maeder ML, Joung JK, Voytas DF (2009) High-frequency modification of plant genes using engineered zinc-finger nucleases. Nature 459:442–425

Triques K, Sturbois B, Gallais S, Dalmais M, Chauvin S, Clepet C, Aubourg S, Rameau C, Caboche M, Bendahmane A (2007) Characterization of *Arabidopsis thaliana* mismatch specific endonucleases: application to mutation discovery by TILLING in pea. Plant J 51:1116–1125

Tsai H, Howell T, Nitcher R, Missirian V, Watson B, Ngo KJ, Lieberman M, Fass J, Uauy C, Tran RK, Khan AA, Filkov V, Tai TH, Dubcovsky J, Comai L (2011) Discovery of rare mutations in populations: TILLING by sequencing. Plant Physiol 156:1257–1268

Uauy C, Paraiso F, Colasuonno P, Tran RK, Tsai H, Berardi S, Comai L, Dubcovsky J (2009) A modified TILLING approach to detect induced mutations in tetraploid and hexaploid wheat. BMC Plant Biol 9:115

Vasil IK (2005) Tissue cultures of maize. Maydica 50:361–365

Wahid MB, Abdullah SNA, Henson IE (2005) Oil palm—Achievements and potential. Plant Prod Sci 8:288–297

Wang D-K, Sun Z-X, Tao Y-Z (2006) Application of TILLING in plant improvement. Acta Genetica Sinica 33:957–964

Wang N, Wang Y, Tian F, King GJ, Zhang C, Long Y, Shi L, Meng J (2008) A functional genomics resource for *Brassica napus*: development of an EMS mutagenized population and discovery of FAE1 point mutations by TILLING. New Phytol 180:751–765

Wang T, Uauy C, Till B, Liu C-M (2010) Tilling and associated technologies. J Integr Plant Biol 52:1027–1030

Wang Y, Li J (2006) Genes controlling plant architecture. Curr Opin Biotech 17:123–129

Weil CF (2009) TILLING in grass species. Plant Physiol 149:158–164

Weil CF, Monde R-A (2007) Getting the point—mutations in maize. Crop Sci 47:S60–S67

Xin Z, Wang ML, Barkley NA, Burow G, Franks C, Pederson G, Burke J (2008) Applying genotyping (TILLING) and phenotyping analyses to elucidate gene function in a chemically induced sorghum mutant population. BMC Plant Biol 8:103–116

Yan ZY, Spencer PS, Li ZX, Liang YM, Wang YF, Wang CY, Li FM (2006) *Lathyrus sativus* (grass pea) and its neurotoxin ODAP. Phytochemistry 67:107–121

Part III
Section 3

Chapter 7
Neglected Oil Crop Biotechnology

Sharad Tiwari and Sunil Kumar

Abstract Global food security has become increasingly dependent on only a handful of crops cultivated intensively leading to crop replacement and a massive reduction in the number of species and diversity of crops. This poses a threat to local and global food security because the replaced indigenous crops are often essential for low input agriculture, have unique nutritional value, and contain diversity of locally adapted genotypes with resistance to a wide array of biotic and abiotic stresses. Most of these plant species are important locally or regionally only, and are known as 'minor', 'neglected', 'underexploited' or 'underutilized' crops. Like many other crops, production of oilseeds has not improved significantly due to their susceptibility to pests, sensitivity to abiotic stresses and low nutrient use efficiency. An approach for meeting the increasing demand for vegetable oils will be to introduce new or underutilized oilseed crops that are more suited for cultivation on less fertile land that do not support production of major oilseed crops. A need also exists for dedicated non-food oilseed crops that can be used for metabolic engineering of novel oil compositions for industrial applications. A number of oilseeds have recently received attention for their potential to fill one or more of these niches. These include Ironweed (*Vernonia galamensis*), crambe (*Crambe abyssinica*), desert mustard (*Lesquerella fendleri*), niger (*Guizotia abyssinica*), camelina (*Camelina sativa*), the Ethiopian mustard (*Brassica carinata*) and Sesame (*Sesamum indicum*). In this chapter emphasis has been given to current biotechnology research and progress for the improvement of these neglected oil crops. Agricultural biotechnology is creating new tools to tackle the problems of crop improvement, rural poverty, employment and income generation by helping to enhance farm productivity and production, improve quality, and explore marketing opportunities in newer ways. Technology like tissue culture provides the means for the culture of protoplasts, ovules and embryos used to create new genetic variation by overcoming reproductive barriers between distantly related crop species and haploid production by the culture of anthers and microspores to shorten the selection cycle in a breeding programme. Characterization of genetic

S. Tiwari (✉) · S. Kumar
Biotechnology Centre, Jawaharlal Nehru Agricultural University, Jabalpur 482004, India
e-mail: shtiwari@gmail.com

S. M. Jain and S. Dutta Gupta (eds.), *Biotechnology of Neglected and Underutilized Crops*, DOI: 10.1007/978-94-007-5500-0_7,

diversity by molecular markers is important for devising effective sampling and conservation strategies. Molecular markers can also be used to certify varieties, to determine the presence or absence of diseases and development of linkage maps for identifying quantitative trait loci and marker assisted selection. Transferred genes through genetic engineering may contribute to a range of properties, including resistance/tolerance to biotic and abiotic factors, improved nutritional status and better management options.

Abbreviations

NAA	α- napthaleneacetic acid
BA	Benzyl adenine
MS	Murashige and Skoog
TDZ	Thidiazuron
2-ip	N^6-[2-isopentenyl] adenine
KN	6-furfurylaminopurine
IBA	Indole-3-butyric acid
2, 4-D	2, 4-dichlorophenoxyacetic acid
AFLP	Amplified fragment length polymorphism
MAS	Marker assisted selection
EST	Expressed sequence tags
DGAT	Diacylglycerol acyltransferase
GISH	Genomic in situ hybridization

7.1 Introduction

An unintended consequence of the intensive agriculture has been a massive reduction in the number of species and diversity of crops. This process of crop replacement is a threat to local and global food security because the replaced indigenous crops are often essential for low input agriculture, have unique nutritional and cultural value, and contain diversity of locally adapted genotypes with resistance to a wide array of biotic and abiotic stresses. Global climate change and degradation of once productive lands have further heightened the demand for crops that perform well in harsh and/or changing environments (http://www.botany.ubc.ca/noug/).

Global food security has become increasingly dependent on only a handful of crops. Over 50 % of the global requirement for proteins and calories are met by just three—maize, wheat and rice. Only 150 crops are traded on a significant global scale. Yet, surveys indicate there are over 7,000 plant species across the world that are cultivated or harvested from the wild for food (www.underutilized-species.org). Most of these plant species are important locally or regionally only, and are known as 'minor', 'neglected', 'underexploited' 'orphan' species or 'underutilized' crops. These terms are often used interchangeably to characterize the range of plant species that are the focus of this chapter. Neglected crops are those grown primarily in their centers of origin or centers of diversity by traditional farmers, where they are still important for the subsistence of local communities. Some species may be globally distributed, but tend to occupy special niches in the local ecology and in production and consumption systems. While these crops continue to be maintained by cultural preferences and traditional practices, they remain inadequately characterized and neglected by research and conservation. In contrast, underutilized crops were once more widely grown but are falling into disuse for a number of reasons. Farmers and consumers are using these crops less because they are in some way not competitive with other crop species in the same agricultural system. The decline of these crops may erode the genetic base and prevent the use of distinctive useful traits in crop adaptation and improvement. Biotechnological approaches may be employed in these neglected crops to overcome slow pace of their improvement (Tiwari et al. 2011).

Agricultural biotechnology is creating new tools to tackle the problem of rural poverty, employment and income generation by helping to enhance farm productivity and production, improve quality, and explore marketing opportunities in newer ways. This offers a considerable promise as a means of improving food security and reducing pressures on the environment. Biotechnology is a rapidly developing field that, in an attempt to meet the current and emerging challenges facing agriculture—such as poor nutrition, unstable and limited food production, and restricted fuel availability—has received considerable attention for the improvement of major crops. In fact, agricultural biotechnology can be considered to cover at least four areas of work: (i) tissue culture and micropropagation, (ii) molecular marker characterization of genetic diversity, (iii) marker assisted selection (MAS), genomics and the related disciplines of proteomics and metabolomics; and (iv) the production of transgenic crops (FAO 2004).

7.1.1 Tissue Culture and Micropropagation

Tissue culture provides the means through in vitro techniques for the culture of protoplasts, anthers, microspores, ovules and embryos used to create new genetic variation in breeding lines, often via haploid production to overcome reproductive barriers between distantly related crop relatives. Micropropagation is in vitro vegetative multiplication through somatic embryogenesis or organogenesis being used to clone large numbers of plants from genotypes of particularly desirable characteristics, allowing these types to be distributed and used more widely. The culture of single cells and meristems is being effectively used to avoid diseases from germplasm, for in vitro transfer of breeding material and conservation of germplasm through cryopreservation. Cell and tissue culture has also produced somaclonal and gametoclonal variants with crop improvement potential.

7.1.2 Molecular Marker Characterization of Genetic Diversity

Characterization of genetic diversity by molecular markers is important for devising effective sampling strategies: e.g., in order to determine diverse material for pre-breeding programmes. It may also be important for developing conservation strategies: e.g., in rationalizing ex situ germplasm collections. Molecular markers can also be used to certify varieties, determine the presence or absence of diseases and assess the reproductive biology of species, among other applications. There is a wide range of markers with different characteristics available. Common markers include: isozymes, random amplified polymorphic DNAs (RAPDs), amplified fragment length polymorphisms (AFLPs), simple sequence repeats, or microsatellites (SSRs); and, more recently, SSRs from expressed sequence tagged sites, derived from transcribed DNA (EST-SSRs).

7.1.3 Marker-Assisted Selection and Genomics

Since 1990s, significant attention has been given to the development of linkage maps for identifying quantitative trait loci (QTL), which represent statistical associations between markers and genes that control a proportion of the variation of a trait. By establishing an association, markers can be used to understand complex traits and assist in selection; when combined with more traditional breeding methods, Marker-Assisted Selection (MAS) has great potential to accelerate genetic improvement. Genomics, a 'second-generation' biotechnology, the ultimate goal of which is to identify all genes and their functions in an organism, has burgeoned over the last decade. By revealing gene sequence similarities and common arrangements of genes (synteny), genomics raises the prospect of information gathered on one species benefiting work on other less researched taxa. Genomics involves a wide range of activities, including: the production of expressed sequence tags (ESTs), genome sequencing, gene function determination, comparative analysis (exploring synteny, cross-identification of candidate genes, etc.), physical mapping and the discipline of bioinformatics. The information gathered is then incorporated into selection and breeding programmes. Genomics, proteomics and metabolomics are disciplines that, among others, can be combined together into a biotechnology meta-analysis, the basis of 'systems biology'.

7.1.4 Genetic Engineering and the Production of Transgenic Crops

Genetic Engineering (GE) is the use of recombinant DNA and asexual gene transfer methods to alter the structure or expression of specific genes and traits in an organism. Active research in this area has been ongoing since 1980s. The product of GE, a transgenic, is one that has been transformed by the insertion of one or more transgenes from another, often unrelated organism. Transferred genes may theoretically contribute to a range of properties, including: resistance/tolerance to biotic and abiotic factors, improved nutritional status and better management options. In future, GE holds the potential for the production of nutraceuticals, oral vaccines and new biofuels in plants. To date, however, commercial deployment of GE has been limited to a narrow range of traits only, herbicide tolerance and insect resistance being the most important (James 2009).

Despite considerable discussion regarding the potential uses of biotechnology for meeting global agricultural challenges, practical deployment for underutilized plants is currently limited. There are relatively simple applications such as tissue culture that do already realize some significant benefits in few crops. However, more advanced techniques may result in significant value in the future (Dowson et al. 2009). It is clear that biotechnology is not a panacea for promoting underutilized species, and the cost and effort involved in realizing successful interventions will often be much greater than many researchers have first considered, with time scales of decades rather than years being the norm (CGIAR 2006).

Vegetable oils have historically been a valued commodity for food use and to a lesser extent for non-edible applications such as detergents and lubricants. Many plants are cultivated for their oil content which is used for cooking and industrial purposes. The production of oilseeds has not improved significantly due to their susceptibility to pests, pathogens and sensitivity to abiotic stresses and low nutrient use efficiency. For qualitative and quantitative improvement, the breeder has to rely upon the extent of genetic variability present in the base population. The lack of genetic variability has been considered a major limiting factor to the progress made in the improvement of these crops. The desired goals can be achieved by incorporating additional genetic variability in the existing germplasm. However, in addition to the conventional methods of plant breeding, like introduction, selection and hybridization, recent advances in biotechnology have opened several new possibilities for the creation of genetic variability and selection of desired traits.

Vegetable oils consist principally of energy-dense triacylglycerols that are composed of three fatty acids bound to a glycerol back bone. These molecules are an important source of calories in human and animal diets and are also used in the preparation of margarines, salad oils and fried foods. The energy density of triacylglycerols has also made these molecules an attractive source of biodiesel that is produced by trans-esterification of their component fatty acids. Furthermore, the immense diversity of fatty acid structures that can be found in triacylglycerols in the plant kingdom opens up opportunities for the use of vegetable oils in a variety of bio-based industrial formulations, including lubricants and drying oils. Biotechnological improvement of fatty acid composition will certainly lead to additional demand for vegetable oils in food, feed, and bio-based industrial materials. One of the most notable recent breakthroughs in oilseed biotechnology is the complex metabolic engineering of oilseeds to produce fish oil-type omega-3 poly unsaturated fatty acids. Progress has also been made in the identification of genes for the synthesis and metabolism of novel fatty acid structures, including fatty acids with hydroxy and epoxy residues that are well-suited for industrial applications such as lubricants, plasticizers, and nylon precursors. These genes have typically been isolated from species with limited agronomic potential and transferred to established oilseed crops to generate vegetable oils with new functionality (Lu et al. 2011).

An approach for meeting the increasing demand for vegetable oils is to introduce new or underutilized oilseed crops that are more suited for cultivation on less fertile land or in arid and semi-arid climates that do not support production of major oilseed crops, such as soybean and canola. A need also exists for dedicated nonfood oilseed crops that can be used for metabolic engineering of novel oil compositions for industrial applications. The use of such crops would preclude the real or perceived risks of mixing industrial and food traits that could arise, for example if soybean was used for production of industrial oils with novel fatty acid compositions (e.g. hydroxy fatty acids). A number of oilseeds have recently received attention for their potential to fill one or more of these niches. These include Ironweed (*Vernonia galamensis*), crambe (*Crambe abyssinica*), desert mustard (*Lesquerella fendleri*), niger (*Guizotia abyssinica*), camelina (*Camelina sativa*), the Ethiopian mustard (*Brassica carinata*) and Sesame (*Sesamum indicum*).

In this chapter, a detailed account of biotechnological approaches is given in context with major neglected oil crop species listed as neglected or underutilized (NUS) species by three organizations that are active in NUS crop promotion in the tropics and subtropics: the Global Facilitation Unit for Underutilized Species (http://www.underutilized-species.org/), the International Centre for Underutilized Crops (http://www.icuc-iwmi.org/) and Bioversity International (http://www.bioversityinternational.org/). These institutions help to set priorities based on the potential impacts of promotion in addressing agricultural challenges, while also considering issues such as cultural importance, the cost of intervention, the feasibility and sustainability of activities, and ethical concerns.

7.2 *Vernonia galamensis* (Cass.) Less.

Vernonia galamensis is an annual plant in the Asteraceae family, known for its use as an oilseed. This species, often called 'Ironweed', is a new potential industrial oilseed crop, which originates from Eastern Africa. It is the largest source of vernonia oil, which is rich in a useful epoxy fatty acid called vernolic acid and is used to make plastics, rubbery coatings, and drying agents. Use of this oil as a replacement for traditional plasticizers and binders in the production of paints and PVC shows promise as a method of reducing smog pollution (www.underutilized-species.org). It produces high quantities of epoxy fatty acids (at least 60 %) in a trivernolin form, useful in the reformulation of oil-based paints to reduce emission of volatile organic compounds (Perdue et al. 1986). About 38 % of the Vernonia seed is oil of which about 72 % is vernolic acid. Vernonia has "reactive diluents" oil properties to serve as solvents that become part of the dry paint surface and do not evaporate to pollute air. Other potential markets for the fatty acids include plasticizers, additives in polyvinyl chloride (PVC), coatings, cosmetic, and pharmaceutical applications (Carlson et al. 1981). Preliminary investigations also showed that the meal after seed oil extraction is a valuable source of crude protein (43.75 %); it also consists of crude fiber (10.90 %), ash (9.50 %) and the carbohydrate fraction (6.57 %) with sucrose (2.36 %), fructose (1.90 %) and glucose (0.77 %). The major mineral elements, calcium (11.08 mg/g), potassium (14.18 mg/g), magnesium (6.90 %) and high phosphorus (644 mg/g) not only meet the nutritional requirements but also are higher than in most other oilseeds (Ologunde et al. 1990). The lipase activity found in the ungerminated seed and the characteristics that the lipase shows make *Vernonia galamensis* an attractive oilseed crop not only as an industrial oil source but also as a source of low-cost lipase (Ncube and Read 1995).

In 1950, the Agricultural Research Service (ARS) of US Department of Agriculture (USDA) made an extensive search to identify plants not competing with existing crops as new sources of industrial raw materials (Perdue 1988). Among many species examined *V. galamensis*, native to East Africa, emerged as one of the most important due to its high triacylglycerols (TAGs) contents with vernolic acid (cis-12-epoxyoctadeca-cis-9-enoic acid) (Baye et al. 2005). Ethiopian Vernonia has the highest oil content, up to 41.9 % with up to 80 % vernolic acid. Apart from Ethiopia it is also grown in many other African regions as an industrial oilseed. Its seed production is poor when it is grown outside of the equatorial region hence, cultivation in other places is not always economically worthwhile. Further studies in the breeding of this species to generate more productive varieties may show promise. Biotechnological studies conducted to this crop is scanty, despite its importance, it has remained a neglected oil crop so far.

7.2.1 Plant Tissue Culture

Belay et al. (1989) in their preliminary work reported Vernonia immature seed callus production to ascertain its oil producing potential. They initiated callus from immature seeds of Vernonia on semisolid MS medium supplemented with 0.05 mg/l 2,4-D. For oil extraction and subsequent chemical analysis, they selected approximately 22 weeks old callus tissues from the fourth subculture on the initiation medium. However, they could not find vernolic acid formation at this stage in the callus.

7.2.2 Molecular Marker Analysis

RAPD analysis among ten selected Ethiopian *V. galamensis* lines showed moderate genetic diversity as shown by the resulting four different groupings using 13 RAPD primers. OPA10 was most informative one. This suggested that for a given number of Vernonia lines tested for polymorphism about 40 % would show polymorphism or genetic variations (Ramalema et al. 2010).

7.2.3 Genomics

Limited commercial supply of epoxy fatty acids, such as vernolic acid has commenced considerable interest for genetic engineering of oil crops to produce high levels of this acid (Kinney 2002). However, there is only limited information as to how plants such as *V. galamensis* accumulate high levels of vernolic acid. Thus, a better understanding of the mechanism for the effective channeling or selective accumulation of vernolic acid into triacylglycerols felt necessary.

Molecular techniques have been used to clone the genes encoding epoxygenase enzyme responsible for vernolic acid synthesis in *V. galamensis* using RT-PCR, heterologous probes and Northern blotting. Seither et al. (1997) isolated two cDNA clones from the cDNA library constructed at a stage in seed development postulated to have abundant transcripts for the epoxygenase. These clones shared highest homology to plant cytochrome P-450s and they reported that epoxygenase in the *V. galamensis* was a cytochrome P-450. In their experiment, total RNA was isolated from seeds, reverse transcribed and a cDNA library was constructed. A PCR was performed with cDNAs by using a highly degenerate 5′ primer designed for the heme binding motif shared among cytochrome P-450s and 3′ oligo dT primer for the poly-A tail. The products were TA cloned and sequenced and a putative candidate was used for screening from cDNA library. A full length clone of 1.6 kb was obtained with highest homology to several plant cytochrome P-450s. The cDNA was excised and used to probe a Northern blot of total RNA from leaf and three stages of seed development. The expression was most abundant in the mid-mature seed. The same cDNA library was also screened using a probe for allene oxide synthase (AOS) from flax. This heterologous probe facilitated the isolation of a cDNA which showed highest homology to several plant cytochrome P-450s. Subsequently, the *V. galamensis* epoxygenase gene was cloned by Hitz (1998). However, transgenic *Arabidopsis* and soybean with epoxygenases gene expressed only low levels of vernolic acid in mature seeds (Hatanaka et al. 2004; Kinney 2002).

Seed microsome assays of *V. galamensis* demonstrated that diacylglycerol acyltransferase (DGAT), an enzyme for the final step of triacylglycerol synthesis, has a strong substrate preference for vernolic acid bearing substrates including acyl-CoA and diacylglycerol (Hatanaka et al 2003). There are two classes of DGATs known as DGAT1 and DGAT2. Yu et al. (2008) reported the isolation, characterization, and functional analysis of two DGAT1 cDNAs from *V. galamensis* (VgDGAT1a and VgDGAT1b). VgDGAT1a and VgDGAT1b were expressed in all plant tissues examined with highest expression in developing seeds. Enzymatic assay using isolated microsomes from transformed yeast showed that VgDGAT1a and VgDGAT1b have the same DGAT activity levels and substrate specificities. They concluded that the two VgDGAT1s are functional, but not likely to be responsible for the selective accumulation of vernolic acid in *V. galamensis* seed oil.

Triacylglycerol (TAG) is the main storage lipid in plants. DGAT1, DGAT2 (Acyl-CoA: diacylglycerol acyltransferase) and PDAT (phospholipid:

diacylglycerol acyltransferase), encoded by three separate gene families, are all capable of catalyzing the final acylation step during TAG synthesis. Li et al. (2010a) investigated the expression patterns of DGAT1, DGAT2 and PDAT in relation to the accumulation of oil and epoxy and hydroxy fatty acid in developing seeds of *V. galamensis*, Euphorbia lagascae, Stokesia and castor that accumulate high levels of these fatty acids in comparison with soybean and *Arabidopsis*. The expression patterns of DGAT1, DGAT2 and the PDAT were consistent with all three enzymes playing a role in the high epoxy or hydroxy fatty acid accumulation in developing seeds of these plants. PDAT and DGAT2 transcript levels were present at much higher levels in developing seeds of Vernonia than in soybeans or *Arabidopsis*. DGAT1 appeared to be a major enzyme for seed oil accumulation at least in Arabidopsis and soybeans. For the epoxy and hydroxy fatty acid accumulating plants, DGAT2 and PDAT also showed expression patterns consistent with a role in the selective accumulation of these unusual fatty acids in seed oil.

7.3 *Crambe abyssinica* Hochst. ex R.E. Fries

Crambe abyssinica, commonly known as "Abyssinian mustard or Abyssinian Kale" is an oilseed crop and belongs to the Brassicaceae family. It is a native to the Mediterranean region and has a short life cycle; 40–50 days for flowering and 75–90 days for seed maturity. One of the most important advantages of this crop is that it does not outcross with any food oil seed crops (Wang and Peng 1998), thus eliminating the problem of gene flow. Besides, the seed oil of Crambe contains 55–60 % erucic acid that makes the oil non-edible, but can be used as starting material for producing oils used in chemical industry. Crambe has thus been considered as a very promising industrial oilseed crop in recent years and it has already been commercially cultivated on small scale with an acceptable yield potential. *C. abyssinica* is a fast growing high biomass crop with significant potential for biofuel production and for phytoremediation of heavy metal contaminated soils and sediments.

There is growing interest in Europe and the United States for the use of crambe as a renewable industrial feedstock, biofuels, lubricants and bioplastics. Crambe seed meal contains 45–58 % protein with a well-balanced amino acid content, with especially high in lysine and methionine levels and thus could be used as high value feedstock protein. However, crambe seed meal has high levels of sulfur-containing glucosinolates known to cause toxicity from ingestion by swine and poultry. Therefore, reducing the levels of expression of the key genes in the glucosinolate biosynthesis pathway could decrease the glucosinolate contents and make crambe seed meal palatable to ruminant animals. Developing and producing quality feed will not only be advantageous to the health of livestock, which will improve the meat quality and thus will also have indirect positive effects on human nutrition.

7.3.1 Tissue Culture

The lack of regeneration protocols for Crambe has hampered the development of transgenic Crambe plants. Gao et al. (1998) reported a regeneration frequency of 45 % from single cell culture, but without further transformation attempts. Li et al. (2010b) achieved a similar regeneration frequency of 43.7 % using hypocotyls of *C. abyssinica* cv. Galactica as explants cultured on a Murashige and Skoog medium supplemented with various plant growth regulators (PGRs). Among the different PGR combinations tested, 10 μM thidiazuron (TDZ) and 2.7 μM α-naphthaleneacetic acid (NAA) promoted the highest frequency of regeneration. However, this regeneration frequency is still very low for developing an efficient transformation system since the regeneration frequency is often dramatically reduced after *Agrobacterium* infection. Li et al. (2011) reported an efficient regeneration protocol for Crambe in which the regeneration frequency reached over 95 % using hypocotyls as explants. They investigated the effects of N-source, C-source, $AgNO_3$, cultural conditions as well as the concentration and combination of plant growth regulators (PGR) on the regeneration frequency of *C. abyssinica*. The results showed that all these factors, especially the N-source and PGR concentrations and combinations, played an important role in shoot regeneration. Among all the factors tested, the combination of using hypocotyls from *C. abyssinica* cv. galactica, the Lepiovre basal medium supplemented with 16 g l^{-1} glucose, 0.5 g l^{-1} $AgNO_3$, 2.2 mg l^{-1}TDZ, 0.5 mg l^{-1} NAA, 2.5 g l^{-1} Gelrite, seeds germinated in dark for 3 days and explants cultured in light, gave the best regeneration frequency (over 95 %). The results also suggest that reducing the content of NH_4^+ or keeping a suitable NO_3^-/NH_4^+ ratio in the regeneration medium would be crucial to Crambe shoot regeneration. Chhikara et al. (2011) described the development of an efficient method of plant regeneration through indirect shoot organogenesis from hypocotyl explants and transformation.

7.3.1.1 Protoplast Fusion

PEG-induced asymmetric somatic hybridization between *B. napus* and *C. abyssinica* was accomplished with the fusion of UV-irradiated mesophyll protoplasts of *C. abyssinica* cv 'Carmen' and cv 'Galactica' fused with hypocotyl protoplasts of different genotypes of *B. napus* cv 'Maplus' and breeding line '11502'. Shoot regeneration frequency varied between 6.1 and 20.8 % among the different doses of UV-irradiation, ranging from 0.05 to 0.30 J/cm^2. In total, 124 shoots were regenerated, of which 20 asymmetric somatic hybrids were obtained and verified by nuclear DNA content and AFLP analysis. AFLP data showed that some of the characteristic bands from *C. abyssinica* were present in the hybrids (Wang et al. 2003). The chromosomes of *B. napus* and *C. abyssinica* origin could be clearly discriminated by genomic in situ hybridization (GISH) in mitotic and meiotic cells. Analysis of cleaved amplified polymorphic sequence (CAPS) markers derived from the fae1 gene showed novel patterns different from the *B. napus* recipient in some hybrid offspring (Wang et al. 2004). The investigation into the fertility of asymmetric somatic hybrids indicated that the fertility increased with increasing UV-doses. All of the hybrids were cultured to full maturity, and could be fertilized and set seeds after self-pollination or backcrosses with *B. napus*. The analysis of the fatty acid composition in the seeds showed significantly greater amounts of erucic acid than *B. napus*. These studies indicated the use of UV-irradiation for the induction of asymmetric somatic hybrids to promote the fertility of the hybrids and their sexual progeny via chromosomal elimination. Protoplast fusion technique also facilitates the introgression of exotic genetic material into crop species.

7.3.1.2 Development of Intergeneric Hybrids Through Tissue Culture

An intergeneric hybrid between *B. juncea* × *C. abyssinica* was obtained for the first time through the conventional crossing method combined with ovary culture, when *C. abyssinica* was used as a paternal parent. The hybridity was confirmed by morphology, cytology and isozyme analysis (Wang and Peng 1998). The intergeneric hybrid from a cross between *B. chinensis* and *C. abyssinica* was observed with $2n = 55$ chromosomes in the original progenies. After several generations of in vitro propagation by tissue culture, the chromosomes of the intergeneric hybrid were remarkably reduced, varying from 25 to 28, averaged at 26. The reduction of chromosomes in the hybrid and the high numbers of bivalents were possibly due to the chromosome of *C. abyssinica* eliminating and the genome of *B. chinensis* doubling in the hybrid cells (Tang et al. 2006).

7.3.1.3 Microspore Culture

Wang et al. (2006) regenerated twenty-seven microspore-derived plants from the *B. napus*–*C. abyssinica* monosomic addition lines obtained from the F2 progeny of the asymmetric somatic hybrid. Fourteen seedlings were determined to be diploid plants ($2n = 38$) arising from spontaneous chromosome doubling, while 13 seedlings were confirmed as haploid plants. Doubled haploid plants produced after a treatment with colchicines. The lines are potentially useful for molecular genetic analysis of novel *C. abyssinica* genes or alleles contributing to traits relevant to oilseed rape breeding.

7.3.1.4 Somatic Embryogenesis

Palmer and Keller (2011) investigated somatic embryogenesis in *C. abyssinica* cv. Prophet using cotyledon, hypocotyl and root explants from 8-day-old seedlings cultured with levels of NAA and 2,4-D ranging from 2.2 to 39.0 μM, combined with 6-benzyladenine (BA) to achieve an auxin:cytokinin ratio of 20:1. Callus formation frequency for cotyledon and hypocotyl explants was 100 % for levels of 2,4-D from 4.5 to 33.9 μM. The response was similar with NAA levels of 13.0 to 39.0 μM. Root explants were less responsive. When calluses were transferred to a medium containing 0.56 μM each of thidiazuron and BA with 1.0 μM indole-3-butyric acid (IBA), somatic embryos were induced. Embryos were induced from calluses grown on media containing either 11.3 μM 2,4-D or 13.0 μM NAA, or higher. On a medium without plant growth regulators, embryos were induced but at a much lower frequency. For all three explants, a combination of 22.6 μM 2,4-D and 26.0 μM NAA was optimal for embryogenic callus induction. Hypocotyl-derived calluses were superior to cotyledon- and root-derived calluses for embryo induction. The best embryo formation response was with medium containing 5.0–6.0 % sucrose. The highest average number of embryos per callus (36) was obtained from hypocotyl calluses from medium with 22.6 μM 2,4-D. Somatic embryos germinated best on half-strength B5 or MS medium with 3 % sucrose, and plantlets were successfully established under greenhouse conditions. The results indicate that high levels of auxins are required for the induction of embryogenic calluses from explants of *C. abyssinicia*, while cytokinins are critical for somatic embryo formation.

7.3.2 Genetic Transformation

Since, a high efficiency genetic transformation system for crambe has recently been developed, it can effectively be manipulated for lower erucic acid content for palatability, high erucic acid for pharmaceuticals, as well as for the production of functionalized fatty acids, increased biomass and seed oil yield for bioenergy, bioplastics or other industrial purposes.

Li et al. (2010a, b) evaluated six *Agrobacterium* stains, each harbouring the cloning vector containing the neomycin phosphotransferase (nptII) and β-glucuronidase (gus) genes. EHA101 and AGL-1 yielded the highest transformation frequencies of 1.3 and 2.1 %, respectively. They successfully recovered putative transgenic lines and confirmed as transgenic by Southern blot analysis. Subsequently, *Agrobacterium*-mediated transformation of hypocotyls of cv. Galactica with constructs harbouring the wax synthase and fatty acid reductase genes have also successfully recovered confirmed transgenic plants carrying these transgenes.

Recently, Chhikara et al. (2011) developed an efficient *Agrobacterium*-mediated transformation system for *C. abyssinica* cv. BelAnn using β-glucuronidase (gus), bacterial arsenate reductase (ArsC), and γ-glutamylcysteine synthetase (γ-ECS) genes with the highest transformation efficiency. Pre-cultured hypocotyl explants were infected with *A. tumefaciens* strain LBA4404 harboring binary vector pCAMBIA1300 containing gus, ArsC, and γ-ECS genes under the control of CaMV35S, leaf-specific SRS1p and constitutive Act2p promoters, respectively. Following co-cultivation and selection, regenerated shoot buds were sub-cultured on MS medium containing GA3 for shoot elongation. Elongated shoots were transferred to root induction medium for 1 week. Semi-quantitative RT-PCR analysis confirmed the expression of mRNA transcripts for gus, ArsC and γ-ECS genes in T0 generation transgenic plants. Histochemical assays showed the gus expression in both vegetative and reproductive tissues of stably transformed T1 generation plants. Germinating seeds from T1 transgenic plants grown on MS medium containing hygromycin revealed a 3:1 Mendelian inheritance pattern for each transgene. This method achieved an overall frequency of 50–70 % regeneration and 6.7–8.3 % transformation with three different gene constructs.

7.3.3 Molecular Marker Analysis of Genetic Diversity and Relatedness

Crambe is an 'Old World' genus with a disjunct distribution among the four major centers of species diversity. A phylogenetic analysis of nucleotide sequences of the internal transcribed spacers (ITS) of the nuclear ribosomal repeat conducted with 27 species of Crambe and 18 related genera using weighted and unweighted parsimony supported Crambe as a monophyletic genus with three major lineages (Francisco-Ortega et al. 1999). *C. abyssinica* (n = 45) is most closely related to *C. hispanica* L. (n = 30) and *C. glabrata* DC. (n = 15). The species complex extends throughout the Mediterranean region, Ethiopia and East Africa. *C. abyssinica* is endemic to Ethiopia, *C. glabrata* to Spain, Portugal and Morocco, and *C. hispanica* is distributed in the Mediterranean region and Middle East. Warwick and Gugel (2003) compared genetic relationships among *C. abyssinica*, *C. hispanica* and *C. glabrata* and attempted a taxonomic separation of them using traditional morphological traits, agronomic and seed quality data, chromosome number, and various molecular data sets including RAPD data, chloroplast (cpDNA) restriction site data and ITS sequence data for the internal transcribed spacer region of the nuclear ribosomal DNA. The three species could be distinguished most reliably by a chromosome number. cpDNA restriction site data and ITS sequence data, two relatively conserved DNA data sets, supported the recognition of *C. glabrata* as a distinct species separate from the *C. hispanica*/*C. abyssinica* accessions. Both RAPD data and field evaluation data revealed greater amounts of genetic variation in *C. hispanica* compared with accessions of *C. abyssinica.* They found that *C. glabrata* was genetically distinct for all data sets and warrants separate species status.

7.3.4 Genomics

7.3.4.1 Organeller Genomics

Southern blot hybridization techniques used to examine the chloroplast DNA (cpDNA) sequences present in the mitochondrial DNAs (mtDNAs) of *C. abyssinica* and other related species of crucifer family led to the conclusion that DNA has been transferred sequentially from the chloroplast to the mitochondrion during crucifer evolution and there cpDNA sequences can persist in the mitochondrial genome over long periods of evolutionary time (Nugent and Palmer 1988).

7.3.4.2 Organ-Specific Expression of Highly Divergent Thionin Variants

Most thionins of higher plants are toxic to various bacteria, fungi, and animal and plant cells. The only known exception is the seed-specific thionin 'crambin' of the crucifer *C. abyssinica*. Crambin has no net charge, is very hydrophobic and exhibits no toxicity. The existence of a large number of novel and highly variable thionin variants in Crambe has been deduced from cDNA sequences that were amplified by the polymerase chain reaction (PCR) from RNA of seeds, leaves and cotyledons. While the deduced amino acid sequences of the thionin domains of most of these thionin precursor molecules are highly divergent, the two other domains are conserved. Most of the predicted thionin variants are positively charged. The presence of positively charged residues in the thionin domains consistently correlates with the presence of a negatively charged residue in the C-terminal amino acid extension of the various thionin precursors. The different thionin variants are encoded by distinct sets of genes and are expressed in an organ-specific manner (Schrader-Fischer and Apel 1994).

7.3.4.3 Functional Characterization of the Fatty Acid Elongase Gene

A genomic fatty acid elongation 1 (FAE1) clone corresponding to a 1521-bp open reading frame, which encodes a protein of 507 amino acids was isolated from *C. abyssinica*. In yeast cells expression of CrFAE led to production of new very long chain monounsaturated fatty acids such as eicosenoic and erucic acids. Seed-specific expression in Arabidopsis thaliana resulted in up to a 12-fold increase in the proportion of erucic acid. On the other hand, in transgenic high-erucic *B. carinata* plants, the proportion of erucic acid was as high as 51.9 % in the best transgenic line, a net increase of 40 % compared to wild type. These results indicate that the CrFAE gene encodes a condensing enzyme involved in the biosynthesis of very long-chain fatty acids utilizing monounsaturated and saturated acyl substrates, with a strong capability for improving the erucic acid content (Mietkiewska et al. 2007).

7.3.4.4 Genomics for Phytoremediation

Arsenic contamination is widespread throughout the world and this toxic metalloid is known to cause cancers of organs such as liver, kidneys, skin, and lungs in human. Similarly, chromium pollution is a serious environmental problem with few cost-effective remediation strategies available. *C. abyssinica* accumulates significantly higher levels of arsenic and chromium as compared to other species of the Brassicaceae family. Thus crambe, a fast growing high biomass crop, is an ideal candidate for phytoremediation of heavy metals contaminated soils. In order to understand the pathways involved in heavy metals metabolism and detoxification in plants, a PCR-Select Suppression Subtraction Hybridization (SSH) approach was employed to identify the differentially expressed transcripts in crambe plants under arsenate stress (Paulose et al. 2010) and chromium stress (Zulfiqar et al. 2011). A total of 105 differentially expressed subtracted cDNAs were sequenced from plants under the arsenate stress, which were found to represent 38 genes. These genes encode proteins functioning as antioxidants, metal transporters, reductases, enzymes involved in the protein degradation pathway, and several novel uncharacterized proteins. Upon chromium exposure, plants revealed a total of 72 differentially expressed subtracted cDNAs representing 43 genes. The subtracted cDNAs suggest that Cr stress significantly affects pathways related to stress/defense, ion transporters, sulfur assimilation, cell signaling, protein degradation, photosynthesis and cell metabolism. The transcripts corresponding to the subtracted cDNAs showed strong upregulation by arsenate and chromium stress as confirmed by the semi-quantitative RT-PCR. These studies revealed novel insights into the plant defense mechanisms and the regulation of genes and gene networks in response to heavy metal toxicity and further characterization of differentially expressed genes may enable the engineering of non-food high-biomass plants for phytoremediation of heavy metal contaminated soils and sediments.

7.4 *Lesquerella fendleri* L.

Lesquerella fendleri L. (Gray) S. Wats (commonly known as "Fendler's bladderpod" or "yellowtop" and desert mustard) is a member of the Brassicaceae and is an important plant producing seed oil high in hydroxy fatty acids (Carlson et al. 1990; Skarjinskaia et al. 2003). The plant is native to Arizona, New Mexico, Colorado, Utah, Texas, and Mexico. It is cultivated for the seed which yields up to 28 % oil rich in hydroxy fatty acids and 22 % protein used as supplement for livestock. Oils high in hydroxy fatty acids can replace castor oil, which is used extensively in industrial applications including cosmetics, plastics and coatings (Reed et al. 1997; Dykinga 1999). Within this genus, *Lesquerella fendleri* L. is a good candidate for domestication because it has the highest agronomic potential, low seed dormancy and low fruit dehiscence (Thompson and Dierig 1994; Ploschuk et al. 2003).

7.4.1 Plant Tissue Culture

Most tissue culture efforts made with *L. fendleri* were directed towards the establishment of an efficient transformation protocol. Wang et al. (2008) developed a protocol for regeneration and *Agrobacterium*-mediated genetic transformation of

L. fendleri. Initially they induced calli from hypocotyls and cotyledons on MS fortified with 0.5 mg l-1 BA, 1 mg l-1 NAA and 1 mg l-1 2,4-D, followed by co-cultivated for 2–3 days in darkness on MS supplemented with 0.5 mg l-1 BA, 0.2 mg l-1 NAA and 100 μmol l-1 As together with *Agrobacterium tumefaciens* strain EHA 105/pCAMBIA1301 harboring gene construct. Following co-cultivation, calli transfected by *A. tumefaciens* were transferred to MS with 0.5 mg l-1 BA, 0.2 mg l-1 NAA, 500 mg l-1 Cef and 10 mg l-1 hygromycin. After 4 weeks the resistant regenerants were transferred to MS with 0.5 mg l-1BA, 0.2 mg l-1 NAA, 500 mg l-1 Cef and 25 mg l-1 hygromycin for further selections. With this approach, they obtained 22.70 % the average regeneration frequency from transfected calli, and 6–13 regenerated shoots per callus.

7.4.1.1 Ovule Culture

Germplasm evaluations revealed several other *Lesquerella* species having significantly elevated hydroxy fatty acid contents or different types of hydroxy fatty acids other than that found primarily in *L. fendleri.* These wild species may be suitable donors for possible novel gene introgression by interspecific hybridization. Several barriers to interspecific hybridization exist. Saprophytic self-incompatibility prevents fertilization and silique development following pollination. Tomasi et al. (2002) obtained interspecific hybrids of *L. fendleri* with *L. auriculata*, *L. pallida* and *L. lindheimeri* utilizing ovule culture, producing sufficient F1 hybrid plants for further breeding purposes.

7.4.1.2 Protoplast Culture and Fusion

Intertribal *Brassica napus* (+) *Lesquerella fendleri* hybrids were produced by Skarzhinskaya et al. (1996) through polyethylene glycol-induced fusions of *B. napus* hypocotyl and *L. fendleri* mesophyll protoplasts. In the symmetric fusion experiments, protoplasts from the two materials were fused without any pretreatments. While in asymmetric fusion experiments, X-ray irradiation at doses of 180 and 200 Gy were used to limit the transfer of the *L. fendleri* genome to the hybrids and significantly decrease growth and differentiation of non-fused *L. fendleri* protoplasts. In total, 128 regenerated plants were identified as intertribal somatic hybrids on the basis of morphological criteria. Nuclear DNA analysis performed on 80 plants, using species specific sequences, demonstrated that 33 plants from the symmetric fusions and 43 plants from the asymmetric fusions were hybrids. X-ray irradiation of *L. fendleri* protoplasts increased the possibility of obtaining mature somatic hybrid plants with improved fertility. From the symmetric fusions 2 plants could be fertilised and set seeds after cross-pollination with *B. napus*. From the asymmetric fusions, 9 plants could be selfed as well as fertilised when backcrossed with *B. napus*. A total of 6 plants were found to have different chromosome numbers.

Transferring of *L. fendleri* genetically transformed plastids to *Brassica napus* plants has been achieved with the somatic hybridization method in which the protoplasts of *B. napus* chlorophyll-deficient plants were fused with gamma-irradiated protoplasts of *L. fendleri* transplastomic plants (Nitovs'ka et al. 2006). A total of 59 green hybrid colonies were isolated while shoot regeneration was observed in two cell lines and only one yielded morphologically normal plants. PCR and isozyme analyses showed that the plants were transplastomic cybrids containing *B. napus* nuclei and *L. fendleri* transformed chloroplasts. Recently, Asymmetric intergeneric hybrid plants were obtained through the protoplast fusion between *Orychophragmus violaceus* and *L. fendleri* (Ovcharenko et al. 2011). *L. fendleri* carried chloroplasts transformed with the fused aadA16gfp gene construct, conferring streptomycin–spectinomycin resistance and UV-induced green fluorescence. The somatic hybrids were selected using the properties of spectinomycin-induced plastid defects in "albino" *O. violaceus* plants (chloroplast recipient) combined with the γ-irradiation-induced inactivation of nuclei in plastid donor *L. fendleri.* The morphology and esterase isozyme pattern of the hybrid plant as well as the results of the PCR analysis of internal transcribed spacer of nuclear ribosomal DNA proved that the regenerated hybrids carried *O. violaceus* nuclei, while PCR amplification of the atpB-rbcL spacer and aadA16gfp gene fragments confirmed the presence of the transformed *L. fendleri* chloroplasts in these plants. Expression of the fused aadA16gfp gene construct was further confirmed by sodium dodecylsulfate-polyacrylamide gel electrophoresis analysis and the resistance of the obtained plants to both streptomycin and spectinomycin.

7.4.1.3 Cell Suspension Culture

In the context of plant molecular studies, model systems are useful for circumventing the lengthy time frames associated with plant development and the limited availability of working materials for analysis. Rapid growth and easy maintenance of suspension cell cultures can provide a constant supply of relevant fresh material for analysis (Biesaga-Koscielniak et al. 2008). *L. fendleri* suspension culture could be useful for studying the effects of different culture conditions, including exogenously applied phytohormones or metabolic inhibitors, as well as mutagenic agents, on lipid metabolic pathways.

In a protocol described for suspension culture system in *L. fendleri* by Kharenko et al. (2011), the hypocotyl segments from 12-day-old seedlings were incubated for 7 weeks on a medium for callus initiation (MS basal medium with B5 vitamins, 3.0 % sucrose, 1.0 mg/l 2,4-D and 0.1 mg/l kinetin at a pH of 5.8. For initiating suspension culture, 50 ml MS basal medium with 1.0 mg/l 2,4-D, 0.1 mg/l kinetin, B5 vitamins and 3.0 % sucrose was inoculated with 350 mg of 28-day-old callus and the flasks placed on an orbital shaker at 145 rpm were maintained at 24 °C with a photoperiod of 16 h at a light intensity of 30 $\mu E\ m^{-2}\ s^{-1}$. The suspension was sub-cultured every 14 days by transferring 10 ml of the suspension to 40 ml of fresh medium of the same composition. Characterization of the lipid content in cell suspension culture showed a different range of fatty acids accumulating in the cells and predominant in the culture medium. Subsequently, the effect of application of abscisic acid (ABA), which modulates lipid accumulation, was assessed. Exogenously applied ABA was taken up by the cells and metabolized via the conjugation pathway, resulting in the accumulation of ABA-glucose ester. The cell line was responsive to exogenous ABA, resulting in increased cellular lipid content and increased accumulation of lipids in the culture medium. This novel *L. fendleri* suspension culture offers a valuable model system for efficient characterization of mechanisms associated with ABA-induced accumulation of lipids.

7.4.2 Molecular Markers

In an AFLP analysis performed by Du et al. (2008) to characterize 27 intertribal sexual F1 hybrids of *Brassica napus* (2n = 38) cultivars and *L. fendleri* (2n = 12) and their progenies obtained through the crossing between as a pollen parent. Analysis revealed that bands absent in *B. napus*, novel for two parents and specific for *L. fendleri* appeared in all F1 plants and their progenies. Some progenies showed the modified fatty acid profiles with higher levels of linoleic, linolenic, eicosanoic and erucic acids than those of *B. napus* parents. The occurrence of these partial hybrids with phenotypes, genomic and fatty acid alterations possibly resulted from the chromosome elimination and doubling accompanied by the introgression of *L. fendleri* DNA segments and genomic reorganization.

7.4.3 Genetic Transformation

Theoretically, the oil and lesquerolic acid content in *L. fendleri* can be increased through genetic engineering. The efforts made for improving the oil content and the quality have been described below.

7.4.3.1 Biolistic Approach for Plastid Transformation

Plastid transformation is an alternative to nuclear gene transformation. It is possible to achieve biological containment of agronomic traits such as resistance to herbicides and insects, to obtain high-value products through modification of the cellular metabolism and for the expression of recombinant proteins. The challenge of plastid transformation is that the plastid genome is present in many copy numbers and each genome should be altered in order to obtain a genetically stable line.

Skarjinskaia et al. (2003) developed a plastid transformation protocol for *L. fendleri,* a species with a high capacity for plant regeneration in tissue culture.

Transformation vector pZS391B carried an aadA16gfp marker gene conferring streptomycin–spectinomycin resistance and green fluorescence under UV light. Biolistic transformation of 51 *Lesquerella* leaf samples, followed by spectinomycin selection, yielded two transplastomic clones. The AAD–GFP fusion protein, the marker gene product, was localized to chloroplasts by confocal laser microscopy. Fertile plants and seed progeny were obtained in line Lf-pZS391B-1. In the 51 samples, a large number (108) of spontaneous mutants were identified. In five of the lines spectinomycin resistance was localized to a conserved stem structure by sequencing 16S rRNA genes. Success in *L. fendleri*, a wild oilseed species, extends plastid transformation beyond *Arabidopsis thaliana* in the Brassicaceae family.

7.4.3.2 Agrobacterium Mediated Transformation

A protocol for *Agrobacterium*-mediated genetic transformation of *L. fendleri* was developed by Wang et al. (2008). Calli were first induced from hypocotyls and cotyledons and then co-cultivated with *A. tumefaciens* strain EHA105/pCAMBIA1301 that harbored genes for uidA (GUS) and hygromycin resistance. They confirmed transgenic plants by polymerase chain reaction analysis, GUS histochemical assay and genomic Southern blot hybridization.

7.4.4 Genomics

An oleate 12-hydroxylase gene LFAH12, from *L. fendleri* was isolated on the basis of nucleotide sequence similarity to an oleate hydroxylase gene from *Ricinus communis* (Broun et al. 1998). Transgenic studies showed its expression restricted to seeds, but not in leaves or roots. However, hydroxylase activity was detectable in crude extracts of vegetative tissues. The discrepancy between the presence of activity and the lack of hydroxy fatty acids suggests selective removal and breakdown of hydroxy fatty acids in vegetative organs. High levels of LFAH12 mRNA accumulation did not lead to correspondingly high levels of protein accumulation, suggesting that accumulation of the hydroxylase may be controlled post-transcriptionally. Expression of the *L. fendleri* gene in transgenic plants of a fad2 mutant of Arabidopsis, which is deficient in cytoplasmic oleate delta 12 desaturase activity, resulted in partial suppression of the mutant phenotype in roots. Thus, unlike the hydroxylase from *R. communis*, the *L. fendleri* enzyme had both hydroxylase and desaturase activities.

L. fendleri seed oil contains up to 60 % hydroxy fatty acids, nearly all of which is the 20-carbon hydroxy fatty acid lesquerolic acid. However, lesquerolic acid is formed by the elongation of the 18-carbon hydroxy fatty acid, ricinoleic acid. To identify a gene encoding the enzyme involved in hydroxy fatty acid elongation, an *L. fendleri* genomic DNA library was screened using the coding region of the Arabidopsis Fatty Acid Elongation1 gene as a probe (Moon et al. 2001). A gene LfKCS3 with a high sequence similarity to a known very long-chain fatty acid condensing enzymes, was isolated. LfKCS3 transcripts accumulated only in the embryos of *L. fendleri* and first appeared in the early stages of development. Transgenic expression studies in *Arabidopsis* confirmed that LfKCS3 condensing enzyme specifically catalyzes elongation of 18-carbon hydroxy fatty acids. In an another experiment, the LfKCS45 gene encoding a root-specific condensing enzyme with a high sequence similarity to known 3-ketoacyl-CoA synthases of the membrane-bound fatty acid elongase was isolated from *L. fendleri* (Moon et al. 2004). Reverse transcription-PCR experiments showed that the LfKCS45 gene is expressed only in root tips.

Seed oil is stored mostly as triacylglycerol (TAG). Although, the pathways and key genes involved in the hydroxyl fatty acids (HFA) synthesis have been elucidated, how they contribute to the accumulation of HFA in TAG is poorly understood. In order to understand how HFA synthesis is regulated in *L. fendleri*, Chen et al. (2011) examined the changes in fatty acid composition and gene expression during seed development from 7 days after pollination (DAP) to desiccation (49 DAP). They examined the expression patterns of three key genes involved in fatty acid synthesis during seed development. Using real-time polymerase chain reaction, the transcript level of the three lipid genes, LFAH12 (bifunctional oleate 12-hydroxylase: desaturase), LfKCS3 (3-ketoacyl-CoA synthase) and LfFen1 (oleate 12-desaturase) were quantified. While all of these genes displayed a bell-shaped expression pattern with a peak at 35 DAP and a sharp decline at 42–49 DAP, they had different expression levels during early seed development and maximum inductions. The results will advance our understanding of regulatory mechanisms underlying synthesis and accumulation of HFAs, which is useful to developing and implementing the effective genetic approaches for enhancing HFA production in *L. fendleri* and other oilseeds.

7.5 *Guizotia abyssinica* (L.f.) Cass.

Niger or Noug is an oil-seed crop, indigenous to Ethiopia and holds significant promise for improving rural livelihoods in Sub-Saharan Africa. It is of economic significance not only for domestic consumption in the countries where it is grown, but also as an export commodity to North America and Europe, where it is mainly sold as bird-feed under its English name Nigerseed (or "Thistle seed"). The species is used in intercropping systems, grows on poor but also extremely wet soils, and contributes to soil conservation. While not fully domesticated, and suffering from low yields and susceptibility to insect herbivores, it contributes up to 50 % of the Ethiopian oil-seed crop. Niger belongs to the Compositae/Asteraceae family and is closely related to sunflower. It differs from domesticated sunflower mainly due to its high level of branching, numerous flower heads and small seeds. The oil content of niger seed varies from 30 to 50 %. The fatty acid composition is typical for seed oils of the compositae family with linoleic acid being a dominant component. Niger has been categorized as a 'Neglected and Underutilized Species (NUS)' in an effort to draw attention to this crop and highlight the importance of research on NUS (http://www.botany.ubc.ca/noug/).

Niger is also grown as a minor oil crop in India, Kenya, Uganda, Sudan, Malawi and some other African countries. Besides edible purposes, niger seed oil is utilized for the manufacture of soaps, paints and lubricants. The protein rich meal obtained after oil extraction is used as feed or for manure. Niger plant has an extremely low harvest index. Some problems hampering the realization of the full potential of niger are the low-yielding capacity of its cultivars and a susceptibility to diseases. In addition, self-incompatibility causes serious difficulty for inbred line development and maintenance which has impeded the improvement of the crop by conventional plant breeding techniques (Getinet and Sharma 1996).

7.5.1 Plant Tissue Culture

In vitro technology could serve as an alternative means for genetic upgrading and its application largely depends on the reliable plant regeneration system. Niger has been the subject of numerous cell/tissue culture studies. Regeneration has been reported from cotyledons and hypocotyl (Ganapathi and Nataraja 1993; Nikam and Shitole 1993; Adda et al. 1993, 1994), leaves (Sujata 1997; Jadimath et al. 1998; Kumar et al. 2000) and seedling explants (Nikam and Shitole 1997).

Naik and Murthy (2010) obtained regeneration from suspension cultures via somatic embryogenesis. Establishment of embryogenic suspension cultures has a great potential to aid crop improvement and is also suitable for in vitro selection of variants especially selection of salt tolerant, disease/toxin resistant and cold tolerant lines in crop plants.

Homozygous lines obtained through anther and microspore culture can be used for hybridization and crop improvement. Induction of embryogenesis from cultured anthers (Adda et al. 1993; Murthy et al. 2000; Hema and Murthy 2008) and plant regeneration from unpollinated ovule cultures (Bhat and Murthy 2007, 2008) has been reported in niger. The effects of amino acids (arginine, aspargine, cysteine, glutamine, glycine and proline) and polyamines (putrescine and spermidine) to enhance embryogenesis and plant regeneration from cultured anthers of niger cv. Ootacamund was also reported (Hema and Murthy 2008). Sujata (1997) was successful in effective maintenance of male sterile plants through culture of leaves from a male sterile plant developed through gamma irradiation.

7.5.2 Genetic Transformation

The introduction of specific desirable genes into niger can be achieved by genetic engineering. The prerequisite for a successful gene transfer of desirable traits is the establishment of an efficient transformation protocol. Murthy et al. (2003)

developed an efficient protocol for *Agrobacterium*-mediated genetic transformation of niger using hypocotyl and cotyledon explants from in vitro grown seedlings. They found that cotyledon was a better explant for transformation with a transformation frequency of 15 % while it was only 3 % in case of hypocotyls.

7.5.3 Molecular Techniques

Crop improvement through breeding depends on the magnitude of the genetic diversity and the extent to which this diversity is utilized. Characterization and evaluation are important to enhance the inherent value to conserved germplasm. The accessible collections of diverse cultivated as well as wild germplasm accessions have made great contribution to crop improvement. Molecular markers and their excellent attributes prove to be extremely useful for the assessment of genetic diversity as well as for identification and maintenance of germplasm collections. AFLP and RAPD markers were used to provide the estimates of the comparative genetic variation within and among populations of various *Guizotia* taxa with the goal of conserving and utilizing their genetic diversity (Geleta et al. 2007a). The genetic diversity analysis for Ethiopian niger was reported for the first time by Geleta et al. (2007b). Using RAPD analysis, they revealed the extent of genetic diversity among 70 populations of niger collected from 11 regions of Ethiopia, representing all its growing regions of the country. Ninety-seven percent of the loci studied were revealed to be polymorphic for the whole data set. UPGMA cluster analysis showed that most of the populations were clustered according to their place of origin. However, some populations were distant from the majority and seem to have unique genetic properties. They concluded that the crop has a wide genetic basis that may be used for the improvement of the species through the conventional breeding and/or the marker assisted selection. In a similar approach, Nagella et al. (2008) reported genetic diversity of selected Indian niger germplasm accessions of different origin and pedigree background through the use of RAPD markers. It was the first attempt to estimate genetic variability among the Indian niger cultivars using molecular markers. An ISSR marker analysis with merely five primers for genetic diversity within and among the population of 37 accessions from Ethiopia (Petros et al. 2007) amplified a total of 118 genomic DNA fragments of which 106 were polymorphic (89.83 %). Among ISSR markers UBC 888 was found most informative. The total genetic diversity and the coefficient of genetic differentiation suggested more variability within the populations than among them.

Fluorescence in situ hybridization (FISH) on somatic chromosome preparations of niger using a DNA probe of the 18S-5.8S-26S rRNA genes including the transcribed and non-transcribed spacer sequences revealed a maximum of six major and two minor signal sites of rDNA. The positions of the FISH signals coincided with the sites of the nucleolar organizer regions and their adjacent C-banded heterochromatin when present (Dagne et al. 2000).

Molecular techniques are being widely used in systematic and phylogenetic studies to measure the genetic relatedness based on DNA sequences variation (Soltis et al. 1998). The internal transcribed spacers (ITS) of the nuclear ribosomal RNA genes have been among the most widely used sequences for DNA sequence variation studies. However, in spite of the small number of species in the genus *Guizotia*, there are no DNA sequences available for systematic purposes. Bekele et al. (2007) drew phylogenetic inferences using the information from ITS sequence generated for five species of Guizotia and related this to the previous understanding and unresolved problems of the genus. They found that *G. scabra,* ssp. *Scabra, G. scabra* ssp. *schimperi* and *G. villosa* have contributed to the origin of *G. abyssinica,* the cultivated species of the genus. Based on their findings they suggested that the present composition of the species of genus *Guizotia* and the subtribe the genus presently placed in should be redefined.

7.5.4 Genomics

In order to develop genomic tools and resources for population genetic studies, phylogeographic and evolutionary analyses, research on mating systems, studies of gene flow between the crop and its wild relatives, as well as to aid modern breeding efforts in a non-model organism like niger, generation of a library of ESTs is often the first step. EST databases can be used for many different purposes, including genome wide studies of gene expression and selection, the study of gene family evolution or simply for providing sequence data for molecular marker development (Bouck and Vision 2007). The development of Simple Sequence Repeat markers (SSRs) from ESTs has become the method of choice for many researchers, as it is a more time- and cost-efficient alternative to more traditional approaches, such as library construction, enrichment, and screening.

Recently, Dempewolf et al. (2010) developed a library of the expressed sequence tags, the microsatellite markers using EST sequences, and the chloroplast genome sequence aimed to establish genomic tools and resources for niger. The EST library consisted of 25 711 Sanger reads, assembled into 17 538 contigs and singletons, of which 4781 were functionally annotated using the *Arabidopsis* Information Resource (TAIR). From the EST library, they selected 43 microsatellites and then designed and tested primers for their amplification. These microsatellite markers can be utilized to study the level and partitioning of the genetic diversity in niger more closely, and arrive at a better understanding of phylogeographic patterns. There is a wealth of genomic resources available in niger's closest crop relative, sunflower (*Helianthus annuus* L.), owing to its global importance as an oilseed crop and its status as model species for research on speciation. By comparing sequences of chloroplast genome of niger with the plastid genomes of sunflower and lettuce, they were able to assess the level of Compositae chloroplast genome divergence at a finer scale as compared to previous analyses. This further enhanced the understanding of Compositae plastid genome evolution.

As the chloroplast is an important source of markers for phylogenetic and phylogeographic analyses, making the full chloroplast genome sequence of niger available empowers researchers to assess the usefulness of a wide range of chloroplast DNA markers for such studies in niger, and the family Compositae as a whole. Based on chloroplast sequence comparison, they did not find large rearrangements between the niger and the sunflower chloroplast genomes and only 1.8 % sites showed a sequence divergence between the two species. They also identified 34 tRNAs, 4 rRNA sequences and 80 coding sequences including one region (trnH-psbA) with 15 % sequence divergence between niger and sunflower. This divergence may prove to be particularly useful for phylogeographic studies in niger and its wild relatives.

7.6 *Camelina sativa* (L.) Crantz

The crucifer oilseed species *Camelina sativa* (L.) Crantz variously known as camelina, false flax, gold of pleasure, German sesame or Siberian oilseed is of a particular interest amongst all the under-exploited oilseed crops of the family Brassicaceae. It is native to Northern Europe and to Central Asian areas (www.underutilized-species.org). It is one of a number of species subject to the renewed interest as a possible alternative oil crop. It has agronomic low-input features with 35–40 % oil content in seed and an unusual fatty acid composition with high levels of alpha-linolenic acid vis-à-vis unusually high cholesterol and brassicasterol content (188 and 133 ppm) as compared to other vegetable oils. Although high cholesterol and presence of eicosenoic acid (15 %) pose a hurdle for its approval as food oil, the presence of omega-3-fatty acids makes its oil unique and nutritionally rich. A cold-pressed meal of *Camelina* after oil extraction contains 10–14 % oil by weight and 40 % protein with lower glucosinolate levels, making it a desirable animal feed. With a variety of non-food usages of the oil as drying oil and in environmentally safe painting and coating applications, minimal agronomic input requirement for cultivation makes it a potential crop for use as bio-fuel without interfering with the edible oil trade and competition for available resources (Agarwal et al. 2010).

7.6.1 Plant Tissue Culture

There have been very few published in vitro studies on *Camelina*. This species has previously been used as a fusion partner in somatic hybridization studies with other Brassica species (Narasimhalu et al. 1994; Hansen 1998; Sigareva and Earle 1999; Jiang et al. 2009). The first report of its culture establishment and regeneration in vitro came from Tattersall and Millam (1999). They successfully established in vitro systems for the regeneration of shoots from leaf explants which were more efficient for the regeneration of a root and shoots than hypocotyls. It was found that for regeneration from leaf tissue the use of auxin (NAA) alone in the medium above a level of 0.54 μM resulted in root or callus growth. Cytokinin, in the form of BA alone failed to induce regeneration, but a combination of 4.44 μM BA and 0.54 μM NAA induced shoot regeneration at the rates over 10.0 shoots per explant. The rooting response of explants was increased from a control level of 26.4 to 46.7 % by the addition of 5.4 μM NAA. Regenerated shoots were successfully transplanted to soil and flowered and set seeds normally.

7.6.1.1 Protoplast Fusion

Protoplasts of *Camelina* have been used in somatic hybridization with *Brassica* species; however, in all cases, the focus of these studies was to improve the *Brassica* species. Black spot, caused by *Alternaria brassicae* and *A. brassicicola,* is an important disease in all the *Brassica oleracea* vegetables. Sufficient resistance to the pathogen is not found within the species or in the species that readily cross to *B. oleracea. Camelina* is highly resistant to *Alternaria spp.* and has, in addition, other desirable characters for the improvement of *B. oleracea.* Protoplast fusion was performed for the production of intertribal somatic hybrids between rapid cycling *B. oleracea* (tribe Brassiceae), which has good regenerability and *C. sativa* (tribe Sisymbrieae) by polyethylene glycol (PEG) treatment, as a step towards the transfer of resistance to this disease into *Brassica* vegetable crops (Hansen 1998). The *B. oleracea* fusion partner was inactivated by treatment with iodoacetate. *C. sativa* has poor regenerability hence, no pretreatment was needed for this species. The protoplasts were cultured using a feeder layer system. A total of 2903 calli were isolated from the fusions. Fourteen of these initiated shoots, i.e., 0.5 % regeneration frequency. Approximately 110 shoots were excised from 6 of these calli and transferred to rooting medium. Rooted plantlets grew vigorously in vitro and flowering was frequently observed. However, establishment of rooted shoots in soil was unsuccessful. Hybrid identity was confirmed by intermediate shoot morphology, RAPD marker analysis, and flow cytometric estimation of nuclear DNA content. In a similar experiment carried out by Sigareva and Earle (1999) rooted plants grew in soil up to 4–5 weeks, and some produced sterile flowers. Two of three hybrids tested showed a high level of resistance to *A. brassicicola.* Resistance was correlated with the induction of high levels of the phytoalexin ‘camalexin’ 48 h after inoculation, as in the resistant Camelina fusion partner. In contrast, susceptible somatic hybrids produced much lower levels of camalexin.

Intertribal somatic hybrids between *Brassica napus* and *Camelina sativa* were also developed by protoplast electrofusion (Jiang et al. 2009). Hybrid identity of the regenerants was determined using flow cytometric analysis of nuclear DNA content and simple sequence repeat (SSR) marker analysis. Three hybrids exhibited specific bands for *B. napus* and *C. sativa.* These hybrids showed intermediate leaf, flower and seed morphology compared with the two parental species.

The seeds of these three hybrids had a modified fatty acid profile, indicating higher levels of linolenic and eicosanoic acids than those of *B. napus*. Their results suggest that somatic hybridization offers opportunities for transferring the entire genomes between *B. napus* and *C. sativa* in improving rapeseed breeding.

7.6.1.2 Microspore Culture

Doubled haploidy (DH) protocols have been used in breeding programs for many years to develop improved crop varieties. In order for doubled haploidy to be effective in a breeding program, an efficient microspore culture protocol is required. The conditions leading to the induction and development of microspore-derived embryos vary depending on the species, and therefore doubled haploidy methods have to be determined for each species. A number of factors influence microspore embryogenesis including genotype, stage of microspore development, donor plant growing conditions, media composition, and culture conditions. Microspore-derived embryos have been produced from *C. sativa* (Ferrie and Bethune 2011). The microspores from buds of 1–3 mm in length were isolated and purified in full-strength B5 extraction medium and cultured in NLN medium with 12.5 % sucrose and 12.5 % polyethylene glycol 4000 (PEG) without glutamine, at a density of 10,000 microspores per milliliter. Glutamine was added to the cultures 72 h after extraction to give a final concentration of 0.8 g/L. The microspore cultures were maintained at 24 °C in dark. After 28 days the generated embryos were transferred to light for continued development and to allow the embryos to become green. After 5–10 days, the embryos were plated on solid medium (½ strength B5, 1 % sucrose, 0.8 % agar, pH 5.8) and kept at 22 °C under 16 h photoperiod. The highest embryogenic frequency achieved was 38 microspore-derived embryos from 100,000 microspores. After approximately 6–8 weeks, regenerated plantlets with well developed roots and shoots were transferred to greenhouse for development.

7.6.2 Genetic Transformation

The high percentage of polyunsaturated fatty acids makes camelina oil more susceptible to oxidation and thus is undesirable for fuel and other industrial applications. Therefore, it is necessary to modify camelina oils to find a role for this potential crop in the world oilseed market. *Camelina* has great potential to become a biotechnological platform for genetically engineered products. There have been limited research activities on camelina biotechnology. However, like many cruciferous oil producing plants such as *Arabidopsis thaliana* and *Brassica napus*, *Camelina* is also amenable to transformation. Efficient methods for *Camelina* genetic transformation would facilitate a series of experiments designed to enhance the fatty acid profile of this species.

7.6.2.1 Transformation via Tissue Regeneration

In vitro *Agrobacterium*-mediated gene transfer involves the introduction of a transgene into appropriate plant tissue and regeneration of the tissue into a whole plant. This method has been widely and successfully used with many dicot and monocot crops. However, transformation by tissue culture can be time-consuming and generally very particular to the skills of the researcher performing the transformation. Kuvshinov et al. (2002, 2004) were pioneers to demonstrate efficient transformation in this crop using hypocotyls, cotyledon and leaf explants through the *Agrobacterium* method. Among explants leaf segments showed highest transformation efficiency. They selected and regenerated the transformed plants on MS medium supplemented with variable concentrations of auxins and cytokinins.

7.6.2.2 In Planta Transformation

In order to improve oil quality and other agronomic characters, Lu and Kang (2008) developed an efficient and simple in planta transformation method to generate transgenic camelina plants. The method included *Agrobacterium*-mediated inoculation of plants at early flowering stage along with a vacuum infiltration procedure. They used a fluorescent protein (DsRed) as a visual selection marker, which allowed them to conveniently screen mature transgenic seeds from a large number of untransformed seeds. Using this method, over 1 % of transgenic seeds could be obtained. Genetic analysis revealed that most of transgenic plants contain a single copy of transgene. In addition, to demonstrate that camelina can be effectively used to produce genetically engineered products, they transformed camelina seeds with a castor fatty acid hydroxylase (FAH12) gene. Fatty acid methyl ester analyses by gas chromatography indicated that all red seeds analyzed accumulated novel fatty acids, which had been previously identified in transgenic FAH 12 Arabidopsis as ricinoleic acid, the major component of castor oil, and three other hydroxy fatty acids: densipolic acid; lesquerolic acid; and auricolic acid. Red fluorescent seeds confirmed the transformation successfully expressing the castor FAH12 gene. They concluded that this low-cost oilseed crop, *C. sativa*, has great utility as an economical platform for a plethora of genetically engineered industrial and pharmaceutical products.

7.6.2.3 Floral Dip Method

Nguyen et al. (2011) described successful transformation using floral dip method resulting into greater than 1 % transformation efficiency. They transformed *Camelina* plants through contacting the plants to a dipping solution comprising Agrobacterium, a sugar, and a nonionic surfactant. Their method does not require a vacuum infiltration step.

Among above three methods in planta *Agrobacterium* -mediated gene transfer has advantages over tissue culture intensive methods. For example, in planta methods do not require performance by a specialist, and less equipment, labor and reagents are needed to obtain transformed plants. Thus, there is minimal somaclonal variation as compared to that typically encountered with tissue culture.

7.6.3 Molecular Markers

Camelina is an alternative oilseed crop species with limited information about the origin and diversity of available germplasm. The first study involving the use of molecular marker in this crop was reported by Vollmann et al. (2005). They evaluated a representative set of 41 accessions selected based on oil content, protein content and phenotypic data from a set of 130 *Camelina* germplasms, using random amplified polymorphic DNA (RAPD) analysis. Of 24 primers, 15 were polymorphic producing a total of 30 marker loci. Genetic distance estimates between the 41 accessions were calculated, based both on RAPD polymorphism and on seed quality characteristics, and dendrograms were generated for comparison. Similarities were found between the two different clustering approaches and grouping was partly in agreement with pedigree information or geographic origin.

During construction of a genetic linkage map of camelina, Gehringer et al. (2006) screened a total of 256 amplified fragment length polymorphism (AFLP) primer combinations for polymorphisms between the two mapping parents 'Lindo' and 'Licalla'. Among those, 44 primer combinations with the highest rate of polymorphism were used to genotype the 181 single seed descent lines. In addition, they also screened a set of 400 publicly available *Brassica* SSR primers in the parental lines. The majority of the SSR primers did not amplify loci in *C. sativa*; however eight polymorphic SSR markers were identified of which four could be integrated into the genetic map.

In an another study, amplified fragment length polymorphism (AFLP) fingerprinting revealed high levels of diversity within the 53 accessions of camelina in order to investigate the role of geographical origin in genetic variation and fatty acid content and a link between ecogeography and both origin and key oil traits (Ghamkhar et al. 2010). The accessions were categorized by principal coordinate analysis using molecular marker data, enabling identification of links between geographical distribution and these categories. Their results clearly confirmed that camelina oil quality characteristics are strongly influenced by environmental factors. The unprecedented high genetic diversity in this group of accessions offers an excellent opportunity to investigate valuable genes for successful adaptation of camelina to specific ecogeographical conditions such as drought.

7.6.4 Quantitative Trait Loci Analysis

Notwithstanding its potential for oil production, there is limited molecular and genomic information on this crop. In addition, a limited amount of molecular and sequence information is available for *C. sativa*. In contrast to the vegetable and oilseed *Brassica* species, almost no information was available prior to the studies of Gehringer et al. (2006) with regard to the genomic make up of *C. sativa* and the genetic control of complex agronomic traits in this species. They constructed a genetic linkage map and used this map for Quantitative Trait Loci (QTL) studies. Gehringer et al. (2006) constructed a genetic map for *C. sativa* using amplified fragment length polymorphism (AFLP) and *Brassica* simple sequence repeat (SSR) markers, in a population of recombinant inbred lines that were developed, through single-seed descent, from a cross between 'Lindo' and 'Licalla', two phenotypically distinct parental varieties. At first, a chromosome number of $2n = 40$ was confirmed in all ten mitotic metaphases from each of the parental genotypes. Accordingly, the linkage map was constructed, which contained 157 AFLP markers and 3 *Brassica* SSR markers, on a total of 20 linkage groups, corresponding to $n = 20$. The map covered a total length of 1385.6 cm, with an average marker interval of 8.6 cm. A moderate level of DNA sequence conservation between *C. sativa* and the *Brassica* A, B and C genomes was demonstrated by the ability of 55 out of 406 tested *Brassica* SSR primer combinations to amplify microsatellite loci in *C. sativa* showing monomorphic amplification products, indicating partial genome homoeology with the *Brassica* species. A genetic map of Camelina will prove to be a valuable tool for future genomics-assisted improvement of this crop.

In *C. sativa* QTL studies were restricted due to the absence of a genetic map for this species which is a prerequisite for QTL detection. Gehringer et al. (2006) used the constructed genetic map with the data from field trials with different fertilization treatments (0 and 80 kg N/ha) at multiple locations over 3 years, to localize QTLs for agronomic characters including seed yield, oil content, 1000-seed weight (TSW), and plant height. QTLs were localized in the genetic map by composite interval mapping (CIM). They detected a total of eight significant QTLs for oil content; four for seed yield; two each for TSW, linoleic acid, linolenic acid and eicosenic acid; one each for plant height, oleic acid and erucic acid, on 12 different linkage groups. The major QTL for oil content was detected on LG4, which also co-localizes with a QTL for seed yield, may be a promising target for simultaneous marker assisted improvement of seed yield and oil content. A *Brassica* SSR marker that in oilseed rape is linked to a QTL for erucic acid biosynthesis and oil content was the nearest marker to a QTL for oleic acid, linoleic acid, eicosenic acid and oil content in C. sativa. Some yield QTLs were found only with the N0 treatment, and might represent loci contributing to the competitiveness of false flax in low-nutrient soils. Their results represent a starting point for future marker-assisted breeding.

Understanding the *Camelina sativa* genome is essential if agronomic properties are to be improved through molecular assisted breeding, mutation breeding, and/or genetic manipulation. For example, modification of the oil composition for superior biodiesel is a natural goal for this oilseed crop. Target genes for modification could therefore include Fatty acid desaturase 2 (FAD2), a membrane bound delta-12-desaturase which converts oleic acid to linoleic acid, and Fatty acid elongase 1 (FAE1) which sequentially adds 2 carbon units to 18 carbon fatty acid CoA conjugates, resulting in very long chain fatty acids.

7.6.5 Genomics

C. sativa is a member of the family Brassicaceae, and thus is a relative of both the genetic model organism Arabidopsis thaliana and the oilseed crop *Brassica napus*. The close relationship between *C. sativa* and *Arabidopsis* makes the *Arabidopsis* genome an ideal reference point for the development of genetic and genomic tools in *C. sativa*.

Manipulation of genes affecting traits of interest requires knowledge of their duplication status. Whole genome duplication is particularly relevant because it is common in plants, and because in the case of allopolyploidy it results in two or three independent copies of each gene. Based on their experiments involving FAD2 and FAE1 genes, Hutcheon et al. (2010) reported that *C. sativa* is a hexaploid, whose oil composition is likely influenced by more than one functional copy of both genes. As a first step to characterize genes involved in fatty acid biosynthesis, they determined the copy number of FAD2 and FAE1 by Southern blot analysis. PCR amplified fragments, consisting of conserved regions of FAD2 and FAE1 obtained using designed primers based on *Arabidopsis* genomic sequence, were used as probe. Results of the Southern blots revealed three bands in *C. sativa* for both FAD2 and FAE1, whereas hybridization revealed only a single band in Arabidopsis for both genes. These results suggest that FAD2 and FAE1 occur in at least three copies in *C. sativa*, while they are single copy in *Arabidopsis*. In many species fatty acid genes have been found to be multi-copy; therefore Hutcheon et al. (2010) verified their result using blot hybridization of the gene LEAFY (LFY), which is known to be single copy in a wide variety of species from several plant families. Three bands were observed following hybridization with the LFY probe of the same blot as was used for FAD2 and FAE1, suggesting LFY also exists as three copies in *C. sativa*. Transcript studies resulted that all three copies of both CsFAD2 and CsFAE1 were expressed in developing seeds, and sequence alignments showed the presence of previously described conserved sites, suggesting that all three copies of both genes could be functional. The regions downstream of CsFAD2 and upstream of CsFAE1 demonstrated colinearity with the *Arabidopsis* genome. In addition, the three expressed haplotypes were observed for six predicted single-copy genes in 454 sequencing analysis and results from flow cytometry indicated that the DNA content of *C. sativa* is approximately three-fold that of diploid *Camelina* relatives. Based on their results, they proposed that *C. sativa* be considered an allohexaploid.

In order to characterize the largely unexplored genome of *C. sativa*, Galasso et al. (2010) developed a new version of the cTBP (combinatorial tubulin-based polymorphism) method based on intron-length polymorphism (ILP), to rapidly characterize the β-tubulin gene family. *C. sativa* β-tubulin gene family of is composed of at least 20 different β-tubulin isotypes, named CsTUB1 through CsTUB20. The method, named h-TBP, allows the rapid cloning of the β-tubulin genomic sequences that encompass the two introns, invariantly present at fixed positions within the coding region of the vast majority of the plant species. The β-tubulin sequences cloned by h-TBP also comprised part of exon1 and exon3 and the whole sequence of exon2. Comparison of the β-tubulin exon sequences of *C. sativa* with those of *Arabidopsis* thaliana, the closest relative among crucifers, defines distinct groups of putative orthologous genes. Analysis of the *C. sativa* β-tubulin intron sequences reveals some molecular features that can provide the first hints for the understanding of intron plasticity and evolution.

It is desirable to increase the monounsaturated oleic acid (18:1), and to decrease polyunsaturated fatty acids (PUFA), linoleic (18:2) and α-linolenic (18:3) acids, in camelina oils to improve oxidative stability. 18:1 desaturation is mainly controlled by the microsomal oleate desaturase encoded by the FAD2 gene. Three FAD2 genes, designated CsFAD2-1 to 3, were identified in *Camelina* by Kang et al. (2011). Functional expression of these genes in yeast confirmed that they all encode microsomal oleate desaturases. Although the three CsFAD2 genes share very high sequence similarity, they showed different expression patterns. Expression of CsFAD2-1 was detected in all the tissues examined, including developing seed, flower, as well as in vegetable tissues such as leaf, root, and stem. Transcripts of CsFAD2-2 and CsFAD2-3 were mainly detected in developing seeds, suggesting their major roles in storage oil desaturation in seed. The introns of the three CsFAD2 genes, which showed greater sequence variations, may provide additional resources for designing molecular markers in breeding. They also demonstrated the roles of CsFAD2 in PUFA synthesis by the mutant analysis and antisense gene expression in camelina seed.

7.7 *Brassica carinata* A. Braun

Brassica carinata A. Braun (Ethiopian mustard), considered to have originated in Ethiopia, is among the oldest oil crops cultivated in Ethiopia, however, hardly cultivated in other parts of the world. It is used both as a vegetable and as an oil seed crop. Oil from the wild species is high in erucic acid, which is toxic, though there are some cultivars that contain very little erucic acid and can be used as food. The seed can also be crushed and used as a condiment. It is an amphidiploid species (2n = 34, BBCC) derived from the diploid species *Brassica nigra* (2n = 16, BB) and *Brassica oleracea* (2n = 18, CC). Owing to its drought and heat tolerance, the crop is now being considered as an alternative to *B. napus* and *B. juncea* in drier areas and has been evaluated as a potential oilseed crop in the United States, Canada, India, Italy and Spain (Teklewold and Becker 2006a, b; Warwick et al. 2006). It is not only an important source of edible oil, but also known to be tolerant to heavy metals and is a potential candidate for phytoremediation. Cultivars of yellow-seeded *B. carinata* are currently being developed in North America for the biodiesel and fish feed biorefinery markets (Li et al. 2009).

7.7.1 *Tissue Culture*

Protocol of plant regeneration in *B. carinata* was developed by Narasimhulu and Chopra (1987) aimed to creating somaclonal variation for plant type and adaptability, so that this species can fit into cropping systems in Indian agriculture. They assessed the response of cotyledonary and stem explants for callus induction and shoot regeneration on MS and B5 basal media containing different combinations of auxin and cytokinin concentrations. MS medium supplemented with BA and NAA favoured callus induction. Supplementing MS with combinations of BA and IAA, as also with BA alone, regenerated shoots from the explants with a high frequency. The frequency of shoot regeneration and the mean number of shoots per explant were higher in cotyledons than in stem explants on identical growth regulator combinations. On B5 medium, supplemented with BA (2 mg/l) and IBA (0.4 mg/l), compact callus was produced which regenerated shoots on transfer to medium containing BA (0.8 mg/l). Jain et al. (1988) found MS medium with zeatin (1.0 mg 1-1) and IAA (0.1 mg l-1) to be best for shoot organogenesis on which the cotyledonary explants invariably underwent callusing followed by multiple shoot formation, which could be separated and subcultured for further propagation. Number of shoots per cotyledon explant cultured varied from 0 to as many as 50. Shoot organogenesis also declined with the reduction in photoperiod from continuous light to 16 h. The shoots were easily rooted during prolonged incubation on the same medium and whole plants could be regenerated and grown to maturity. Genotypic differences among carinata accessions for regeneration were common. Further, species-specific responses for in vitro shoot regeneration from cotyledon explants of three basic diploid species of *Brassica*, *B. campestris* (AA), *B. nigra* (BB), *B. oleracea* (CC) and their amphidiploids *B. juncea* (AABB), *B. napus* (AACC) and *B. carinata* (BBCC) have also been observed; in particular *B. carinata* showed less regeneration frequencies than the parental diploid species (Narasimhulu and Chopra 1988). In an another study, immature stem segments of seven different genotypes of *B. carinata* produced shoots with variable frequencies when cultured in MS medium with BAP and picloram at 0.2 mg/l each (Narasimhulu et al. 1992b). They observed that 'Line 171' produced shoots with 100 % efficiency from both cut ends of the explant.

Hypocotyl explants from 6 to 7-day-old (but not younger or older) seedlings cultured on medium containing combinations of 2 mg l-1 BA and 0.01 mg l-1 NAA or 4 mg l-1 kinetin and 0.01 mg l-1 2,4-D regenerated shoots at 100 % frequency (Yang et al. 1991). Explants showed higher regeneration capacity at the distal end than the proximal end, and the upper segment of hypocotyl was more regenerative than the lower one. Regenerants were successfully rooted on half-strength growth regulator-free medium, acclimatized and developed into normal, fertile plants.

7.7.1.1 Anther and Microspore Culture

Isolated microspores from a cultivar of *B. carinata* was cultured in modified Nitsch and Nitsch (NN) medium supplemented with 13 % (W/V) sucrose, 0.05 mg/l benzyladenine (BA) and 1.00 mg/l naphthaleneacetic acid (NAA). Embryogenic responses were observed at cultured temperatures ranging from 22 to 32 °C. The highest frequency of embryos occurred at 30 °C and 7–54 embryos per anther (approx. 17,000 microspores per anther) developed (Chuong and Beversdorf 1985). A split temperature culture regime of incubation at 32 °C for 3 days followed by incubation at 25 °C resulted in both high embryo yields and a high percentage of normal embryos. Plantlet development from microspore-derived embryos appeared to be influenced by both medium and culture conditions.

Pollen embryogenesis occurred in the anther cultures of two genotypes of *B. carinata* following pretreatment of anthers at 35 °C for 3 or 6 days that was essential for the induction of androgenesis on growth regulator-free culture medium. A combination of sucrose and glucose was found better than sucrose alone. However, none of the pollen embryos germinated normally. Complete plants were raised through adventitious bud differentiation from their hypocotyls (Arora and Bhojwani 1988).

7.7.1.2 Protoplast Culture and Fusion

Protoplasts isolated from hypocotyls of three-day-old seedlings of *B. carinata* cv R-2128 were cultured in a modified Nitsch and Nitsch liquid medium containing 13 % sucrose, 0.4 % Ficoll, 0.25 mg/l BA, 0.5 mg/l NAA and 0.5 mg/l 2,4-D (Chuong et al. 1987). After 4–6 weeks developing microcalli were approximately 0.5 mm in diameter were transferred onto MS medium containing 3 % sucrose, 0.4 % agarose, 200 mg/l casein hydrolysate, 5 mg/l BA and 0.5 mg/l NAA, pH 5.7. Approximately 20 % of the calli transferred to this medium produced plantlets. Narasimhulu et al. (1992a) reported rapid and efficient plant regeneration in protoplasts isolated from hypocotyls of 7-d-old seedlings of three genotypes of *B. carinata* after enzymatic digestion in cellulase R-10 (0.5 %) and pectolyase Y-23 (0.025 %). The protoplasts were stabilized with 0.4 M mannitol used as osmoticum, and were cultured in darkness in Kao's liquid medium containing 0.4 M glucose and the growth regulators 2,4-D (1.0 mg/l), NAA (0.1 mg/l) and zeatin riboside (0.5 mg/l). These were transferred to 16 h photoperiod conditions after 3 days of dark culture, and the medium was diluted to reduce the osmoticum on the seventh and tenth days of culture. Developed microcolonies, upon transfer to MS agarose medium with 2,4-D (0.1 mg/l), BAP (1 mg/l) and 0.1 M sucrose, proliferated further to produce callus clumps. The plating efficiency of the three genotypes varied from 1 to 2 %. Calli 2–3 mm in diameter were further transferred to MS agarose plates with zeatin (2 mg/l) where they produced shoot buds and shoots with frequencies ranging from 22.5 to 74.2 % for the three genotypes. The shoots were rooted in medium with IBA (1 mg/l) and were then established in soil. The total time required for the protoplast-to-plant development was 8–10 weeks.

Protoplasts isolated from cotyledons of *B. carinata*, underwent sustained division when cultured at 5.0 × 104 ml-1 in modified 8p medium (KM8P) with 1.0 % (w/v) Seaplaque agarose. Cell colonies produced callus when agarose droplets, in which the protoplasts had been embedded, were transferred to K8 medium with 0.6 % (w/v) Type I agarose on day 16, giving a plating efficiency of 1.6 %. It was found that 70 % of the protoplast derived-tissues produced shoot buds after subculture to MS medium containing 3.0 % (w/v) sucrose, 1.125 mgl-1 BAP, 0.035 mgl-1 GA and 0.6 % (w/v) Type I agarose, resulting in shoot formation from 1.1 % of the protoplasts originally plated. Regenerated shoots developed prolific root systems when placed on hormone-free MS medium with 1.0 % (w/v) sucrose and 0.6 % (w/v) Type I agarose (Jaiswal et al. 1990).

Jourdan and Salazar (1993) obtained 64 hybrid plants in two fusion experiments in an attempt to resynthesize *Brassica carinata* (BBCC) by the protoplast fusion between *B. nigra* (BB) and *B. oleracea* (CC) and identified them to be true hybrids by isoenzyme analysis, nuclear DNA content, chromosome number, and intermediate morphology. Of these plants 56 % were normal amphidiploids with 2n = 34 chromosomes and a DNA content equivalent to that of natural *B. carinata*. The remaining plants were polyploid, morphologically abnormal, and infertile. The majority of the hybrids contained both chloroplasts and mitochondria from *B. nigra*, but some plants combined chloroplast and mitochondria from the different progenitors. Hybrids with a DNA content equivalent to that of *B. carinata* had a wide range of male fertility (4–98 %), but consistently low female fertility. However, only a few selfed seed could be produced, but these germinated and grew into vigorous plants.

Intergeneric protoplast fusion has also been attempted to transfer Alternaria blight resistance from *Camelina sativa* into *B. carinata* (Narasimhulu et al. 1994). Polyethylene glycol mediated fusion between protoplasts from etiolated hypocotyls of *B. carinata* and mesophyll protoplasts of *C. sativa* resulted to 6.8 % mean frequency of heterokaryons. Three hybrid shoots were regenerated, each from a single fusion derived callus but these shoots failed to produce roots capable of withstanding transplantation. Confirmation of hybridity was obtained from the morphology of in vitro produced leaves, somatic chromosome number in leaf tips, and restriction fragment length polymorphism for a nuclear rDNA probe. Analysis for organelle constitution using RFLPs indicated that the hybrid contained chrloroplasts derived from *Camelina* and mitochondria from the cultivated *Brassica* species.

7.7.2 Genetic Transformation

Narasimhulu et al. (1992b) selected immature stem segments of 'Line 171' for genetic transformation using a non-oncogenic *Agrobacterium tumefaciens* containing plasmid PCV 730, binary vector carrying resistance genes for kanamycin and hygromycin. A co-cultivation period of 4 days with a bacterial concentration of approximately 2.5 × 10 cells/ml, followed by a recovery period of 2 days, produced transformed shoots that could be selected and rooted in the presence of kanamycin at 15 mg/l. Transformation was confirmed by neomycin phosphotransferase assay and Southern blot analysis. Seed analysis of transformed plants indicated that kanamycin resistance was inherited in the progeny.

Cotyledonary petioles and hypocotyl explants were used for *Agrobacterium*-mediated transformation with a construct containing the selectable marker genes, neomycin phosphotransferase II, phosphinothricin acetyl transferase and the reporter gene β-glucuronidase, under the control of a tandem 35S promoter (Babic et al. 1998). Although transformation was achieved with both cotyledonary petioles and hypocotyls, cotyledonary petioles responded best, with 30–50 % of the explants producing GUS-positive shoots after selection on 25 mg/l kanamycin. Direct selection on L-phosphinothricin also produced resistant shoots at a lower frequency (1–2 %).

B. carinata offers an attractive alternative for the production of recombinant proteins using oleosin technology. Hirudin, a blood anticoagulant protein from leeches was produced in *B. carinata* seeds using oleosin as a carrier (Chaudhary et al. 1998). Cotyledonary petioles were infected with *Agrobacterium* strains containing oleosin-glucuronidase (pCGNOBPGUS-A) or oleosin-hirudin (pCGN-OBHIRT) constructs. Polymerase chain reaction and neomycin phosphotransferase II enzyme assays confirmed the presence of the fusion genes in plants regenerating under selection. The fusion polypeptides were correctly expressed and targeted to the oil-bodies of the seeds with high fidelity (ca. 90 %). Recombinant protein was purified from all other cellular protein by a simple flotation process and cleaved from oil-bodies using the endoprotease, factor Xa. Hirudin activity was measured using a colorimetric thrombin inhibition assay and an activity in the range of 0.2–0.4 antithrombin units per milligram of oil-body protein was detected.

Eicosapentaenoic acid (EPA) plays an important role in many aspects of human health. Recently, Cheng et al. (2009) successfully expressed two novel genes, an 18-carbon ω3 desaturase (CpDesX) from *Claviceps purpurea* and a 20-carbon ω3 desaturase (Pir-ω3) from *Pythium irregulare* in zero-erucic acid B. carinata with EPA levels in transgenic seed of this line reaching up to 25 %. They also reported that conlinin1 promoter from flax functioned reasonably well in *B. carinata,* and can serve as an alternative to the napin promoter from *B. napus*.

7.7.3 Molecular Markers

As PCR techniques have developed over the last 15 years, the wealth of new DNA marker technologies have arisen which have facilitated the analysis of genetic relationships in crops species as an important component of crop improvement. It helps to analyze genetic variability of cultivars, select parental materials for hybridization for making new genetic recombination select inbred parents or tester for maximizing heterotic response and identify materials that should be maintained to preserve maximum genetic diversity in germplasm sources. However, Information on genetic diversity and genetic relationships among genotypes of *B. carinata* and marker assisted selection is currently limited.

7.7.3.1 Random Amplified Polymorphic DNA

Geographic diversity is a potent source of allelic diversity. The extent of genetic diversity among Forty-three germplasm accessions of Ethiopian mustard from five different countries, comprising 29 accessions from eight different geographic regions of Ethiopia and 14 exotic accessions from Australia, Pakistan, Spain, and Zambia were analyzed using random amplified polymorphic DNA (RAPD) technique (Teklewold and Becker 2006a). A set of 50 primers yielded a total of 275 polymorphic bands allowing an unequivocal separation of every Ethiopian mustard accession. The usefulness of the 50 RAPD primers in measuring heterozygosity and distinguishing accessions was variable such that polymorphic information content (PIC) varied from 0.05 to 0.40, band informativeness (BI) from 0.05 to 0.65 and primer resolving power (RP) from 0.15 to 6.83. Jaccard's similarity coefficients ranged from 0.44 to 0.87 indicating the presence of a high level of genetic diversity. On the average, Australian and Ethiopian accessions were the most similar while, Spanish and Zambian accessions were the most distant ones. In another investigation the extent and structure of genetic variation in sixty-one accessions of *B. carinata* from 49 collection sites in Tanzania using RAPD markers resulted in 88 % variation among accessions, 4 % among regions and 8 % within accessions (Volis et al. 2009).

RAPD has also been applied to investigate heterosis in 36 F1 individuals, generated from crosses among nine inbred lines representing seven different geographic regions of Ethiopia. The nine parents along with their 36 F1s were evaluated using 14 phenotypic traits and 182 RAPD markers. The analysis depicted low correlation between phenotypic and molecular distances and it was concluded that parental distances estimated from phenotypic traits better predicted heterosis, F1 performance and GCA than distances estimated from RAPD markers (Teklewold and Becker 2006b).

7.7.3.2 Simple Sequence Repeats Markers

The presence of high levels of sinigrin in the seeds represents a serious constraint for the commercial utilization of Ethiopian mustard meal. Introgression of genes for low glucosinolate content from *B. juncea* into *B. carinata* was attempted through back crossing (Marquez-Lema et al. 2008). BC1F1 seed from crosses between double zero *B. juncea* line Heera and *B. carinata* line N2-142 was produced and further back cross generations were advanced. Forty-three BC1F4 derived lines were selected and subjected to a detailed phenotypic and molecular evaluation to identify lines with low glucosinolate content and genetic proximity to *B. carinata* using sixteen phenotypic traits and 80 Simple Sequence Repeats (SSR) markers. Eight BC1F4 derived lines were very close to N2-142 both at the phenotypic and molecular level. Three of them, with average glucosinolate contents from 52 to 61 micromoles g − 1, compared to 35 micromoles g − 1 for Heera and 86 micromoles g − 1 for N2-142 were selected.

Marquez-Lema et al. (2010) first time reported on the transferability and amplification quality of microsatellite (SSR) markers of the public domain in *B. carinata*. They studied the amplification of a set of 73 SSRs from *B. nigra* and *B. napus* in *B. carinata*, and compared the results with those obtained in the amplification of the same markers in other *Brassica* species. This set of SSRs from *B. nigra* (B genome) and *B. napus* (AC genome) led to the identification of the 3 basic genomes of the *Brassica* species tested. 94.3 % of the SSR markers from *B. nigra* and 97.4 % of those from *B. napus* amplified SSR-specific products in *B. carinata*. Very high-quality amplification with a strong signal and easy scoring in *B. carinata* was recorded for 52.8 % of the specific loci from *B. nigra* SSRs and 59.3 % of the specific loci from *B. napus* SSRs, compared to 66.7 % in *B. nigra* and 62.8 % in *B. napus*. These high-quality transferable SSR markers may provide an efficient and cost-effective platform to advance in molecular research in *B. carinata*.

7.7.3.3 Amplified Fragment Length Polymorphisms

Amplified Fragment Length Polymorphisms (AFLP) has been used to evaluate patterns and levels of genetic diversity in sixty six *B. carinata* germplasms in western Canada with comparison to twenty *B. juncea* and seven *B. nigra* accessions (Warwick et al. 2006). A total of 296 AFLP bands were generated from four primer pair combinations and scored for presence/absence in all the accessions of three species. Based on the analysis, *B. carinata* was found less genetically diverse than the other two species. The differences in diversity were evident in the proportion of polymorphic loci within each species: 23, 35 and 50 % for *B. carinata*, *B. nigra* and *B. juncea*, respectively. AFLPs proved to be useful for fingerprinting cultivars as two primer pair combinations were sufficient to uniquely identify all the accessions of *B. carinata*.

7.7.4 Genomics

Since *Brassica* species represent the closest crop plant relatives to the model plant *Arabidopsis thaliana*, significant progress will be achieved in the coming years through integration of candidate gene approaches in crop brassicas, using the detailed information now available for the *Arabidopsis* genome. The integration of information from the model plant with the increasing supply of data from physical mapping and sequencing of the diploid *Brassica* genomes will undoubtedly give great insight into the genetics underlying both simple and complex traits in oilseed crops of *Brassica*.

7.7.4.1 Genome Studies

Studies involving chloroplast (ct) DNA fragment patterns generated by digestion with fifteen restriction endonucleases from the three elementary *Brassica* species (*B. nigra*, *B. oleracea* and *B. campestris*) and the three amphiploid *Brassica* species (*B. carinata*, *B. napus* and *B. juncea*) showed that in all species restriction sites for enzymes with GC-rich recognition sequences were less frequent and not as variable as for those with AT-rich sequences. The ct DNA fragment patterns of *B. carinata* were virtually identical to those of *B. nigra* indicating its origin and little alteration since the origin of this amphiploid (Erickson et al. 1983). Flow cytometry has been used to estimate 2C nuclear DNA content in parents and interspecific F1 hybrids from the crosses between *Brassica campestris*, *B. carinata*, *B. juncea* and *B. napus* obtained through in vitro ovary and ovule culture (Sabharwal and Dolezel 1993). It was found that in comparison with the A genome, the B and the C genomes of *Brassica* contained 26.9 and 43.9 % more DNA, respectively. The Flow cytometric analysis of nuclear DNA content might be a useful tool in *Brassica* breeding to distinguish interspecific hybrids containing various genome combinations.

7.7.4.2 Seed Coat and Seedling Leaf Pigmentation

Flavonoid differences between near-isogenic lines of yellow- and brown-seeded *B. carinata* were used to identify a genetic block in seed coat and seedling leaf pigment biosynthesis (Marles et al. 2003). Seed coat pigment in the brown-seeded line consisted of proanthocyanidins (condensed tannins), while anthocyanin was absent. Where as dihydroquercetin, dihydrokaempferol, quercetin and kaempferol accumulated only in the mature seed coat of the yellow-seeded line, indicating dihydroflavonol reductase (DFR) as an element of genetic control in pigment biosynthesis. The DFR transcripts from the developing seed coat in the yellow-seeded line were absent or less abundant at 5–30 days after pollination compared to transcript levels in the brown-seeded line. When grown at 25/20 °C (day/night) temperature, seedling leaves of the yellow-seeded line also exhibited a reduced expression of DFR and contained less anthocyanin compared to the respective tissues from the plants of the brown-seeded line. Cooler (18/15 °C) growing temperatures affected seedling leaf pigmentation, mature seed coat colouration and DFR expression in the yellow-seeded line while, the brown-seeded line tissues were unaffected by these temperature changes. The results were suggestive of a temperature-sensitive regulator of DFR in the yellow-seeded line of *B. carinata* which ultimately affects the formation of pigments in the seedling leaves and in the mature seed coats.

7.7.4.3 Antisense Repression and Silencing of the Endogenous FAD2 Gene

Erucic acid and its derivatives represent important industrial feedstock compounds, and, in this regard, there is an increasing demand for the production of high erucate oils. With a goal to develop high erucic acid *B. carinata* lines with the increased proportions of erucic acid and very long-chain fatty acids, the expression of the endogenous FAD2 gene was manipulated using co-suppression and antisense approaches. Both methods resulted in transgenic lines exhibiting decreased proportions of polyunsaturated C18 fatty acids (18:2 + 18:3) and concomitant and significantly increased proportions of 18:1, 22:1 and total very long-chain fatty acids (Jadhav et al. 2005).

The 3′-UTR of the FAD2 gene was cloned by PCR and used to prepare an intron-spliced hairpin RNA (ihpRNA) construct. Compared to that of the wild type (control) background, this construct, when expressed in *B. carinata*, resulted in a high degree of FAD2 gene silencing accompanied by strong increases of up to 16 and 10 % in oleic acid and erucic acid proportions, respectively (Mietkiewska et al. 2008). The increase in 18:1 was accompanied by a concomitant proportional reduction in 18:2. Further experiment involving transformation of *B. carinata* with combination of ihpRNA with *Crambe abyssinica* FAE gene under the control of seed specific napin promoter resulted in an even greater increase in erucic acid proportions, by up to 16 % in T1 segregating seeds as compared to that of the control.

7.7.4.4 Genomics for Phytoremediation by *B. carinata*

Metal salt contamination of soils is a serious environmental problem with potential harmful consequences to agriculture and human health. About 20 % of the world's cultivated land and nearly half of all irrigated lands are affected by salinity. In nature, plant species possess a range of mechanisms involved in the detoxification of metals, allowing some to survive better than others under metal stress. Understanding these mechanisms will impact strongly on the success of developing salt tolerant crops.

Recently, *B. carinata* has been reported as a promising phytoextractor for Zn, Cu, Ni and Pb for the purpose of phytoremediation in multiply metal contaminated soils without suffering a significant biomass reduction (Purakayastha et al. 2008). Isolation and characterization of novel defence-related genes induced by copper, salicylic acid, methyl jasmonate, abscisic acid and pathogen infection have been reported in this species (Zheng et al. 2001).

Exposure of *B. carinata* seedlings to the increasing concentrations of a non-physiological ion, lithium, showed significant effects on the germination rate, root length, chlorophyll content and fresh weight in brown-seeded and yellow-seeded near-isogenic lines (Li et al. 2009). The lipid and phenolic composition dramatically changed in brown-seeded seedlings after lithium exposure. In contrast, the yellow-seeded plants maintained the same phenolic and lipid composition before and after exposure to lithium and did not tolerate the high metal concentrations tolerated by the brown-seeded line. Microarray analysis using *B. napus* 15000 expressed sequence tags (EST) array indicated a total of 89 genes in the brown-seeded line and 95 genes in the yellow-seeded line. These genes differentially expressed more than 20-fold and 1083 genes with more than 2-fold after the treatment of *B. carinata* seedlings with lithium chloride. The putative functions of the differentially expressed genes included proteins involved in defense, primary metabolism, transcription, transportation, secondary metabolism, cytochrome P450, as well as proteins with unknown functions. From the results of this study, *B. carinata* brown-seeded germplasm showed an ability to survive under moderately high concentrations of lithium chloride (>150 mM) and has some potential in phytoremediation of lithium-contaminated water and soil.

7.7.5 Proteomics

In order to understand the biochemical basis for the observed resistance against fungal pathogen *Leptosphaeria maculans* in the plants generated by an interspecific cross between the highly susceptible *B. napus* and the highly resistant *B. carinata,* changes in the leaf protein profiles of hybrid lines were investigated (Subramanian et al. 2005). Two-dimensional electrophoresis followed by tandem mass spectrometry led to the identification of proteins unique to the susceptible (five proteins) and resistant genotypes (seven proteins) as well as those that were differentially expressed in the resistant genotype 48 h after a challenge with the pathogen (twenty eight proteins). The proteins identified as being unique in the resistant plant material included superoxide dismutase, nitrate reductase, and carbonic anhydrase. Photosynthetic enzymes (fructose bisphosphate aldolase, triose phosphate isomerase and sedoheptulose bisphosphatase), dehydroascorbate reductase, peroxiredoxin, malate dehydrogenase, glutamine synthetase, N-glyceraldehyde-2-phosphotransferase, and peptidyl-prolyl cis–trans isomerase were observed to be elevated in the resistant genotype upon pathogen challenge. They further validated the increased levels of the antioxidant enzyme superoxide dismutase by spectrophotometric and in-gel activity assays.

7.8 *Sesamum indicum* L.

Sesame has been described as the most ancient oilseed crop in the world and is regarded as queen of oilseeds, perhaps for its resistance to oxidation and rancidity, even when stored at ordinary ambient air temperatures (Bedigian and Harlan 1986). The importance of sesame lies in the quality of the oil, the presence of antioxidants sesamin and sesmolin, its antiquity and use in religious rituals in India, Egypt and Persian region. The world production is estimated at 3.66 million tones with Asia and Africa producing 2.55 and 0.95 million tons, respectively (Anonymous 2008). The major sesame growing countries are India, China, Myanmar and Sudan. Unfortunately, average world yield of sesame is still low at 0.46 ton ha^{-1} (FAO 2005). The area and production of this crop is declining in the traditional areas. Despite the potential for increasing the production and productivity of sesame, there are a number of challenges inhibiting sesame production and productivity. Among many production constraints, the most important include a lack of improved cultivars and a poor seed supply system. In addition, there are severe biotic stresses, such as bacterial blight (*Xanthomonas campestris* pv. sesami), phyllody (Mycoplasma-like organism), Fusarium wilt (*Fusarium oxysporum*), Powdery mildew (*Oidium erysiphoides*), Alternaria leaf spot (*Alternaria sesame*) and Cercospora leaf spot (*Cercospora sesame*) (Daniel 2008). The above mentioned constraints to the productivity of sesame pose the need of concerte efforts for sesame crop improvement.

7.8.1 Plant Tissue Culture

Plant tissue culture technology has been available to plant breeders for nearly four decades and has been extensively employed for the crop improvement in several oil seed crops. However, very little information is available on sesame. It is found to be highly recalcitrant in nature. The first reported study on tissue culture in Sesame was that of Lee et al. (1985) from shoot tip culture and George et al. (1987) from different parts of sesame. Effects of explants and hormone combinations on callus induction were studied by Kim et al. (1987) in order to in vitro selection of herbicide tolerant lines of sesame. However, successful plant regeneration from herbicide tolerant calli was achieved by Chae et al. (1987). The effect of growth regulators on organ cultures (Kim and Byeon 1991) and its combination with cold pretreatment and genotype in anther culture (Lee et al. 1988) of sesame was investigated. Micropropagation has been achieved from shoot tip (Rao and Vaidyanath 1997a), nodal explants (Gangopadhyay et al. 1998) and leaf (Sharma and Pareek 1998) cultures. Somatic embryos were obtained from zygotic embryos (Ram et al. 1990) and seedling-derived callus (Mary and Jayabalan 1997; Xu et al. 1997) with low conversion frequencies in callus cultures. Indirect adventitious shoot regeneration from hypocotyl and/or cotyledon explants has also been reported but at low frequencies (Rao and Vaidyanath 1997b; Takin and Turgut 1997; Younghee 2001). Bhaskaran and Jayabalan (2006) reported standardization of a reproducible morphogenesis, micropropagation and callus induction protocol in cultivated varieties of sesame. Influence of macronutrients, plant growth hormones and genotype on adventitious shoot regeneration from cotyledon explants in sesame was reported by Were et al. (2006). High-frequency plant regeneration through direct adventitious shoot formation from de-embryonated cotyledon segments of sesame was achieved by Seo et al. (2007). Chattopadhyaya et al. (2010) established an efficient protocol for shoot regeneration from sesame internodes using the transverse thin cell layer (tTCL) culture method. Abdellatef et al. (2010) evaluated the in vitro regeneration capacity of sesame cultivar exposed to culture media containing ethylene inhibitors such as cobalt chloride and silver nitrate and found growth promotive effects due to reduction in ethylene concentration or inhibition of ethylene action.

A simple and efficient protocol for producing an inter-specific hybrid between *Sesamum alatum* and *S. indicum* through ovule culture has been optimized (Rajeswari et al. 2010). Direct organogenesis was successfully achieved when the ovules, excised from 7-day-old capsules from the cross *S. alatum* × *S. indicum*, were cultured on MS medium containing 8.8 μM benzylaminopurine (BAP), 2.8 μM indole acetic acid (IAA) and 1712.3 μM glutamine. The regenerants produced roots on half strength MS medium supplemented with 0.27 μM NAA. Phenotypically, the hybrid plants were intermediate to those of parents for majority of the traits. Peroxidase and esterase isozymes were found to be useful in the identification of hybrid plants. Further, screening against phyllody disease under greenhouse conditions revealed that the hybrids were moderately resistant.

7.8.2 Genetic Transformation

The yield potential of this crop is very low when compared with major oil seed crops due to early senescence and extreme susceptibility to biotic and abiotic stress factors including photosensitivity (Rao et al. 2002). Wild species of sesame possess genes for resistance to biotic and abiotic stresses (Joshi 1961; Kolte 1985; Brar and Ahuja 1979; Weiss 1971). However, introgression of useful genes from wild species into cultivars via conventional breeding has not been successful due to post-fertilization barriers. The only option left for improvement of sesame is to transfer genes from other sources through genetic transformation techniques. However, the main obstacle to genetic transformation is the recalcitrant nature of sesame to in vitro regeneration (Baskaran and Jayabalan 2006). There are very few reports on shoot regeneration, with low frequencies in a few genotypes from cotyledon and/or hypocotyl explants (Rao and Vaidyanath 1997a; Taskin and Turgut 1997; Younghee 2001; Were et al. 2006; Seo et al. 2007). Somatic embryos have also been induced from hypocotyl-derived calluses, but no plant regeneration was achieved (Mary and Jayabalan 1997). Although sesame has been shown to be susceptible to *Agrobacterium tumefaciens*, but no transformed shoot/plant was recovered (Taskin et al. 1999). For the first time, Yadav et al. (2010) reported conditions for establishing an *A. tumefaciens*-mediated transformation protocol for generation of fertile transgenic sesame plants. This was achieved through the development of an efficient method of plant regeneration through the direct multiple shoot organogenesis from cotyledon explants and the establishment of an optimal selection system.

Hairy root cultures using *Agrobacterium rhizogenes* have been successfully established (Ogasawara et al. 1993; Jin et al. 2005). In order to investigate the possible quinone derivative intermediate of Anthraquinones in sesame, hairy root culture was induced by Furumoto et al. (2007) through direct infection of axenic seedlings with *Agrobacterium rhizogenes* ATCC 15834. The established hairy root clone (SI-16) was subcultured in a phytohormone-free B5 liquid medium containing 2 % sucrose at 25 °C on a rotary shaker at 80 rpm in the dark at intervals of 14–18 days. After 4 weeks of inoculation, the secreted pigments were separated from the medium by gravity filtration. The same hairy root culture was further used to explain the biosynthetic origin of 2-geranyl-1,4-naphthoquinone and its biogenetically related anthraquinone administering 13C labeled glucose to the hairy root culture (Furumoto and Hoshikuma 2011). Using hypocotyl and cotyledon explants from sesame seedlings, hairy root cultures were established and cDNA coding for Dehydroascorbate reductase (DHAR) (Chun et al. 2007) and peroxidase (Chun et al. 2009) were characterized and cloned from the roots. The frequency of sesame hairy root formation was higher from hypocotyl than cotyledon explant. It was also found that DHAR and peroxidase genes were differentially expressed in distinct tissues of sesame plant.

7.8.3 Molecular Techniques

DNA markers provide a powerful tool for genetic evaluation and marker-assisted breeding of crops, and especially for cultivar identification. Among the different types of molecular markers, random amplified polymorphic DNA (RAPD) markers are particularly useful for the assessment of genetic diversity because of their simplicity, speed and relatively low cost (Nybom 2004). RAPD markers have been used extensively in several crops including cucumber (Horejsi and Staub 1999), potato (Demeke et al. 1996) and pepper (Prince et al. 1995). Abdellatef et al. (2008) used RAPD markers to characterize 10 germplasm collections from Sudan to investigate the genetic diversity of selected sesame germplasm accessions from different origin and pedigree background.

Diversity estimates in cultivated plants provide a rationale for conservation strategies and support the selection of starting material for breeding programs. The diversity measures applied to crops usually have been limited to the assessment of genome polymorphism at the DNA level. Occasionally, selected morphological features are recorded and the content of key chemical constituents is determined, but unbiased and comprehensive chemical phenotypes have not been included systematically in diversity surveys. Laurentin et al. (2008) assessed metabolic diversity in sesame by non-targeted metabolic profiling and elucidated the relationship between metabolic and genome diversity in sesame and observed different patterns of diversity at the genomic and metabolic levels, which indicates that selection plays a significant role in the evolution of metabolic diversity in sesame. Earlier Laurentin and Karlovsky (2007) and Ali et al. (2007) successfully used AFLP to distinguish cultivars of sesame to elucidate the genetic relationship among genotypes.

The determination of genetic differences among crop genotypes has become the primary need to grant patent and the protection of Plant Breeder's Rights (PBR). Sharma et al. (2009) characterized 16 sesame genotypes by employing RAPD and ISSR markers and suggested that putative variety specific RAPD and ISSR markers could be converted to Co-dominant Sequence Characterized Amplified Region/Sequence Tagged Site (SCAR/STS) markers to develop robust variety specific markers. The comparative analysis for the genetic diversity of sesame has been carried out using agro-morphological and molecular markers such as RAPD among twelve sesame populations collected from three regions in Cambodia and Vietnam (Pham et al. 2011) and ISSR among eighteen genotypes of sesame collected from various agro-climatic regions of Iran along with six exotic genotypes from the Asian countries (Parsaeian et al. 2011). ISSR analyses with merely 13 ISSR primers for genetic variation among them revealed 170 loci, of which 130 (76.47 %) were polymorphic. A high genetic variation was revealed both by agro-morphological and molecular markers within and among the sesame populations. Although both agro-morphological and RAPD markers were found to be useful in genetic diversity analysis in sesame, their combined use would give superior results. Further, the parental lines for hybridization should be selected on the basis of genetic diversity rather than the geographical distribution.

In another approach, sequence-related amplified polymorphism (SRAP) was used by Zhang et al. (2011) for the analysis of 67 sesame cultivars from the major sesame producing areas of China. A total of 561 bands were amplified using 21 SRAP random primer pairs, with 265 of them were polymorphic, resulting in a polymorphism ratio of 47.2 %. The average genetic similarity coefficient and the genetic distance of the 67 cultivars were 0.9104 and 0.0706, respectively, indicating limited genetic diversity and narrow genetic basis. The genetic basis of landraces was found wider than that of bred cultivars.

7.8.4 Marker Assisted Selection

Uzun et al. (2003) were first to identify a molecular marker linked to an agronomically important trait in sesame. They identified an AFLP marker linked to the closed capsule mutant trait in sesame using bulked segregant analysis (BSA) approach on segregating progenies of a cross between the closed capsule mutant line ‘cc3’, and the Turkish variety ‘Muganli-57’. They tested a total of 72 primer combinations to screen for linkage to the trait, but only one closely linked AFLP marker was identified. The linkage was confirmed by analysing the AFLP profile from single plants. They suggested that this marker had a potential to accelerate breeding programmes aimed at modifying unwanted side-effects of the closed capsule mutation through marker-assisted selection.

The first report on molecular tagging of the dt gene regulating determinate growth habit in sesame came from Uzun and Cagirgan (2009). The development of determinate cultivars has become an objective of high priority in sesame breeding programmes. They investigated RAPD and inter simple sequence repeat (ISSR) techniques for the development of molecular markers for this induced mutant character. Using the F2 segregating population and bulked segregant approach, they were able to detect two ISSR marker loci originated from a (CT)8AGC primer. They proposed that this marker would be potentially useful for assisting sesame breeding programmes through marker assisted selection and can facilitate the integration of determinate growth habit into new genetic backgrounds.

7.8.5 Genomics

In spite of extensive efforts to develop new sesame varieties by conventional and mutational breeding, the lack of a non-shattering sesame variety is one of the major barriers to obtaining high yield of sesame seeds (Yermanos et al. 1972; Ashri 1987). In addition, after oil extraction, the remaining meal corresponding to 50 % of seed dry weight is wasted or used for feeding poultry. Therefore, identification of novel genes involved in biosynthesis of sesame-specific flavor or lignans and understanding the metabolic pathways from photosynthates toward storage oil are desirable as an aid to improve the quality and quantity of oil in sesame cultivars. Sesame is an important oil crop, but limited transcriptomic and genomic data are currently available. In addition, a shortage of sesame molecular markers limits the efficiency and accuracy of genetic breeding. High-throughput transcriptomic sequencing is essential to generate a large transcriptome sequence dataset for gene discovery and molecular marker development. Expressed Sequence Tags (ESTs) generated by large-scale single-pass cDNA sequencing have proven valuable for the identification of novel genes in specific metabolic pathways. cDNA clones encoding seed-specific stearoyl-acyl carrier protein desaturase (Yukawa et al. 1996) and metallothionein-like protein (Chyan et al. 2005) have been isolated and their expression analysis revealed the maximum accumulation of RNA transcripts in the seeds.

Suh et al. (2003) in order to elucidate the metabolic pathways for lignans in developing sesame seeds and to identify genes involved in accumulation of storage products and in the biosynthesis of antioxidant lignans, obtained 3328 Expressed Sequence Tags (ESTs) from a cDNA library of 5–25 days old immature sesame seeds. ESTs were clustered and analyzed by the BLASTX or FASTAX program against the GenBank NR and Arabidopsis proteome databases. They carried out a comparative analysis between developing sesame and Arabidopsis seed ESTs for gene expression profiles during development of green and non-green seeds. Analyses of these two seed EST sets helped to identify similar and differential gene expression profiles during seed development, and to identify a large number of sesame seed-specific genes. Seed-specific expression of several candidate genes was confirmed by northern blot analysis. They identified EST candidates for genes possibly involved in biosynthesis of sesame lignans, sesamin and sesamolin, and suggested a possible metabolic pathway for the generation of cofactors required for synthesis of storage lipid in non-green oilseeds.

Sesame seed has been recognized as a nutritional protein source owing to its richness in methionine. Storage proteins have been implicated in allergenic responses to sesame consumption. Two abundant storage proteins, 11S globulin and 2S albumin, constitute 60–70 and 15–25 % of total sesame proteins, respectively. Two gene families separately encoding four 11S globulin and three 2S albumin isoforms were identified in a database search of 3328 expressed sequence tag (EST) sequences from maturing sesame seeds (Hsiao et al. 2006). Full-length cDNA sequences derived from these two gene families were completed by PCR using a maturing sesame cDNA library as the template. The amino acid compositions of these deduced storage proteins revealed that the richness in methionine is attributed mainly to two 2S albumin isoforms and partly to one 11S globulin isoform.

Leaves of *Sesamum* spp. are used as leafy vegetables in Nigeria and many tropical areas around the world. Sesame leaves are also being used in Japan as a new food material containing functional components. Sesamin, a major lignan constituent of sesame seed, is considered responsible for a number of beneficial health effects in humans. Hata et al. (2010) scrutinized genotypic differences in sesamin content of two Japanese sesame varieties that differ in seed sesamin content (Higher in 'Gomazou' and lower in 'Kin-goma'). The expression of the sesamin biosynthetic gene CYP81Q1 was analysed through Quantitative RT-PCR analysis of stem and leaf samples performed with the Real-time PCR system relative to the expression of the reference gene 18S rRNA. The gene expression was found to be considerably higher in 'Gomazou' than in 'Kin-goma', indicating that genotypic difference of CYP81Q1 gene expression is one of the important factors affecting leaf sesamin contents. They further reported that the CYP81Q1 gene expression and sesamin content in leaves are photoperiod dependent (Hata et al. 2011) and concluded that cultivation of sesame under continuous light enables high-yield production of sesame leaves containing distinctively high levels of sesamin.

Recently introduced technique, Illumina paired-end sequencing is a fast and cost-effective approach to gene discovery and molecular marker development in non-model organisms. Sesame transcriptomes from five tissues were sequenced using this technology (Wei et al. 2011) leading to generation of 86,222 unigenes with an average length of 629 bp. Of the unigenes, 46,585 had significant similarity with proteins in the NCBI nonredundant protein database and Swiss-Prot database. In total, 22,003 unigenes were mapped onto 119 pathways using the Kyoto Encyclopedia of Genes and Genomes Pathway database (KEGG). Furthermore, 44,750 unigenes showed homology to 15,460 Arabidopsis genes based on BLASTx analysis against The Arabidopsis Information Resource (TAIR). Among these, 7,702 unigenes were converted into SSR markers (EST-SSR) in which dinucleotide SSRs were the dominant repeat motif (5,166). Randomly selected forty EST-SSR primer pairs successfully amplified DNA fragments and detected significant amounts of polymorphism among 24 sesame accessions.

7.9 Conclusions

Neglected crops are essential to the livelihoods of millions of poor farmers throughout the world. Newer technologies will certainly play their part in the process of improvement, conservation and use strategies. There are already a number of examples which show how useful neglected crops can be, but often only in small research scale activities which needs to be scaled up. Perhaps there is a need for some deliberate determination of the way in which these powerful tools can be best used for such crops. As implied above, there is also a lot of work to be done on the development of sustainable linkages between various organizations, farmers and consumers. It will always be unlikely that any one organization will have the resources to support work on the scale needed for the individual neglected oilseed crops. Thus a major challenge will be to make sustainable networks and filieres. Strengthened community involvement in the management of underutilized crops and deliberate attention to resourcing their needs for new materials (and securing access to existing ones) will provide a basis for some more work on key production issues. The first of these is obviously that of the development of improved materials. Participatory plant breeding approaches including transgenics and molecular approaches may not only be an important element of the work on these crops; it will be the only feasible approach to obtaining improved materials.

Ultimately, we have to recognize that neglected oilseed crops present their own range of problems and opportunities. These are important to many farmers in the ways that are complementary to and are different from their concerns for the major crops. Attempting to copy large crop solutions across these species will help neither in the improved conservation and use of the crops nor the interests of the farmers who grow them. Developing an agenda specific to the crops will have to be recognized as an important and continuing need.

References

Abdellatef E, Sirelkhatem R, Ahmed MMM, Radwan KH, Khalafalla MM (2008) Study of genetic diversity in Sudanese sesame (*Sesamum indicum* L.) germplasm using random amplified polymorphic DNA (RAPD) markers. Afr J Biotechnol 7:4423–4427

Abdellatef E, Ahmed MMM, Daffalla HM, Khalafalla MM (2010) Enhancement of adventitious shoot regeneration in sesame (*Sesamum indicum* L.) cultivar promo KY using ethylene inhibitors. J Phytol 2:61–67

Adda S, Reddy TP, Kavi Kishor PB (1993) Somatic embryogenesis and organogenesis in *Guizotia abyssinica*. In Vitro Cell Dev Biol-Plant 30:104–107

Adda S, Reddy TP, Kavi Kishor PB (1994) Androclonal variation in niger (*Guizotia abyssinica* Cass). Euphytica 79:59–64

Agarwal A, Pant T, Ahmed Z (2010) *Camelina sativa*: a new crop with biofuel potential introduced in india. Curr Sci 99:1194–1195

Ali GM, Yasumoto S, Katsuta MS (2007) Assessment of genetic diversity in sesame (*Sesamum indicum* L.) detected by amplified fragment length polymorphism markers. J Biotechnol 10:12–23

Anonymous (2008) World Sesame situation, American Sesame Growers Association. http://www.sesamegrowers.org/worldstatusofsesame.htm. Accessed 15 Jan 2008

Arora R, Bhojwani SS (1988) Production of androgenic plants through pollen embryogenesis in anther cultures of *Brassica carinata* A. Braun. Biol Plant 30:25–29

Ashri A (1987) Sesame. In: Robbelen G, Downey RK, Ashri A (eds) Oil crops of the world: their breeding and utilization. McGraw-Hill, New York

Babic V, Datla RS, Scoles GJ, Keller WA (1998) Development of an efficient *Agrobacterium*-mediated transformation system for *Brassica carinata*. Plant Cell Rep 17:183–188

Baskaran P, Jayabalan N (2006) In vitro mass propagation and diverse callus orientation on *Sesamum indicum* L.-an important oil plant. J Agric Technol 2:259–269

Baye T, Becker HC, Witzke-Ehbrecht SV (2005) *Vernonia galamensis*, a natural source of epoxy oil: variation in fatty acid composition of seed and leaf lipids. Ind Crop Prod 21:257–261

Bedigian D, Harlan JR (1986) Evidence for cultivation of Sesame in the ancient world. Econ Bot 40:137–154

Bekele E, Geleta M, Dagne K, Jones AL, Barnes I, Bradman N, Thomas MG (2007) Molecular phylogeny of genus *Guizotia* (Asteraceae) using DNA sequences derived from ITS. Genet Resour Crop Evol 54:1419–1427

Belay S, Rier JP, Ayorinde FO (1989) Preliminary observation of the chemical composition of callus derived from immature seeds of *Vernonia galamensis*, Var. Ethiopica, Gilbert. J Am Oil Chem Soc 66:828

Bhat JG, Murthy HN (2007) Factors affecting in vitro gynogenic haploid production in niger (*Guizotia abyssinica* (L. f.) Cass.). Plant Growth Regul 52:241–248

Bhat JG, Murthy HN (2008) Haploid plant regeneration from unpollinated ovule cultures of niger (*Guizotia abyssinica* (L. f.) Cass.). Russ J Plant Physiol 55:241–245

Biesaga-Koscielniak J, Koscielniak J, Filek M, Janeczko A (2008) Rapid production of wheat cell suspension cultures directly from immature embryos. Plant Cell Tiss Organ Cult 94:139–147

Bouck A, Vision T (2007) The molecular ecologist's guide to expressed sequence tags. Mol Ecol 16:907–924

Brar GS, Ahuja L (1979) Sesame: its culture, genetics, breeding and biochemistry. Annu Rev Plant Sci 1:245–313

Broun P, Boddupalli S, Somerville C (1998) A bifunctional oleate 12-hydroxylase: desaturase from *Lesquerella fendleri*. Plant J 13:201–210

Carlson KD, Schneider WJ, Chang SP, Princen LH (1981) *Vernonia galamensis* seed oil: a new source for epoxy coatings. In: Pryde EH, Princen LH, Mukherjee KD (eds) New sources of fats and oils. American Oil Chemists' Society, Champaign

Carlson KD, Chaudhry A, Bagby MO (1990) Analysis of oil and meal from *Lesquerella fendleri* seed. J Am Oil Chem Soc 67:438–442

CGIAR (2006) Enhancing the delivery of genomics research outcomes. Genomics research in the CGIAR: effective means of establishing platforms for genetic research. Science council of the consultative group for international agricultural research, the secretariat, Rome, Italy

Chae YA, Park SK, Anand IJ (1987) Selection in vitro for herbicide tolerant cell lines of *Sesamum indicum* 2: selection of herbicide tolerant calli and plant regeneration. Kor J Plant Breed 19:75–80

Chattopadhyaya B, Banerjee J, Basu A, Sen SK, Maiti MK (2010) Shoot induction and regeneration using internodal transverse thin cell layer culture in *Sesamum indicum* L. Plant Biotechnol Rep 4:173–178

Chaudhary S, Parmenter DL, Moloney MM (1998) Transgenic *Brassica carinata* as a vehicle for the production of recombinant proteins in seeds. Plant Cell Rep 17:195–200

Chen GQ, Lin JT, Lu C (2011) Hydroxy fatty acid synthesis and lipid gene expression during seed development in *Lesquerella fendleri*. Ind Crop Prod 34:1286–1292

Cheng B, Wu G, Vrinten P, Falk K, Bauer J, Qiu X (2009) Towards the production of high levels of eicosapentaenoic acid in transgenic plants: the effects of different host species, genes and promoters. Transgenic Res. doi:10.1007/s11248-009-9302-z

Chhikara S, Dutta I, Paulose B, Jaiwal PK, Dhankher OP (2011) Development of an *Agrobacterium*-mediated stable transformation method for industrial oilseed crop Crambe abyssinica 'BelAnn'. Ind Crop Prod. doi:10.1016/j.indcrop.2011.07.021

Chun JA, Lee WH, Han MO, Lee JW, Yi YB, Goo YM, Lee SW, Bae SC, Cho KJ, Chung CH (2007) Molecular and biochemical characterizations of dehydroascorbate reductase from sesame (*Sesamum indicum* L.) hairy root cultures. J Agric Food Chem 55:6067–6073

Chun JA, Lee JW, Yi YB, Park GY, Chung CH (2009) Induction of hairy roots and characterization of peroxidase expression as a potential root growth marker in sesame. Prep Biochem Biotechnol 39:345–359

Chuong PV, Beversdorf WD (1985) High frequency embryogenesis through isolated microspore culture in *Brassica napus* L. and *B. carinata* Braun. Plant Sci 39:219–226

Chuong PV, Pauls KP, Beversdorf WD (1987) Protoplast culture and plant regeneration from *Brassica carinata* Braun. Plant Cell Rep 6:67–69

Chyan CL, Lee TT, Liu CP, Yang YC, Tzen JT, Chou WM (2005) Cloning and expression of a seed-specific metallothionein-like protein from sesame. Biosci Biotechnol Biochem 69:2319–2325

Dagne K, Cheng B, Heneen WK (2000) Number and sites of rDNA loci of *Guizotia abyssinica* (L. f.) Cass. as determined by fluorescence in situ hybridization. Hereditas 132:63–65

Daniel E (2008) Investigation of the genetic variability among land races of sesame from Ethiopia. MA Thesis, University of Hohenhein, Stuttgart, Germany

Dawson IK, Hedley PE, Guarino L, Jaenicke H (2009) Does biotechnology have a role in the promotion of underutilised crops? Food Policy 34:319–328

Demeke T, Lynch DR, Kawchuk LM, Kozub GC, Armstrong JD (1996) Genetic diversity of potato determined by random amplified polymorphic DNA analysis. Plant Cell Rep 15:662–667

Dempewolf H, Kane NC, Ostevik KL, Geleta M, Barker MS, Lai Z, Stewart ML, Bekele E, Engels JM, Cronk QB, Rieserberg LH (2010) Establishing genomic tools and resources for *Guizotia abyssinica* (L.f.) Cass.—the development of a library of expressed sequence tags, microsatellite loci, and the sequencing of its chloroplast genome. Mol Ecol Resour 10:1048–1058

Du XZ, Ge XH, Zhao ZG, Li ZY (2008) Chromosome elimination and fragment introgression and recombination producing intertribal partial hybrids from *Brassica napus* × *Lesquerella fendleri* crosses. Plant Cell Rep 27:261–271

Dykinga J (1999) A storybook future for lesquerella? Agric Res Mag 47:14–15

Erickson LR, Straus NA, Beversdorf WD (1983) Restriction patterns reveal origins of chloroplast genomes in *Brassica* amphiploids. Theor Appl Genet 65:201–206

FAO (2004) Preliminary review of biotechnology in forestry, including genetic modification. Forest genetic resources working paper FGR/59E. Forest resources development service, Forest resources division, Food and Agriculture Organization of the United Nations, Rome, Italy

FAO (2005) Food and Agricultural Organisation of the United Nations. FAOSTAT Database. http:/apps.fao.org/default.htm

Ferrie AMR, Bethune TD (2011) A microspore embryogenesis protocol for *Camelina sativa*, a multi-use crop. Plant Cell Tiss Org Cult 106:495–501

Francisco-Ortega J, Fuertes-Aguilar J, Gómez-Campo C, Santos-Guerra A, Jansen RK (1999) Internal transcribed spacer sequence phylogeny of *Crambe* L. (Brassicaceae): molecular data reveal two old world disjunctions. Mol Phylogenet Evol 11:361–380

Furumoto T, Hoshikuma A (2011) Biosynthetic origin of 2-geranyl-1,4-naphthoquinone and its related anthraquinone in a *Sesamum indicum* hairy root culture. Phytochem 72:871–874

Furumoto T, Ohara T, Kubo T, Kawanami Y, Fukui H (2007) 2-Geranyl-1,4-naphthoquinone, a possible intermediate of anthraquinones in a *Sesamum indicum* hairy root culture. Biosci Biotechnol Biochem 71:2600–2602

Galasso I, Manca A, Braglia L, Martinelli T, Morello L, Breviario D (2010) h-TBP: an approach based on intron-length polymorphism for the rapid isolation and characterization of the multiple members of the β-tubulin gene family in *Camelina sativa* (L.) Crantz. Mol Breed. doi:10.1007/s11032-010-9515-0

Ganapathi TR, Nataraja K (1993) Effect of auxins and cytokinins on plant regeneration from hypocotyls and cotyledons of niger (*Guizotia abyssinica*). Biol Plant 35:209–215

Gangopadhyay G, Poddar R, Gupta S (1998) Micropropagation of sesame (*Sesamum indicum* L.) by in vitro multiple shoot production from nodal explants. Phytomorphology 48:83–90

Gao HB, Wang Y, Gao F, Luo P (1998) Studies on the plant regeneration from single cell culture of *Crambe abyssinica*. Hereditas (Beijing) 20:50–52

Gehringer A, Friedt W, Lühs W, Snowdon RJ (2006) Genetic mapping of agronomic traits in false flax (*Camelina sativa* subsp. sativa). Genome 49:1555–1563

Geleta M, Bryngelsson T, Bekele E, Dagne K (2007a) AFLP and RAPD analyses of genetic diversity of wild and/or weedy *Guizotia* (Asteraceae) from Ethiopia. Hereditas 144:53–62

Geleta M, Bryngelsson T, Bekele E, Dagne K (2007b) Genetic diversity of *Guizotia abyssinica* (L. f.) Cass. (Asteraceae) from Ethiopia as revealed by random amplified polymorphic DNA (RAPD). Genet Resour Crop Evol 54:601–614

George V, Bapat A, Rao PS (1987) In vitro Multiplication of Sesame (*Sesamum indicum*) through tissue culture. Ann Bot 60:17–21

Getinet A, Sharma SM (1996) Niger, *Guizotia abyssinica* (L. f.) Cass. Promoting the conservation and use of underutilized and neglected crops. International Plant Genetic Resources Institute, Rome, pp 1–59

Ghamkhar K, Croser J, Aryamanesh N, Campbell M, Kon'kova N, Francis C (2010) Camelina (*Camelina sativa* (L.) Crantz) as an alternative oilseed: molecular and ecogeographic analyses. Genome 53:558–567

Hansen LN (1998) Intertribal somatic hybridization between rapid cycling *Brassica oleracea* L. and *Camelina sativa* (L.) Crantz. Euphytica 104:173–179

Hata N, Hayashi Y, Okazawa A, Ono E, Satake H, Kobayashi A (2010) Comparison of sesamin contents and CYP81Q1 gene expressions in aboveground vegetative organs between two Japanese sesame (*Sesamum indicum* L.) varieties differing in seed sesamin contents. Plant Sci 178:510–516

Hata N, Hayashi Y, Okazawa A, Ono E, Satake H, Kobayashi A (2011) Effect of photoperiod on growth of the plants, and sesamin content and CYP81Q1 gene expression in the leaves of sesame (*Sesamum indicum* L.). Environ Exp Bot. doi:10.1016/j.envexpbot.2011.07.004

Hatanaka T, Yu K, Hildebrand DF (2003) Cloning and expression of a Vernonia and Euphorbia diacylglycerol acyltransferase cDNAs. In: Murata N, Yamada M, Nishida I, Okuyama H, Sekiya J, Hajime W (eds) Advanced research on plant lipids. Kluwer Academic Publishers, Dordrecht

Hatanaka T, Shimizu R, Hildebrand DF (2004) Expression of a Stokesia laevis epoxygenase gene. Phytochemistry 65:2189–2196

Hema BP, Murthy HN (2008) Improvement of in vitro androgenesis in niger using amino acids and polyamines. Biol Plant 52:121–125

Hitz WD (1998) Fatty acid modifying enzymes from developing seeds of *Vernonia galamensis*. US Patent 5846784, December 8

Horejsi T, Staub JE (1999) Genetic variation in cucumber (*Cucumis sativus* L.) as assessed by random amplified polymorphic DNA. Genet Resour Crop Evol 46:337–350

Hsiao ES, Lin LJ, Li FY, Wang MM, Liao MY, Tzen JT (2006) Gene families encoding isoforms of two major sesame seed storage proteins, 11S globulin and 2S albumin. J Agric Food Chem 54:9544–9550

Hutcheon C, Ditt RF, Beilstein M, Comai L, Schroeder J, Goldstein E, Shewmaker CK, Nguyen T, Rocher JD, Kiser J (2010) Polyploid genome of *Camelina sativa* revealed by isolation of fatty acid synthesis genes. BMC Plant Biol. doi:10.1186/1471-2229-10-233

Jadhav A, Katavic V, Marillia EF, Giblin EM, Barton DL, Kumar A, Sonntag C, Babic V, Keller WA, Taylor DC (2005) Increased levels of erucic acid in *Brassica carinata* by co-suppression and antisense repression of the endogenous FAD2 gene. Metab Eng 7:215–220

Jadimath VG, Murthy HN, Pyati AN, Kumar HGA, Ravishankar BV (1998) Plant regeneration from leaf cultures of *Guizotia abssinica* (Niger) and *Guizotia scabra*. Phytomorphology 48:131–135

Jain RK, Chowdhury JB, Sharma DR, Friedt W (1988) Genotypic and media effects on plant regeneration from cotyledon explant cultures of some *Brassica* species. Plant Cell Tiss Org Cult 14:197–206

Jaiswal SK, Hammatt N, Bhojwani SS, Cocking EC, Davey MR (1990) Plant regeneration from cotyledon protoplasts of *Brassica carinata*. Plant Cell Tiss Org Cult 22:159–165

James C (2009) Global status of commercialized biotech/GM crops. ISAAA briefs 41. ISAAA, Ithaca

Jiang JJ, Zhao XX, Tian W, Li TB, Wang YP (2009) Intertribal somatic hyprids between *Brassica napus* and *Camelina sativa* with high linolenic acid content. Plant Cell Tiss Org Cult 99:91–95

Jin UH, Chun JA, Han MO, Lee JW, Yi YB, Lee SW, Chung CH (2005) Sesame hairy root cultures for extra-cellular production of a recombinant fungal phytase. Prog Biochem Biophys 40:3754–3762

Joshi AB (1961) Sesamum. A monograph. Indian Central Oilseed Committee, Hyderabad

Jourdan P, Salazar E (1993) *Brassica carinata* resynthesized by protoplast fusion. Theor Appl Genet 86:567–572

Kang J, Snapp AR, Lu C (2011) Identification of three genes encoding microsomal oleate desaturases (FAD2) from the oilseed crop *Camelina sativa*. Plant Physiol Biochem 49:223–229

Kharenko OA, Zaharia LI, Giblin M, Čekić V, Taylor DC, Palmer CD, Abrams SR, Loewen MC (2011) Abscisic acid metabolism and lipid accumulation of a cell suspension culture of *Lesquerella fendleri*. Plant Cell Tiss Org Cult 105:415–422

Kim HY, Byeon GH (1991) Effect of growth regulators on organ cultures of sesame. J Subtrop Agric Res Dev 8:93–103

Kim MK, Park SK, Chae YA (1987) Selection in vitro for herbicide tolerant cell lines of *Sesamum indicum*: effects of explants and hormone combinations on callus induction. Kor J Plant Breed 19:70–94

Kinney AJ (2002) Perspectives on the production of industrial oils in genetically engineered oilseeds. In: Kuo TM, Gardner HW (eds) Lipid biotechnology. Marcel Dekker Inc., New York

Kolte SJ (1985) Disease of annual edible oil seed crops. Vol II: Rapeseed-mustard and sesame diseases. CRC Press, Boca Raton

Kumar HG, Murthy HN, Jadimath VG, Sheelavantmath SS, Pyati AN, Ravishankar BV (2000) Direct somatic embryogenesis and plantlet regeneration from leaf explants of niger, *Guizotia abyssinica* (L.f.) Cass. Ind J Exp Biol 38:1073–1075

Kuvshinov V, Kanerva A, Koivu K, Pehu E, Kuvshinova S (2002, 2004) A transformation system in *Camelina sativa*. Patents WO 02/38779 A1 and US 2004/0031076 A1

Laurentin H, Karlovsky P (2007) AFLP fingerprinting of sesame (*Sesamum indicum* L.) cultivars: identification, genetic relationship and comparison of AFLP informativeness parameters. Genet Resour Crop Evol 54:1437–1446

Laurentin H, Ratzinger A, Karlovsky P (2008) Relationship between metabolic and genomic diversity in sesame (*Sesamum indicum* L.). BMC Genomics 9:250–260

Lee JI, Park YH, Park YS, Kim BG (1985) Propagation of sesame (*Sesamum. indicum* L.) through shoot tip culture. Kor J Plant Breed 17:367–372

Lee SY, Kim HS, Lee YT, Park CH (1988) Effect of growth regulators, cold pretreatment and genotype in anther culture of sesame (*Sesamum indicum* L.). Res Rep Rural Dev Adm Biotechnol 30:74–79

Li X, Gao P, Gjetvaj B, Westcott N, Gruber MY (2009) Analysis of the metabolome and transcriptome of *Brassica carinata* seedlings after lithium chloride exposure. Plant Sci 177:68–80

Li R, Yu K, Hildebrand DF (2010a) DGAT1, DGAT2 and PDAT expression in seeds and other tissues of epoxy and hydroxy fatty acid accumulating plants. Lipids. doi:10.1007/s11745-010-3385-4

Li X, Ahlman A, Yan X, Lindgren H, Zhu LH (2010b) Genetic transformation of the oilseed crop *Crambe abyssinica*. Plant Cell Tiss Org Cult 100:149–156

Li X, Ahlman A, Lindgren H, Zhu LH (2011) Highly efficient in vitro regeneration of the industrial oilseed crop *Crambe abyssinica*. Ind Crop Prod 33:170–175

Lu C, Kang J (2008) Generation of transgenic plants of a potential oilseed crop *Camelina sativa* by *Agrobacterium* -mediated transformation. Plant Cell Rep 27:273–278

Lu C, Napier JA, Clemente TE, Cahoon EB (2011) New frontiers in oilseed biotechnology: meeting the global demand for vegetable oils for food, feed, biofuel, and industrial applications. Curr Opin Biotechnol 22:252–259

Marles MAS, Gruber MY, Scoles GJ, Muir AD (2003) Pigmentation in the developing seed coat and seedling leaves of *Brassica carinata* is controlled at the dihydroflavonol reductase locus. Phytochemistry 62:663–672

Marquez-Lema A, Velasco L, Perez-Vich B (2010) Transferability, amplification quality, and genome specificity of microsatellites in *Brassica carinata* and related species. J Appl Genet 51:123–131

Márquez-Lema A, Fernández-Martínez JM, Pérez-Vich B, Velasco L (2008) Development and characterisation of a *Brassica carinata* inbred line incorporating genes for low glucosinolate content from *B. juncea*. Euphytica 164:365–375

Mary RJ, Jayabalan N (1997) Influence of growth regulators on somatic embryogenesis in sesame. Plant Cell Tiss Org Cult 49:67–70

Mietkiewska E, Brost JM, Giblin EM, Barton DL, Taylor DC (2007) Cloning and functional characterization of the fatty acid elongase 1 (FAE1) gene from high erucic *Crambe abyssinica* cv. Prophet. Plant Biotechnol J 5:636–645

Mietkiewska E, Hoffman TL, Brost JM, Giblin EM, Barton DL, Francis T, Zhang Y, Taylor DC (2008) Hairpin-RNA mediated silencing of endogenous FAD2 gene combined with heterologous expression of Crambe abyssinica FAE gene causes an increase in the level of erucic acid in transgenic *Brassica carinata* seeds. Mol Breed 22:619–627

Moon H, Smith MA, Kunst L (2001) A condensing enzyme from the seeds of *Lesquerella fendleri* that specifically elongates hydroxy fatty acids. Plant Physiol 127:1635–1643

Moon H, Chowrira G, Rowland O, Blacklock BJ, Smith MA, Kunst L (2004) A root-specific condensing enzyme from *Lesquerella fendleri* that elongates very-long-chain saturated fatty acids. Plant Mol Biol 56:917–927

Murthy HN, Kumar AHG, Paek KY (2000) Anther culture of niger. Kor J Plant Tiss Cult 27:353–358

Murthy HN, Jeong JH, Choi YE, Paek KY (2003) *Agrobacterium*-mediated transformation of niger [*Guizotia abyssinica* (L. f.) Cass.] using seedling explants. Plant Cell Rep 21:1183–1187

Nagella P, Hosakatte NM, Ravishankar KV, Hahn E, Paek K (2008) Analysis of genetic diversity among Indian niger [*Guizotia abyssinica* (L. f.) Cass.] cultivars based on randomly amplified polymorphic DNA markers. Electron J Biotechnol 11:1–5

Naik PM, Murthy HN (2010) Somatic embryogenesis and plant regeneration from cell suspension culture of niger (*Guizotia abyssinica* Cass.). Acta Physiol Plant 32:75–79

Narasimhulu SB, Chopra VL (1987) Plant regeneration from callus culture of *Brassica carinata* A. Br. and its implication to improvement of oilseed *Brassica*. Plant Breed 99:49–55

Narasimhulu SB, Chopra VL (1988) Species specific shoot regeneration response of cotyledonary explants of Brassicas. Plant Cell Rep 7:104–106

Narasimhulu SB, Kirti PB, Prakash S, Chopra VL (1992a) Rapid and efficient plant regeneration from hypocotyl protoplasts of *Brassica carinata*. Plant Cell Rep 11:159–162

Narasimhulu SB, Kirti PB, Prakash SV, Chopra VL (1992b) Shoot regeneration in stem explants and its amenability to Agrobacterium-mediated gene transfer in *Brassica carinata*. Plant Cell Rep 11:359–362

Narasimhulu SB, Kirti PB, Bhatt SR, Prakash S, Chopra VL (1994) Intergeneric protoplast fusion between *Brassica carinata* and *Camelina sativa*. Plant Cell Rep 13:657–660

Ncube I, Read JS (1995) Evaluation of *Vernonia galamensis* lipase (acetone powder) for use in biotechnology. Ind Crop Prod 3:285–292

Nguyen T, Liu X, Derocher J (2011) Floral dip method for transformation of *Camelina*. US Patent Application 20110145950. Date 16-06-2011

Nikam TD, Shitole MG (1993) Regeneration of niger (*Guizotia abyssinica* Cass.) CV Sahyadri from seedling explants. Plant Cell Tiss Org Cult 32:345–349

Nikam TD, Shitole MG (1997) In vitro plant regeneration from callus of niger (*Guizotia abyssinica* Cass.) cv. Sahyadri. Plant Cell Rep 17:155–158

Nitovs'ka IO, Shakhovs'kyĭ AM, Cherep MN, Horodens'ka MM, Kuchuk MV (2006) Construction of the cybrid transplastomic *Brassica napus* plants containing *Lesquerella fendleri* chloroplasts. Tsitol Genet 40:3–11

Nugent JM, Palmer JD (1988) Location, identity, amount and serial entry of chloroplast DNA sequences in crucifer mitochondrial DNAs. Curr Genet 14:501–509

Nybom H (2004) Comparison of different nuclear DNA markers for estimating intraspecific genetic diversity in plants. Mol Ecol 13:1143–1155

Ogasawara T, Chiba K, Tada M (1993) Production and high yield of napthoquinone by a hairy root culture of *Sesamum indicum*. Phytochem 33:1095–1098

Ologunde MO, Ayorinde FO, Shepard RL (1990) Chemical evaluation of defatted *Vernonia galamensis* meal. J Am Oil Chem Soc 67:92–94

Ovcharenko O, Momot V, Cherep N, Sheludko Y, Komarnitsky I, Rudas V, Kuchuk N (2011) Transfer of transformed *Lesquerella fendleri* (Gray) Wats chloroplasts into *Orychophragmus violaceus* (L.) by protoplast fusion. Plant Cell Tiss Org Cult 105:21–27

Palmer CD, Keller WA (2011) Somatic embryogenesis in *Crambe abyssinica* Hochst. ex R.E. Fries using seedling explants. Plant Cell Tiss Org Cult 104:91–100

Parsaeian M, Mirlohi A, Saeidi G (2011) Study of genetic variation in sesame (*Sesamum indicum* L.) using agro-morphological traits and ISSR markers. Genetika 47:359–367

Paulose B, Kandasamy S, Dhankher OP (2010) Expression profiling of *Crambe abyssinica* under arsenate stress identifies genes and gene networks involved in arsenic metabolism and detoxification. BMC Plant Biol 10:108

Perdue RE (1988) Systematic botany in the development of *Vernonia galamensis* as a new industrial oilseed crop for the semi-arid tropics. Symb Bot Ups 28:125–135

Perdue RE, Carlson KD, Gilbert MG (1986) *Vernonia galamensis* potential new crop source of epoxy acid. Econ Bot 40:54–68

Petros Y, Merker A, Zeleke H (2007) Analysis of genetic diversity of *Guizotia abyssinica* from Ethiopia using inter simple sequence repeat markers. Hereditas 144:18–24

Pham TD, Geleta M, Bui TM, Bui TC, Merker A, Carlsson AS (2011) Comparative analysis of genetic diversity of sesame (*Sesamum indicum* L.) from Vietnam and Cambodia using agro-morphological and molecular markers. Hereditas 148:28–35

Ploschuk EL, Cerdeiras G, Windauer L, Dierig DA, Ravetta DA (2003) Development of alternative *Lesquerella* species in Patagonia (Argentina): potential of *L. angustifolia*. Ind Crop Prod 18:1–6

Prince JP, Lackney VK, Angeles C, Blauth JR, Kyle MM (1995) A survey of DNA polymorphism within the genus *Capsicum* and the fingerprinting of pepper cultivars. Genome 38:224–231

Purakayastha TJ, Viswanath T, Bhadraray S, Chhonkar PK, Adhikari PP, Suribabu K (2008) Phytoextraction of zinc, copper, nickel and lead from a contaminated soil by different species of *Brassica*. Int J Phytoremediation 10:61–72

Rajeswari S, Thiruvengadam V, Ramaswamy NM (2010) Production of interspecific hybrids between *Sesamum alatum* Thonn and *Sesamum indicum* L. through ovule culture and screening for phyllody disease resistance. S Afr J Bot 76:252–258

Ram R, Catlin D, Romero J, Cowley C (1990) Sesame: new approaches for crop improvement. In: Janick J, Simon JE (eds) Advances in new crops. Timber Press, Portland

Ramalema SP, Shimelis H, Ncube I, Kunert KK, Mashela PW (2010) Genetic analysis among selected vernonia lines through seed oil content, fatty acids and RAPD DNA markers. Afr J Biotechnol 9:117–122

Rao KR, Vaidyanath K (1997a) Callus induction and morphogenesis in sesame (*Sesamum indicum* L.). Adv Plant Sci 10:21–26

Rao KR, Vaidyanath K (1997b) Induction of multiple shoots from seedling shoot tips of different varieties of Sesamum. Ind J Plant Physiol 2:257–261

Rao KR, Kavi Kishor PB, Vaidyanath K (2002) Biotechnology of sesame-an oil seed crop. Plant Cell Biotechnol Mol Biol 3:101–110

Reed DW, Taylor DC, Covello PS (1997) Metabolism of hydroxy fatty acids in developing seeds in the genera *Lesquerella* (Brassicaceae) and *Linum* (Linaceae). Plant Physiol 114:63–68

Sabharwal PS, Dolezel J (1993) Interspecific hybridization in *Brassica*: application of flow cytometry for analysis of ploidy and genome composition in hybrid plants. Biol Plant 35:169–177

Schrader-Fischer G, Apel K (1994) Organ-specific expression of highly divergent thionin variants that are distinct from the seed-specific crambin in the crucifer *Crambe abyssinica*. Mol Gen Genet 245:380–389

Seither C, Avdiushko S, Hildebrand D (1997) Isolation of cytochrome P-450 genes from *Vernonia galamensis*. In: Williams JP, Khan MU, Lem NW (eds) Physiology, biochemistry and molecular biology of plant lipids. Kluwer Academic Publishers, Dordrecht

Seo HY, Kim YJ, Park TI, Kim HS, Yun SJ, Park KH, Oh MK, Choi MY, Paik CH, Lee YS, Choi YE (2007) High-frequency plant regeneration via adventitious shoot formation from deembryonated cotyledon explants of *Sesamum indicum* L. In Vitro Cell Dev Biol Plant 43:209–214

Sharma M, Pareek LK (1998) Direct shoot bud differentiation from different explants of in vitro regenerated shoots in sesame. J Phytol Res 11:161–163

Sharma SN, Kumar V, Mathur S (2009) Comparative analysis of RAPD and ISSR markers for characterization of sesame (*Sesamum indicum* L) genotypes. J Plant Biochem Biotechnol 18:266–271

Sigareva MA, Earle ED (1999) Camalexin induction in intertribal somatic hybrids between *Camelina sativa* and rapid-cycling *Brassica oleracea*. Theor Appl Genet 98:164–170

Skarjinskaia M, Svab Z, Maliga P (2003) Plastid transformation in *Lesquerella fendleri*, an oilseed Brassicacea. Transgenic Res 12:115–222

Skarzhinskaya M, Landgren M, Glimelius K (1996) Production of intertribal somatic hybrids between *Brassica napus* L. and *Lesquerella fendleri* (Gray) Wats. Theor Appl Genet 93:1242–1250

Soltis DE, Soltis PS, Doyle JJ (1998) Molecular systematics of plants II: DNA sequencing. Kluwer Press, Dordrecht

Subramanian B, Bansal VK, Kav NNV (2005) Proteome-level investigation of *Brassica carinata*-derived resistance to *Leptosphaeria maculans*. J Agric Food Chem 53:313–324

Suh MC, Kim MJ, Hur CG, Bae JM, Park YI, Chung CH, Kang CW, Ohlrogge JB (2003) Comparative analysis of expressed sequence tags from *Sesamum indicum* and *Arabidopsis thaliana* developing seeds. Plant Mol Biol 52:1107–1123

Sujatha M (1997) In vitro adventitious shoot regeneration for effective maintenance of male sterile niger (*Guizotia abyssinica* (L.f.) Cass.). Euphytica 93:89–95

Tang TZ, Niu YZ, Shui HX (2006) Cytological observation on intergeneric hybrid between *Brassica chinensis* and *Crambe abyssinica*. Yi Chuan 28:189–194

Taskin KM, Turgut K (1997) In vitro regeneration of sesame (*Sesamum indicum* L.). Turk J Bot 21:15–18

Taskin KM, Ercan AG, Turgut K (1999) *Agrobacterium tumefaciens*—mediated transformation of sesame (*Sesamum indicum* L.). Turk J Bot 23:291–295

Tattersall A, Millam S (1999) Establishment and in vitro regeneration studies of the potential oil crop species *Camelina sativa*. Plant Cell Tiss Org Cult 55:147–149

Teklewold A, Becker HC (2006a) Geographic pattern of genetic diversity among 43 Ethiopian mustard (*Brassica carinata* A. Braun) accessions as revealed by RAPD analysis. Genet Resour Crop Evol 53:1173–1185

Teklewold A, Becker HC (2006b) Comparison of phenotypic and molecular distances to predict heterosis and F1 performance in Ethiopian mustard (*Brassica carinata* A. Braun). Theor Appl Genet 112:752–759

Thompson AE, Dierig DA (1994) Initial selection and breeding of *Lesquerella fendleri* L., a new industrial oil seed. Ind Crop Prod 2:91–106

Tiwari S, Kumar S, Gontia I (2011) Biotechnological approaches for sesame (*Sesamum indicum* L.) and niger (*Guizotia abyssinica* L.f. Cass.). Asia-Pac J Mol Biol Biotechnol 19:2–9

Tomasi P, Dierig D, Dahlquist G (2002) An ovule culture technique for producing interspecific *Lesquerella* hybrids. In: Janick J, Whipkey A (eds) Trends in new crops and new uses. ASHS Press, Alexandria

Uzun B, Cagirgan MI (2009) Identification of molecular markers linked to determinate growth habit in sesame. Euphytica 166:379–384

Uzun B, Lee D, Donini P, Cagirgan MI (2003) Identification of a molecular marker linked to the closed capsule mutant trait in sesame using AFLP. Plant Breed 122:95–97

Volis S, Mendlinger S, Shulgina I, Oluoch M (2009) Genetic diversity in Tanzanian accessions of *Brassica carinata* A. Braun. Int J Plant Breed 3:86–91

Vollmann J, Grausgruber H, Stift G, Dryzhyruk V, Lelley T (2005) Genetic diversity in camelina germplasm as revealed by seed quality characteristics and RAPD polymorphism. Plant Breed 124:446–453

Wang Y, Peng L (1998) Intergeneric hybridization between *Brassica* species and *Crambe abyssinica*. Euphytica 101:1–7

Wang Y, Sonntag K, Rudloff E (2003) Development of rapeseed with high erucic acid content by asymmetric somatic hybridization between *Brassica napus* and *Crambe abyssinica*. Theor Appl Genet 106:1147–1155

Wang Y, Snowdon RJ, Rudloff E, Wehling P, Friedt W, Sonntag K (2004) Cytogenetic characterization and fae1 gene variation in progenies from asymmetric somatic hybrids between *Brassica napus* and *Crambe abyssinica*. Genome 47:724–731

Wang Y, Sonntag K, Rudloff E, Wehling P, Snowdon RJ (2006) GISH analysis of disomic *Brassica napus–Crambe abyssinica* chromosome addition lines produced by microspore culture from monosomic addition lines. Plant Cell Rep 25:35–40

Wang W, Wang C, Huang BL, Huang B (2008) *Agrobacterium tumefaciens*-mediated transformation of *Lesquerella fendleri* L., a potential new oil crop with rich lesquerolic acid. Plant Cell Tiss Org Cult 92:165–171

Warwick SI, Gugel RK (2003) Genetic variation in the *Crambe abyssinica–C. hispanica–C. glabrata* complex. Genet Resour Crop Evol 50:291–305

Warwick SI, Gugel RK, McDonald T, Falk KC (2006) Genetic variation of Ethiopian mustard (*Brassica carinata* A. Braun) germplasm in Western Canada. Genet Resour Crop Evol 53:297–312

Wei W, Qi X, Wang L, Zhang Y, Hua W, Li D, Lv H, Zhang X (2011) Characterization of the sesame (*Sesamum indicum* L.) global transcriptome using Illumina paired-end sequencing and development of EST-SSR markers. BMC Genomics 12:451

Weiss EA (1971) Castor, sesame and safflower. Leonard Hill, London

Were BA, Gudu S, Onkware AO, Carlsson AS, Welander M (2006) In vitro regeneration of sesame (*Sesamum indicum* L.) from seedling cotyledon and hypocotyl explants. Plant Cell Tiss Org Cult 85:235–239

Xu ZQ, Jia JF, Hu ZD (1997) Somatic embryogenesis in *Sesamum indicum* L. cv. nigrum. J Plant Physiol 150:755–758

Yadav M, Chaudhary D, Sainger M, Jaiwal PK (2010) Agrobacterium tumefaciens-mediated genetic transformation of sesame (*Sesamum indicum* L.). Plant Cell Tiss Org Cult 61:543–551

Yang MZ, Jia SR, Pua EC (1991) High frequency plant regeneration from hypocotyl explants of *Brassica carinata* A. Br. Plant Cell Tiss Org Cult 24:79–82

Yermanos DM, Hemstreet S, Saleeb W, Huszar CK (1972) Oil content and composition of the seed in the world collection of sesame introductions. J Am Oil Chem Soc 49:20–23

Younghee K (2001) Effects of BA, NAA, 2, 4-D and $AgNO_3$ treatments on callus induction and shoot regeneration from hypocotyl and cotyledon of sesame (*Sesamum indicum* L.). J Kor Soc Hortic Sci 42:70–74

Yu K, Li R, Hatanaka T, Hildebrand D (2008) Cloning and functional analysis of two type 1 diacylglycerol acyltransferases from *Vernonia galamensis*. Phytochemistry 69:1119–1127

Yukawa Y, Takaiwa F, Shoji K, Masuda K, Yamada K (1996) Structure and expression of two seed-specific cDNA clones encoding stearoyl-acyl carrier protein desaturase from sesame, *Sesamum indicum* L. Plant Cell Physiol 37:201–205

Zhang YX, Sun J, Zhang XR, Wang LH, Che Z (2011) Analysis on genetic diversity and genetic basis of the main sesame cultivars released in China. Agric Sci China 10:509–518

Zheng ZF, Uchacz TM, Taylor JL (2001) Isolation and characterization of novel defence-related genes induced by copper, salicylic acid, methyl jasmonate, abscisic acid and pathogen infection in *Brassica carinata*. Mol Plant Pathol 2:159–169

Zulfiqar A, Paulose B, Chhikara S, Dhankher OP (2011) Identifying genes and gene networks involved in chromium metabolism and detoxification in *Crambe abyssinica*. Environ Pollut 159:3123–3128

Chapter 8
Prospects for Quinoa (*Chenopodium Quinoa* Willd.) Improvement Through Biotechnology

Eric N. Jellen, Peter J. Maughan, Daniel Bertero and Hassan Munir

Abstract Quinoa (*Chenopodium quinoa* Willd., $2n = 4x = 36$) is an Andean broadleaf seed and vegetable crop of ancient origin. Quinoa represents one or two botanical varieties of a much broader tri-species complex native to North and South America and dominated by weedy forms of pitseed goosefoot (*C. berlandieri* Moq.) and avian goosefoot (*C. hircinum* Schrad.). This biological species complex includes at least two extant domesticated forms of *C. berlandieri* subsp. *nuttaliae* (Safford) H.D. Wilson and Heister: Mexican huazontle and chia roja. Within quinoa itself the two main limitations to the crop's improvement and dissemination are restricted access to cultivated Highland Andean germplasm and heat-stress susceptibility in the best agronomic types from the southern Altiplano. These limitations underscore the importance of the exotic gene pool for future quinoa breeding. A sophisticated tool box of DNA-based genetic markers and genomic resources has been developed to facilitate gene transfer from exotic sources in pre-breeding and accelerate the process of breeding elite cultivars. In addition, quinoa physiology and agronomy research have identified promising strategies and potential gene targets for improving yield, heat tolerance, maturity, and other traits critical to the expansion of quinoa into temperate and subtropical lowland production environments. The existing political climate in key areas of the Andean region, while unfavorable for the application of transgenic breeding approaches, should encourage accelerated efforts to incorporate MAS strategies in quinoa breeding.

E. N. Jellen (✉) · P. J. Maughan
Department of Plant and Wildlife Sciences, Brigham Young University,
275 WIDB, Provo, UT 84602, USA
e-mail: jellen@byu.edu

D. Bertero
Departamento de Produccion Vegetal, Facultad de Agronomia, Universidad de Buenos Aires,
Avenue San Martin 4453, C1417DSE, Buenos Aires, Argentina

H. Munir
Department of Crop Physiology, University of Agriculture, Faisalabad 38040, Pakistan

S. M. Jain and S. Dutta Gupta (eds.), *Biotechnology of Neglected and Underutilized Crops*, DOI: 10.1007/978-94-007-5500-0_8,

Abbreviations

AFLP	Amplified fragment length polymorphism
BAC	Bacterial artificial chromosome
BAFI	Ezra T. Benson Agriculture and Food Institute
BVP	Basic vegetative phase
CIP	International Potato Center (Lima, Peru)
EST	Expressed sequence tag
FCB	Financial costbenefit ratio
FISH	Fluorescent in situ hybridization
GA	Gibberellic acid
GLAI	Green leaf area index
GM	Genetically modified
IGS	Intergenic spacer
IPAR	Incidental photosynthetically active radiation
LAI	Leaf area index
MAB	Marker assisted breeding
MAS	Marker assisted selection
NOR	Nucleolar organizer region
NTS	Non-transcribed spacer or 5S rRNA gene
PAR	Photosynthetically active radiation
PPS	Photoperiod sensitivity
RAPD	Random amplified polymorphic DNA
RFLP	Restriction fragment length polymorphism
RIE	Radiation interception efficiency
RIL	Recombinant inbred line
RUE	Radiation use efficiency
SNP	Single nucleotide polymorphism
SOS1	Salt Overly Sensitive 1 (gene)
SRA	Sequence read archive
SSR	Simple sequence repeat
TPAR	Transmitted photosynthetically active radiation

8.1 Introduction

Quinoa is a pseudocereal of Andean origin whose mounting popularity is due to its superb nutritional characteristics. These attributes include good seed protein quantity (7.5–22.1 %; Tapia et al. 1979); excellent protein quality with respect to the eight essential dietary amino acids (Repo-Carrasco et al. 2003); and oil content around 4.6 %, of which upwards of 80 % consists of unsaturated fatty acids (Repo-Carrasco et al. 2003). Since quinoa is produced as a dryland crop on alkaline soils surrounding salt flats (*salares*) on the arid southern Altiplano at elevations exceeding 3,500 m, quinoa is attractive to countries with similar environments that want to increase agricultural productivity while tapping into the growing international commodity market for quinoa seed. The existence of subtropical- and lowland temperate-adapted quinoa landraces, for example in the torrid mid-low elevation valleys of the Bolivian *Yungas* and the Mediterranean climate of the central Chilean coast, suggests that suitable quinoa production environments could be found across much of the globe. However, the development of high-yielding, agronomically desirable cultivars adapted to a wide range of environments will require combining agronomic characteristics of improved highland Andean germplasm with environmental adaptation traits of local varieties, semi-improved strains from similar environments, or even local weedy strains of quinoa's allied weed complex. Important prerequisites to the development of quinoa adaptation and selective breeding programs include the following: (1) elucidation of the genome constitution of the species; (2) delineation of the genetic resource pool for improving quinoa; and (3) investigation of the physio-agronomic characteristics of the crop.

8.1.1 Quinoa Genome Structure

Quinoa and its close allotetraploid New World relatives have a 2*n* chromosome number of 36, with a relatively small genome ranging in size from 724 to 967 Mbp (Stevens et al. 2006; Palomino et al. 2008). Recent research indicates that one of its two sub genomes bears similarity for dispersed repetitive sequences to a complex of phenotypically and ecologically diverse New World diploid species, while the other traces back to an Old World diploid that also contributed to the formation of Eurasian allohexaploid *C. album* (Kolano et al. 2011). Although Palomino et al. (2008) reported karyotypic evidence for two active pairs of nucleolar organizing region (NOR) -bearing chromosomes in pitseed goosefoot and quinoa, Kolano et al. (2008) found only one locus hybridized with NOR probe using fluorescent in situ hybridization (FISH). Furthermore, this remnant NOR locus in quinoa and two subspecies of pitseed goosefoot was associated with the Eurasian subgenome. Maughan et al. (2006) described minor DNA sequence variation between quinoa and pitseed goosefoot within the NOR intergenic spacer (IGS). Other DNA sequence variation has been described for the two subgenome-variant orthologs of the *SALT OVERLY SENSITIVE* 1 (*SOS*1) gene in quinoa (Maughan et al. 2009a). Extensive sequencing of introns 16 and 17 of *SOS*1 in a number of cultivated and weedy *Chenopodium* genotypes of various ploidy levels and genomic constitutions has confirmed the Eurasian and New World origins of the two constituent subgenomes of quinoa, pitseed goosefoot, and avian goosefoot (Jellen et al. unpublished).

8.1.2 Genome Relationships Based on Crossability

Cross-compatibility of the three New World tetraploid taxa (*C. berlandieri*, *C. hircinum*, *C. quinoa*) is evidence that they are closely related. In the first published study on hybridization among 4*x* New World cultigens, Heiser and Nelson (1974) reported obtaining quinoa x huazontle hybrids and that crosses between light-seeded parents produced dark-seeded F_1 progeny. Although Wilson and Heiser (1979) reported that quinoa X huazontle hybrids manifested sterility, Wilson's later 1980 study found that quinoa X pitseed goosefoot and quinoa X huazontle hybrids were fertile. It is interesting to note that $4x \times 6x$ crosses involving pitseed goosefoot and *C. bushianum* had 71–97 % pollen stainability, whereas in quinoa X *C. bushianum* hybrids the figure was 1–8 % (Wilson 1980). Since *C. bushianum* is a North American hexaploid carrying the two subgenomes of pitseed goosefoot plus an additional subgenome related to the Eurasian *C. album* group (unpublished results), the chromosomes of quinoa, therefore have diverged somewhat from those of their North American ancestor. Nevertheless, the cross-compatibility of quinoa and pitseed goosefoot was verified in a later study by Wilson and Manhart (1993), who reported isozyme-based evidence that an excess of 30 % of progeny of pitseed goosefoot plants free-living around cultivated quinoa fields in Washington were inter-taxa hybrids; these hybrids also manifested partial fertility and heterosis. At BYU we produced highly heterotic, partially self-fertile hybrids between *C. berlandieri* var. *macrocalycium* and highland quinoa cv. 'Ingapirca'. These hybrids had the black-pericarp trait from their wild parent and their main impediment to full seed set was heat-induced pollen mortality—apparently a dominant trait transmitted from their quinoa parent (Jellen unpublished). An F_2 population of 115 plants derived from this cross displayed mostly normal phenotypes with 19 evidencing hybrid breakdown at the seedling stage, including two tri-whorled seedlings, one with gray-green coloration, and dwarfed phenotypes (Fig. 8.1). We should also note that Wilson (1981), while reporting isozyme-based evidence for abundant gene flow among South American allotetraploid populations (their "subsection Cellulata complex"), failed to detect evidence for gene flow between lowland avian goosefoot and sympatric populations of exotic *C. album*. This confirms the notion that chromosome number differences present a barrier to gene exchange between 4*x* quinoa and 6*x* lambsquarters: an important consideration favoring the deployment of transgenic quinoa in areas of the world where *C. album* is a prominent and aggressive weed.

8.1.3 Tools for Quinoa Genome Analysis

Modern plant breeding programs in all major, and increasingly in many minor, crops are using molecular markers to facilitate and accelerate selection of superior genotypes. Marker-assisted selection (MAS) has progressed from low-resolution isozymes to technically challenging restriction fragment length polymorphisms (RFLPs) to a whole suite of markers based on polymerase chain reaction (PCR) assays. This latter group of markers generally require smaller amounts of template DNA than RFLPs, are less expensive, and are easier to assay in the laboratory; however, the most powerful, codominant, and high-throughput-capable PCR-based

Fig. 8.1 Segregating F_2 seedlings from the *C. berlandieri* var. *macrocalycium* (BYU 803) X quinoa cv. 'Ingapirca' cross. Note the appearance of mostly vigorous plants but some exhibiting stunted growth or hybrid breakdown. One seedling has a tri-whorled phenotype

markers—SSRs and SNPs—require up-front DNA sequence information. The adaptation of novel methods for DNA sequencing via genome reduction, first described by Maughan et al. (2009b) for *Amaranthus hypochondriacus* and lately applied successfully in *C. quinoa* (unpublished), has allowed for the identification of these types of DNA sequence-based markers in quinoa while circumventing the need for expensive whole-genome sequence data.

8.1.3.1 Molecular Markers

While quinoa is of increasing importance as an alternative crop species, only a few genomic studies have been reported in the literature for quinoa and its close relatives. Initial efforts were aimed at elucidating the genetic relationships among the quinoa ecotypes. Wilson (1988b) constructed the first phylogenetic tree of *Chenopodium* species based on isozyme data. The data supported the hypothesis that the Altiplano was the center of diversity (origin) for quinoa. Follow-up studies, conducted using seed protein variation and morphological markers (Wilson 1988a; Fairbanks et al. 1990), reconfirmed Wilson's initial conclusion. Fairbanks et al. (1993) were the first to use DNA markers in quinoa. They used random amplified polymorphic DNA (RAPD) markers to detect DNA polymorphisms among different quinoa accessions. Bonifacio (1995) also used RAPDs as a way to identify true hybrids from intergeneric crosses. Ruas et al. (1999) used RAPDs to investigate the relationship among 19 *Chenopodium* species, and found that accessions clustered according to their species classifications.

Maughan et al. (2004) produced the first genetic linkage map of quinoa. The map was based on 80 F_2 individuals from a cross between a Chilean lowland ecotype and an Altiplano ecotype. The map consisted of 230 amplified fragment length polymorphism (AFLP) markers, spanned 1,020 cM and consisted of 35 linkage groups. Mason et al. (2005) expanded the genetic resources for quinoa by developing the first set of simple sequence repeat or microsatellite (SSR) markers for use in quinoa. Through sequencing 1,276 clones from enriched CA, ATT, and ATG repeat DNA libraries they identified 208 were polymorphic SSRs on a panel of 31 diverse quinoa accessions. Christensen et al. (2007) used 35 of the these SSR markers to assess the level genetic diversity among 152 quinoa accessions representing the USDA and CIP-FAO collections and to test four hypotheses regarding quinoa's center of diversity, Highland and Lowland clustering patterns, origin of Lowland varieties and the origin of domestication. Jarvis et al. (2008) reported the development of an additional 216 new SSR markers and a new linkage map based on a recombinant inbred line (RIL) population. The new linkage map included the genomic location of two 11S seed storage protein loci, the nucleolar organizing region (NOR) and 275 molecular markers, including 200 SSR markers. Unfortunately the map is still incomplete as it consists of 38 linkage groups (LGs) spanning just 913 cM ($2n = 36$). Segregation distortion was observed in the mapping population for several marker loci, indicating possible chromosomal regions associated with selection or gametophytic lethality. Fuentes et al. (2009) utilized 20 of the newly developed SSR loci to develop 20 fluorescently multiplexed sets of markers, which he utilized to assess the genetic diversity patterns in Chilean quinoa germplasm. Vargas et al. (2011) screened 424 of these quinoa-based SSRs for amplification and polymorphism in cañahua (*C. pallidicaule* Aellen), a poorly studied, annual subsistence crop of the high Andes that is a close diploid relative of quinoa, and reported that 34 amplified and were clearly polymorphic within cañahua germplasm. The number of alleles per marker identified ranged from 2 to 8 with an average of 3.5 alleles per marker locus.

In addition to SSRs, single nucleotide polymorphism (SNP) markers have been developed in quinoa. The first report was from Coles et al. (2005), who utilized 424 expressed sequence tags (ESTs) to identify 51 SNPs from 20 EST sequences, of which 38 were single nucleotide changes and 13 were insertion/deletion changes. More recently, Maughan et al. (unpublished) used a genomic reduction based on restriction-site conservation and 454-pyrosequencing (Maughan et al. 2009b) to discover >5,000 SNPs from a pooled quinoa library of four mapping parents (Table 8.1). Approximately 500 of these SNPs have been validated and are being mapped using the KBioscience KASPar genotyping chemistry as detected on the Fluidigm integrated-fluidic chip (Fluidigm Corp., South San Francisco, CA). We note that use of the nano-fluidic technology for SNP genotyping significantly reduces the overall data point cost to <US $0.05/dtp—a cost per data point that should enable marker assisted selection in many orphaned crops. A single 96.96 Fluidigm integrated fluidic chip is capable of producing 9,216 genotypic data points in a single run (~3 h) with little technical expertise (Fig. 8.2).

Table 8.1 Description of the C. quinoa populations used for single nucleotide polymorphism discovery and the number of SNPs, and type of SNPs identified in each populations

					SNP type (%)					
Population	Mapping generation	Population size	SNP discovered	Unique contigs[a]	A/C	A/G	A/T	C/G	C/T	G/T
KU2 X 0654	F2:9	73	1,344	776	10.9	28.8	13.5	4.1	32	10.1
NL-6 X 0654	F2:9	72	1,872	1,043	11.5	29.2	12.2	5.2	30.8	10.8
NL-6 X Chucapaca	F2:9	86	1,005	588	11.8	26.3	12.2	5.4	30.1	13.6
Average:			1,407	802	11.4	28.1	12.6	4.9	31.0	11.5

Only contigs larger than 200 bp were included in the SNP discovery analysis. SNP were called only if coverage at the base was $\geq$6X, the frequency of the minor allele was >20 and 100 % of the alleles called within a parental line were identical

[a]Number of contigs with at least one SNP

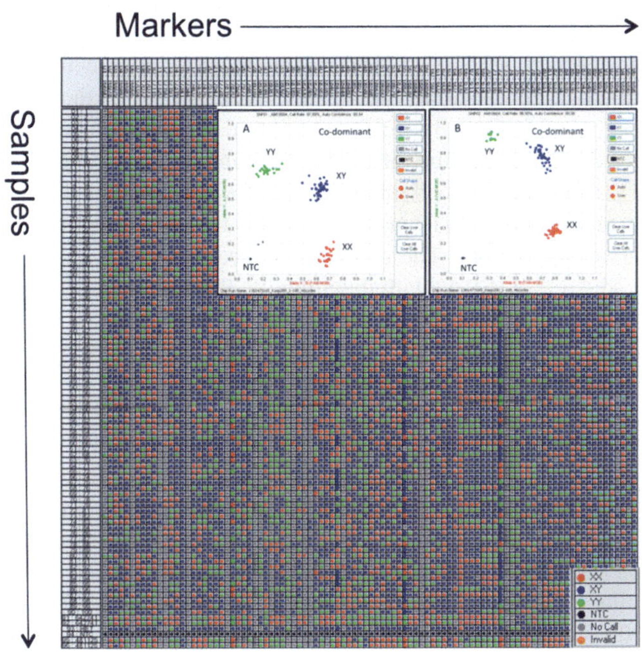

Fig. 8.2 Example of SNP assays using the KASPar^{TM} genotyping chemistry on the Fluidigm access array in the F_2 mapping population. *The background* shows the genotyping across the 96.96 IFC chip (96 DNA samples on the vertical, 96 SNP assays on the horizontal). *Inset panels A and B* show individual SNP loci. No template controls (*NTC*) are identified at the origin of each *Cartesian graph*

8.1.3.2 Expressed Sequence Tag Libraries

Expressed Sequence Tag (EST) sequences are partial sequences from transcribed cDNAs that reflect expressed genes in a given tissue type at a specific point of development. Made publically available, EST sequences facilitate gene discovery, genetic marker development, and homology searches with sequences from other organisms. Collections of these sequences can also provide researchers with a rapid and cost-effective tool to analyze transcriptome changes via DNA microarray analysis or RNAseq experiments. Coles et al. (2005) reported the development of an immature seed and floral expressed sequenced tag (EST) libraries. A total of 424 cDNA clones derived from immature seed and floral tissue were sequenced and analyzed for homology with known gene sequences. Three hundred and eleven of the clones were identified as homologues of known plant proteins while 75 did not share significant homology with any protein in the queried databases. More recently, Bajgain et al. (unpublished, personal communication) completed the first large scale sequencing (454-pyro sequencing and Sanger sequencing) and annotation of the quinoa transcriptome, including deep sequencing of eight development stages of the seed transcriptome, as well as the root, leaf and flower transcriptomes collected from plants grown under differing salinity concentrations. De novo assembly of the reads resulted in 63,151 contig consensus sequences with an average read length of 530 bp. Significant BLASTx alignments with the non-redundant protein database were identified 26,302 (42 %) of the consensus contig sequences ($\leq$1e-15). The Blast2GO suite, used to assign Gene Ontology (GO) annotations to the EST assembly, identified GO annotations for 22,249 (35.23 %) contigs. The contigs were placed into three categories, including 16,358 in the category biological process, 17,410 in the molecular function category, and 22,262 in cellular component category. The raw read sequences have been uploaded to the Sequence Read Archive (SRA) with the Study ID SRP007561.1.

8.1.3.3 Bacterial Artificial Chromosome Libraries

Another tool of importance for future genomic studies in quinoa is a bacterial artificial chromosome (BAC) library. BAC libraries are critical for identifying full-length genomic sequences, correlating genetic and physical maps, comparative genomics and are often vital first step towards whole genome sequencing projects. BAC Libraries have been developed for a number of plant species of major and minor economic/agricultural impact; examples include durum wheat (Cenci et al. 2003), coffee (Noir et al. 2004), amaranth (Maughan et al. 2008), and ginseng (Hong et al. 2004).

Stevens et al. (2006) reported the development of a BAC library for quinoa, which was estimated to represent approximately 9X coverage of the haploid quinoa genome. The library contains 26,880 clones from *BamH*I digests and 48,000 clones from *EcoR*I digests. The utility of the library was recently shown for gene discovery. Maughan et al. (2009a) used BAC clones to fully characterize two homoeologous *SOS*1 loci (*cq*SOS1A and *cq*SOS1B) from quinoa, including full-length cDNA sequences, complete genomic sequences, relative expression levels, FISH analysis, and a phylogenetic analysis of *SOS*1 genes from 13 plant taxa. The *SOS*1 gene encodes a plasma membrane Na+/H+ antiporter that play an important role in germination and growth of plants in saline environments.

Balzotti et al. (2008) reported the genomic and cDNA sequences for two 11S genes representing two orthologous loci from the quinoa genome. The nucleotide sequence of the entire gene and flanking regions were obtained directly from BAC clones using a primer walking strategy. Important features of the genes and the proteins they encode were described based on a comparison with homologous 11S sequences from other 49 other plant species. Gene expression and protein accumulation were determined via real time RT-PCR and SDS-PAGE analysis.

8.1.3.4 Cytological Markers

Repetitive sequences are useful for studying evolutionary relationships between species. They help to elucidate the origin of polyploids, allow for the following of chromosomal rearrangements through evolution, and provide insight into the mechanisms of polyploidy genome evolution. Kolano et al. (2008, 2011) described chromosomal organization of several novel repetitive DNA sequences isolated from the quinoa genome. Fluorescence in situ hybridization (FISH) analysis with one of the repetitive DNA clones (18–24 J) in quinoa and the closely related allotetraploid *C. berlandieri* Moq. evidenced hybridization signals that were mainly present on 18 chromosomes, suggesting that the repeat sequence is subgenome specific. A second analyzed sequence, 12–13P, localized exclusively in pericentromeric regions of each chromosome of quinoa, suggesting that the 12–13P sequence constitutes a major part of the repetitive centromeric DNA of *C. quinoa*.

Maughan et al. (2006) sequenced the NOR-IGS and 5S rDNA non-transcribed spacers (NTS) from five quinoa and one *C. berlandieri* accession. IGS sequences revealed length differences due to insertion/deletions (indels), differing numbers of repeat copies, and other rearrangements among the accessions. NTS sequencing revealed two sequence classes, likely representing one locus from each of the genomes in allotetraploid quinoa.

8.2 Quinoa Genetic Resources

Although domesticated quinoa is a highly polymorphic, variable crop in its own right, quinoa breeders face two substantial obstacles to expanding production into new areas: heat susceptibility and germplasm availability. The greatest numbers of superior agronomic types—those of the Altiplano—are susceptible to heat-induced pollen mortality when cultivated at lower elevations. This is presumably a function of domestication and selection in an environment above 3,000 m elevation where daytime temperatures rarely rise above 25–30 °C. Risi and Galwey (1984), citing Tapia et al. (1980), classified cultivated quinoa into five ecotypes: Valley (2,000–4,000 m elevation from central Peru northward); Altiplano (northern Altiplano surrounding Lake Titicaca); Salares (southern Altiplano); Sea level (Chilean coastal region, day length-insensitive); and Subtropical (from the Bolivian Yungas).

Within quinoa itself, approximately 130 accessions are publicly available through the USDA-ARS repository at Ames, Iowa, with an additional 30–40 in the CIP-FAO International Quinoa Nursery, though the vast majority of quinoa accessions in germplasm repositories are not publicly available. Yung as genotypes—which presumably would harbor heat tolerance—are notably absent from the USDA collection. Nonetheless, the USDA germplasm collection has been a tremendous resource for quinoa introduction and adaptation studies in countries like Pakistan and in efforts to validate sets of quinoa genetic markers (Mason et al. 2005; Christensen et al. 2007; Jarvis et al. 2008). Bonifacio (2003) reviewed initial work by the Bolivian government and the PROINPA Foundation to collect and preserve quinoa germplasm; these efforts have gained significant momentum, including assembly of a core collection, with the construction of an advanced germplasm repository at Quipaquipani on the northern Altiplano near Viacha, Bolivia. Ortiz et al. (1998, 1999) described a similar quinoa collection under the stewardship of the National University of the Altiplano in Puno, Peru. Other quinoa repositories are maintained by governmental agricultural entities and by several universities in the Andean region, most notably the UNALM (La Molina, Peru), UNSAAC (Cuzco, Peru), UNTA (Oruro, Bolivia), and UAP (Iquique, Peru).

Since the *nuttaliae* domesticates of Mexico, pitseed goosefoot, and avian goosefoot are cross-compatible with quinoa, they can be considered components of the primary gene pool for quinoa improvement. The problems of quinoa heat intolerance and restrictions to Andean germplasm access outlined above underscore the potential value of these other taxa, especially pitseed goosefoot. This taxon (*C. berlandieri* subsp. *berlandieri*) is found across a very wide range of mostly disturbed environments throughout North America. These environments include torrid lowland subtropics of the Gulf Coast (*C. berlandieri* var. *boscianum*); interior plains, temperate to subarctic forests, mountains and saline desert basins (*C. berlandieri* var. *zschackei*); arroyos of the Colorado Plateau and Sonoran Desert (*C. berlandieri* var. *sinuatum*); and forested shores of the Northeast Coast (*C. berlandieri* var. *macrocalycium*). The authors have also observed a potentially unique, brightly pigmented botanical variety as a weed in intercropped maize fields (milpas) in the Sierra Madre Oriental of Puebla State, Mexico. Throughout its range, pitseed goosefoot is often misidentified as Eurasian lambsquarters and vice versa, though the two species can usually be differentiated by the alveolate, adhering pericarp that lends a "honeycombed" appearance to the achenes of *C. berlandieri*, in contrast to the disadhering pericarp and shiny seeds of the utriculate *C. album* fruits. Like lambsquarters, pitseed goosefoot is a highly successful adventitious weed of human and natural disturbances, commonly encountered at construction sites, pastures, streambeds, field margins, and occasionally roadsides, throughout its native range as well as in other parts of the world. Avian goosefoot (*C. hircinum*) has slightly different plant morphology versus pitseed goosefoot, the primary differences being deeply lobed leaves and fetid odor in the former. Avian goosefoot is likewise encountered in disturbed, nitrogen-rich soils, though its range is more restricted to temperate and subtropical low lands of the southern cone of South America.

8.3 Quinoa Physiology and Agronomy: Targets for Crop Improvement

8.3.1 Physiological Targets for Quinoa Improvement

In spite of quinoa becoming a familiar food for an expanding number of consumers all over the world, scientific knowledge about the species is still fragmentary. This is most evident when it comes to knowledge about the physiological basis of yield determination and plant adaptation to environment. Because quinoa is able to grow in harsh environments, adaptation to low temperatures, water deficits and salinity have been the focus of several studies (Jensen et al. 2000; Jacobsen et al. 2005, 2007; Koyro and Eisa 2008; Jacobsen et al. 2009; Winkel et al. 2009; González et al. 2011; Ruffino et al. 2010; Hariadi et al. 2011; Orsini et al. 2011; Ruiz Carrasco et al. 2011). However, the environmental control of development and growth under abiotic stress-free environments has been less explored (i.e. Bertero et al. 1999a; Bois et al. 2006; Ruiz and Bertero 2008).

Investigations of yield determination, from a physiological standpoint, have been greatly facilitated since the adoption of methodologies using a top-down approach (Passioura 1981; Slafer and Savin 2006) and as researchers have increasingly recognized the existence of a critical period for yield determination—the crop cycle time window in which yield is most sensitive to environmental factors (Fischer 1985; Slafer and Savín 2006; Slafer et al. 2006). Using a top-down approach, crop yield is explained as the product of final crop aerial biomass and harvest index—the proportion of biomass partitioned to grains.

This approach was used to evaluate how quinoa yield is determined in a multi-environment trial involving a diverse set of 24 cultivars tested in 14 inter-tropical sites under irrigation and fertilization (Bertero et al. 2004). The relative contributions of genotype (G) effects versus genotype by environment interaction (GxE) effects to total yield variation were similar to those found in other crop adaptation studies, and emphasized the need for selection to be conducted within each environmental group (environments that are similar in the manner in which they discriminate among genotypes). Four environmental groups were distinguished: one from the cold Andean highlands; two from tropical valleys of moderate altitude (in Peru as well as Nairobi, Kenya); and one from warmer, low altitude environments (in Brasilia, Brazil and Gia loc, Vietnam). There were four genotypic groups, each having similar patterns of response across a range of environments: from Sea Level (of central Chile origin or ancestry); the Inter-Andean valleys; the Northern Highlands around Lake Titicaca between Bolivia and Peru; and the Southern Bolivian Highlands. These groups showed a strong correspondence to quinoa adaptation groups previously proposed by Tapia et al. (1979). The environmental groups not only affected genotypic rankings across environments, but also yield determinations, with potential implications for breeding strategies. Within the cold Andean environments, there was a positive association ($R^2 = 0.61$) between crop yield and total aerial biomass, but a non-significant one with harvest index.

Total biomass depended on both cycle duration (days from sowing to maturity) and crop growth rate ($gm^{-2}\ d^{-1}$), with $R^2 = 0.80$ and 0.92, respectively. The highest biomass was reached by genotypes from the valleys, but they suffered a yield penalty due to their long cycle, which forced grain filling to coincide with lower temperatures and decreased harvest index. Yield determination was very different in the low altitude environments of Brazil and Vietnam, with a positive association between yield and harvest index ($R^2 = 0.72$) but a barely significant one with biomass ($R^2 = 0.23$). Variation in biomass was not associated with crop duration, only with crop growth rate ($R^2 = 0.74$). Part of these differences between environments can be found in the range of variation for the different variables: from ~400 to 1,400 $g\ m^{-2}$ in biomass, 110–190 days in cycle duration and 3–8 $g\ m^{-2}\ d^{-1}$ in crop growth rate in the Highlands; while values for the low altitude environments ranged from ~600 to 1,000 $g\ m^{-2}$, 85–105 days and 7–11 $g\ m^{-2}\ d^{-1}$ for the same variables. Longer durations and reduced growth rates in the cold environments can be attributed to the lower temperatures, while the narrow range of variation in crop duration at lower altitudes can be interpreted as the consequence of crops experiencing higher temperatures under short photoperiods. These results also highlight the relevance of crop growth rate variation for biomass determination in both types of environment. The importance of harvest index in warmer environments points to variation in capacity to set grains under those conditions. There are reports of quinoa sterility due to high temperatures (Johnson and McCamant 1987), but sterility could also be associated with variation in cultivars' sensitivity to seed growth inhibition by high temperatures (Bertero et al. 1999b), which will be discussed later.

8.3.2 Environmental Control of Development and Genetic Variation in Responses

Variation in duration of development is a key aspect of crop adaptation to environment (Evans 1993) and also for quinoa, as mentioned in the previous section. Responses to environment have been studied from germination to seed filling in this crop. Jacobsen and Bach (1998) studied the influence of temperature on seed germination rate in a cultivar selected in Denmark from a Chilean (Sea Level-type) source population. They found a base temperature (T_b, threshold for temperature responses) of 3 °C with an optimum temperature (associated with the maximum rate of development) between 30 and 35 °C. Seed reached 100 % germination in 30°Cd^{-1} (thermal time units), meaning that under high temperatures and adequate humidity all seeds germinate in around one day.

Variation in germination responses to temperature was studied in 10 Bolivian cultivars of quinoa by Bois et al. (2006). They reported variation in T_b and time to 50 % germination, T_b ranged from 0.24 to−1.97 °C, several degrees lower than those reported by Jacobsen and Bach (1998). It is interesting that lower temperature thresholds appear to characterize cultivars originating from colder environments (the Bolivian Highlands compared with South Chile). Similar tendencies in T_b were observed for time to flowering when comparing cultivars from Highland and Sea Level origin (Bertero et al. 1999a). Variation between cultivars was more evident in seeds incubated at 2 °C, with times to 50 % germination ranging from 45 to 67 h. Quinoa can be sown as early as late winter-early spring in South Bolivia (Joffre and Acho 2008), so variation in capacity for crop establishment under low temperatures deserves attention.

Development from emergence to maturity is affected most by two environmental factors in quinoa: temperature and photoperiod. Experiments conducted under controlled environments using nine quinoa cultivars with origins from Colombia to Southern Chile demonstrated that all cultivars have a facultative short-day response for duration of emergence to flowering; furthermore, all phases of development—emergence to flower initiation, flower initiation to anthesis and anthesis to physiological maturity—are sensitive to photoperiod (Bertero et al. 1999a, 1999b). For time to visible floral buds, a non-destructive surrogate of floral initiation, two parameters of photoperiodic responses—the basic vegetative phase (BVP) and photoperiod sensitivity (PPS)—appeared to be more useful to explain differences between cultivars. The BVP is the minimal duration to flowering, found under short days in a short day plant, and PPS is the change in phase duration per unit change in photoperiod, expressed in thermal time units (°Cd) to account for temperature effects. Both parameters varied along a temperature gradient: the tropical cultivar 'Nariño' from Colombia exhibited the longest BVP duration and PPS values [700 Cd and 65°Cd h^{-1} (T_b = 1.5 °C)], while the lowest values were observed in the southern Chilean (Sea Level) cultivar 'Baer' [380 °Cd and 12 °Cd h^{-1} (T_b = 3.4 °C); Bertero 2003]. Early flowering was also observed in cultivars from the Peruvian and Bolivian highlands, as an adaptation to the short vegetative period experienced in those environments. The genetic basis of this variation is still unknown.

Another factor that appears to play an important role in the control of development, at least in varieties from the Bolivian Highlands, is water stress. Geerts et al. (2008a) reported a 30 days (from 65 to 95) delay in time to first anthesis with increases in water deficit, while a similar stress can accelerate maturity if occurring during seed development. This finding holds several implications. Prolonged dry spells can occur during the cropping season coinciding with flowering and seed filling in this environment (García et al. 2007). Flowering is very sensitive to stress (García 2003) but also is part of the critical period for yield determination (Bertero and Ruiz 2008; Mignone and Bertero 2008); therefore, postponing flowering could act as an escape mechanisms if that means exposing flowering to a more favorable water availability condition later in the season. If a positive association between actual yield and variation in flowering delay (if variation could be found) in the event of transient water deficit could be established, then selection for this trait could be relevant for this type of environment under rain-fed conditions. Perhaps the most determinant limitation for quinoa phenological adaptation to temperate environments is linked to photoperiod and temperature sensitivity during seed filling. There is a temperature x photoperiod interaction affecting seed filling, being strongly inhibited by the combination of long days and high temperatures (Bertero et al. 1999b). While some Andean cultivars can be cultivated and mature at high latitudes (Carmen 1984; Risi and Galwey 1991), being limited only by crop duration, seed set is strongly inhibited in mid latitudes when flowering occurs under long days and high temperatures. In the multi-environment quinoa trial, all temperate testing sites were excluded from the analysis because tropical adapted cultivars produced a high amount of vegetative biomass but failed in producing grain (Bertero et al. 2004). Interesting for adaptation to temperate environments is that this sensitivity appears not to occur in cultivars of Sea Level origin and in some from the Highlands. High temperatures also appear to explain the poor adaptation of varieties from the Chilean and Bolivian highland to altitudes of around 2,500 m in Colorado, USA, while they had a good performance at 2,800 m in other localities from the same state (Johnson and Mc Camant 1987).

Other developmental processes are those related to leaf appearance. The phyllochron (thermal time between the appearance of two successive leaves on the main stem) is affected by both temperature and photoperiod in quinoa, though temperature effects are more pronounced in relative terms (Bertero et al. 2000). Variation in the phyllochron exhibits a pattern similar to that of time to flowering, such that late-flowering plants are also the ones with the longest phyllochron (and hence lowest rate of leaf appearance), while the opposite is observed in accessions from the Highlands and South Chile. This indicates that in short-season environments, such as those from the southern Altiplano, plants flower in a shorter time but can partially compensate for this reduction in the time available to capture resources by producing a higher number of leaves per unit time than cultivars from warmer and more humid environments. An interesting finding is that the phyllochron is shorter (25 %) in new varieties selected in the Bolivian highlands compared with traditional landraces (Bois et al. 2006)—perhaps a consequence of selection for higher crop growth rate and biomass production.

The interaction between duration of developmental phases and leaf appearance depends on its temporal coordination. This was explored in a group of cultivars from Sea Level origin by Ruiz and Bertero (2008). Time to first anthesis and time to end of leaf appearance on the main stem have an almost one to one correlation, but a more detailed analysis showed that cv. NL-6, from Holland (Jacobsen 2003) is able to extend leaf appearance beyond the date of first anthesis, perhaps conferring an advantage to this cultivar in terms of radiation capture. As a rule of thumb, early to flower plants pay a yield potential penalty associated with the shorter time available to capture resources (aerial and subterranean), as the positive association between crop biomass and duration indicates (Bertero et al. 2004). Biomass growth is also a function of variation in crop growth rate, however, and faster leaf appearance could lead to faster leaf area development, higher radiation interception and growth. This suggests a quinoa improvement strategy based upon the design of cultivars that reach similar biomass values with shorter cycles or higher biomass with similar cycles (Padilla and Otegui 2005). Partial overlapping between leaf appearance and reproductive development is also an interesting option and, in fact, quinoa leaf area development and panicle growth are partially simultaneous (Ruiz and Bertero 2008).

8.3.3 Crop Carbon Balance and Yield Determination

For cultivars growing under optimal conditions (no water or nutrient limitation), radiation is the driving force of crop growth. Biomass accumulation depends on the amount of incident radiation intercepted by the crop canopy and on radiation use efficiency (RUE, g aerial biomass per unit of intercepted radiation, gm^{-2} MJ PAR^{-1}). Intercepted radiation during the crop cycle, besides crop duration and daily incident radiation, depends on radiation interception efficiency (RIE), which is determined by the Green Leaf Area Index (GLAI, the photosynthetically active leaf surface per unit soil surface) and the light attenuation coefficient (k). This coefficient relates transmitted and incidental PAR to LAI and is defined by the equation $k = -\ln$ (TPAR/IPAR)/LAI (Flenet et al. 1996).

Total plant leaf area exhibited higher values the later the time to flowering in a comparison of four Sea Level quinoa cultivars (Ruiz and Bertero 2008) but cv. 'NL-6' reached similar values than cv. 'RU-5' (from England) despite its shorter cycle because of faster leaf growth early in development. Early growth can be very slow in quinoa, particularly under low temperatures, affecting its capacity to compete with weeds and capture radiation. Slow initial growth can be an advantage in the dry highlands, where weed pressure is low and plants rely on soil water reserves during initial growth, but a higher early vigor is needed in more humid, fertile environments. Selection for early vigor was explored by breeding for wheat adaptation to water limited Mediterranean environments in Australia (Richards et al. 2002) conferring higher transpiration to evapotranspiration ratio and water use efficiency. In a comparison of 15 quinoa cultivars early vigor (measured as plant leaf area 10 days after emergence) exhibited a strong negative association with time from emergence to appearance of the first pair of leaves, and a positive one to plant biomass and leaf area 20 and 36 days after emergence (unpublished results).

Besides GLAI, radiation interception also depends on the light attenuation coefficient k. A single value of 0.59 described variation between cultivars, years and densities (Ruiz and Bertero 2008) similar to those of cereals like wheat and maize or legumes like soybean. Interestingly, k values increased with increasing density, perhaps because of a more even distribution of plants and hence of the foliage. Plants also need to reach high interception efficiencies early in development to maximize radiation capture and keep them at least until the end of the critical period for yield determination. Under high densities (50 plants m^{-2}) and for a spring sowing in Buenos Aires, Argentina, crops reached full radiation interception (RIE > 0. 95) 37 days after emergence (at the start of reproductive growth) and maintained it at least until day 60, covering the whole reproductive and flowering phase (Gómez et al. 2011); however, further research is necessary to determine whether these high k and RIE values can be reached under less optimal conditions.

Finally, variation in RUE could also play a role in biomass accumulation. RUE was clearly different between early stages of development and subsequent stages close to flowering in Sea Level quinoa cultivars (Ruiz and Bertero 2008), when a two-fold increase could be recorded (from 1.25 to 2.68 g MJ IPAR$^{-1)}$. RUE changes occurred in GLAI values from 0.61 to 1.38 and RIE values from 0.33 to 0.51 when leaf area both in main stem and branches was increasing and it was concluded that light distribution within the canopy is the major cause of observed changes. Variation between cultivars was detected, but differences could be explained in terms of the unequal weighting of cumulative intercepted radiation before and after a minimum GLAI was reached. RUE estimates during the second stage were within the typical range of C_3 crops and compare well with those of cereals like wheat (Abbate et al. 1997), or barley (Bingham et al. 2007) and also with that of sunflower for rapid growth phase prior to anthesis (Trapani et al. 1992) but the low initial values can be an important factor in limiting early growth. In another experiment (Gómez et al. 2011) no breakpoint in RUE was detected for a crop growing at high density and sown in spring and values were similar to that in the second stage (2.24 g MJ IPAR^{-1}), perhaps because fast growth and high densities caused plants to remain for a short time in the first RUE stage. Low RUE values could contribute to the low initial growth observed in quinoa, perhaps more in cold and dry environments where GLAI evolution is expected to be slow. No other reports of variation in RUE at the crop level in varieties from Andean origin are available.

The first analysis of variation in yield sensitivity to environment during the plant cycle was conducted by Garcia (2003) using plants of a Bolivian highland variety grown in pots and exposed to water stress treatments centered at different phenological phases. The highest reductions were found for stress applied between bud formation (start of reproductive growth) and the milky grain stage (average reduction 39 %, similar for stresses applied at the bud formation, flowering and milky grain stages period), while seed weight was only affected by treatments applied at the milky grain stage (38 %). Yield reduction was caused by changes in seed number for the first two sensitive stages and by seed weight reduction in the later. Maximum sensitivity to stress during flowering and early seed filling (milky grain stage) was later confirmed by Geerts et al. (2008b). In an analysis using Sea Level cultivars, seed number was found to be more sensitive to growth during the flowering phase (from first anthesis to the end of anthesis), less sensitive to growth during seed filling and not linked to growth during previous phases (Bertero and Ruiz 2008), based on the correlation between growth per phase and seed number in a range of cultivars, sowing dates and plant densities. Crop biomass increment during flowering explained variation in seed number ($R^2 = 0.71$) and this is the phase when most panicle growth (excluding seeds) occurs during the crop cycle; no difference between cultivars in partitioning to reproductive structures was detected. Finally, in an experiment in which shading was applied at different phenological stages using two cultivars, yield and seed number were reduced by treatments applied only after first anthesis (Mignone and Bertero 2008) with highest reduction in seed number (up to 50 %) and lower in seed weight (less than 20 %). All these works indicate that the critical period for yield determination starts late in quinoa, probably not earlier than first anthesis. Maximizing growth and reproductive partitioning during this period (perhaps not later than the milky grain stage) should be the focus of breeding efforts.

One avenue to increase yield potential was explored by the artificial manipulation of reproductive partitioning in quinoa. Average harvest indices are low in this crop (i.e. 0.30, Bertero et al. 2004) and similar to those of wheat and rice before the green revolution (Sakamoto and Matsuoka 2004). Given the previous history of crop breeding (wheat and rice, for instance), an increase in yield is expected to be achieved by affecting gibberellic acid (GA) metabolism, and thus manipulating plant height with the underlying hypothesis that a reduction in competition between stems and panicles for photo-assimilates will result in higher seed number and yield, as a consequence of increased reproductive partitioning. Using Paclobutrazol, a GA synthesis inhibitor applied at the start of reproductive growth, plant height decreased from 197 to 138 cm, yield increased from 517 in controls to 791 g m^{-2} in treated plants, seed number raised from 308 to 432 thousand seeds m^{-2}, and harvest index increased from 0.282 to 0.398 g g^{-1}. This manipulation did not reduce total aerial biomass or seed weight, and root biomass and lateral roots were increased (Gómez et al. 2011).

8.3.4 Other Targets

Other advances in the search for genetic variability in relation to abiotic stress responses have appeared in the literature in recent years. Ruiz-Carrasco et al. (2011) evaluated the mechanisms involved in salinity tolerance in four lowland quinoa genotypes and were able to relate variation in responses to salinity to the pattern of expression of salt-transporter genes related to ionic homeostasis. Expression of CqNHX1, a tonoplast-localized vacuolar Na^+/H^+ antiporter, was strongly up-regulated at 300 mM NaCl in comparison to control plants in shoots of salt-tolerant genotypes, but not in a sensitive genotype. However, CqSOS1 (a plasma membrane antiporter) was strongly induced in shoots of one tolerant genotype but remained unaltered in a salt-sensitive one. Physiological responses to salt stress were measured in *Chenopodium quinoa* by Morales et al. (2011) in a greenhouse experiment. High levels of trigonelline, a known osmoprotectant was found to accumulate in the high salt treatment suggesting a key role in salt tolerance of quinoa. The expression profiles of genes involved in salt stress (*SOS*1, *NHX*1, *TIP*2, *BADH* and *GAPDH*) showed constitutive expression in leaf tissue and up-regulation in root tissue in response to salt stress. These data suggest that quinoa tolerates salt through a combination of salt exclusion and accumulation mechanisms.

Variation in frost resistance was explored under controlled environments using two cultivars—one from the highlands and one from a temperate valley—with higher resistance in the one from the highlands (Jacobsen et al. 2005, 2007). Frost avoidance through supercooling capacity (a decrease in tissue temperature below the freezing point without the occurrence of extracellular freezing of water) linked to soluble sugar accumulation is the main mechanisms proposed to explain frost resistance and variation between cultivars and cold acclimatization conditions. As damage by frost is higher in late developmental stages, selection for earliness can also act as an escape mechanism (Jacobsen et al. 2005).

Concerning responses to water availability González et al. (2011) compared eight Bolivian and two Peruvian cultivars grown in the field in a dry valley of Northwest Argentina for gas exchange parameters and isotope composition. There was 160 mm of rain and 250 more mm were applied as irrigation during the crop cycle, thus representing a condition of moderate water deficit for the species (Geerts et al. 2009). Variation in carbon isotope discrimination, an estimator of long-term water use efficiency (WUE; Araus et al. 2002) was positively related to yield and negatively to intrinsic instant WUE (net CO_2 assimilation/leaf conductance ratio) measured during seed growth. Under the moderate deficits experienced, plants with higher conductance were able to reach high photosynthetic rates and had high discrimination values, while stomata closed early during the day in some cultivars having higher WUE but also lower productivity.

Carbon isotope discrimination is an easy-to-measure trait that can be manipulated by selective breeding (Condon et al. 2004). There are different associations between it and yield due to environment, however, so selection for high discrimination ability can be a valuable approach under transient or moderate deficits or in irrigated crops, while selection for higher WUE might be valuable for more limiting conditions, in particular when plants rely on stored water reserves (Tambussi et al. 2007). Consequently, it would be interesting to evaluate the association between carbon isotope discrimination, biomass accumulation, and yield of quinoa under more intense water deficits like those experienced in the Bolivian highlands under rain-fed conditions (Vacher 1998). Constitutive low osmotic potentials (but limited osmotic adjustment), partial stomatal closure that maintains a moderate photosynthesis but strongly reduces transpiration, low total weight/dry weight ratios, and small thick-walled cells that allow large losses of water without loss of turgor are some of the processes or anatomical characteristics conferring stress tolerance under severe water deficits in quinoa (Vacher 1998; Jensen et al. 2000; Jacobsen et al. 2009). However, genetic variation for these traits has not yet been explored in a range of cultivars from diverse environmental origins.

Genotypic variation for nitrogen fertilizer responses is an area that has recently begun to be explored by the authors (Hassan and Jellen unpublished). Under controlled greenhouse conditions in a preliminary experiment, growth and photosynthetic parameters in five quinoa genotypes were evaluated under five N regimes ranging from deficit to surfeit and genotypes were identified that had two distinct responses. One cultivar group exhibited a "normal" response, consisting of a progressive increase in growth characteristics with increasing N availability and extreme stress under N deficit. A second cultivar group exhibited a "low-N tolerance" response with superior growth under N deficit but reduced growth under excess N conditions. A segregating recombinant inbred line (RIL) population derived from the cross between "normal" and "low-N tolerance" parents has been genotyped using SNP markers (see Sect. 8.1.3.1) and is now being evaluated phenotypically to identify genomic regions controlling these responses to varying N regimes.

The successful establishment of a crop depends on the availability of high quality seeds. Two factors affecting seed quality, especially in humid environments, are the loss of seed viability during conservation and the risk of pre-harvest sprouting (vivipary). Quinoa is an orthodox species (tolerant to desiccation and able to survive for a long time under low humidity and temperature), but these conditions are not met when it is cultivated in mesic to humid temperate environments. Fortunately, variation in seed longevity was detected in a comparison of Sea Level and Highland cultivars, with slower rate of deterioration under conditions of accelerated aging in some varieties (one from Temuco and other from Chiloé, Chile) of Sea Level origin (Castellión 2008; Castellión et al. 2010). Pre-harvest sprouting occurs due to environmental conditions surrounding the maternal plant that favor germination, such as a prolonged period of high humidity or rain and moderate to warm temperatures. Risk of sprouting for quinoa is low in its traditional area of cultivation, where it matures after the end of the rainy season (García et al. 2007), but coincidence with rains around maturity is known to affect seed quality (Jacobsen and Bach 1998). Two sources of resistance to sprouting were recently characterized by Ceccato et al. (2011). Both accessions exhibited dormancy during seed development and maturation in the humid pampas of Argentina. One of those accessions, originating from Chiloe, Chile (and the same exhibiting a slower rate of seed deterioration under conservation) expressed a high level of dormancy within the temperature range from 17 to 24 °C and under 11.5–14.5 h photoperiods during seed development.

Munir (2011) found that germination percentage of quinoa seed declines precipitously after 5–6 months storage and therefore seed inviability can have a dramatic impact on seed yield. Seed stored at room temperature exhibited the same viability as seed stored under cold, dry, and/or refrigerated conditions. An extensive seed-priming program to enhance vigor and viability is also in progress with some preliminary successes (personal communication). Among the quinoa introductions Munir (2011) examined in the subtropical lowland Punjabi environment under winter cultivation conditions, PI 634919, PI 596293, Ames 13730, Ames 13737, and Danish Q-52 were the most reliable and stable in terms of their performance across all testing environments. Grain yield of these accessions ranged between 2 and 2.8 t ha^{-1} (Munir 2011).

One agronomic character of extreme interest to the authors is seed size. The genetic mapping population mentioned in mentioned in Sect. 8.1.2 above, derived from a cross between large-seeded weedy *C. berlandieri* var. *macrocalycium* and Ecuadorian *C. quinoa* cv. 'Ingapirca', is in process of advancement to study this trait. We expect this population to exhibit transgressive segregation for seed size, as well as other interesting seed morphology traits like testa width, coloration, seed dormancy and vivipary resistance, etc. (Jellen and Maughan unpublished).

8.3.5 Conclusions Regarding Quinoa Physiology

Most advances in physiological knowledge about quinoa were made in the last 15 years, and many in the last five. As seen in the examples given, genotypic variability for traits of interest has been little explored, even less their genetic basis or heritability. Some areas of extreme agronomic interest, like the control of seed size and responses to nutrient availability—an area where only empirical information on yield responses to nitrogen fertilization has been published (Jacobsen et al. 1994; Berti et al. 2000; Schulte auf'm Erley et al. 2005)—are currently being explored.

8.4 Agronomic Potential for Quinoa in New Environments

8.4.1 Overview of Quinoa Introduction Efforts in the Eastern Hemisphere

The introduction of new crops becomes a focal subject for the crop science community almost any time that a region of the world experiences a scarcity of food. New crop introduction, including acclimatization for better yields and economic benefits in low productivity systems or under severe abiotic or biotic stress conditions, remained a major strategy during the latter half of the previous century (Jacobsen 2003). Many crops from the Western Hemisphere were included in this crop introduction model. Sunflower, soybean, potato, maize and upland cotton are a few of the native crops from the Americas that have been looked to as sources of food, feed and fiber security to human populations in the Eastern Hemisphere (Rauf et al. 2010). Quinoa has been one of the "minor" crops that fit into this crop introduction scenario (Bhargava et al. 2006).

Quinoa introduction into the Eastern Hemisphere can be traced back to the late 1950s (Tapia et al. 1979; Jacobsen 2003). England was the first site of its introduction in the post-Hispanic era and quinoa's outstanding nutritional value was the primary motive for initial cultivation efforts in the United Kingdom. Later, quinoa's adaptability and tolerance to extreme climatic stresses, especially its salinity tolerance, increased the appeal of the crop to European agronomists. Denmark was the second site of European quinoa introduction during 1986–1987 and quinoa breeding research—based primarily on reselection from heterogeneous populations—led to the development in the later years of new quinoa varieties, both in England at Cambridge University and in Denmark at Royal Agriculture University. Almost simultaneously, a team of scientists started work in Holland on quinoa variety development (Limburg and Mastebroek 1996). By 1993, quinoa was widely recognized for its quality, drought tolerance and nutrition, though it was considered to be a relatively unimproved crop (Jacobsen and Denis-Ramirez 1986; Risi and Galwey 1989; Galwey 1989; Jacobsen and Stølen 1993). With the release of the first variety in Holland in 1995, the breeding work in Europe began to focus more on obtaining higher yields and superior plant "ideotypes" (Limburg and Mastebroek 1996; Mastebroek and Limburg 1996; Mastebroek and Marvin 1997). Quinoa adaptation research also began to focus more on phenology and plant development under the new climatic conditions (Jacobsen and Stølen 1993).

Quinoa's successful introduction in different regions of Northern Europe inspired the United Nations Food and Agriculture Organization (FAO) and non-governmental organizations like the Ezra T. Benson Food and Agriculture Institute

(BAFI) to perform a set of comprehensive, worldwide assessments of quinoa using publicly available germplasm. Vietnam, Kenya, Egypt, Turkey, Morocco, Greece, Austria, Poland, Czech Republic, Sweden, Norway, Italy, Spain, Finland, England, Canada, and Australia were included among the group of test countries representing diverse terrestrial environments for assessing the crop's adaptability (Mujica et al. 2001; Jellen et al. 2005; Iliadis et al. 1997, 2001; Keskitalo 1997; Ohlsson 1997). Furthermore, the cultivation of quinoa was successfully attempted in the Indian Subcontinent nations of Pakistan and India (Bhargava et al. 2007; Munir and Basra 2010; Munir 2011). However, one serious problem reported by Munir et al. (2011) was with viability of exotic quinoa seed lots, due either to improper handling during transport or prolonged postharvest storage under high temperatures and/or humidity (Bertero 2003; Betero and Ruiz 2010).

The demonstration of quinoa's tolerance to drought and salinity stresses was the next success, in the first decade of the 21st Century. Drought tolerance was assessed in a variety of ways, not only to understand the basic mechanism behind it, but also to model this crop for dry, arid regions and to search out seed and foliar remedies for drought mitigation (Jensen et al. 2000; Jacobsen 2003; Jacobsen et al. 2009; Kamman et al. 2011). The long-term commercial production of large-seeded Royal Quinoa varieties ('Achachino', 'Hilo', 'Kellu', 'Lipeña', 'Ollague', 'Sayaña', 'Toledo', 'Uyuni', etc.) around the shores of the high salt flats (*salares*) of Uyuni and Lake Titicaca motivated investigations of salinity tolerance. As mentioned above in Sect. 1.1.3.3, characterization of the *SOS*1 gene, which encodes a plasma membrane Na^+/H^+ antiporter protein, was one of the first steps toward understanding the molecular basis of quinoa's salinity tolerance (Maughan et al. 2009a). Hence, quinoa can be viewed as a potential gene source for inducing salinity tolerance in other crops (Prado et al. 2000; Jacobsen 2003; Hariadi et al. 2011; Maughan et al. 2009b).

Yields of quinoa grain vary between 650 and 2,200 kg ha^{-1} in Bolivia and 560–1,780 kg ha^{-1} in Peru, the two principal quinoa-producing countries (FAOSTAT 2008). In new regions of quinoa introduction, the potential is altogether astonishing. Kenya in Africa reportedly had the highest grain yields (4,000 kg ha^{-1}), followed by Greece (3,960 kg ha^{-1}) and Italy (2,280 kg ha^{-1}; Mujica et al. 2001). In Denmark, the grain yield varies between 1.5 and 2 t ha^{-1}, while in Vietnam the average yield was reportedly between 1,125 and 1,685 kg ha^{-1}. In Pakistan, a maximum of 2,800 kg ha^{-1} grain yield was obtained, which if compared with the local staple yields approximately half the comparative financial cost-benefit ratio (FCB)—proving that quinoa could be a future cash crop of the country (unpublished).

8.4.2 Quinoa Introduction Efforts in Pakistan

Pakistan is a developing country with an agrarian economy. Agriculture contributes 20.6 % to its GDP and the vast majority of that is derived from only a few crops (ESP 2009). The country contains a wide variety of ecological zones, including snowcapped mountains in the north, coastal deserts in the south along the Arabian

Sea, and the vast irrigated plains of Punjab and Sindh. The country's disproportionate dependence on maize, cotton, sugarcane, and wheat crops has increased the risk factor, such that failure of only one of these crops is sufficient to shatter country's economy. Moreover, the flow of the five major rivers of the province is at risk because of an ongoing conflict in their headwaters with neighboring India. One-third of the total arable land is salt-affected. Half of the population is food insecure and 57 % of children under the age of 14 are malnourished (ESP 2009).

Crop diversity is one potential solution for minimizing the risk of catastrophic failure of one of Pakistan's major crop. Quinoa is a climate-resilient crop having the potential to resist sudden climatic changes in addition to biotic stresses; therefore, it has become a subject for introduction in Pakistan (Jacobsen 2003; Munir 2011). During the winter of 2008, a total of 25 accessions were obtained from the USDA Agricultural Research Station at Ames, Iowa. Seed lots were examined for vigor and viability and ten accessions with more than 90 % germination and better vigor status were selected for field trials. Research stations at Faisalabad, Chakwal and Bahawalpur were selected for this purpose and accessions were compared on the basis of grain yield. A number of agronomic and growth parameters such as leaf area, leaf number, panicle number, 1,000 grain count, number of tillers, main panicle weight, sub panicles' weight and number, grain yield, biological or forage yield and viability of produced seed were recorded. Biochemical parameters were also measured, including leaf chlorophyll *a* and *b* content, carotenoids, total phenolic content, protein content, enzymatic activity of catalase, peroxidase and superoxide dismutase, along with Na and K content. Based on preliminary success at all test sites, refinement of field production methods was initiated by planting all ten accessions at four dates (September 15, October 15, November 15, and December 15) at a single station in Faisalabad during three consecutive winter seasons. The mid-November and mid-December plantings performed significantly better than the others and grain obtained was of better size and higher germination rate when compared on yearly basis.

In addition, quinoa accessions were tested in order to define optimum stand establishment parameters. Direct field-sown quinoa performed significantly better than quinoa that was transplanted from polythene nursery bags. Optimization of nitrogen doses for quinoa is also in its third year and determination of its irrigation requirements is moving forward. Since 2008, there have been no serious pest or disease infestations.

Seed multiplication is also among the priority areas and the fact that sufficient seed has been obtained over the course of three years of extensive experimentation is encouraging for the formation of quinoa breeding projects. A total of three sets of quinoa varietal plantings are in progress in the field and mutation breeding has been initiated (Munir unpublished).

8.5 Political Scenario of Quinoa Biotechnology

Quinoa is a sacred crop, the "mother grain", among Aymara- and Quechua-speaking natives of the Andes. It should therefore not be surprising that there is tremendous opposition to, as well as ignorance about, genetic crop modification among these peoples. This situation is reminiscent of indigenous Mexican opposition to "contamination" of their sacred native landraces by genetically modified maize. Ironically, both quinoa and maize owe their existence and status as food crops to the ingenious genetic modification efforts—via selection and hybridization—of the ancestors of these New World natives. Nevertheless, despite the opposition of large segments of their populations to transgenic crops, the Andean governments have taken a favorable approach to biotechnology within the framework of marker-assisted breeding (MAB). These are pragmatic responses to increasing consumer demand, both domestically and internationally, and concomitant increases in the price of quinoa. The other consideration regarding transgenic quinoa is that a considerable portion of the current market demand is organic.

8.6 Conclusions

Quinoa is a remarkable crop in many respects, and worldwide recognition of its excellent nutritional qualities has resulted in a dramatic increase in consumer demand for this commodity. Unfortunately, supply coming from the Andean countries has lagged behind demand, threatening a future of chronic shortages and skyrocketing quinoa prices in the marketplace. Consequently, we should expect to see increasing efforts to expand quinoa production into new areas, as we have observed in Pakistan. Although genetically modified quinoa may not be the short- or medium-term solution to increase production in the "GM-phobic" and germplasm exchange-restricted Andean countries, non-transgenic biotechnological solutions such as MAS and MAB—particularly using molecular markers and abundant germplasm resources publicly available in North America—are tremendously promising avenues for future quinoa breeding research. The physiological considerations outlined in this chapter are important guidelines for this effort, since the variation and phenotypic plasticity of quinoa and its close relatives simultaneously represent opportunities and obstacles to geographic expansion of this crop.

Successful quinoa introduction efforts under a broad range of conditions in Pakistan provide encouragement for similar efforts to expand quinoa production in other at-risk areas of the Developing World. However, while quinoa has the potential to yield enormously on marginal lands irrigated with brackish water, the market—as well as subsistence farm families—will have little tolerance for a crop that is susceptible to periodic episodes of complete yield loss from unseasonal heat waves, or to germination failure due to improper harvesting and postharvest storage. Based on historical observations in other crops, we should also assume that pests and diseases will adapt to quinoa as the crop becomes established in new areas of production. The best way to anticipate these future catastrophes is to ensure that quinoa breeding programs are established in these new production regions; that quinoa breeders found their programs upon broad germplasm bases; and that they utilize, whenever possible, the powerful tools of genomics and biotechnology to streamline the quinoa variety development pipeline.

References

Abbate PE, Andrade FH, Culot JP et al (1997) Grain yield in wheat: effects of radiation during spike growth period. Field Crops Res 54:245–257

Araus JL, Slafer GA, Reynolds MP, Royo C (2002) Plant breeding and drought in C3 cereals: what should we breed for? Ann Bot 89:925–940

Balzotti MRB, Thornton JN, Maughan PJ et al (2008) Expression and evolutionary relationships of the *Chenopodium quinoa* 11s seed storage protein gene. Int J Pl Sci 169:281–291

Bertero HD (2003) Response of developmental processes to temperature and photoperiod in quinoa (*Chenopodium quinoa* W.). Food Rev Int 19:87–97

Bertero HD, Ruiz RA (2008) Determination of seed number in sea level Quinoa (*Chenopodium quinoa* Willd.) cultivars. Eur J Agron 28:186–194

Bertero HD, Ruiz RA (2010) Reproductive partitioning in sea level quinoa (*Chenopodium quinoa* Willd.) cultivars. Field Crops Res 118:94–101

Bertero HD, King RW, Hall AJ (1999a) Modelling photoperiod and temperature responses of flowering in quinoa (*Chenopodium quinoa* W.). Field Crops Res 63:19–34

Bertero HD, King RW, Hall AJ (1999b) Photoperiod-sensitive phases of development in quinoa (*Chenopodium quinoa* W.). Field Crops Res 60:231–243

Bertero HD, King RW, Hall AJ (2000) Photoperiod and temperature effects on the rate of leaf appearance in quinoa (*Chenopodium quinoa* W.). Austral J Pl Physiol 27:349–356

Bertero HD, de la Vega AJ, Correa G et al (2004) Genotype and genotype-by-environment interaction effects for grain yield and grain size of quinoa (*Chenopodium quinoa* Willd.) as revealed by pattern analysis of international multi-environment trials. Field Crops Res 89:299–318

Berti M, Wilckens R, Hevia F et al (2000) Fertilización nitrogenada en quinoa (*Chenopodium quinoa* Willd.). Ciencia Invest Agraria 27:81–90

Bhargava A, Shukla S, Ohri D (2006) Karyotypic studies on some cultivated and wild species of *Chenopodium* (Chenopodeaceae). Genet Res Crop Evol 53:1309–1320

Bhargava A, Shukla S, Ohri D (2007) Genetic variability and interrelationships among various morphological and quality traits in quinoa (*Chenopodium quinoa* Willd.). Field Crops Res 101:104–116

Bingham IJ, Blake J, Foulkes MJ, Spink J (2007) Is barley yield in the UK sink limited? I. Post-anthesis radiation interception, radiation-use efficiency and source–sink balance. Field Crops Res 101:198–211

Bois JF, Winkel T, Lhomme JP et al (2006) Response of some Andean cultivars of quinoa (*Chenopodium quinoa* Willd.) to temperature: effects on germination, phenology, growth and freezing. Eur J Agron 25:299–308

Bonifacio A (1995) Interspecific and intergeneric hybridization in chenopod species.MS thesis. Brigham Young University, USA

Bonifacio A (2003) *Chenopodium* sp.: genetic resources, ethnobotany, and geographic distribution. Food Rev Int 19:1–7

Carmen ML (1984) Acclimatization of quinoa (*Chenopodium quinoa* Willd.) and canihua (*Chenopodium pallidicaule* Aellen) to Finland. Ann Agric Fenn 23:135–144

Castellión ML (2008) Procesos de deterioro y mecanismos de protección y reparación involucrados en la pérdida diferencial de la viabilidad durante el almacenamiento en semillas de *Chenopodium quinoa* Willd. Ph D thesis. University of Buenos Aires, Argentina

Castellión M, Matiacevich S, Buera P, Maldonado S (2010) Protein deterioration and longevity of quinoa seeds during long-term storage. Food Chem 121:952–958

Ceccato D, Bertero HD, Batilla D (2011) Environmental control of dormancy in quinoa (*Chenopodium quinoa* Willd.). Two potential sources of pre-harvest sprouting tolerance. Seed Sci Res 21:133–141

Cenci A, Chantret N, Kong X et al (2003) Construction and characterization of a half million clone BAC library of durum wheat (*Triticum turgidum* ssp *durum*). Theor Appl Genet 107:931–939

Christensen SA, Pratt DB, Pratt C et al (2007) Assessment of genetic diversity in the USDA and CIP-FAO international nursery collections of quinoa (*Chenopodium quinoa* Willd.) using microsatellite markers. Plant Genet Resources: Charact Util 5:82–95

Coles ND, Coleman CE, Christensen SA et al (2005) Development and use of an expressed sequenced tag library in quinoa (*Chenopodium quinoa* Willd.) for the discovery of single nucleotide polymorphisms. Plant Sci 168:439–447

Condon AG, Richards RA, Rebetzke GJ, Farquhar GD (2004) Breeding for high water-use efficiency. J Exp Bot 55:2447–2460

ESP (2009) Economic survey of Pakistan. Minister of Finance, Government of Pakistan, 2:17–19

Evans LT (1993) Crop evolution, adaptation and yield. Cambridge University, Cambridge

Fairbanks DJ, Burgener KW, Robison LR et al (1990) Electrophoretic characterization of quinoa seed proteins. Plant Breed 104:190–195

Fairbanks DJ, Waldrigues A, Ruas CF et al (1993) Efficient characterization of biological diversity using field DNA extraction and random amplified polymorphic DNA markers. Rev Brasil Genet 16:11–22

FAOSTAT(2008)http://www.fao.org/economic/ess/publications-studies/statistical-yearbook/fao-statistical-yearbook-2007-2008/g-human-welfare/en/. Accessed 5 Jan 2012

Fischer RA (1985) Number of kernels in wheat crops and the influence of solar radiation and temperature. J Agric Sci 105:447–461

Flenet F, Kiniry J, Board J, Westgate M, Reicosky DC (1996) Row spacing effects on light extinction coefficients of corn, sorghum, soybean and sunflower. Agron J 88:185–190

Fuentes FF, Martinez EA, Hinrichsen PV et al (2009) Assessment of genetic diversity patterns in Chilean quinoa (*Chenopodium quinoa*Willd.) germplasm using multiplex fluorescent microsatellite markers. Conserv Genet 10:369–377

Galwey NW (1989) Quinoa. Biologist 36:267–274

García M (2003) Agroclimatic study and drought resistance analysis of quinoa for an irrigation strategy in the Bolivian Altiplano. Ph D thesis. Katholieke Universiteit Leuven, Belgium

García M, Raes D, Jacobsen SE, Michel T (2007) Agroclimatic constraints to rainfed agriculture in the Bolivian Altiplano. J Arid Environ 71:109–121

Geerts S, Raes D, Garcia M et al (2008a) Indicators to quantify the flexible phenology of quinoa (*Chenopodium quinoa* Willd.) in response to drought stress. Field Crops Res 108:150–156

Geerts S, Raes D, Garcia M et al (2008b) Introducing deficit irrigation to stabilize yields of quinoa (*Chenopodium quinoa* Willd.). Eur J Agron 28:427–436

Geerts S, Raes D, Garcia M et al (2009) Modeling the potential for closing quinoa yield gaps under varying water availability in the Bolivian Altiplano. Agric Water Man 96:1652–1658

Gómez MB, Aguirre Castro P, Mignone C, Bertero HD (2011) Can yield potential be increased by manipulation of reproductive partitioning in quinoa? Evidence from giberellic acid synthesis inhibition using paclobutrazol. Funct Plant Bio 38:420–430

González JA, Bruno M, Valoy M, Prado FE (2011) Genotypic variation of gas exchange parameters and leaf stable carbon and nitrogen isotopes in ten quinoa cultivars grown under drought. J Agron Crop Sci 197:81–93

Hariadi Y, Marandon K, Tian Y et al (2011) Ionic and osmotic relations in quinoa (*Chenopodium quinoa* Willd.) plants grown at various salinity levels. J Exp Bot 62:185–193

Heiser CB, Nelson C (1974) On the origin of the cultivated chenopods. Genet 78:503–505

Hong CP, Lee SJ, Park JY et al (2004) Construction of a BAC library of Korean ginseng and initial analysis of BAC-end sequences. Mol Genet Genomics 271:709–716

Iliadis C, Karyotis T, Mitsibonas T (1997) Research on quinoa (*Chenopodium quinoa* Willd.) and amaranth (*Amaranthus caudatus*) in Greece. In: Proceedings of COST workshop 24–25 Oct 1997, CPRO-DLO, Wageningen, pp. 85–91

Iliadis C, Karyotis T, Jacobsen SE (2001) Adaptation of quinoa under xerothermic conditions and cultivation for biomass and fibre production. In: Jacobsen SE, Portillo Z (eds) International potato centre (CIP) memorias, Primer Taller Internacional sobre Quinua—Recursos Geneticos y Sistemas de Produccion, UNALM, Lima, Peru, pp 371–378, 10–14 May 1999

Jacobsen SE (2003) The worldwide potential of quinoa (*Chenopodium quinoa* Willd.). Food Rev Int 19:167–177
Jacobsen SE, Bach AP (1998) The influence of temperature on seed germination rate in quinoa (*Chenopodium quinoa* Willd.). Seed Sci Tech 26:515–523
Jacobsen SE, Denis-Ramirez AR (1986) Quinoa-en sydamerikansk proteinafgrode. Agrologisk Tidsskrift Marken 12:23–25
Jacobsen SE, Stølen O (1993) Quinoa—morphology and phenology and prospects for its production as a new crop in Europe. Eur J Agron 2:19–29
Jacobsen SE, Liu F, Jensen CR (2009) Does root-sourced ABA play a role for regulation of stomata under drought in quinoa (*Chenopodium quinoa* Willd.)? Sci Hort 122:281–287
Jacobsen SE, Jorgensen I, Stølen O (1994) Cultivation of quinoa (*Chenopodium quinoa*) under temperate climatic conditions in Denmark. J Agric Sci 122:47–52
Jacobsen SE, Monteros C, Christiansen JL et al (2005) Plant responses of quinoa (*Chenopodium quinoa* Willd.) to frost at various phenological stages. Eur J Agron 22:131–139
Jacobsen SE, Monteros C, Corcuera et al (2007) Frost resistance mechanisms in quinoa (*Chenopodium quinoa* Willd.). Eur J Agron 26:471–475
Jarvis DE, Kopp OR, Jellen EN et al (2008) Simple sequence repeat marker development and genetic mapping in quinoa (*Chenopodium quinoa* Willd.). J Genet 87:39–51
Jellen EN, Benlhabib O, Maughan PJ et al (2005) Introduction of the Andean crop quinoa in Morocco. Am Soc Agron Proceed. 6–9 Nov 2005, Salt Lake City, UT., USA, P7560
Jensen CR, Jacobsen SE, Andersen MN et al (2000) Leaf gas exchange and water relation characteristics of field quinoa (*Chenopodium quinoa* Willd.) during soil drying. EurJ Agron 13:11–25
Joffre R, Acho J (2008) Quinua, descanso y tholars en el sur de Altiplano Boliviano. Rev Habitat 75:38–43
Johnson D, McCamant J (1987) Quinoa research and development 1987. Sierra Blanca Associates, Denver p 7
Kamman CI, Linsel S, Gobling JW, Koyro HW (2011) Influence of biochar on drought tolerance of *Chenopodium quinoa* Willd. and on soil-plant relations. Pl Soil. doi:10.1007/s1110401107715
Keskitalo M (1997) Quinoa (Chenopodiumquinoa Willd.). In: Proceedings of COST workshop, 24–25 Oct 1997, CPRO-DLO, Wageningen, The Netherlands
Kolano B, Plucienniczak A, Kwasniewski M, Maluszynska J (2008) Chromosomal localization of a novel repetitive sequence in the *Chenopodium quinoa* genome. J Appl Genet 49:313–320
Kolano B, Gardunia BW, Michalska M et al (2011) Chromosomal localization of two novel repetitive sequences isolated from the *Chenopodium quinoa* Willd. Genome. Genome 54:710–717
Koyro HW, Eisa SS (2008) Effect of salinity on composition, viability and germination of seed of *Chenopodium quinoa* Willd. Pl Soil 302:79–90
Limburg H, Mastebroek HD (1996) Breeding high yielding lines of *Chenopodium quinoa* Willd. with saponinfree seed. In: Proceedings of COST workshop, 22–24 Feb 1996, Eur Comm EUR 17473/KVL, Copenhagen, Denmark
Mason SL, Stevens MR, Jellen EN et al (2005) Development and use of microsatellite markers for germplasm characterization in quinoa (*Chenopodium quinoa* Willd.). Crop Sci 45:1618–1630
Mastebroek HD, Limburg H (1996) Breeding for harvest security of *Chenopodium quinoa* Willd. In: Proceedings of COST workshop, 22–24 Feb, 1996, Eur Comm EUR 17473/KVL, Copenhagen, Denmark
Mastebroek HD, Marvin HP (1997) Content of sapogenins in leaves and seeds of quinoa (*Chenopodium quinoa* Willd.). In: Proceedings of COST workshop, 24–25 Oct 1997, Wageningen, The Netherlands
Maughan PJ, Bonifacio A, Jellen EN et al (2004) A genetic linkage map of quinoa (*Chenopodium quinoa*) based on AFLP, RAPD, and SSR markers. Theor Appl Genet 109:1188–1195
Maughan PJ, Kolano BA, Maluszynska J et al (2006) Molecular and cytological characterization of ribosomal RNA genes in *Chenopodium quinoa* and *C. berlandieri*. Genome 49:825–839

Maughan PJ, Sisneros N, Luo MZ et al (2008) Construction of an *Amaranthus hypochondriacus* bacterial artificial chromosome library and genomic sequencing of herbicide target genes. Crop Sci 48:S85–S94

Maughan PJ, Turner TB, Coleman CE (2009a) Characterization of salt overly sensitive 1 (SOS1) gene homoeologs in quinoa (*Chenopodium quinoa* willd.). Genome 52:647–657

Maughan PJ, Yourstone SM, Jellen EN, Udall JA (2009b) SNP discovery via genomic reduction, barcoding and 454-pyrosequencing in amaranth. Plant Genome 2:260–270

Mignone C, Bertero D (2008) Relación entre la tasa de crecimiento, el número y peso de granos en el cultivo de quínoa (*Chenopodium quínoa*). XIII Latin American/XXVII Argentinean Pl physiol meeting, Rosario, 21–24 Sep 2008

Morales AJ, Bajgain P, Garver Z, Maughan PJ, Udall JA (2011) Evaluation of the physiological responses of *Chenopodium quinoa* to salt stress. Int J Plant Physiol Biochem 3:219–232

Mujica A, Jacobsen SE, Izquierdo J, Marathee JP (2001) Results of American and European tests of quinoa. FAO, UNA-Puno, CIP, Peru

Munir H (2011) Introduction and assessment of quinoa (*Chenopodium quinoa* Willd.) as a potential climate proof grain crop. Ph D Thesis, University of Agriculture, Faisalabad

Munir H, Basra SMA (2010) Improved germination and seedling vigor enhancement of quinoa (*Chenopodium quinoa* Willd.) grown at different salinity levels. In: Qadir D, Wichelns J, Oster J et al (ed) Sustainable management of saline waters and salt affected soils for agriculture, 2nd Bridging Workshop, ICARDA & IWMI, Aleppo, Syria and Colombo, Sri Lanka, 15–18 Nov 2009

Munir H, Basra SMA, Cheema MA, Wahid A (2011) Phenotypic flexibility in exotic quinoa (*Chenopodium quinoa* Willd.) germplasm for seedling vigor and viability. Pak J Agric Sci 48:255–261

Noir S, Patheyron S, Combes MC, Lashermes P, Chalhoub B (2004) Construction and characterization of a BAC library for genome analysis of the allotetraploid coffee species (*Coffea arabica* l.). Theor Appl Genet 109:225–230

Ohlsson I (1997) Quinoa-a potential crop for Sweden? In: Proceedings of COST Workshop, CPRO-DLO, Wageningen, The Netherlands, 24–25 Oct 1997

Orsini F, Accorsi M, Gianquinto G et al (2011) Beyond the ionic and osmotic response to salinity in *Chenopodium quinoa*: functional elements of successful halomorphism. Funct Plant Biol 38:818–831

Ortiz R, Ruiz-Tapia EN, Mujica-Sanchez A (1998) Sampling strategy for a core collection of Peruvian quinoa germplasm. Theor Appl Genet 96:475–483

Ortiz R, Madsen S, Ruiz-Tapia EN et al (1999) Validating a core collection of Peruvian quinoa germplasm. Genet Res Crop Evol 46:285–290

Padilla JM, Otegui ME (2005) Co-ordination between leaf initiation and leaf appearance in field grown maize (*Zea mays*): genotypic differences in response of rates to temperature. Ann Bot 96:97–1007

Palomino G, Trejo Hernandez L, de la Cruz Torres E (2008) Nuclear genome size and chromosome analysis in *Chenopodium quinoa* and *C. berlandieri* subsp. *nuttaliae*. Euphytica 164:221–230

Passioura JB (1981) The interaction between the physiology and the breeding of wheat. In: Evans LT, Peacock WJ (eds) Wheat science-today and tomorrow. Cambridge University, Cambridge

Prado FF, Gallardo MB, Gonzalez JA (2000) Effect of NaCl on germination, growth and soluble sugar contents in *Chenopodium quinoa* (Willd.) seeds. Bot Bull Acad Sin Hortic 41:27–34

Rauf S, Khan AA, DeSilva T, Naveed A (2010) Consequences of plant breeding on genetic diversity. Int J Plant Breed 4:1–21

Repo-Carrasco R, Espinoza C, Jacobsen S-E (2003) Nutritional value and use of the Andean crops quinoa (*Chenopodium quinoa*) and kaniwa (*Chenopodium pallidicaule*). Food Rev Int 19:179–189

Richards RA, Rebetzke GJ, Condon AG, van Herwaarden AF (2002) Breeding opportunities for increasing the efficiency of water use and crop yield in temperate cereals. Crop Sci 42:111–121

Risi JC, Galwey NW (1984) The *Chenopodium* grains of the Andes: Inca crops for modern agriculture. Adv Appl Biol 10:145–216

Risi JC, Galwey NW (1989) The pattern of genetic diversity in the Andean grain crop quinoa (*Chenopodium quinoa* Willd.), I. Associations between characteristics. Euphytica 41:147–167

Risi JC, Galwey NW (1991) Genotype x environment interaction in the Andean grain crop quinoa (*Chenopodium quinoa*) in temperate environments. Plant Breed 107:141–147

Ruas PM, Bonifacio A, Ruas CF et al (1999) Genetic relationship among 19 accessions of six species of *Chenopodium*L., by random amplified polymorphic DNA fragments (RAPD). Euphytica 105:25–32

Ruffino AMC, Rosa M, Hilal M et al (2010) The role of cotyledon metabolism in the establishment of quinoa (*Chenopodium quinoa*) seedlings growing under salinity. Plant Soil 326:213–224

Ruiz RA, Bertero HD (2008) Light interception and radiation use efficiency in temperate quinoa (*Chenopodium quinoa* Willd.) cultivars. Eur J Agron 29:144–152

Ruiz-Carrasco K, Antognoni F, Coulibaly AK et al (2011) Variation in salinity tolerance of four lowland genotypes of quinoa (*Chenopodium quinoa* Willd.) as assessed by growth, physiological traits, and sodium transporter gene expression. Plant Physiol Biochem 49:1333–1341

Sakamoto T, Matsuoka M (2004) Generating high-yielding varieties by genetic manipulation of plant architecture. Curr Opin Biotech 15:144–147

Schulte auf'm Erley G, Kaul HP, Kruse M, Aufhammer W (2005) Yield and nitrogen utilization efficiency of the pseudocereals amaranth, quinoa, and buckwheat under different nitrogen fertilization. Eur J Agron 22:95–100

Slafer GA, Savín R (2006) Physiology of crop yield. In: Goodman RE (ed) Encyclopedia of plant and crop science. Taylor and Francis, New York

Slafer GA, González FG, Kantolic AG et al (2006) Grain number determination in major grain crops. In: Amarjit SB (ed) Handbook of seed science and technology. Food Products Press, New York

Stevens MR, Coleman CE, Parkinson SE et al (2006) Construction of a quinoa (*Chenopodium quinoa* Willd.) BAC library and its use in identifying genes encoding seed storage proteins. Theor Appl Genet 112:1593–1600

Tambussi EA, Bort J, Araus JL (2007) Water use efficiency in C3 cereals under Mediterranean conditions: a review of physiological aspects. Ann Appl Biol 150:307–321

Tapia M, Alandia S, Cardozo A et al (1979) Quinua y canihua. In: Tapia M (ed) Cultivos andinos. Serie libros y materiales educativos 49. IICA, Bogotá

Tapia ME, Mujica SA, Canahua A (1980) Origen, distibucion geografica y sistemas de produccion en quinua. In: Primera reunion sobre genetica y fitomejoramiento de la quinua. UNTA, IBTA, IICA, Centro de Investigacion Internacional para el Desarrollo, Puno

Trápani N, Hall AJ, Sadras VO, Vilella F (1992) Ontogenic changes in radiation-use efficiency of sunflower (*Helianthus annuus* L.). Field Crops Res 29:301–316

Vacher JJ (1998) Response of two main Andean crops, quinoa (*Chenopodium quinoa* Willd.) and papa amarga (*Solanum juzepczukii* Buk.) to drought on the Bolivian Altiplano: significance of local adaptation. Agric Ecosyst Environ 68:99–108

Vargas A, Elzinga DB, Rojas-Beltran JA et al (2011) Development and use of microsatellite markers for genetic diversity analysis of canahua (*Chenopodium pallidicaule* Aellen). Genet Res Crop Evol 58:727–739

Wilson HD (1980) Artificial hybridization among species of *Chenopodium* sect. Chenopodium. Syst Bot 5:253–263

Wilson HD (1981) Genetic variation among South American populations of tetraploid *Chenopodium* sect. Chenopodium subsect. Cellulata Syst Bot 6:380–398

Wilson HD (1988a) Allozyme variation and morphological relationships of *Chenopodium hircinum* (s.1.). Syst Bot 13:215–228

Wilson HD (1988b) Quinua biosystematics: domesticated populations. Econ Bot 42:461–477

Wilson HD, Heiser CB (1979) The origin and evolutionary relationships of 'Huauzontle' (*Chenopodium nuttaliae* Safford), domesticated chenopod of Mexico. Am J Bot 66:198–206
Wilson HD, Manhart J (1993) Crop/weed gene flow: *Chenopodium quinoa* Willd. and *C. berlandieri* Moq. Theor Appl Genet 86:642–648
Winkel T, Lhomme JP, Nina Laura JP et al (2009) Assessing the protective effect of vertically heterogeneous canopies against radiative frost: the case of quinoa on the Andean Altiplano. Agric Forest Meteor 149:1759–1768

Chapter 9
Biotechnology of *Eruca Sativa* Mill

Susan MH Slater

Abstract *Eruca sativa* Mill is amenable to biotechnological procedures and these initial protocols represent a means of quickly adapting the inherent potential of this species to specific applications and markets. *E. sativa* has potential for food, feed, industrial, and medicinal purposes and is a pool of genetic diversity for the economically important *Brassica* species. Intergeneric hybrids with *Brassica napus, B. rapa,* and *B. juncea* have been produced through either embryo rescue or protoplast fusion. *Brassica* markers are transferrable to *E. sativa* making marker assisted selection possible. Tissue regeneration protocols focus around the use of cotyledons, hypocotyls or cotyledonary node explants. Low concentrations of α-naphthaleneacetic acid (NAA) induce shoot organogenesis in 20 % of cultured explants, whereas somatic embryogenesis protocols focus around the use of 2, 4-dichlorophenoxyacetic acid (2, 4-D) and give a regeneration rate of greater than ~6 %. Double haploid protocols exist for both isolated microspore culture and anther culture. The key feature of these protocols is the need for heat shock to trigger androgenesis. An *Agrobacterium tumefaciens*-mediated transformation protocol based around shoot organogenesis gives a transformation frequency of 1.1 %.

Abbreviations

2, 4-D	2, 4-Dichlorophenoxyacetic acid
BA	6-Benzyladenine
IAA	Indole-3-acetic acid
IBA	Indole-3-butryic acid
NAA	α-Naphthaleneacetic acid

S. MH. Slater (✉)
Crop Development Centre, University of Saskatchewan, 51 Campus Drive, Saskatoon, SK S7N5A8, Canada
e-mail: smhslater@gmail.com

S. M. Jain and S. Dutta Gupta (eds.), *Biotechnology of Neglected and Underutilized Crops*, DOI: 10.1007/978-94-007-5500-0_9,

9.1 Introduction

Eruca sativa Mill or *Eruca vesicaria* subsp. *sativa* (Miller) (taramira, garden rocket or rocket, arugula, roquette, yunjie, white pepper, roca; n = 11) is part of a cytodeme, which includes *E. vesicaria* and *E. pinnatifida,* within the *Brassiceae* family (Fig 9.1a; Pignone and Gomez-Campo 2011; Sun et al. 2004; Warwick 1995; Warwick et al. 2007). It is an erect annual native to south Europe and central Asia (Pignone and Gomez-Campo 2011). *E. sativa* is a self-incompatible, salt tolerant, drought tolerant, cool season crop propagated by seeds or transplants (Bianco 1995; Pignone and Gomez-Campo 2011; Verma et al. 1977). The botany, use, conservation, and breeding of *E. sativa* have been recently reviewed (Pignone and Gomez-Campo 2011) and will be highlighted here.

As a food and fodder, *E. sativa* is used as a vegetable crop where its spicy leaves are eaten raw in salads and snacks or cooked in a variety of meals (Batra and Dhingra 1991; Bianco 1995; Yaniv et al. 1998). The leaves are rich in fiber, iron, and Vitamin C (Bianco 1995). The amino acid composition of *E. sativa* proteins indicates that it has potential as a protein source for edible purposes (Kaushal et al. 1982). The seed cake and entire plant are used as fodder for domestic animals (Fagbenro 2004; Pignone and Gomez-Campo 2011).

E. sativa is grown as an oilseed crop (jamba or taramira oil) in India, Iran, Afghanistan, Pakistan, and Ethiopia (Batra and Dhingra 1991; Bianco 1995). The oil content is 25–47 % and is rich in erucic acid (C22:1) and glucosinolates, with the proportion of oleic to linoleic being similar to *B. napus* (Table 9.1). The high erucic acid content makes *E. sativa* a potential source of industrial oil (Batra and Dhingra 1991; Pignone and Gomez-Campo 2011). This oil is a suitable bio-diesel alternative (Chakrabarti et al. 2011; Li et al. 2009; Li 2011) and has outperformed

Fig. 9.1 **a** W. Canadian field grown *E. sativa,* **b** Shoot regeneration from *E. sativa* hypocotyl segments, **c** *Agrobacterium* transformed shoots of *E. sativa* cultured under selection

Table 9.1 A comparison of oil content between *Brassica* species and *E. sativa*[a]

Species	Days to maturity	Yield (t ha^{-1})	Oil content (%)	Protein content (%)	Erucic (C22:1) (%)	Oleic (C18:0) (%)	Linoleic (C18:1) (%)	Glucosinolates (µmoles/g defatted seed meal)
B. juncea	75–85; 80–90; 140–150		35.6–41.3	18.3–22.0	33.3–49.3	7.4–16.5	14.9–25.6	90.0–125.2
B. carinata			25.0–39.1	20.4–21.0	26.5–38.1	12.3–15.0	19.6–27.8	82.8–122.5
B. napus			37.8–45.0	19.4–20.5	0.4–2.5	46.6–66.9	13.5–26.2	35.1–43.6
B. rapa (B. campestris)	70–90; 60–70	2.71–6.53	38.1–48.8	17.6–19.8	32.2–55.3	8.1–24.0	11.7–20.9	76.1–118.6
E. sativa	93–109	1.95–5.72	23.0–47.6	19.0–41.6	26.7–54.79	11.6–23.4 (2.3 %)	4.23–15.7 (11.9 %)	74.7–120.4

a An amalgamation of data from Chauhan et al. 2010; Fazili et al. 2010; Flanders and Abdulkarim 1985; Kanya and Urs, 1989; Kaushal et al. 1982; Sun et al. 2004; Ugur et al. 2010; Warwick et al., 2007; Yadava et al. 1998 jatropha, castor, and canola oils as a potential source of biodiesel due to its high oxygen content and high combustion rate (Chakrabarti et al. 2011).

The oil and leaf extracts are an effective insect repellant and suppressant suggesting their use as a biological control of crop pests (Khater and Shalaby 2007; Warwick et al. 2007; Yaniv et al. 1998; Zasada and Ferris 2004). The incorporation of the seed-oil meal byproduct or the use of *E. sativa* as a green manure crop decreases soil-borne populations of weeds, nematodes, and pathogenic fungi (Angelini et al. 1998; Riga et al. 2006; Tiyagi and Alam 1995). The glucosinolate concentration and its isothiocyanate derivatives are linked to the nematode suppressant activity (Zasada and Ferris 2004). The usefulness of a crop as a biofumigant depends on shoot yield, as well as glucosinolate, phenol, and ascorbic acid content. *E. sativa* has comparable yield and active compounds to *B. rapa*, *B. juncea*, and *B. napus* cultivars (Antonious et al. 2009).

Eruca oil and plant parts are used in the traditional pharmacopoeia for various purposes including depurative, diuretic, emollient, tonic stimulant, laxative, anti-inflammatory, digestive aid, diabetic control, and phlegm preventative (Pignone and Gomez-Campo 2011; Yaniv et al. 1998). Erucic acid, glucoerucin, isothiocyanate, and sinapin content are linked to the potential health benefits of the oil (Bianco 1995). Isothiocyanate and free erucic acid are linked to antimicrobial, anti-genotoxic, antioxidant, and anti-ulcer effects (Alqasoumi et al. 2009; Alqasoumi 2010; Barillari et al. 2005; Khoobchandani et al. 2010; Lamy et al. 2008; Sarwar et al. 2007). The primary glucosinolate in the seeds is 4-methylthiobutyl glucosinolate (glucoerucin) which has antioxidant properties (Barillari et al. 2005) and is around 109.4–149.2 μmol/g whole seed (Warwick et al. 2007). Glucoerucin is a precursor to 4-methylthiobutylthiocyanate, which is an anti-inflammatory agent active against skin disease (Yehuda et al. 2009). Erucin (4-(methylthio) butyl isothiocyanate), a product of glucoerucin, has selective anti-proliferative activity and therefore cancer preventative potential (Melchini et al. 2009).

E. sativa has a broad market range, wide genetic diversity, extreme environmental habitat, and limited breeding work. The potential for germplasm improvement is high, especially for niche markets. The application of biotechnology techniques to *E. sativa* will quickly advance this species for use in these specific markets.

9.2 *E. sativa* Germplasm Improvement Through Biotechnology

The usefulness of biotechnology protocols in a species is linked to the breeding goals and uses of that species. *E. sativa* is cultivated in the Mediterranean, Middle East, central Asia, north India, and central India. The distinct Mediterranean and Asian varieties are linked to its uses as a vegetable versus an oilseed crop, respectively (Pignone and Gomez-Campo 2011; Warwick et al. 2007). *E. sativa* acreage is low and is grown where rainfall and soil fertility are too low to cultivate cereals, rape, or mustard (Flanders and Abdulkarim 1985; Sun et al. 2004). There is a significant effect of environment and genotype by environment on agronomic and seed quality traits in *E. sativa* (Warwick et al. 2007). There are only a few open-pollinated cultivars (Leskovsek et al. 2008) and a great amount of diversity exists in the gene pool (Bozokalfa et al. 2011). Western Canadian studies indicate that *E. sativa* selections are less susceptible to blackleg fungus then *B. napus* checks (Warwick et al. 2007) indicating that *E. sativa* is a source of disease and pest resistance for *Brassica* (Sun et al. 2004; Warwick et al. 2007). *E. sativa* is reported to have cold, heat, drought, salt and insect tolerance (Ashraf 1994; Kumar et al. 2009; Magrath and Mithen 1997; Sun et al. 2004). When discussing biotechnology protocols in *E. sativa*, a direct comparison to *Brassica* species is useful due to the similarity of the species and the amenability of *Brassica* to biotechnology.

9.2.1 Intergeneric Hybrids of E. Sativa and Brassica Species

Intergeneric hybrids between *B. napus* x *E. sativa* (Dai et al. 2004; Sundberg and Glimelius 1991), *B. rapa* x *E. sativa*, (Agnihotri et al. 1990; Matsuzawa et al. 1999), and *E. sativa* x *B. juncea* (Sikdar et al. 1990) have been produced. Although achievable, crosses between *E. sativa* and all these *Brassica* species are incompatible due to pollen-stigma interactions (Sun et al. 2005a). *E. sativa* is a source of male sterility and sporophytic self-incompatibility genes for *Brassica* species. Traits useful in the development of hybrid systems for *E. sativa* (Matsuzawa et al. 1999; Sharma et al. 1985; Sun et al. 2005b; Verma 1984; Verma et al. 1977). Intergeneric hybridization between *B. rapa* and *E. sativa* yielded a cytoplasmic male sterile line of *B. rapa* containing the cytoplasm genome of *E. sativa* (Matsuzawa et al. 1999).

9.2.2 Genome Maps and Genetic Markers

Genome maps and genetic markers for *E. sativa* are limited and linked to either evolutionary studies or intergeneric hybrids with *Brassica* species. Evolutionary classification of the various members of the *Eruca* genus and related species has been done using restriction site variation of chloroplast DNA (Warwick and Black 1991), inter simple sequence repeats (Egea-Gilabert et al. 2009), and amplified fragment length polymorphisms (Warwick et al. 2007). Inter simple sequence repeats are useful to measure the toxicity of heavy metal agronomic conditions (Al-Qurainy 2010) while markers for a *E. sativa* rDNA intergeneric spacer would be useful for both analyzing intergeneric hybrids or for species evolution (Lakshmikumaran and Negi 1994). Genetic markers developed for *Brassica* breeding programs may be transferable to *E. sativa* breeding programs, since micro-satellite markers developed for *Brassica* species are transferable to *E. sativa* (Plieske and Struss 2001; Yadava et al. 2009).

9.3 Tissue Culture Studies in *E. sativa*

9.3.1 Organogenesis and Somatic Embryogenesis

A species specific tissue regeneration system is required for clonal propagation, introduction of somaclonal variation, use of marker-assisted selection, basic research into the biology of a species, development of a double haploid system, and the development of transformation systems for foreign gene introduction (Cardoza and Stewart 2004). *E. sativa* is amenable to both shoot organogenesis and somatic embryogenesis from a variety of explants, including cotyledons, cotyledonary petioles, hypocotyls, immature seeds, immature embryos and mesophyll protoplast using various media supplemented with either α-naphthaleneacetic acid (NAA), indole-3-acetic acid (IAA), or 2, 4-dichlorophenoxyacetic acid (2, 4-D) and 6-benzyladenine (BA) or kinetin or zeatin (Table 9.2).

Plant regeneration is affected by numerous aspects including: genotype, age and type of the explants used, ethylene inhibitors, and media constituents (Cardoza and Stewart 2004). Initial *E. sativa* tissue regeneration systems used cotyledons or cotyledonary nodes cultured in MS base medium with various concentrations of NAA, BA and kinetin and indicated 2.7 μM NAA or 2.9 μM IAA with 4.4–8.9 μM BA gave a shoot organogenesis at 8–20 % (Batra and Dhingra 1991; Parkash et al. 1989). Parkash et al. (1989) found that an auxin to cytokinin proportion of 0.1–0.25 only worked if a lower concentration of NAA was used. If the concentrations were increased, but the auxin to cytokinin proportions remained the same, organogenesis was not as high. Although BA worked consistently better than kinetin for both cotyledons and cotyledonary nodes, the type of auxin used made a difference in shoot organogenesis from these two explants (Table 9.2; Batra and Dhingra 1991; Parkash et al. 1989). This is different than reported by Slater et al. (2011) where a pulse of a high concentration of two different auxins with a high cytokinin concentration improved shoot regeneration from hypocotyl explants (Fig. 9.1b). The usefulness of combined auxins and an initial high auxin pulse is not well reported in the *Brassica* literature, except for one study which also found that a high auxin in the initial medium improved shoot regeneration from *B. napus* hypocotyl explants (Tang et al. 2011).

Newer regeneration protocols focus on somatic embryogenesis and indicate that 4.5 μM 2, 4-D with or without kinetin in hypocotyl or cotyledonary explants of *E. sativa* will give greater than 6 % regeneration (Table 9.2; Chen et al. 2011; Zhang et al. 2005). Zhang et al. (2005) indicated that the most effective auxin for somatic embryogenesis was 2, 4-D as compared to NAA, IAA, or IBA. BA was a more effective cytokinin than kinetin and somatic embryogenesis was improved if a low level of cytokinin was included (Chen et al. 2011). The efficiency of using hypocotyl versus cotyledonary explants is likely protocol or genotype based (Christey et al. 1997; De Block et al. 1989; Pental et al. 1993). Cotyledonary explants are more efficient for somatic embryogenesis (Chen et al. 2011; Zhang et al. 2005); whereas hypocotyl explants worked best for shoot organogenesis

Table 9.2 Plant regeneration in *E. sativa*

Explant	Explants age (days)	Supplements	Regenerants	Percent regeneration	Reference
Cotyledons	7–8	0.5 NAA + 2.0 BA	Shoot organogenesis	8–21 %	Parkash et al.1989
Cotyledonary nodes	14	0.5 mg/L IAA + 1.0 mg/L BA	Shoot organogenesis	?	Batra and Dhingra 1991
Cotyledon, hypocotyl	5	1.0 mg/L μM 2, 4-D	Somatic embryogenesis	65–84 %	Zhang et al. 2005
Cotyledons, cotyledonary petioles, hypocotyls	7	1.0 mg/L 2, 4-D + 0.1-0.3 mg/L kinetin	Somatic embryogenesis	6–48 %	Chen et al. 2011
Immature embryos	20–30	0.5 mg/L NAA + 5.0 mg/L BA	Somatic embryogenesis and organogenesis	?	Ahloowalia 1987
Immature seeds	4–5	0.1 mg/L NAA + 1.0 mg/L kinetin + 1.0 mg/L GA + 10.0 mg/L casein hydrolysate	Somatic embryogenesis	?	Agnihotri et al. 1990
Immature seeds	6–8	1.0 mg/L NAA + 0.5 mg/L BA	Somatic embryogenesis	?	Dai et al. 2004
Mesophyll protoplasts		0.2 mg/L 2,4-D + 1.0 mg/L BA + 0.1 mg/L GA3	Somatic embryogenesis	11.25 %	Sikdar et al. 1987
Mesophyll protoplasts		0.75 mg/L NAA + 4.0 mg/L BA + 0.1 mg/L GA3	Shoot organogenesis	15.71 %	Sikdar et al. 1987, 1990
Mesophyll protoplasts		0.1 mg/L IAA + 5.0 mg/L BA + 0.5 mg/L zeatin or 0.1 mg/L IAA + 2.0 mg/L zeatin	Shoot organogenesis	80 %	Fahleson et al. 1988
Hypocotyl protoplasts		?	Shoot organogenesis	34 × 5.6 %	Zhang et al. 2008
Isolated microspores		200 mg/L activated charcoal	Embryogenesis	4–69 embryos/2 × 10^5 microspores	Leskovsek et al. 2008
Anthers		Silver nitrate, NAA, 2, 4-D	Embryogenesis	?	Tribulato and Branca 2008
Hypocotyls		*Agrobacterium rhizogenes*	Hairy root cultures	?	Xue et al. 2008
Hypocotyls, cotyledons	6	1.0 mg/L IAA + 0.1 mg/L NAA + 5.0 mg/L BA + *Agrobacterium tumefaciens*	Transformed shoot organogenesis	1.1	Slater et al. 2011

? = unknown

(Slater et al. 2011). As with *Brassica* protocols, the addition of the ethylene inhibitor, silver nitrate, improved efficiency of shoot organogenesis (Mukhopadhyay et al. 1992; Slater et al. 2011)

9.3.2 E. Sativa Protoplasts and Embryo Rescue

The culture of immature seeds and embryos has been important for rescuing intergeneric hybrids between *E. sativa* and *Brassica* species and as a potential basis for genetic modification of the species. Ahloowalia (1987) was one of the first to produce somatic embryos and shoot organogenesis from immature embryos of *E. sativa* (Table 9.2). The culture of immature seeds (pollinated ovaries) from intergeneric crosses with *Brassica* species produced somatic embryos using various NAA and cytokinin concentrations (Agnihotri et al. 1990; Dai et al. 2004).

Intergeneric fertile somatic hybrids were produced by protoplast fusion of *E. sativa* x *B. juncea* (Sikdar et al. 1990) and *E. sativa* x *B. napus* (Fahleson et al. 1988). Somatic cell fusion bypasses the sexual incompatibility barriers between species and widens the gene pool of a domesticated crop. For example, the inclusion of *E. sativa*'s drought, disease and insect resistance into *B. juncea, B. napus, and B. rapa* (Agnihotri et al. 1990; Fahleson et al. 1988; Sikdar et al. 1990). *E. sativa* mesophyll or hypocotyl derived protoplasts are an initial explant for shoot organogenesis or somatic embryogenesis depending on the type of auxin and/or the auxin to cytokinin proportion (Table 9.2).

9.3.3 Haploid Production

Double haploid plants fix all alleles in the homozygous state within one generation. Due to self-infertility, *E. sativa* is heterozygous and the production of double haploids would be extremely advantageous for any breeding program, as well as for applied and basic genetic studies, mutation research, gene mapping, and as targets for gene transformation (Ferrie and Mollers 2011). *E. sativa* is amenable to both anther culture (Tribulato and Branca 2008) and isolated microspore culture (Leskovsek et al. 2008).

Development of anther or isolate microspore culture protocols requires consideration of: donor plant conditions, harvest floral organ conditions, isolation of the anther/microspores, culture and induction of the anthers/microspores, regeneration of the embryos, and doubling of the chromosomes (Ferrie and Caswell 2011; Germana 2011). For *E. sativa*, careful donor care and specific genotypes defines whether embryogenesis occurs (Leskovsek et al. 2008). A bud length of 4–5 mm gives late uninucleate to early binucleate microspores which were cultured on a high sucrose medium containing activated charcoal for microspores or NAA, silver nitrate, and 2, 4-D for anthers (Leskovsek et al. 2008; Tribulato and Branca

2008). For both anther and isolated microspore culture, globular embryos are induced after a 24 h heat shock at 32–35°C (Leskovsek et al. 2008; Tribulato and Branca 2008). The need for activated charcoal in microspore culture indicates that a detrimental substance is either interfering with the triggering of androgenesis or the development of embryos (Leskovsek et al. 2008). The inclusion of a cold pretreatment with the heat shock made no difference to embryogenic potential of microspores (Leskovsek et al. 2008).

9.3.4 Agrobacterium-Mediated Transformation

Transformation protocols start with the questions linked to regeneration systems and include an analysis of the genotype, age and type of the explant, and the media constituents (Cardoza and Stewart 2004). The *Agrobacterium tumefaciens*–mediated transformation protocols for *E. sativa* and *Brassica* species are similar (Table 9.3; Babic et al. 1998; Damgaard and Rasmussen 1991; De Block et al. 1989; Kuvshinov et al. 1999; Mukhopadhyay et al. 1992). Six day old hypocotyl explants were co-cultivated with *A. tumefaciens* on a high auxin: high cytokinin media for 2 days and then transferred to a medium containing lower concentrations

Table 9.3 Media used for *Agrobacterium tumefaciens*—mediated transformation of *Eruca sativa* hypocotyl segments[a]

Medium[b]	Plant growth regulators	Time
Co-cultivation medium	1.0 mg L^{-1} IAA 5.0 mg L^{-1} BA 0.1 mg L^{-1} NAA	2 days
Recovery medium	2.0 mg L^{-1} BA 0.1 mg L^{-1} NAA 5.0 mg L^{-1} silver nitrate Antibiotic	7 days
Regeneration/selection medium	2.0 mg L^{-1} BA 0.1 mg L^{-1} NAA 5.0 mg L^{-1} silver nitrate Antibiotic Selective agent	4–8 weeks
Shoot elongation medium	5.0 mg L^{-1} silver nitrate Antibiotic Selective agent	4–8 weeks
Rooting medium	0.1 mg L^{-1} NAA 5.0 mg L^{-1} silver nitrate Antibiotic Selective agent	2 weeks

a Adapted from Slater et al. 2011

[b] Murashige and Skoog (1962) basal medium with 3 % sucrose and 0.7 % solidifying agent

of auxin and cytokinin along with silver nitrate and antibiotic for several days (Slater et al. 2011). The explants are transferred to a similar medium containing a selective agent where shoots form within 60 days (Fig. 9.1c; Slater et al. 2011).

Brassica protocols use a wide range of explants, although hypocotyl explants are the most commonly used despite being sensitive to co-cultivation with *Agrobacterium* (Poulsen 1996). This sensitivity was not observed in *E. sativa* (Slater et al. 2011). Neither a pre-culture stage nor acetosyringone are required in *E. sativa* transformation; however, the ethylene inhibitor silver nitrate improves shoot regeneration (Mukhopadhyay et al. 1992; Poulsen 1996; Slater et al. 2011). *A. tumefacien* based vectors were attempted in *E. sativa* (Slater et al. 2011). However, the cultivation of leaf explants of *E. sativa* with *A. rhizogenes* to produce hairy root cultures for cryopreservation indicates that *A. rhizogenes* would likely work (Xue et al. 2008).

The regeneration frequency of transformed *E. sativa* shoots was 1.1 % (Slater et al. 2011) which is low compared with *Brassica* species transformation rates (0.5–36 %; Poulsen 1996). This could be due to the greater amount of work that has been done in *Brassica* species, the limited amount of organogenesis protocols tested for *E. sativa*, or the use of kanamycin as a selectable marker (Poulsen 1996). A majority of the transformed shoots (64 %) were with single inserts, which is important for subsequent genetic studies and to decrease the likelihood of gene silencing. These plants were selfed and the genes remained stable for two generations (Slater et al. 2011).

9.4 Conclusions

E. sativa is a drought tolerant, salt tolerant, cold season, disease and pest resistant crop. Breeding work is limited and there is a lot of genetic variability. Although *E. sativa* produces less oil than other oil seed crops, it has a larger range of potential oil content then *Brassica* species (Table 9.1) with a greater potential for cultivar development (Bozokalfa et al. 2011). As well, *E. sativa* has great potential for niche markets or extreme growing conditions.

E. sativa cultivars with reduced amounts of erucic acid and glucosinolate levels are needed to make the oil and meal acceptable for human and animal nutrition (Kanya and Urs 1989; Warwick et al. 2007). Alternatively, cultivars with increased amounts of these anti-nutritive compounds would make the crop more attractive as an industrial oil, medicinal crop, or biofumigant. Increasing glucosinolate content increases *E. sativa*'s usefulness as an antimicrobial and anti-genotoxic compound (Khoobchandani et al. 2010; Lamy et al. 2008; Melchini et al. 2009; Yehuda et al. 2009). Increasing the erucic acid content increases the usefulness of this oil as a biodiesel (Chakrabarti et al. 2011). Increasing both of these compounds would make *E. sativa* oil more useful as a natural, organic insect repellent (Khater and Shalaby 2007) or the meal as a soil fumigant and weed suppressant with low environment impact (Angelini et al. 1998; Tiyagi and Alam 1995).

In *Brassica* species, gene transfer systems have been used to improve oil quality, seed storage protein, secondary metabolite concentrations (glucosinolates), herbicide tolerance, disease and pest resistance, pollination systems, and to provide unique protein production and environment cleansing systems (Poulsen 1996). Given the limited amount of breeding done in *E. sativa* and the potential of this crop for niche markets, the use of regeneration and transformation protocols will advance development of use-specific cultivars in much the same way it has assisted in the development of *Brassica* species cultivars.

References

Agnihotri A, Gupta V, Lakshmikumaran MS, Shivanna KR, Prakash S, Jagannathan V (1990) Production of *Eruca-Brassica* hybrids by embryo rescue. Plant Breed 104:281–289

Ahloowalia BS (1987) Somatic embryogenesis and plant regeneration in *Eruca sativa*. Crop Sci 27:813–814

Alqasoumi S (2010) Carbon tetrachloride-induced hepatotoxicity: protective effect of 'Rocket' *Eruca sativa* L. in Rats. Am J Chin Med 30:75–88

Alqasoumi S, Al-Sohaibani M, Al-Howiriny T, Al-Yahya M, Rafatullah S (2009) Rocket "*Eruca sativa*": a salad herb with potential gastric anti-ulcer activity. World J Gastroenterol 15:1958–1965

Al-Qurainy F (2010) Application of inter simple sequence repeat (ISSR marker) to detect genotoxic effect of heavy metals on *Eruca sativa* (L.). Afr J Biotechnol 9:467–474

Angelini L, Lazzeri L, Galletti S, Cozzani A, Macchia M, Palmieri S (1998) Antigerminative activity of three glucosinolate-derived products generated by myrosinase hydrolysis. Seed Sci Technol 26:771–780

Antonious GF, Bomford M, Vincelli P (2009) Screening *Brassica* species for glucosinolate content. J Environ Sci Health Part B 44:311–316

Ashraf M (1994) Organic substances responsible for salt tolerance in *Eruca sativa*. Biol Plant 36:255–259

Babic V, Datla RS, Scoles GJ, Keller WA (1998) Development of an efficient *Agrobacterium*-mediated transformation system for *Brassica carinata*. Plant Cell Rep 17:183–188

Barillari J, Canistro D, Paolini M, Ferroni F, Pedulli GF, Iori R, Valgimigli L (2005) Direct antioxidant activity of purified glucoerucin, the dietary secondary metabolite contained in rocket (*Eruca sativa* Mill.) seeds and sprouts. J Agric Food Chem 53:2475–2482

Batra A, Dhingra M (1991) Production of plantlets of *Eruca sativa* in vitro. J Phytol Res 4:73

Bianco VV (1995) Rocket, an ancient underutilized vegetable crop and its potential. In: Padulos S (ed) IPGRI: Rocket genetic resources network. International Plant Genetic Resources Institute, pp 35–57

Bozokalfa MK, Esiyok D, Ilbi H, Kavak S, Asciogul TK (2011) Evaluation of phenotypic diversity and geographical variation of cultivated (*Eruca sativa* L.) and wild (*Diplotaxis tenuifolia* L.) rocket plant. Plant Genet Resour 9:454–463

Cardoza V, Stewart CN (2004) Invited Review: *Brassica* biotechnology: progress in cellular and molecular biology. In Vitro Cell Dev Biol Plant 40:542–551

Chauhan JS, Kumar S, Singh KH, Meena SS, Meena ML (2010) Oil and seed meal quality indices of Indian rapeseed-mustard varieties. J Plant Biochem Biotechnol 19:83–86

Chakrabarti MH, Ali M, Baroutian S, Saleem M (2011) Techno-economic comparison between B10 of *Eruca sativa* L. and other indigenous seed oils in Pakistan. Process Saf Environ Prot 89:165–171

Chen K, Zhang XB, Jiang JL, Wang XY (2011) Plantlet regeneration from cotyledon, cotyledon petiole, and hypocotyl explants via somatic embryogenesis pathway in roquette (*Eruca sativa* Mill). Plant Biosyst 145:68–76

Christey MC, Sinclair BK, Braun RH, Wyke L (1997) Regeneration of transgenic vegetable brassicas (*Brassica oleracea* and *B.campestris*) via Ri-mediated transformation. Plant Cell Rep 16:587–593

Dai LJ, Li X, Guan CY, Zhong J (2004) On ovary and embryo culture of hybrid between rapeseeds and rocket salad. J Hunan Agric Univ 30:201–204

Damgaard O, Rasmussen O (1991) Direct regeneration of transformed shoots in *Brassica napus* from hypocotyl infections with *Agrobacterium rhizogenes*. Plant Mol Biol 17:1–8

De Block MD, DeBrouwer D, Tenning P (1989) Transformation of *Brassica napus* and *Brassica oleracea* using *Agrobacterium tumefaciens* and the expression of the *bar* and *neo* genes in the trangenic plants. Plant Physiol 91:694–701

Egea-Gilabert C, Fernandez JA, Migliaro D, Martinez-Sanchez JJ, Vicente MJ (2009) Genetic variability in wild vs. cultivated *Eruca vesicaria* populations as assessed by morphological, agronomical and molecular analyses. Sci Hortic 121:260–266

Fagbenro OA (2004) Soybean meal replacement by roquette (*Eruca sativa* Miller) seed meal as protein feedstuff in diets for African catfish *Clarias gariepinus* (Burchell 1822), fingerlings. Aquacult Res 35:917–923

Fahleson J, Rahlen L, Glimelius K (1988) Analysis of plants regenerated from protoplast fusions between *Brassica napus* and *Eruca sativa*. Theor Appl Genet 76:507–512

Fazili IS, Masoodi M, Ahmad S, Jamal A, Khan JS, Abdin MZ (2010) Interactive effect of sulfur and nitrogen on growth and yield attributes of oilseed crops (*Brassica campestris* L. and *Eruca sativa* Mill.) differing in yield potential. J Plant Nutr 33:1216–1228

Ferrie AMR, Caswell KL (2011) Isolated microspore culture techniques and recent progress for haploid and doubled haploid plant production. Plant Cell Tiss Org Cult 104:301–309

Ferrie AMR, Mollers C (2011) Haploids and doubled haploids in *Brassica* spp. for genetic and genomic research. Plant Cell Tiss Org Cult 104:375–386

Flanders A, Abdulkarim SM (1985) The composition of seed and seed oils of Taramira (*Eruca sativa*). JAOCS 62:1134–1135

Germana MA (2011) Anther culture for haploid and doubled haploid production. Plant Cell Tiss Org Cult 104:283–300

Kanya TCS, Urs MK (1989) Studies on Taramira (*Eruca sativa*) seed oil and meal. JAOCS 66:139–140

Kaushal GP, Sital JS, Bhatia IS (1982) Studies on Taramira seed (*Eruca sativa* Lam.) proteins. J Agric Food Chem 30:431–435

Khater HF, Shalaby AA (2007) Potential of biologically active plant oils for controlling of *Culex pipiens* (*Diptera*: *Culicidae*). Acta Scientiae Veterinariae 35:153–160

Khoobchandani M, Ojeswi BK, Ganesh N, Srivastava MM, Gabbanini S, Matera R, Iori R, Valgimigli L (2010) Antimicrobial properties and analytical profile of traditional *Eruca sativa* seed oil: Comparison with various aerial and root plant extracts. Food Chem 120:217–224

Kumar G, Purty RS, Sharma MP, Singla-Pareek SL, Pareek A (2009) Physiological responses amoung *Brassica* species under salinity stress show strong correlation with transcript abundance for SOS pathway-related genes. J Plant Physiol 166:507–520

Kuvshinov V, Koivu K, Kanerva A, Pehu E (1999) *Agrobacterium tumefaciens*—mediated transformation of greenhouse: grown *Brassica rapa* ssp. *oleifera*. Plant Cell Rep 18:773–777

Lakshmikumaran M, Negi MS (1994) Structural analysis of two length variants of the rDNA intergenic spacer from *Eruca sativa*. Plant Mol Biol 24:915–927

Lamy E, Schroder J, Paulus S, Brenk P, Stahl T, Mersch-Sundermann V (2008) Antigenotoxic properties of *Eruca sativa* (rocket plant), erucin and erysolin in human hepatoma (HepG2) cells towards benzo (α) pyrene and their mode of action. Food Chem Toxicol 46:2415–2421

Leskovsek L, Jakse M, Bohanec B (2008) Double haploid production of rocket (*Eruca sativa* Mill.) through isolate microspore culture. Plant Cell Tiss Org Cult 93:181–189

Li WW (2011) One-step preparation of efficient and reusable SO4(2)/ZrO(2)-based hybrid solid catalysts functionalized by alkyl-bridged organosilica moieties for biodiesel production. Chem Sus Chem 4:744–756

Li S, Wang Y, Dong S, Chen Y, Cao F (2009) Biodiesel production from *Eruca sativa* gars vegetable oil and motor, emissions properties. Renew Energy 34:1871–1876

Magrath R, Mithen R (1997) Introgression of disease resistance genes from *Eruca sativa* into *Brassica napus*. In: ISHS Symposium on *Brassicas*. Tenth crucifer genetics workshop, pp 217

Matsuzawa Y, Mekiyanon S, Kaneko Y, Bang SW, Wakui K, Takahata Y (1999) Male sterility in alloplasmic *Brassica rapa* L. carrying *Eruca sativa* cytoplasm. Plant Breed 118:82–84

Melchini A, Costa C, Traka M, Miceli N, Mithen R, DePasquale R, Trovato A (2009) Erucin, a new promising cancer chemopreventive agent from rocket salads, show anti-proliferative activity on human lung carcinoma A549 cells. Food Chem Toxicol 47:1430–1436

Mukhopadhyay A, Arumugam N, Anandakaumar PB, Pradhan AK, Gupta V, Pental D (1992) *Agrobacterium*-mediated genetic transformation of oilseed *Brassica campestris*: transformation frequency is strongly influenced by the mode of shoot regeneration. Plant Cell Rep 11:506–513

Murashige T, Skoog K (1962) A revised medium for rapid growth and bioassays with tobacco tissue cultures. Physiol Plant 15:473–497

Parkash S, Chowdhury JB, Jain RK (1989) Callus initiation and regeneration potential in different genotypes of *Eruca sativa*. Curr Sci 58:979–980

Pental D, Pradhan AK, Sodhi YS, Mukhopadhyay A (1993) Variation amongst *Brassica juncea* cultivars for regeneration from hypocotyl explants and optimization of conditions for *Agrobacterium*-mediated genetic transformation. Plant Cell Rep 12:462–467

Pignone D, Gomez-Campo C (2011) Eruca In: Kole C (ed) Wild crop relatives: genomic and breeding resources, oilseeds. Springer, Heidelberg

Plieske J, Struss D (2001) Microsatellite markers for genome analysis in *Brassica*. I. development in *Brassica napus* and abundance in *Brassicaceae* species. Theor Appl Genet 102:689–694

Poulsen GB (1996) Genetic transformation of *Brassica*. Plant Breed 115:209–225

Riga E, Pierce F, Collins HP (2006) Performance of arugula (*Eruca sativa*) as a green manure and trap crop for fungal pathogens and parasitic nematode suppression in potato. Proc Am Phytopathol Soc 96:S97

Sarwar AM, Kaur G, Jabbar Z, Javed K (2007) *Eruca sativa* seeds possess antioxidant activity and exert a protective effect on mercuric chloride induced renal toxicity. Food Chem Toxicol 45:910–920

Sharma N, Bajaj M, Shivanna KR (1985) Overcoming self-incompatibility through the use of Lectins and sugars in *Petunia* and *Eruca*. Ann Bot 55:139–141

Sikdar SR, Chatterjee G, Das S, Sen SK (1987) Regeneration of plants from mesophyll protoplast of the wild crucifer *Eruca sativa* Lam. Plant Cell Rep 6:486–489

Sikdar SR, Chatterjee G, Das S, Sen SK (1990) 'Erussica', the intergeneric fertile somatic hybrid developed through protoplast fusion between *Eruca sativa* Lam. and *Brassica juncea* (L.) Czern. Theor Appl Gen 79:561–567

Slater SMH, Keller WA, Scoles G (2011) *Agrobacterium*-mediated transformation of *Eruca sativa*. Plant Cell Tiss Org Cult 106:253–260

Sun W, Guan C, Meng Y, Liu Z, Zhang T, Li X, Yang S, Ling L, Chen S, Zeng X, Wang H (2005a) Intergeneric crosses between *Eruca sativa* Mill. and *Brassica* species. Acta Agron Sin 31:36–42

Sun W, Pan Q, Liu Z, Meng Y, Zhang T, Wang H, Zeng X (2004) Genetic resources of oilseed *Brassica* and related species in Gansu Province, China. Plant Genet Resour 2:167–173

Sun W, Pan Q, Liu Z, Meng Y, Zhang T, Wang H, Zeng X (2005b) Overcoming self-incompatibility in *Eruca sativa* by chemical treatment of stigmas. Plant Genet Resour 3:13–18

Sundberg E, Glimelius K (1991) Effects of parental ploidy level and genetic divergence on chromosome elimination and chloroplast segregation in somatic hybrids within *Brassicaceae*. Theor Appl Genet 83:81–88

Tang GX, Knecht K, Yang XF, Qin YB, Zhou WJ, Cai D (2011) A two-step protocol for shoot regeneration from hypocotyl explants of oilseed rape and its application for *Agrobacterium*-mediated transformation. Biol Plant 55:21–26

Tiyagi SA, Alam MM (1995) Efficacy of oil-seed cakes against plant-parasitic nematodes and soil-inhabiting fungi on mung-bean and chickpea. Bioresource Technol 51:233–239

Tribulato A, Branca F (2008) Haploid plants regeneration of *Eruca sativa* by in vitro anther culture. In: Modern variety breeding for present and future needs. Proceedings of the 18th EUCARPIA general congress, p 259

Ugur A, Suntar I, Aslan S, Orhan IE, Kartal M, Sekeroglu N, Esiyok D, Sener B (2010) Variations in fatty acid compositions of the seed oil of *Eruca sativa* Mill. caused by different sowing periods and nitrogen forms. Pharmacogn Mag 6:305–308

Verma SC (1984) Genic male sterility in *Eruca sativa* cruciferae. Incompat Newsl 16:9

Verma SC, Malik R, Dhir I (1977) Genetics of the incompatibility system in the crucifer *Eruca sativa* L. Proc R Soc Lond B 96:131–159

Warwick S (1995) Rocket, an ancient underutilized vegetable crop and its potential. In: Padulos S (ed) IPGRI: rocket genetic resources network. International plant genetic resources institute, pp 22–34

Warwick SI, Black LD (1991) Molecular systematic of *Brassica* and allied genera (Subtribe Brassicinae, Brassiceae)—chloroplast genome and cytodeme congruence. Theor Appl Genet 82:81–92

Warwick SI, Gugel RK, Gomez-Campo C, James T (2007) Genetic variation in *Eruca vesicaria* (L.) Cav. Plant Genet Resour 5:142–153

Yadava TP, Friedt SW, Gupta SK (1998) Oil content and fatty acid composition of Taramira (*Eruca sativa* L.) genotypes. J Food Sci Technol 35:557–558

Yadava DK, Parida SK, Dwivedi VK, Varshney A, Ghazi IA, Sujata V, Mohapatra T (2009) Cross-transferability and polymorphic potential of genomic STMS markers of *Brassica* species. J Plant Biochem Biotechnol 18:29–36

Yaniv Z, Schafferman D, Amar Z (1998) Tradition, uses and biodiversity of rocket (*Eruca sativa, Brassicaceae*) in Israel. Econ Bot 52:394–400

Yehuda H, Khatib S, Sussan I, Musa R, Vaya J, Tamir S (2009) Potential skin anti-inflammatory effects of 4-methylthiobutylisothiocyanate (MTBI) isolated from rocket (*Eruca sativa*) seeds, vol 35. International Union of Biochemistry and Molecular Biology, Inc, pp 295–305

Xue S-H, Luo X-J, Wu Z-H, Zhang H-L, Wang X-Y (2008) Cold storage and cryopreservation of hairy root cultures of medicinal plant *Eruca sativa* Mill., *Astragalus membranaceus* and *Gentiana macrophylla* pall. Plant Cell Tiss Org Cult 92:251–260

Zasada IA, Ferris H (2004) Nematode suppression with brassicaceous amendments: application based upon glucosinolate profiles. Soil Biol Biochem 36:1017–1024

Zhang T, Cao ZY, Wang XY (2005) Induction of somatic embryogenesis and plant regeneration from cotyledon to hypocotyl explants of *Eruca sativa* Mill. In Vitro Cell Dev Biol-Plant 41:655–657

Zhang CL, Gui XM, Ahu Y, Liu YT, Mao XQ, Lin LB (2008) Plant regeneration and hypocotyl protoplasts isolation of *Eruca sativa*. J S China Agric Univ 29:63–68

Chapter 10
Biotechnology of *Stylosanthes*

Amaresh Chandra

Abstract The genus *Stylosanthes* contains some 40 species largely classified on morphological characteristics and has a monophyletic origin, closely related to *Arachis* with the *S. guianensis* species complex as the most ancient group clearly distinct from the rest of the genus. Many species are important forage legumes and grown both in annual and perennial system of agriculture. Species are grouped in two subgeneric sections, sections *Styposanthes* and *Stylosanthes* based on the presence or absence, respectively, of an axis rudiment, a small appendage at the base of the pod or loment. Most species of *Stylosanthes* are diploid (2n = 20) but polyploid species (2n = 40 and 2n = 60) also exist. Latter are exclusively allopolyploid. Section *Styposanthes* contains both diploid and polyploid species, while species in Section *Stylosanthes* are exclusively diploid. Molecular markers have proved that a tertraploid (4n) is a combination of a diploid (2n) species from section *Stylosanthes* and a diploid (2n) species from Sec. *Styposanthes*. A reasonable numbers of sequence-tagged site (STS) markers have been developed in this genus and used to define the progenitors of tetraploid *S. scabra*, *S. hamata* and only hexaploid species *S. erecta*. Genetic diversity study in this genus is largely based on isozyme, RAPD and STS markers. Lack of sufficient EST sequences and simple sequence repeat (SSR) markers have hampered a desired good linkage map; however, RAPD and few RFLPs based map delivered QTLs for drought and anthracnose disease. Before the introduction of *Stylosanthes seabrana* in year 1998–1999, *S. scabra*, *S. hamata*, *S. guianensis* and *S. humilis* were major species cultivated in Indian tropics and sub-tropics. Though regeneration protocols have been developed for few species, gene transfers through genetic engineering means have been very limited in this genus. Incorporation of leaf meal, stylo meal (as the replacements of concentrate) in feeding system of small and large ruminants,

A. Chandra
Indian Grassland and Fodder Research Institute, Gwalior Road,
Jhansi, Uttar Pradesh 284003, India

A. Chandra (✉)
Division of Plant Physiology and Biochemistry, Indian Institute of Sugarcane Research,
Rae Bareli Road, Lucknow, Uttar Pradesh 226002, India
e-mail: amaresh62@yahoo.com

S. M. Jain and S. Dutta Gupta (eds.), *Biotechnology of Neglected and Underutilized Crops*, DOI: 10.1007/978-94-007-5500-0_10,

poultry ration as well as their conservation in form of fodder banks are some of the new and impacted initiatives have been taken with *Stylosanthes* crop in South Asian countries. Narrow genetic base, insufficient SSR markers, loose genetic map and overall low breeding efforts world wide in *Stylosanthes* invite concerted attention for its improvement by exploiting genomic resources and information being generated with related legumes, vital to this less explored but an important forage legume for better and improved livestock husbandry.

10.1 Introduction

The genus *Stylosanthes* encompasses many legumes of forage value. Generally, *Stylosanthes* is referred as stylo. Persistence gaps in demand and supply of fodder as observed in many countries can be minimized by incorporation of species of *Stylosanthes* in forage production systems. Both plant and animal scientists are routinely involved in identifying, developing and evaluating good quality fodder varieties for better livestock production. Range fodder legumes will have edge over cultivated as later would be inadvertently increasing the input-cost significantly especially being low priority crops. Some important species of *Stylosanthes* namely *S. scabra* (shrubby stylo), *S. hamata* (Caribbean stylo), *S. guianensis* (common stylo), *S. viscosa* (sticky stylo) and *S. seabrana* (Caatinga stylo) have been identified as an important component of programmes sponsored by various government and non-governments agencies and is extensively utilized in pastoral, agropastoral and silvipastoral systems for animal production (Chandra et al. 2006). Due to its ability to restore soil fertility, improve soil physical properties and provide permanent vegetation cover *Stylosanthes* species can also play a vital role in the development of wastelands in many countries including India. It is also considered a better nurse crop in plantation on degraded lands. Areas like improving natural grazing lands, rejuvenating rainfed crop lands and livestock production system have been greatly influenced by use of *Stylosanthes* species (Fig. 10.1). The most specific problems associated with *Stylosanthes* are the limited variations of available germplasm and the susceptibility to anthracnose disease caused by the fungus *Colletotrichum gloeosporioides*.

The *Stylosanthes* species are distributed across the tropical, subtropical and temperate regions of the Americas (mainly South America), tropical Africa and SE Asia (Williams and Gardner 1984). Stylo has been the most economical pasture and forage legume in the tropical regions of the globe. It is widely distributed throughout Brazil, one of the main centers of diversity (Costa and Ferreira 1984). Except for *Stylosanthes fruticosa, S. erecta* and *S. sundaica*, the genus is native to tropical America. *Stylosanthes* species are part of the natural flora in many regions of South and Central America, Mexico and the Caribbean. It was only in 1914 that the economic value of *S. humilis* in agriculture was recognized in Australia. This was followed by *S. guianensis* in 1933 in Brazil (Stace and Edye 1984). Burt and Williams (1975), in their '*Stylosanthes* story', described the rising

Fig. 10.1 Major species of *Stylosanthes*

interest in these two stylo species. The early success with annual *S. humilis* and perennial *S. guianensis* (common and fine-stem stylos) laid the foundation for research and development in stylo. They have a common basic chromosome number of 10 (Cameron 1967). Most *Stylosanthes* species are diploid, but tetra- and hexaploid species are known. *S. hamata* exists in both diploid and tetraploid form, but the tetraploid type is the one widely used for forage (Stace and Cameron 1984). Morphological differentiation appears insufficient for species identification and, as a result, some disagreements remain regarding the species status of much of the genus (Mannetje 1984; Williams et al. 1984). Classified on morphological characteristics, this genus contains some 40 species (Kirkbride and Kirkbride 1985). The genus *Stylosanthes* has a monophyletic origin (Gillies and Abbott 1996) and is closely related to *Arachis* (Lavin et al. 2001), with the *S. guianensis* species complex as the most ancient group, clearly distinct from the rest of the genus. The species of this genus are grouped in two subgeneric sections, sections *Styposanthes* and *Stylosanthes*, based on the presence or absence, respectively, of an axis rudiment, a small appendage at the base of the pod or loment (Mannetje 1984). Most species of *Stylosanthes* are diploid (2n = 20) but polyploid species (2n = 40 and 2n = 60) also exist. Latter are exclusively allopolyploid. Section *Styposanthes* contains both diploid and polyploid species, while species in section *Stylosanthes* are exclusively diploid. In most of the cases molecular markers have proved that a tertraploid (4n) is a combination of a diploid (2n) species from section *Stylosanthes* and a diploid (2n) species from Sec. *Styposanthes* (Liu et al. 1999; Vander Stappen et al. 2002). Before identification and use of *S. seabrana,* four species namely *S. scabra, S. hamata, S. guianensis* and *S. humilis* have been widely cultivated as tropical forage legumes. Among these, *S. scabra* and *S. hamata* are allotetraploids (2n = 2x = 40) and appear to be highly diploidized with regular bivalent formation at meiosis and disomic inheritance (Stace and Cameron 1984). The other 2 species, *S. guianensis* and *S. humilis*, are diploid (2n = 2x = 20). These 4 cultivated species are predominantly self-pollinating with a low but variable degree of out-crossing (Stace 1982).

10.2 Distribution of Genetic Diversity

Geographic information systems (GIS) has tremendously helped in identifying regions with similar ecological profile to target new exploration and collection sites for new useful genetic resources (Burt et al. 1980). Sawkins (1999) studied four species using GIS and showed that *S. guianensis* has the widest geographic distribution as compared to *S. viscosa*, *S. humilis* and *S. capitata*. Vander Stappen et al. (2000) investigated the diversity of Mexican accessions of *S. humilis* and concluded that Mexico may be rich in unique diversity that could be used for the conservation and utilization of *S. humilis* germplasm. Promising *S. scabra* introduction from northeastern Brazil emphasizes further collections of such lines from these regions (Burt et al. 1979). Over 600 accessions of *S. scabra* have been accumulated world wide. More than 90 % of them were collected in Brazil, with Colombia and Venezuela each contributing about another four percent of the collections (Schultze-Kraft et al. 1984). Clustering analysis based on morphological and agronomical (M-A) characteristics grouped *S. scabra* accessions into four varietal types, including continental, Brazilian Coastal, cf. *scabra* and aff. *scabra* types (Maass 1989). Relationship between genetic diversity and geographical distribution of different species of stylo would lead in facilitating a systematic exploitation of the existing collections and identifying regions with high genetic diversity where there are good opportunities to collect novel genotypes. Recently concluded ACIAR, Australia sponsored stylo project (High yielding anthracnose–resistant *Stylosanthes* for agricultural systems) where Australia, India, Brazil, Colombia and China participated, has highlighted the importance of regional evaluation of stylo germplasm in exploring new ecological niches that may be suitable for the more productive and anthracnose resistant germplasm. Being most damaging disease of stylo, a good collection of germplasm and their evaluation have been made against anthracnose (Cameron and Trevorrow 1988).

10.3 Nutritional Assessment

Nutritional evaluation, particularly the amino acid profile revealed that though *S. scabra* cv Seca is anthracnose resistant cultivar, it is poor in quality (low essential amino acids) in comparison to *S. guianensis* cultivars (Guodao et al. 2004). *Stylosanthes* leaf meal (SLM) when supplied in different animal rations especially of ruminants and monogastrics, almost 50 % cut in concentrate supply was observed to get equivalent milk production. This was because of abundance of sulphur containing amino acids in *S. seabrana* over other species of *Stylosanthes* (Fig. 10.2). As success observed in China, role of SLM in significant gains in live weight and the cost of production in boiler chicken were observed in India too. Findings indicated that *Stylosanthes* meal can replace more expensive ingredients in the ration by up to 6 % without any adverse effect on the final product. Skin and shank colour of the boilers were more appealing on rations containing leaf meal (Changjun et al. 2004). Results based on modeling using the available Indian data pertaining to *S. hamata*, *S. scabra* and *S. guianensis* indicated widespread suitability of soils for *S. hamata* and *S. guianensis*. Due to requirement of high rainfall *S. guianensis* is likely to be restricted to coastal regions of Southern India. *S. scabra* was shown to be moderately suitable over a wide area, but less suitable in central and western districts (White et al. 2001). Both transpiration efficiency and carbon isotope discrimination analysis in stylo indicated the importance of these characters along with specific leaf area in identifying the drought tolerant lines (Chandra and Bhatt 2008; Thumma et al. 1998). The causal nature of relationships among these characters has been also substantiated using QTLs approach (Thumma et al. 2001). Hence, wider suitability of stylo species to different types of soils and environments indicate their role in providing nutritional security to ruminants and small animals.

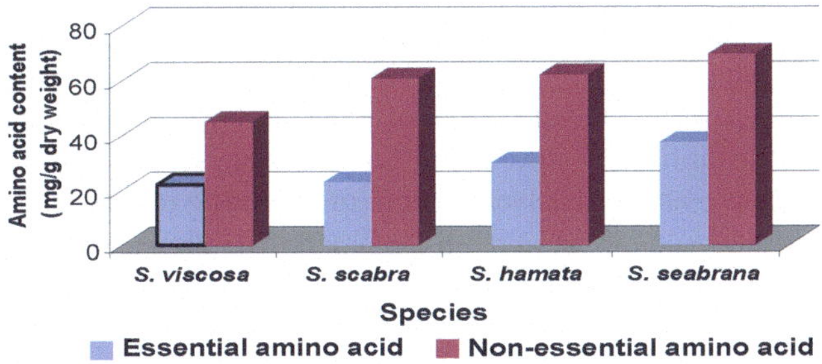

Fig. 10.2 Amino acid profile of *Stylosanthes* species

10.4 Genome Constitution and Details of Major Cultivable Species

Ten basal genomes named A through J have been reported and in the evolution of *S. scabra* (genome AB), *S. seabrana* and *S. viscosa* have been proposed as A and B genome donor respectively. Accessions of *S. seabrana* resemble *S. scabra* in many characters, however utilizing STS markers we have recently identified nine new accessions of *S. seabrana* (Chandra and Kaushal 2009) which not only increased world germplasm pool, also provided options for its suitability testing under various cropping systems. Use of genome-specific molecular markers, such as STS, not only facilitates identification of unknown species/accessions, but also provides information about the genetic relationships among the accessions. Emphasis have been made wherein the impact of *S. seabrana*, being one of the progenitors of *S. scabra*, in improving *S. scabra* for characters such as drought tolerance as it could be possible to breed *S. scabra* using pre-selected better performing progenitors as genetic variations existing in the two diploid progenitors species can be exploited (Tewari and Chandra 2008).

Of late, research on this crop has taken a momentum however the improvement of stylo utilizing molecular tools has not yet made any dent primarily due to non-availability of linkage map and enough co-dominant markers *vis a vis* low level of polymorphism exhibited with different sets of available markers. Therefore, there is a need to develop enough SSR markers to be used for linkage map development and also in variation study to identify lines and for allele mining for abiotic stress adaptation as *S. scabra* have been reported as a hardy species. Testing and evaluation of wide germplasm lines on acid and saline soil which contribute major part of the soil of some countries has to be strengthened. The paramount importance of this crop can be truly assessed when gradual decline in cultivated lands are taken into account. The scenario of limited grasslands world wide is alarming which further deteriorated due to ecological retrogression and increased soil erosion and climate change. Potentially, the vast areas under wastelands, degraded forests etc. can be transformed by introducing pasture and forage species via agropastoral, agroforestry and hortipastoral systems. The greater effort should be driven towards seed industry which will play a major role in any future expansion of *Stylosanthes* utilization schemes. There is a need to improve the contribution of *Stylosanthes* to small and marginal land holder production systems and also its incorporation into mixed farming which hitherto has been highly effective in improving the living standard of such peasants. Great opportunity lies ahead in research and development to suggest its utilization for cut and carry, in situ grazing, leaf meal production under intensive and extensive systems of livestock rearing and livelihood. Each species of *Stylosanthes* as some of them narrated below has its role in different agro-ecozones for its optimum production and use.

Stylosanthes hamata: In comparison to all other species, *S. hamata* have been observed highly diversified species in terms of both adaptation and yield performance including seed production. The main advantage of this species is its land covering capacities to maintain and conserve moisture and that leads to fast growth (Fig. 10.1). Both diploid and tetraploid lines have been reported in this species. However, tetraploid species is more prominently used as forage. Diploid *S. hamata* and *S. humilis* are the two progenitors of tetraploid *S. hamata* (Curtis et al. 1995). This species is annual/short perennial and in nature and flower mostly in month of September and October. It has been observed highly palatable and has been preferred by animals over other species. The minimum level of stickiness present in this species is also one of the reasons causing preference by animals. In comparison to *S. scabra* it is less drought–hardy, however, it is grazing tolerant. Some of the lines were peculiarly identified due to their different habit. However, majority of the lines were observed as herbaceous procumbent. Being highly adaptable species has been extensively used by different government and semi-government agencies for different purposes. This species has been identified suitable for reclaiming hill slopes.

Stylosanthes viscosa: This species is marked by early emergence and highly stickiness of the leaf and stems. This is also characterized by the whitish leaves. The height of this species usually do not cross 50 cm. Being one of the diploid progenitor of *S. scabra*, the potential lines of this species can be used in synthesizing artificial *S. scabra* lines. This species has got less preference however, browser chews them as and when made available.

Stylosanthes scabra: It is most hardy species and well suited under low rainfall conditions. It is tetraploid ($2n = 4x = 40$) and perennial in nature. This particular species has been widely bred and many varieties are reported. The molecular map so far developed using RAPD and some STS markers in *Stylosanthes*, cultivars of this species has been used in generating intra or inter-specific populations. The two diploid progenitors of this species is *S. viscosa* ($2n = 2x = 20$) and *S. seabrana* ($2n = 2x = 20$).

Stylosanthes humilis: This species was once the major stylo species in Australia but due to anthracnose much of the areas was destroyed in 1970s and it is now had little significance in stylo based pasture in Australia (Chakraborty et al. 1996). It is herbaceous and short in height in comparison to *S. hamata* and flower very early in comparison to all other species of *Stylosanthes*. It is also normal in feel and annual in nature. The seed of this species is recognized by the presence of large hook and small size of the seed in comparison to *S. hamata*. Stem is also thin and leaves are pale green in colour. This species has been sidelined due to its susceptibility towards anthracnose and low productivity. Presence of hairs on stems and leaves are also some of the characteristic features of this species.

Stylosanthes guianensis: This species has been found most suitable where humidity and moisture are more. Thus, in India it is found well suited to high rainfall areas of Assam, west Bengal, Maharastra and Andaman and Nicobar islands. It is leafy and provides good quality forage. Due to big leaves it has been identified as a suitable line for the production of leaf meal. In China and Brazil this species is largely used for such purposes. Flowering is also late in this species.

Stylosanthes fruticosa: *S. fruticosa* Alston synonyms *Stylosanthes flavicans* Baker, *Stylosanthes mucronata* Wild, *Stylosanthes bojeri* Vogel and *Arachis fruticosa* Tetz.

This is commonly known as African stylo and amenable to stable mixture with perennial grasses such as *Andropogon gayanus* and *Heteropogon contortus*. In areas protected from grazing and intermittently mown, it generally forms dense sward. Collections have been made from 36 sites of Tamil Nadu, 7 sites of Pondicherry, 29 sites of Andhra Pradesh and 38 sites of Karnataka state of India. Materials collected have shown morphological distinctness as some of them have been found growing to the level of 1800 m from sea level. Though in India the largest collections of *S. fruticosa* accessions is maintained at Indian Grassland and Fodder Research Institute (IGFRI), Regional Research Station (RRS), Dharwad, the molecular characterization to maintain identity and purity for proper conservation and management for better use in breeding and proprietary reasons have not been attempted.

Stylosanthes seabrana: This is a recently identified species first time introduced in India in 1998–99 through ACIAR collaborative project. This is one of the progenitors of hardy *S. scabra,* shown great promise in India and China. Evaluation at different research centers and farmers fields this species has shown great potential in terms of its utility. It is highly nutritious and showed best establishment in different types of soils. It is diploid in nature and has been selected from the original field of *S. scabra*. It is erect in nature, stem are not hardy can be used for leaf meal prodcutions. It is also observed highly palatable. In the very first year of establishment this species was superior in both seed production and green biomass. Apart from materials received from ACIAR project many lines were selected from *S. scabra* growing plots. Molecular characterization of 19 lines of *S. seabrana* showed low levels of polymorphism among lines. The number of fragment shared by *S. seabrana* to *S. scabra* was higher in comparison to *S. viscosa*. The availability of such lines provided an opportunity for the improvement of *S. scabra* by selecting nutritious *S. seabrana* lines. It is also superior in many terms like frost tolerance, high seedling vigour and essential amino acid level. The regeneration potential of this species was also found remarkable in terms of both yield and protein. Species has performed well when it was sown in watershed areas where moisture is better. The added advantage of this species as observed was the better production of seeds in comparison to other species. Animals especially browsers and even in form of cut and carry systems this species has shown excellent promise as this is highly preferred over other species. The flowering of this species is almost synchronized and thus helps in collection of seeds and economically highly viable. Preliminary results indicated the better germination of this species under mild level of salinity. In addition to the presence of high sulphur containing amino acids, erect plant habit with dark green leaves and thin stem makes suitability of this species for leaf meal and stylo meal.

10.5 Plant Systematic Study

In total ten basal genomes (A to J) have been defined in *Stylosanthes* (Liu et al. 1999; Ma et al. 2004). The genus *Stylosanthes* has a monophyletic origin and is closely related to *Arachis*. In 1838 Vogel established the main division of the genus into the sections *Styposanthes* and *Stylosanthes* (Maass 1989), based on the presence of rudimentary secondary floral axis and two inner bracteoles in the former and no such axis and only one inner bracteole in the latter (Kirkbride and Kirkbride 1985; Mannetje 1984). Annual and diploid namely *Stylosanthes guianensis* species complex is the most ancient group. Genome A and C has been identified as natural donor for tetraploid with AABB and AACC genome respectively. As more than one species have the same ancestral genomes, it seems most likely that more than one hybridization event has taken place. Because of the disjunction of the natural distribution of several *Stylosanthes* species independent evolution of similar tetraploid combination has also been suggested. This has led to restriction of the *Stylosanthes* species to the isolated areas and only few species are widely distributed. This scenario proposes utilization of species having good quality attributes to artificially synthesize the species possessing similar level of genome structure for easy and stable survival.

Low genome specificity of sequence tagged site (STS) markers makes the markers most suitable for progenitor analysis in conjunction with morphological attributes in *Stylosanthes* genus. Sequence tagged sites (STS) markers also revealed similar type of inter-species relationships as shown by RAPD. Such type of markers have been found having added advantage of identifying the species, subspecies and genotypes in *Stylosanthes*, with a view to plant conservation and breeding (Nagaich and Chandra 2009; Nagaich et al. 2009; Vander Stappen et al. 1999). These markers have been successfully demonstrated in identifying the diploid progenitors of *S. scabra, S. fruticosa, S. erecta, S. hamata* allotetraploid and hexaploid species. Two diploid progenitors namely *S. viscosa* and *S. seabrana* of *S. scabra* have been identified by Liu and Musial (1997). In the evolution of *S. scabra* (genome AB), *S. seabrana* and *S. viscosa* have been proposed as A and B genome donor respectively (Figs. 10.3 and 10.4). Recently, our efforts have

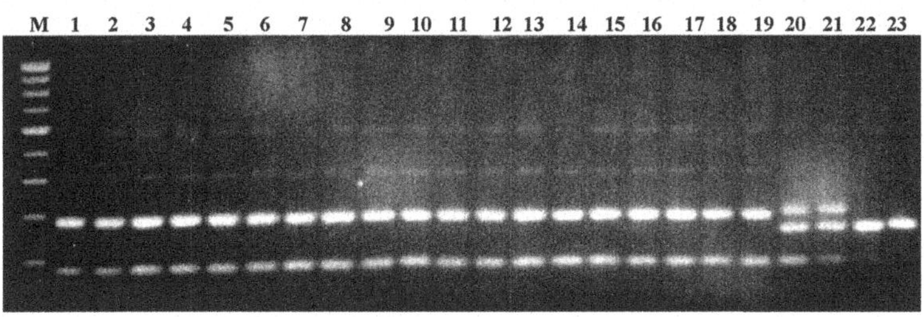

Fig. 10.3 PCR amplification profile obtained using the STS primer SHST1F3/R3 of all *S. seabrana* lines and two accessions of *S. scabra* and *S. viscosa* showing sharing of three bands amplified by *S. scabra*. Lanes 1–10 contains newly identified *S. seabrana* (IG-97-369, IG-97-387, IG-97-391, IG-97-370, IG-97-346, IG-97-355, IG-97-325, IG-97-339, IG-97-384 and IG-97-352) whereas 11–19 already known *S. seabrana* lines (EC-408403, EC-408404, EC-408405, CPI-110372, CPI-105546B, CPI-104710, CPI-2523, CPI-2534 and CPI-2539). Lanes 20 and 21 contains two accessions of *S. scabra* (*cv* Seca and RRR 94–97) and 22 and 23 contains two accessions of *S. viscosa* (CPI-33904 and CPI-33941)

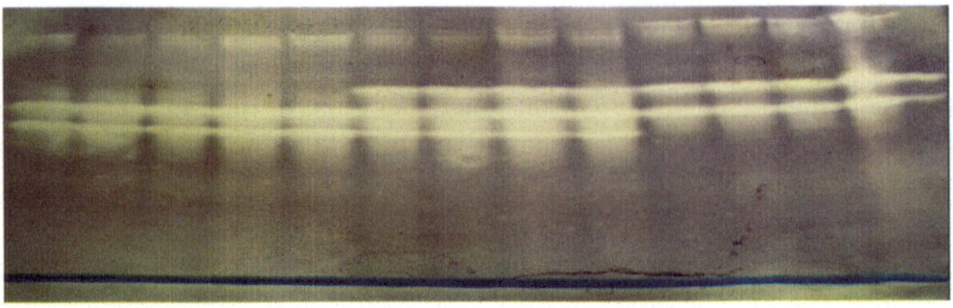

Fig. 10.4 SOD isoform indicating the sharing of bands of two progenitors (*S. viscosa* and *S. seabrana*) of *S. scabra*

made success in identifying the third putative progenitor of only hexaploid species (*S. erecta*) (Ma et al. 2004). Similarly the progenitors of *S. hamata* have been also identified using same markers (Curtis et al. 1995).

The progenitor analysis of *S. scabra* clearly indicated that one of the progenitor of this species is *S. viscosa* belongs to section *Stylosanthes* since it lacks the axis rudiment whereas other diploid progenitor of *S. scabra* i.e., *S. seabrana* belongs to section *Styposanthes* and possessed the axis rudiment. Such characteristic features have tremendously helped in solving problems related species distinction and delimitation (Kirkbride and Kirkbride 1985). Additionally, characters like the presence of prominent hook in *S. hamata* and *S. humilis* seeds also helped in identification of these species. However, as such species identification in *Stylosanthes* based on morphological characteristics is notoriously difficult. There are different views on the species concept and little agreement exists on characteristics most suitable for species identification (Costa and Ferreira 1984; Mannetje 1984). As a result of these difficulties, there have been disagreements over the species status of a large proportion of the *Stylosanthes* genus (Mannetje 1984; Williams and Gardner 1984). This situation has, however, changed dramatically over the last few years due to the development and applications of molecular markers. In addition, the application of chloroplast DNA markers has allowed the identification of maternal donors of majority of the polyploid *Stylosanthes* species (Liu and Musial 2001; Ma et al. 2004). There are reports that a group of accessions of *Stylosanthes,* of special interest for their adaptation to the heavy clay soils, were assessed on the basis of morphological characters and were grouped in three categories and main attributes contributed to the separation of the groups were the presence or absence of a stipule horn bristles, stipule horn lateral bristles, inflorescence bristles and leaf hairs either absent, all over leaf or back of leaf only.

10.6 Molecular Markers and Genetic Diversity Estimate

The accessions of major *Stylosanthes* species have been characterized using both biochemical and molecular markers. When large number of *S. scabra* accessions was characterized some of the accessions though acquired as *S. scabra* found more closeness to the newly identified species *S. seabrana*. The average dissimilarity among Brazilian accessions was much lower than those among Colombian and Venezuelan. The variation among Brazilian materials was more in comparison to other accessions and this was may be due to long distances introductions/dispersions of *S. scabra* accessions with Brazilian genotypes (Liu 1997). Four agronomic important species (*S. scabra, S. hamata, S. humilis, S. guianensis*) have been together analyzed using RAPD markers. Relatively low level of polymorphism (0–16 % of total bands in pair wise comparison) were found within each species, while polymorphism between the species were much higher (up to 46 %) (Kazan et al. 1993; Liu et al. 1999). Low polymorphism (0–2 %) were detected between the individuals of the same cultivar or accessions. The allotetraploid species *S. hamata* and its putative progenitor, *S. humilis* were more akin to each other than *S. scabra* and *S. guianensis*. No variation in RAPD markers was found between the two commercial *S. hamata* cvs Verano and Amiga. Recently Nagaich et al. (2009) used RAPD, IISR, SSR and STS markers in diversity estimate of more than 60 accessions of *S. hamata*. Low variations in seed proteins of *S. seabrana* have been also observed (Chandra 2008) (Fig. 10.5). Suitable RAPD and STS markers have been also reported for genetic estimation of *S. fruticosa*, only species endemic to southern parts of India (Chandra 2007). Cultivar Oxley in *S. guianensis* was considerably different from the other cultivars and accessions of this species (Fig. 10.6). The low level of polymorphism as reported within each species suggested that interspecific crosses may be better approach for the construction of linkage map in *Stylosanthes* (Chandra 2006).

AFLP markers have been also used in studying the variability in *S. humilis* accessions of South American origin and these markers have differentiated all Mexican accessions from the accessions of South American origin. Although most Mexican accessions formed one major group, one accessions clustered with South American gene pool, indicated that Mexico may contain unique sources of *S. humilis* and therefore would merit attention for conservation and maintenance of *S. humilis* germplasm (Vander Stappen et al. 2000). At IGFRI, Jhansi, India nineteen accessions of *S. seabrana* have been evaluated using 30 RAPD markers and seed protein markers indicated sharing of more than 95 % bands among accessions (Tewari and Chandra 2008). In 12 accessions of *S. seabrana* the dissimilarity values have been reported to the level of 0.078. These 19 *S. seabrana* lines were evaluated along with *S. scabra* and *S. viscosa* indicated many bands which were not present in *S. seabrana* though being one of the progenitors of *S. scabra*. Similarly *S. viscosa* has also not shown all bands present in *S. scabra*. The sharing of bands among *S. seabrana* to *S. viscosa* was also not to high level. Very low level of polymorphism have been also observed among the commercial Seca individuals (Chakraborty 2004).

Despite the fact that STS markers have provided a new look in the genetics of *Stylosanthes*, limitation of such markers in diversity analysis and linkage map is known due to its low genome specificity and number. Therefore, the breeding implications in this important genus is lacking as no molecular analysis has been carried out using a robust set of markers in the introduced and assembled

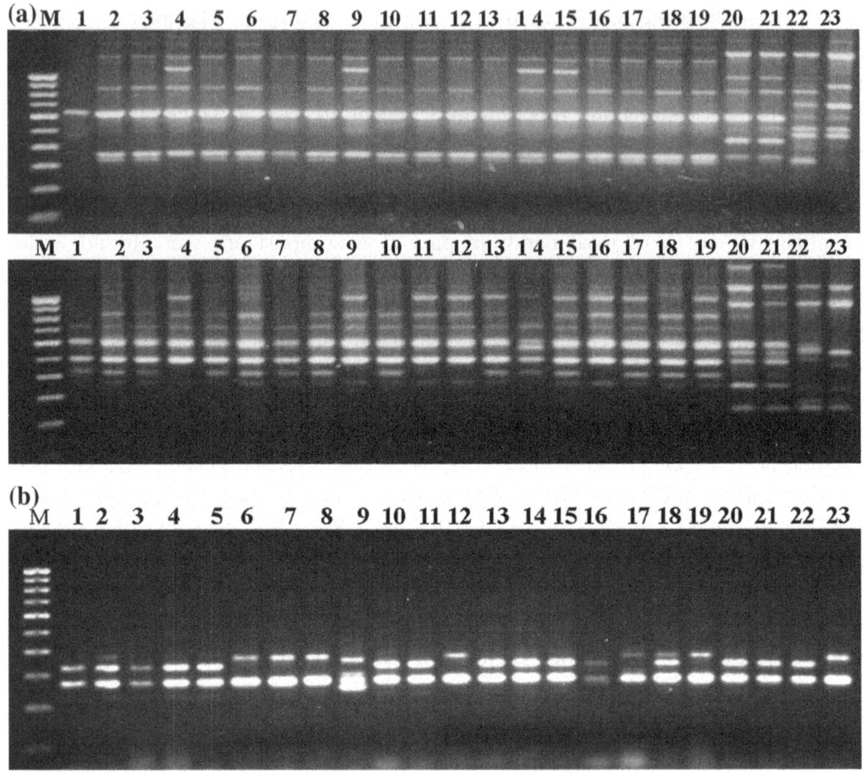

Fig. 10.5 **a** RAPD profile of 19 lines of *S. seabrana* (1–19) along with *S. scabra* (20, 21) and *S. viscosa* (22, 23) with primers OPAE07 and OPQ09 indicating low level of polymorphism among *S. seabrana* lines. **b** Agarose (1.6 %) gel electrophoretic patterns of 23 *S. hamata* accessions (Lanes 1–23) with STS primer pairs SHCAPEF3/R3. First lane has 100 bp DNA ladder as molecular weight marker (M)

accessions from different countries. The development of molecular markers like sequence-tagged microsatellite site (STMS) or simple sequence repeats (SSR) has been very slow in this genus. Till now 38 SSRs markers (both genomic and EST based) from *S. guianensis* (Santos et al. 2009a; Vander Stappen et al. 1999), 18 from *S. capitata* (Billotte et al. 1999; Santos et al. 2009b) and 16 SSR markers developed in silico using genomic and EST sequences of 11 species of *Stylosanthes* (Mace et al. 2008), are available. Recently we have developed 41 EST and genomic based SSR markers in *S. seabrana* (Chandra et al. 2011).

Breeding for disease resistance and other characters would benefit from the development of molecular marker technology. Considerable pathogenic and genetic diversity has been reported in the pathogen population from centre of both diversity and utilization (Weeds et al. 2003). Breeding programs for improvement of stylo were initiated long back by CSIRO Australia, CIAT Colombia and

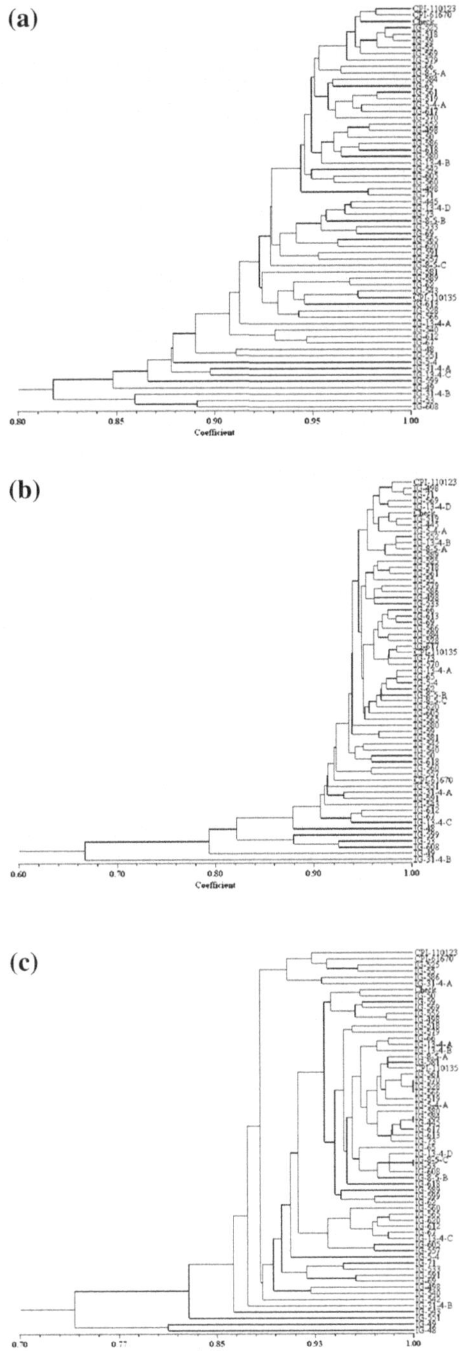

Fig. 10.6 UPGMA dendrogram showing clustering of 63 *S. hamata* accessions based on RAPD (**a**), ISSR (**b**), and STS (**c**) data

EMBRAPA Brazil in addition to their germplasm enrichment activities (Cameron et al. 1997). Only two released cultivars i.e., cv Siran (Cameron et al. 1996) and cv. Estilosantes Campo Grande (Miles and Lascano 1997) have been bred to date. Though classical taxonomic treatments of *Stylosanthes* based on some aspects of the floral and fruit morphology provided basic taxonomic relationships of the species, the molecular techniques offers a number of advantages in deciphering such relationships. Variation detected by the molecular analysis of DNA can be quantified, provides many characters, and is not subject to environmental effects. The level of variations and estimate of genetic diversity and at the same time the production potential of different species and cultivars required to be assessed in larger perspective so that the available germplasm can be targeted for different zones and climate of the country.

10.7 Abiotic and Biotic Stresses and Impact of Biotechnology

Perennial stylo like *S. scabra* depicted various adaptive strategies to cope with very severe water deficits. *S. scabra* over other species exhibit substantial osmotic adjustment that contribute to the maintenance of turgor at low tissue water potentials (Kelemu et al. 2004) and are able to root deeply and extract water at water potentials considerably lower than the −1.5 MPa normally associated with wilting point. Accumulation of biochemicals ultimately led to better osmotic adjustment and thus the maintenance of turgor at low tissue water potential. This makes the perennial species of stylo in general to survive in dry seasons of tropics. Variation in drought tolerance observed on the basis of different biochemical attributes in different *Stylosanthes* accessions indicated some of the *S. scabra* lines especially Rate Reducing Resistance (RRR) as a good source of materials for drier regions (Chandra et al. 2004). Additionally, transpiration efficiency (TE) or water use efficiency and carbon isotope discrimination (Δ) study indicated negative relationship between TE and $\Delta(r = -0.71)$. In this case also perennial *S. scabra* cv Seca maintain the highest TE under both control and stress treatments (Thumma et al. 1998). Using *S. scabra* 93116 × *S. scabra* cv Fitzroy F_2 population, the causal nature of different drought responsive traits as well as relationship between TE and Δ was confirmed through QTL analysis (Thumma et al. 2001). Genetic linkage map developed with intra-specific population (*S. scabra* cv Fitzroy × CPI 93116) used with traits for QTL a significant correlation of Δ was observed with TE and biomass production. Most of the QTLs for TE and Δ were observed on linkage group 5 and 11. Similarly, QTL for specific leaf area (SLA), transpiration and biomass productivity traits were clustered on linkage group13 and 14. The causal nature of relationship between TE and Δ was observed as at the coincident markers between TE and Δ, high alleles of TE were associated with low alleles of Δ (Thumma et al. 2001).

In another set of experiment where twenty genotypes of *Stylosanthes* consisting four species were evaluated under rain fed condition employing biochemical and physiological attributes indicated that *S. scabra* genotypes were more tolerant to drought over other lines as evidenced by high leaf thickness, proline accumulation, content of sugars and chlorophyll and nitrate reductase activity (Chandra et al. 2004). Thicker leaves (low SLA) and ability to possess high level of osmotic potential under water stress makes most of the *S. scabra* RRR lines suitable for dry zones. Of these, *S. scabra* RRR94–97 and RRR94–86 were most promising as they showed better response in terms of NR activity, content of soluble protein and soluble sugar as well as other physiological characteristics (Chandra et al. 2004; Chandra and Bhatt 2008) (Fig. 10.7).

In early 1970s, it was estimated that approximately 2 million ha colonized by sown and naturalized Townsville stylo (TS, *S. humilis*) in Australia was destroyed by the major disease of stylo, anthracnose, caused by *C. gloeosporioides* and it is now of little significance in stylo pastures (Chakraborty et al. 1996). The demise of TS was a devastating event in the historical development of stylo pastures in Australia. This led to the identification and characterization of suitable parents and selective breeding for the development of durable resistant cultivars. Though a reasonable number of cultivars in different countries (Australia, Colombia, Brazil, China and USA) have been released, only limited numbers of lines have so far been identified in India. When *S. scabra* cv. Seca (Oram 1990) was released in Australia in 1976 it was highly resistant to all known strains of *C. gloeosporiodes* (Irwin and Cameron 1978), by 1982 a new race virulent on Seca was identified (Davis et al. 1984). The latest two cultivars of *S. seabrana* Primar and Unica have started showing disease as new races have evolved within 5 years of their release (Trevorrow et al. 1998). Since its introduction in India, at least four pathogen race clusters were observed aggressive to highly aggressive on the *S. seabrana* accession 105546B in India (Ramesh et al. 2004). However, widespread damage has not been recorded at field sites in India on this species (Ramesh et al. 2004). Thus, careful and continued monitoring of the pathogen population is necessary to detect early signs of severe damage following the introduction of new cultivars. Two mapping populations viz., *S. scabra* 93116 × *S. scabra* cv Fitzroy and *S. scabra* cv Seca × *S. fruticosa* were used to develop map along with phenotypic data for anthracnose disease to identify QTLs. With the first population a

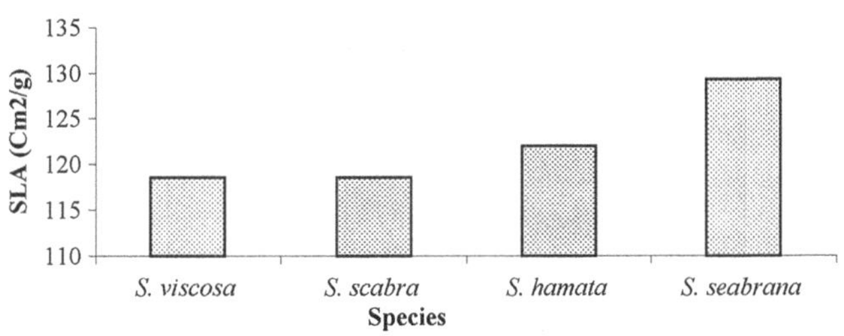

Fig. 10.7 Variation in specific leaf area in four species of stylo

map was constructed using 151 RAPD markers. Six QTLs linked with anthracnose resistance have been identified at different levels of LOD scores located on six chromosomes.

Two genetically distinct biotypes of *C. gloeosporioides* infecting *Stylosanthes* spp. have been identified in Australia, which have clonally descended from two separate introductions (Irwin and Cameron 1978; Manners and He 1997). Biotype A causes disease on most of the species whereas biotype B manifests as a blight symptom on leaf and stem of *S. guianensis*. Apart from these two biotypes there are strains in South America and Africa that do not clearly belong to either biotype (Chakraborty et al. 2004). Despite the relatively short history of the *Stylosanthes* in India the anthracnose pathogen has diversified and needs regular monitoring of the pathogen population to detect early appearance of new races especially when new species i.e, *S. seabrana* which is close to native *S. fruticosa* (Liu and Musial 2001) has been introduced. Various forms of resistance to stylo anthracnose have been reported from breeding programs conducted at CSIRO, Australia (Cameron et al. 1997) and in South America by CIAT (Miles and Grof 1997). Oligomeric resistance has been detected in some accessions of *S. guianensis* (Cameron and Irwin 1983) and *S. scabra* (Irwin et al. 1986); race specific is found in *S. viscosa* accession CPI 33941 (Irwin et al. 1986); quantitative resistance believed to be due to more than one gene has been observed in some accessions of *S. guianensis*, (Miles and Lenne 1984) *S. scabra* (Chakraborty et al. 1990) and in *S. hamata* (Iamsupasit et al. 1991). Results employing eight parents suggested that the resistance is controlled by several genes and is quantitatively inherited (Iamsupasit et al. 1995).

10.8 Genetic Engineering and Its Prospects

Recombinant DNA technology and gene transfer systems allows the isolation and introduction of various genes from one source to other and therefore many genes of agronomical value have been targeted. Genetic engineering not only widens the pool of useful genes (by removing the species barriers encountered in traditional plant breeding methods) for use in biotic and abiotic stress management, but also allows the use of several desirable genes in a single event, thus shrinking the time needed to incorporate novel genes into an elite plant background. No transgenics bearing genes of economical importance so far have been reported in genus *Stylosanthes*, however, the genes against anthracnose resistant and drought tolerance are of the prime importance. In *S. guianensis* the expression of SgNCED1 (9-cis-epoxycarotenoid dioxygenase) has been observed to be induced in both leaves and roots. Both dehydration and salt stress induced the expression of SgNCED1 strongly and rapidly (Yang and Guo 2007). However, in lack of clear understanding of the molecular basis of host-pathogen interactions in *Stylosanthes* applicability of such technology is quite limited so as to modify plants with durable resistance to a wide range of pathogens as reported in other crops (Dixon et al. 1996; Martin et al. 2003). Nevertheless, attempts have been made to increase the digestibility and resistance to anthracnose through genetic engineering approach by inserting anti-fungal protein in stylo (Rae et al. 1997). Lignin is one of the major components of fiber in tropical forages and its high level has been found to be associated with poor digestibility and low nutritional value. In this direction, efforts have been made in suppressing one of the enzymes of lignin biosynthesis in *S. humilis* and further the difference in concentration of lignin levels has been reported (Rae et al. 1997).

10.9 Tissue Culture and Its Impact on *Stylosanthes* Research

As with other crop species, transformation and regeneration protocols have been developed for *Stylosanthes* species which is prerequisite for any molecular manipulation study (Kelemu et al. 2001; Manners 1987, 1988; Manners and Way 1989; Quecini et al. 2002; Sarria et al. 1994). In past, over other species *S. guianensis* was the main target for such work. Recently, Kumar and Chandra (2010) reported the regeneration protocols for *S. hamata* and *S. seabrana* via callus induction from cotyledonary and hypocotyl explants (Fig. 10.8). Direct plant regeneration via multiple shoot induction in *S. seabrana* have been reported by Kumar and Chandra (2009). This protocol would be very useful for *Agrobacterium*-mediated genetic transformation of *Stylosanthes* with the gene(s) of interest. Vieira et al. (1990) reported plant regeneration from protoplast in *S. guianensis*, *S. macrocephala*, and *S. scabra*, while Valarini et al. (1997) reported plant regeneration from

Fig. 10.8 Callus induction and plantlet regeneration in *S. seabrana*

leaf-derived callus culture of *S. scabra*. BAP and NAA in different concentrations have been used for callus induction as well as plant regeneration in stylo by most of the workers (Consoli et al. 1996; Dornelas et al. 1991; Godwin et al. 1987; Meijer 1982; Meijer and Broughon 1981). The standardized protocol for in vitro plant regeneration would be very useful in genetic transformation of the stylo species, thus making molecular manipulation of the genus feasible.

10.10 Utilization of Stylo and Grazing System

Though stylo adapts best to light texture soils the relative importance of different abiotic constraints on the growth, survival and productivity of it depends on the physiological adaptation of species. The tolerance of the crop towards a reasonable level of drought and providing quality fodder and also enriching the soil, stylo has been recommended for different use. The recent research in the direction of quality and amino acid profile also indicated the use of stylo, in stylo meal and leaf meal for poultry and as a replacement of concentrate. This crop has also provided opportunities to several seed growers to improve their livelihood. In contrast to Australia where native grasses are usually intolerant of the increased grazing pressure and that follows introduction of *Stylosanthes* cultivars, here in India the employed stylo are preferred and in few years it disappears. Dominance of stylo in Australian pasture is one of the main concerns and that causes introduction of many improved grasses and controlled grazing. Improved grasses like *Cenchrus ciliaris*, *Bothriochloa pertusa* are better able to persist with *Stylosanthes* under this increased grazing pressure. The improved grasses are more competitive where phosphorus availability is moderate to high; in the case of *C. ciliaris,* legume may disappear from the pasture at the moderate to high levels of soil P. The fertilizer management has made more competitiveness of stylo with different grasses and in particular ratio of *Cenchrus* and stylo has yielded sustainable production.

Successful intercropping of *S. hamata* with cereals such as sorghum, maize and bajra under rainfed conditions has been shown to increase grain yield of cereals by 6–26 % at various sites in India (Ramesh et al. 1997). Compared with cereal monoculture, residue from legume-cereal intercropping improves animal nutrition due to high nitrogen content. Use of legume leys helps maintain soil cover, provide additional forage and improves soil nutritional and physical properties. Many framers in southern part of the country have started using stylo on their bunds to conserve the soil moisture in addition to soil erosion. The underlying aim must be to retain enough seed of stylo in the soil seedbank so that once it is established; it can be used for one or more years without reseeding.

More than 50 % of the land in India is subject to land degradation of one or another. As a pioneering colonizer, stylo establishes well on poor soils under dryland conditions, even when the top soil is severely eroded. Additionally, it can fix 20 to >200 kg N/ha/year depending on the agronomic, edaphic and environmental conditions (Vallis and Gardener 1984). Its ability to improve bulk density, infiltration rate and water holding capacity makes it a useful species in the conservation, stabilization and sustainable development of land and water resources, especially in fragile environments. Due to these properties many Government agencies have identified this crop to improve degraded and wastelands.

10.11 Conclusion and Prospects

Being narrow genetic base and susceptibility to anthracnose disease, it is quite pertinent that the combining of elite stylo lines not only generate suitable materials having durable resistance genetic resources but also make available good back up cultivars of stylo if new pathogen race seriously damage such elite lines. Therefore, such efforts should continue. The screening of materials under different agro-climatic zones are required to be continued so that the genetic vigour that is existing can be adapted easily under diverse environment. Use and exploitation of weather data and GIS should be promoted for targeting new ecological niches. More and enough number of germplasm may be required for evaluation at different established stylo sites so that the suitable genotypes can be identified for problem soils i.e., salt affected and acidic soils. Therefore, a program on such line has to be initiated and implemented. The efficiency and number of polymorphic molecular markers has to be geared up and employed in targeted breeding programs as the crop in general is less divergent. The development and use of SSR markers would be one step ahead in this direction. Anthracnose disease caused by the fungus *C. gloeosporioides* disease was first recorded at Deodoro in Brazil in 1937 on *S. humilis* (Anon 1937) and is now widespread in all countries where this legume is grown. The reliance on susceptible to moderately resistant cultivars poses a potentially serious threat from anthracnose. Identification and development of germplasm with increased resistance and wider geographical adaptations are important research priorities.

Apart from the anthracnose threat the limited number of potential germplasm availability necessitated adaptations of the new breeding strategies to increase the overall production of stylo. The potential of *S. seabrana* for tropical and subtropical regions with clay and heavy soils, cool winters and distinct wet-dry seasonal conditions directed to use of this species in developing new breeding approach. The one could be based on the finding that it is the one of the progenitors of hardy and drought tolerant *S. scabra* which in turn elucidated the evolution of one of the most important *Stylosanthes* species, *S. scabra* may lead to important impacts on the efforts of improving *S. scabra*. It should be possible to artificially synthesize *S. scabra* using pre-selected *S. viscosa* and *S. seabrana* accessions. These artificial *S. scabra* genotypes could be used directly or more likely, be used in breeding programs. By doing so the genetic variation existing in the two diploid progenitors species would become available in improving the allotetraploid *S. scabra*. So far developed map and linked markers with anthracnose resistance also provide the opportunity to use them after converting them in STS or SCAR in breeding programs. The phylogenetic relationships among different *Stylosanthes* species also provided sufficient clue to use them in making new breeding strategies. The resistance available in some elite lines of *S. scabra* as well as further evaluation of these and some additional lime is required to select putative replacement cultivars especially where *S. scabra* is a productive pasture component. The molecular genetical approaches have provided the means to broaden the resistance base of elite lines. The new resistance source identified through molecular markers can be backcross into elite lines so that the back-cross selection would have broad based resistance to anthracnose.

Though tolerance of abiotic constraints (low soil fertility, low pH, frost, flooding and drought) can be readily identified in diverse germplasm collections, the challenge to combine either required tolerance or tolerance(s) of one or more constraints with other attributes (e.g., forage yield, quality and seed yield) is a formidable task. A long term breeding project aimed at retaining desirable attributes (frost tolerance and drought resistance) of subtropical fine-stem stylo (*S. guianensis* var intermedia) while improving forage yielding ability and broadening environmental adaptation was eventually abandoned (Cameron et al. 1997). However, the recent finding by Thumma et al. (2001) where identified QTLs for physiological and production traits would permit pyramiding of the different physiological components of drought tolerance in the coming years breeding program of *Stylosanthes.*

Leaf meal production from *S. guianensis* in China is an important success story. Studies in India have established the economic feasibility of using dried stylo leaf meal as feed concentrate to supply protein and other nutrients for poultry and pigs. With the available post-harvest processing technologies (Davis et al. 1984; Trevorrow et al. 1998), it has been considered as one of the best forage legumes for development of edible/oral vaccine for animals (Manners and He 1997).

Sowing *Stylosanthes* in tropical grass pastures has improved its quality and tremendously increased the animal production in many countries. Additionally, it has the potential to improve tropical and subtropical environments. Despite the major initiatives taken by IGFRI, Jhansi, India as reported in special issue on Stylo–Research and Development, no cultivar has yet been formally released at national level. Of late, a variety (Phule Kranti) of *S. seabrana* has been released at state level from Rahuri (Maharastra) and the concerted efforts placed at IGFRI led to the identification of 8 promising accessions of *S. seabrana* for their trials by AICRP on forage crops for evaluation and release of a variety at national level. In absence of released and notified varieties, two Australian cultivars *S. scabra* cv. Fitzroy and *S. hamata* cv. Verano are widely grown in India. Due to vulnerability of these two cultivars towards anthracnose, the commercial seed production of these cultivars suffered seriously. Of late, introduction of several new improved germplasm including newly identified species, cultivars and bred materials from Australia, Colombia, Brazil and Ethiopia and their evaluation at selected centers under different agro-climatic conditions improved the existing scenario as many lines including *S. scabra* RRR as well as newly introduced species *S. seabrana* has shown great promise for diverse agro-climatic zones (Chandra 2010). Because of concerted efforts which was largely generated from the recently concluded ACIAR stylo project and background study during the period of early 80s *Stylosanthes* has been considered as the most important tropical legume which not only improve the soil fertility but also provide nutritive forage (Pathak et al. 2002).

In general microsatellite markers are considered as marker of choice owing to their capacity to detect high variability *vis-à-vis* being co-dominant in nature, making them powerful tools for population genetical analyses of *Stylosanthes* should be continued. To strengthen the molecular breeding work it is necessary that sufficient SSRs must be developed in this genus. Our reports along with other studies in *Stylosanthes* species, clearly establish that microsatellite markers could be reliably used for assessing genetic diversity, identification of agronomically valuable and diverse germplasm for use in linkage mapping and genetic enhancement of specific traits in *Stylosanthes*.

References

Anon (1937) Informaceos sobro algumas plantas forrageiras. Ministerio da Agricultra: Rio de Janeiro

Billotte N, Lagoda PJR, Risterucci AM, Baurens FC (1999) Microsatellite-enriched libraries: applied methodology for the development of SSR markers in tropical crops. Fruits 5:277–288

Burt RL, Williams WT (1975) Plant introduction and the *Stylosanthes* story. Aust Meat Res Committee Rev 25:1–26

Burt RL, Isbell RF, Williams WT (1979) Strategy of evaluation of a collection of tropical herbaceous legumes from Brazil and Venezuela. I. Ecological evaluation at the point of collection. Agro-Ecosyst 5:99–117

Burt RL, Pengelly BC, Williams WT (1980) Network analysis of genetic resources data: III. The elucidation of plant/soil/climate relationships. Agro-Ecosyst 6:119–127

Cameron DF (1967) Chromosome number and morphology of some introduced *Stylosanthes* species. Austral J Agric Res 18:375–379

Cameron DF, Irwin JAG (1983) Inheritance of resistance to anthracnose disease in *Stylosanthes guianensis*. In: Proceeding of 8th Australian plant breeding conference, Adelaide, South Australia, pp 243–245

Cameron DF, Trevorrow RM (1988) Yield of lines bred to improve anthracnose resistance in *Stylosanthes scabra*. In: Michalk DL, Prately JE (eds) Agronoimy-growing greener future. Proceedings of the 9th Australian society of agronomy conference, Wagga Wagga

Cameron DF, Edye LA, Chakraborty S, Manners JM, Liu CJ (1996) An integrated program to improve anthracnose resistance in *Stylosanthes*-a review (online). http://www.regional.org.au/au/asa/1996/contributed/112cameron.htm

Cameron DF, Trevorrow RM, Liu CJ (1997) Recent advances in studies of anthracnose of *Stylosanthes*. II. Approaches to breeding for anthracnose resistance in *Stylosanthes* in Australia. Trop Grasslands 31:424–429

Chakraborty S (2004) Anthracnose disease of *Stylosanthes*. In: Chakraborty S (ed) High yielding anthracnose–resistant *Stylosanthes* for agricultural systems. ACIAR, Australia

Chakraborty S, Cameron DF, Lupton J (1996) Management through improved understanding-A case history of *Stylosanthes* anthracnose in Australia. Pasture and Forage Crop Pathology. Miscellaneous Publication, American Society of Agronomy, Crop science society of America, and soil science society of America, 677 S. Segoe Rd, Madison, WI 53711, USA

Chakraborty S, Pettitt AN, Boland RM, Cameron DF (1990) Field evaluation of quantitative resistance to anthracnose in *Stylosanthes scabra*. Phytopath 80:1147–1154

Chakraborty S, Ramesh CR, Kexian Y (2004) *Colletotrichum gloeosporioides* diversity at centers of utilization in Australia, China and India. In: Chakraborty S (ed) High yielding anthracnose–resistant *Stylosanthes* for agricultural systems. ACIAR, Australia

Chandra A (2006) A partial genetic map of *Stylosanthes* developed using inter-specific F2 cross. Cytologia 71:353–358
Chandra A (2007) Molecular assessment of variation in *Stylosanthes fruticosa* accessions employing STS and RAPD markers. Cytologia 72:251–258
Chandra A (2008) Genetic relationship among accessions of *Stylosanthes hamata* based on seed proteins profile. Ind J Genet Plant Breed 68:208–212
Chandra A (2010) Research and promises of range fodder legume Stylo. NAAS News 10:6–8
Chandra A, Tiwari KK, Nagaich D, Dubey N, Kumar S, Roy AK (2011) Development and characterization of microsatellite markers from tropical forage *Stylosanthes* species and analysis of genetic variability and cross-species transferability. Genome 54:1016–1028
Chandra A, Bhatt RK (2008) Carbon isotope discrimination function analysis and drought tolerance of stylo species grown under rain-fed environment. Acta Physiol Plant 30:63–69
Chandra A, Pathak PS, Bhatt RK (2006) *Stylosanthes* research in India: prospects and challenges ahead. Curr Sci 90:915–921
Chandra A, Pathak PS, Bhatt RK, Dubey A (2004) Variation in drought tolerance of different *Stylosanthes* accessions. Biol Plant 48:457–460
Chandra A, Kaushal PK (2009) Identification of diploid *Stylosanthes seabrana* lines from existing germplasm of *S. scabra* utilizing STS markers and flow-cytometry, and their molecular characterization. Mol Biotechnol 42:282–291
Changjun B, Guodao L, Dongjum W, Krishna D, Qudratullah S, Prasad VLK, Rama Rao SV, Parthasarthy R, Ramesh CR, Balagopal R, Gopalan A (2004) *Stylosanthes* leaf meal for animal industries in China and India. In: Chakraborty S (ed) High yielding anthracnose –resistant *Stylosanthes* for agricultural systems. ACIAR, Australia
Consoli L, Vieira MLC, de Souza Jr CL, Garcia AAF (1996) Tissue culture effects on quantitative traits in *Stylosanthes guianensis* (Leguminosae). Braz J Genet 19:469–474
Costa NMS, Ferreira MB (1984) Some Brazilian species of *Stylosanthes*. In: Edye LA, Stace HM (eds) The biology and agronomy of *Stylosanthes*. Academic, Sydney
Curtis MD, Cameron DF, Manners JM (1995) Molecular evidence that diploid *Stylosanthes humilis* and diploid *Stylosanthes hamata* are the progenitors of allotetraploid *Stylosanthes hamata* cv. Verano. Genome 38:344–348
Davis RD, Irwin JAG, Cameron DF (1984) Variation in virulence and pathogenic specialization of *Colletotricum gloeosporioides* isolates from *Stylosanthes scabra* cvs. Fitzroy and Seca. Aust J Agric Res 35:653–662
Dixon RA, Lamb CJ, Masoud S, Sewett VJH, Paiva NL (1996) Metabolic engineering prospects for crop improvement through the genetic manipulation of phenylpropanoid biosynthesis and defense response- a review. Gene 179:67–71
Dornelas MC, Vieira MLC, de Souza Jr CL (1991) Fonte e idade de explante, radiação luminosa e combinações de NAA e BAP na indução de calos em *Stylosanthes scabra* Vog. Pesq Agropec Bras 26:1901–1909
Gillies ACM, Abbott RJ (1996) Phylogenetic relationships in the genus *Stylosanthes* (Leguminosae) based upon chloroplast DNA variation. Plant Syst Evol 200:193–211
Godwin ID, Gordon GH, Cameron DF (1987) Plant regeneration from leaf-derived callus cultures of the tropical pasture legume *Stylosanthes scabra* Vog. Plant Cell Tiss Org Cult 9:3–8
Guodao L, Changuin B, Dongiun W, He Huaxuan (2004) *Stylosanthes* cultivars in China: their development and performance. In: Chakraborty S (ed) High yielding anthracnose–resistant *Stylosanthes* for agricultural systems. ACIAR, Australia
Iamsupasit N, Cameron DF, Chakraborty S, Gordon G, Irwin JAG, Davis RD (1991) Glasshouse and field evaluation of quantitative resistance to *Colletotricum gloeosporioides* in tetraploid accessions of *Stylosanthes hamata*. Aust J Agric Res 42:429–439
Iamsupasit N, Cameron DF, Cooper M, Chakraborty S, Edye LA (1995) Inheritance of anthracnose resistance in the tropical pasture legume *Stylosanthes hamata*. Aust J Agric Res 46:1353–1364
Irwin JAG, Cameron DF (1978) Two diseases in *Stylosanthes* spp. caused by *Colletotricum gloeosporioides* in Australia and pathogenic specialization within one of the causal organisms. Aust J Agric Res 29:305–317

Irwin JAG, Cameron DF, Lenne JM (1986) Anthracnose problems with Stylosanthes. In: Murtagh GJ, Jones RM (eds) Proceedings of 3rd australian conference on tropical pastures. Tropical Grasslands Society of Australia, Brisbane

Kazan K, Manners JM, Cameron DF (1993) Genetic variation in agronomically important species of *Stylosanthes* determined using random amplified polymorphic DNA markers. Theor Appl Genet 85:882–888

Kelemu S, Guixi H, Segura G (2001) *Stylosanthes guianensis* plants transformed with a rice chitinase gene confer resistance to foliar blight disease. Phytopathology 91:47

Kelemu S, Miles JW, Rao IM (2004) Biotic and abiotic constraints to *Stylosanthes* production. In: Chakraborty S (ed) High yielding anthracnose–resistant *Stylosanthes* for agricultural systems. ACIAR, Australia

Kirkbride JH, de Kirkbride CG (1985) Typification of *Stylosanthes* (Leguminosae) and its sections. Taxon 36:455–458

Kumar S, Chandra A (2009) Direct plant regeneration via multiple shoot induction in *Stylosanthes seabrana*. Cytologia 74:391–399

Kumar S, Chandra A (2010) *In vitro* plantlet regeneration in *Stylosanthes* spp via callus induction from cotyledonary and hypocotyls explants. Natl Acad Sci Lett 33:289–297

Lavin M, Pennington RT, Klitgaard BB, Spent JI, Cavalcante de Lima H, Gasson PE (2001) The dalbergiod legumes (Fabaceae): delimitation of pantropical monophyletic clade. Am J Bot 88:503–533

Liu CJ, Musial JM (1997) S*tylosanthes* sp. aff. *S. scabra:* a putative diploid progenitor of *Stylosanthes scabra* (Fabaceae). Plant Syst Evol 208:99–105

Liu CJ, Musial JM (2001) The application of chloroplast DNA clones in identifying maternal donors for polyploid species of *Stylosanthes*. Theor Appl Genet 102:73–77

Liu CJ, Musial JM, Thomas BD (1999) Genetic relationships among *Stylosanthes* species revealed by RFLP and STS analyses. Theor Appl Genet 99:1179–1186

Liu CJ (1997) Geographical distribution of genetic variation in *Stylosanthes scabra* revealed by RAPD analysis. Euphytica 98:21–27

Ma ZY, Chandra A, Musial JM, Liu CJ (2004) Molecular evidence that *Stylosanthes angustifolia* is the third putative diploid progenitor of the hexaploid *S. erecta* (Fabaceae). Plant Syst Evol 248:171–176

Maass BL (1989) Die tropische Weideleguminose *Stylosanthes scabra* vog.—variabilitat, Leistungsstand und Moglichkeiten zuchteischer Verbesserung. Landbauforschung Voikenrode, Sonderheft 97:1–189

Mace ES, Varshney RK, Mahalakshmi V, Seetha K, Gafoor A, Leeladevi Y, Crouch JH (2008) In silico development of simple sequence repeat markers within the aeschynomenoid/dalbergoid and genistoid clades of the Leguminosae family and their transferability to *Arachis hypogaea*, groundnut. Plant Sci 174:51–60

Manners JM (1987) Transformation of *Stylosanthes* spp. using *Agrobacterium tumefaciens*. Plant Cell Rep 6:204–207

Manners JM (1988) Transgenic plants of tropical pasture legume *Stylosanthes humilis*. Plant Sci 55:61–68

Manners JM, He C (1997) Recent advances in studies of anthracnose of *Stylosanthes*. IV. Molecular approaches to studies of *Colletotrichum gloeosporioides* causing anthracnose of *Stylosanthes* in Australia. Trop Grasslands 31:435–444

Manners JM, Way H (1989) Efficient transformation with regeneration of the tropical pasture legume *Stylosanthes humilis* using *Agrobacterium rhizogene* and Ti plasmid binary vector system. Plant Cell Rep 8:341–345

Mannetje L (1984) Considerations on the taxonomy of the genus *Stylosanthes*. In: Stace HM, Edye LA (eds) The biology and agronomy of *Stylosanthes*. Academic, Sydney

Martin GB, Bogdanove AJ, Sessa G (2003) Understanding the functions of plant disease resistance proteins. Ann Rev Plant Biol 54:23–61

Meijer EGM, Broughon WJ (1981) Regeneration of whole plants from hypocotyl-, root-, and leaf-derived tissue cultures of the pasture legume *Stylosanthes guianensis*. Physiol Plant 52:280–284

Meijer EGM (1982) High-frequency plant regeneration from hypocotyl- and leaf-derived tissue cultures of the tropical pasture legume *Stylosanthes humilis*. Physiol Plant 56:381–385

Miles JW, Grof B (1997) Recent advances in studies of anthracnose of *Stylosanthes*. III. *Stylosanthes* breeding approaches in South America. Trop Grasslands 31:430–434

Miles JW, Lenne JM (1984) Genetic variation within a natural *Stylosanthes guianensis*, *Colletotricum gloeosporioides*, host-pathogen population. Aust J Agric Res 35:211–218

Miles JW, Lascano CE (1997) Status of *Stylosanthes* development in other countries. I. *Stylosanthes* development and utilization in South America. Trop Grasslands 31:454–459

Nagaich D, Chandra A (2009) Assessment of Genetic Diversity and Identification of Informative Molecular Markers for Germplasm Characterization in Caribbean Stylo (*Stylosanthes hamata*). J Plant Biochem Biotech 18:257–260

Nagaich D, Kumar S, Chandra A (2009) Genome-specific STS markers- a tool for species testing of important forage range legume Stylo. Natl Acad Sci Lett 32:327–331

Oram RN (1990) Register of Australian herbage plant cultivars, 3rd edn. CSIRO, Melbourne

Pathak PS, Ramesh CR, Faruqui SA, Bhatt RK, Chandra A, Gangaiah B (2002) Stylo research and development in India: a technical report. CATAS, China

Quecini VM, de Oliveira CA, Alves AC, Vieiraet MLC (2002) Factors influencing electroporation-mediated gene transfer to *Stylosanthes guianensis* (Aubl.) Sw. protoplasts. Genet Mol Bio 25:73–80

Rae AL et al (1997) Genetic engineering to improve *Stylosanthes*. Trop Grasslands 31:128

Ramesh CR, Mal Bhag, Hazara CR, Sukanya DH, Ramamurthy V, Chakraborty S (1997) Status of *Stylosanthes* development in other countries. III. *Stylosanthes* development and utilization in India. Trop Grasslands 31:467–475

Ramesh CR, Melkania MP, Desale JS et al (2004) Regional evaluation of *Stylosanthes* germplasm in India. In: Chakraborty S (ed) High yielding anthracnose –resistant *Stylosanthes* for agricultural systems. ACIAR, Australia

Santos MO, Karia CT, Resende RM, Chiari L, Jungmann L, Zucchi MI, Souza AP (2009a) Isolation and characterization of microsatellite loci in the tropical forage legume *Stylosanthes guianensis* (Aubl.) Sw. Conserv Genet Res 1:43–46

Santos MO, Sassaki RP, Chiari L, Resende RMS, De Souza AP (2009b) Isolation and characterization of microsatellite loci in the tropical forage *Stylosanthes capitata* Vogel. Mol Ecol Resour 9:192–194

Sarria R, Calderón A, Thro AM, Torres E, Mayer JE, Roca WM (1994) *Agrobacterium*-mediated transformation of *Stylosanthes guianensis* and production of transgenic plants. Plant Sci 96:119–127

Sawkins MC (1999) Geographic and genetic studies of *Stylosanthes* Sw. species. Ph D thesis at University of Birmingham, England

Schultze-Kraft R, Reid R, Williams RJ, Coradin L (1984) The existing *Stylosanthes* collection. In: Stace HM, Edye LA (eds) The biology and agronomy of *Stylosanthes*. Academic, Sydney

Stace HM (1982) Breeding systems in Stylosanthes. I. Observations of outcrossing in *S. scabra* at an alcohol dehydrogenase locus. Australian J Agricul Res 33:87–96

Stace HM, Cameron DF (1984) Cytogenetics and the evolution of *Stylosanthes*. In: Stace HM, Edye LA (eds) The biology and agronomy of *Stylosanthe*. Academic, Sydney

Stace HM, Edye LA (1984) The biology and agronomy of *Stylosanthes*. Academic, Sidney

Tewari S, Chandra A (2008) Genetical assessment of diploid progenitors of *Stylosanthes scabra* by Isozyme, RAPD and STS markers: a possible strategy for improvement of drought tolerant allo-tetraploid *S. scabra* species. Euphytica 162:39–50

Thumma BR, Naidu BP, Chandra A, Cameron DF, Bahnisch LM, Liu CJ (2001) Identification of causal relationships among traits related to drought resistance in *Stylosanthes scabra* using QTL analysis. J Exp Bot 355:203–214

Thumma BR, Naidu BP, Cameron DF, Bahnisch LM (1998) Transpiration efficiency and its relationships with carbon isotope discrimination under well-watered and water-stressed conditions in *Stylosanthes scabra*. Aust J Agric Res 49:1039–1045

Trevorrow RM, Chakraborty S, Jansen PI (1998) Anthracnose resistance in *Stylosanthes seabrana*. In: Michalk DL, Pratley JE (eds) Agronomy-growing greener future. Proceedings of the 9th Australian society of agronomy conference, Wagga, NSW

Valarini MJ, Otsuk IP, Vieira MLC (1997) Changes in N_2 fixation in *Stylosanthes scabra* derived from tissue culture. Braz J Genet 20:713–716

Vallis I, Gardener CJ (1984) Nitrogen input into agricultural systems by *Stylosanthes*. In: Stace HM, Edye LA (eds) The biology and agronomy of *Stylosnathes*. Academic, Australia

Vander Stappen J, Gama Lopez S, Davila P, Volckaert G (2002) Molecular evidence for the hybrid origin of a new endemic species of *Stylosanthes* Sw. (Fabaceae) from the Mexican Yucatan Peninsula. Bot J Lin Soc 234:27–51

Vander Stappen J, Weltjens I, Gama-Lopez S, Volckaert G (2000) Genetic diversity in Mexican *Stylosanthes humilis* as revealed by AFLP, compared to variability of *S. humilis* accessions of South American origin. Euphytica 113:145–154

Vander Stappen J, Weltjens I, Volckaert G (1999) Microsatellite markers in *Stylosanthes guianensis*. Mol Ecol 8:514–517

Vieira LMC, Jones B, Cocking EC, Davey MR (1990) Plant regeneration from protoplasts isolated from seedling cotyledons of *Stylosanthes guianensis*, *S. macrocephala* and *S. scabra*. Plant Cell Rep 9:289–292

Weeds PL, Chakraborty S, Fernandes CD, d'A Charchar MJ, Ramesh CR, Kexian Y, Kelemu S (2003) Genetic diversity in *Colletotrichum gloeosporioides* from *Stylosanthes* spp. At centers of origin and utilization. Phytopathology 93:176–185

White N, Chakraborty S, Fernandes CD et al (2001) Selection of high yielding and anthracnose resistant *Stylosanthes* for Brazil, India and China. In: Proceeding of XIX international grassland congress. Pinacicaba, Brazil

Williams RJ, Reid R, Schultze-Kraft R, Sousa Costa NM, Thomas BD (1984) Natural distribution of *Stylosanthes*. In: Stace HM, Edye LA (eds) The biology and agronomy of *Stylosanthes*. Academic, New York

Williams J, Gardner CJ (1984) Environmental constraints to growth and survival of *Stylosanthes*. In: Stace HM, Edye LA (eds) The biology and agronomy of *Stylosanthes*. Academic, Sydney

Yang J, Guo Z (2007) Cloning of a 9-cis-epoxycarotenoid dioxygenase gene (SgNCED1) from *Stylosanthes guianensis* and its expression in response to abiotic stresses. Plant Cell Rep 26:1383–1390

Chapter 11
Biotechnology of *Miscanthus*

S. J. Dalton

Abstract *Miscanthus x giganteus* is a natural hybrid C4 grass genotype of great size and of a proven utility for biomass cropping, but its growing range is restricted by cold susceptibility. New requirements for fermentability and many other characteristics have also arisen over the last 10 years. However, the *Miscanthus x giganteus* genotype is not very easily included in breeding programmes because it is a sterile triploid hybrid and cannot produce seed. The genetic resources of the parental species *M. sinensis* and *M. sacchariflorus* and related species are being collected, studied and analysed using many new genomic and transcriptomic molecular tools. Breeders have selected new cultivars from within the genetic pool of *Miscanthus sinensis* and have also created new *Miscanthus x giganteus* and other interspecific hybrids. There is also progress in creating new intergeneric hybrids with close relatives such as sugarcane and sorghum. Initially the main purpose of biotechnology research was to develop cheaper micro-propagation methods for *Miscanthus x giganteus*, because rhizome propagation was so expensive. More recently, methods of in vitro polyploidy have been developed in the hybrid and two parental species, which will allow the creation of new hybrid combinations and the exploitation of the greater size of polyploids. Genetic transformation by particle bombardment and via *Agrobacterium* has also been achieved relatively recently and is now being applied to several characteristics potentially involved with fermentation for ethanol production.

Abbreviations

ABA	Abscisic acid
AFLP	Amplified fragment length polymorphism
AOA	Alpha-aminoxyacetic acid

S. J. Dalton (✉)
Institute of Biological, Environmental and Rural Studies, Aberystwyth University, Wales, UK
e-mail: snd@aber.ac.uk

S. M. Jain and S. Dutta Gupta (eds.), *Biotechnology of Neglected and Underutilized Crops*, DOI: 10.1007/978-94-007-5500-0_11,

BAC	Bacterial artificial chromosome
BAP	6-benzylaminopurine
Bar	Basta resistance gene
CH	Casein hydrolysate
2,4-D	2,4 dichlorophenoxy acetic acid
DH	Doubled haploid
DHPS	Sulfonamide herbicide resistance gene
DMSO	Dimethylsulphoxide
EST	Expressed sequence tag
FAEA	Ferulic acid esterase gene
FAE	Ferulic acid esterase
G418	Geneticin
GBS	Genotyping-by-sequencing
GA3	Giberellic acid
GFP	Green fluorescent protein
Gfp	Green fluorescent protein gene
GUS	β-glucuronidase
Gus	β-glucuronidase gene
HB	Holley and Baker medium
Hpt	Hygromycin resistance gene
IAA	Indole acetic acid
IBA	Indole-3-butyric acid
2Ip	2-isopentenyladenine
KIN	Kinetin
MAS	Marker assisted selection
MES	2-(N-morpholino) ethanesulfonic acid
MET	Methionine
MS	Murashige and Skoog medium
NAA	Napthelene acetic acid
nptII	Neomycin phosphotransferase gene
PBZ	Paclobutrazol
PHB	Poly-β-hydroxybutyric acid
pinII	Potato proteinase II gene
PPT	Phosphinothricin
PVP	Polyvinylpyrrolidone
QTL	Quantitative trait loci
xyn2	Xylanase gene
RAD	Restriction site associated DNA
RAPD	Random amplification of polymorphic DNA
RFLP	Restriction fragment length polymorphism
SNP	Single nucleotide polymorphism
SSR	Microsatellite simple sequence repeat
TDZ	Thidiazuron
2,4,5-T	2,4,5 trichlorophenoxyacetic acid
UidA	β-glucuronidase gene

11.1 Introduction

Fifteen *Miscanthus* (Anderss.) species of grass are endemic to the temperate and tropical regions of south-east and East Asia and were traditionally used for animal feed and thatching (Stewart et al. 2009). Several were introduced to Europe as ornamental grasses in the 19th century. Taxonomically the *Miscanthus* genus belongs to the grass tribe *Andropogonae,* subtribe *Saccharinae* which also includes maize, sorghum and sugarcane (Hodkinson et al. 2002a). The useful species of the *Miscanthus* genus comprise *Miscanthus sinensis*, a generally diploid plant with many tillers and high seed set and *Miscanthus sacchariflorus* which is generally tetraploid and with thicker stems and rhizomatous habit (Fig. 11.1). *Miscanthus x giganteus* is an allotriploid, sterile, rhizomatous natural hybrid between the two species and is of great size and vigour. It was found in southern Japan in the 1930s by one collector at Yokahama (Olsen) and by another collector (Honda) in the Kuma river basin. They called the plants, which are now considered to be the same taxon if not the same genotype, Clone Aksel Olsen and *Miscanthus x ogiformus* cv Honda, respectively, but the hybrid is now universally known as *Miscanthus x giganteus*. Triploid hybrid seeds have recently been found on *Miscanthus sacchariflorus* plants in the Kuma region, but their incidence is rare (Nishiwaki et al. 2011).

Miscanthus x giganteus was introduced into Europe as a striking garden plant (Greef and Deuter 1993), but interest in the potential of this genotype and others as bioenergy crops emerged in the 1980s as evidence of climate change accumulated and there are presently at least five major *Miscanthus* breeding programmes outside Asia (Clifton-Brown et al. 2013). The *Miscanthus x giganteus* genotype is widely grown in the European Union on a small scale, principally in the UK, Italy, Ireland and France and amounting to 22–29 KHa in 2008 (AEBIOM 2011). Another triploid hybrid 'Illinois' clone was a 1960s accession grown in the

Fig. 11.1 Plants of *Miscanthus sinensis* 'Goliath', *M. sacchariflorus* and *M. x giganteus* 'Illinois'

Botanic garden in Chicago and is similar, but distinguishable from *Miscanthus x giganteus* 'Honda', based on molecular phylogeny (John Clifton-Brown, pers comm). This clone is becoming increasingly grown in Canada and in America, particularly in Illinois as a higher-yielding alternative to switchgrass (*Panicum virgatum*) and large scale plantings are proposed in four other states amounting to about 81 KHa by 2014 (USDA 2011).

New *Miscanthus* tillers grow each spring, then flower, senesce and die, re-allocating nutrients back into the rhizomes as they do so. The dead straw is harvested annually for combustion. There is interest too in using *Miscanthus* fibres for paper, thatch, textiles and construction materials, the vegetative leaves for fermentation products and the plant itself for bio-remediation (Fowler et al. 2003). However, a major drawback is the high cost of establishing a crop from vegetatively cloned rhizomes and the slow establishment over 3–4 years, although the crop typically lasts 10–15 years.

The collection of genetic resources of *M. sinensis*, *M. sacchariflorus*, related species and any potential hybrids is now a central part of breeding activity. Their genomic evaluation through the creation of mapping populations, genetic maps and quantitative trait loci, to more recent advances with new generations of molecular markers, is facilitating sophisticated marker-assisted breeding and reducing the need for conventional agronomic screening. In vitro tissue culture and other biotechnology techniques have also been developed and genetic manipulation is starting to be applied with genes of interest. The in vitro characteristics of the *M. x giganteus* genotype have been well researched, particularly in Denmark, but latterly all over Europe and in the United States. Initially this effort was principally to establish micro-propagation as a cheaper alternative to conventional propagation. The approaches used were direct micro-propagation in which rapid tiller multiplication was induced or through callus induction followed by the regeneration of numerous plants.

As a crop currently used for direct combustion and gasification, the initial breeding requirements have been to increase biomass and carbon capture by whatever means and to widen the growing range (Clifton-Brown et al. 2013). To this end *Miscanthus* breeders have created a new range of *Miscanthus x giganteus* triploid hybrid cultivars such as 'Amuri' (Allen 2008), 'Nagara' 'Lake Erie' and 'MBS1002' (Deuter 2011a, b, c),'Freedom' (Jennings 2011) and 'No. 1', 'No. 2', and 'No. 3' Xiang Hybrid *Miscanthus* (Zili et al. 2010). Other interspecific *Miscanthus* hybridisations are also possible (Deuter 2000). There is also interest in *M. sinensis* alone, not least because the crop could be seed-sown, although conversely there is concern over possible invasiveness, because *Miscanthus* seeds are so small and easily spread. A new seed sown cultivar Powercane ™ is currently in field trials (Mendel Biotechnology Inc. 2012).

There has also been research on chromosome doubling to exploit polyploidy, both for the expected increase in size and to enable new and different hybrids to be created. Naturally occurring triploids of *Miscanthus sinensis* were identified as early as 1985 (Matumura et al. 1985) and a triploid *M. sinensis* ornamental garden plant 'Goliath' has been grown as a more cold tolerant alternative to

M. x giganteus for biomass (Christian and Haase 2001). Similar *M. sinensis* triploids such as 'Gofal' have been recovered in crossing programmes (Deuter 2000). The creation of doubled haploids to fix characteristics has also been attempted (Głowacka and Jeżowski 2009).

Breeders would also like to manipulate the crop for fermentation products such as biogas and ethanol, with the potential to use genetic manipulation to reduce the lignin content and to alter sugar composition. Interest in this field is growing in Japan and China, where these useful species originated. Other traits of interest include herbicide tolerance and other stress tolerances such as heat, cold, drought and salt. Greater tolerance to cold and drought would in particular extend the growing range of *Miscanthus x giganteus* itself (Zub and Brancourt-Hulmel 2010) and genetic manipulation of this genotype would be instantly exploitable, although perhaps not in Europe at the present time. However, disease and pest resistance is currently hardly considered in these remarkably healthy species. Other potential targets are more efficient nitrogen utilisation, larger root biomass, greater photosynthetic efficiency, improved recycling of nutrients during senescence, and the delay or inhibition of flowering to concentrate resources on vegetative growth. There is also a programme at Texas A&M University to create intra-generic hybrids of *Miscanthus* with sugarcane and sorghum species (Rooney et al. 2010). However, as with other species, the core requirement for the application of many biotechnology techniques is the development of reliable and efficient plant tissue culture and regeneration systems.

11.2 Establishment of In Vitro Cultures

11.2.1 Species, Genotypes, Surface Sterilisation and Explants

As an established cultivar, the original *Miscanthus x giganteus* genotype has been the most studied of all *Miscanthus* material for in vitro culture, but that is beginning to change with the emergence of new cultivars and breeders lines (Table 11.1). Shoot apices, node buds and rhizome-derived apical shoots and lateral buds have been used to establish sterile shoots in culture, while these explants and immature inflorescences, leaf sections, seeds and anthers have been used for callus induction. Above ground tillers containing shoot apices and young leaves, or stem sections containing node buds or immature inflorescences are the most easily sterilised tissue (Table 11.2). Lewandowski 1997 gave contamination rates of less than 3 % for inflorescence explants. However, below ground rhizome sections are difficult to clean and sterilise and their high contamination rates have precluded their general use until now (Lewandowski 1997).

In our laboratory we have used the more rigorous sterilisation protocol routinely used for tillers and seeds of other grasses (Dale 1980) to sterilise rhizome sections (Fig. 11.2).

Table 11.1 *Miscanthus* species and genotypes used for experiments

Species	Genotypes	Explants	Description	Purpose	Reference
Miscanthus x giganteus	*Miscanthus x ogiformis* Honda 'Giganteus'	In vivo and in vitro shoot apices, leaf sections, immature inflorescences	Widely grown cultivar	Propagation, micro-propagation	Nielsen et al. 1993, 1995; Lewandowski 1997; Holme et al. 1997; Petersen et al. 1999
	Miscanthus x giganteus, 23 different clones	In vivo shoot apices and immature inflorescences	Breeders lines	Propagation	Christian and Haase 2001
	Miscanthus x giganteus G1 and G3	Axillary buds	Widely grown cultivars	Propagation	Pepó and Tóth 2005
	M.x giganteus	Immature inflorescences,	Widely grown cultivar	Propagation	Płażek et al. 2007, Płażek and Dubert 2010; Głowacka et al. 2010a; Gubisova et al. 2013
	M.x giganteus (Greef et Deu)	Immature inflorescences	Widely grown cultivar	Polyploidy	Głowacka et al. 2010a, b, c; Yu et al. 2009
	M.x giganteus	Immature inflorescences, flowers, ovaries, pollen	Widely grown cultivar	Propagation	Kim et al. 2010
	M.x giganteus	Noded stem sections	Widely grown cultivar	Micro-propagation	Rothrock 2010
	M.x giganteus	Immature inflorescences	Widely grown cultivar	Genetic manipulation	Trieu 2010
	'Honda' and 'Illinois'	In vitro shoot apices	Widely grown cultivars	Micro-propagation, GM	Dalton (unpublished)
Miscanthus sacchariflorus	(Maxim) Benthet Hook	Mature embryos, anthers, immature inflorescences and stems, shoot apices	–	Genetic manipulation	Zili et al. 2004
	'Robustus'	Immature inflorescences	Cold tolerant ornamental	Genetic manipulation	Trieu 2010
	MSac1, 'Robustus' (2x and 4x)	In vitro shoot apices	Breeders line, ornamental	Micro-propagation, GM Polyploidy	Dalton (unpublished)

(continued)

Table 11.1 (continued)

Species	Genotypes	Explants	Description	Purpose	Reference
Miscanthus sinensis	MS-88-110, SW217, M78, 92-123-II-8, 92-126-II-10, 92-132-II-4, 92-134-II-5, 93-230, 93-245, 93-246-1, MS-90-12, MS-90-6	In vivo and in vitro shoot apices, leaf sections, immature inflorescences	Breeders lines	Polyploidy	Petersen et al. 2002, 2003
	Ms17 (Japan), Ms21 cv 'Silberfeder', Ms1 x Ms18, Ms7 x Ms18	Anthers	Cultivar and breeders lines	Doubled haploidy	Głowacka and Ježowski 2009
	Ms10 European hybrid, Ms17 (Japan), Ms20 cv'Grosse Fontane'	Immature inflorescences	Cultivars and breeders lines	Propagation	Głowacka et al. 2010a
	Ms10, Ms11European hybrids, Ms16(Japan) Ms19 cv 'Goliath', Ms21 cv 'Silberfeder'	Immature inflorescences	Cultivars and breeders lines	Polyploidy	Głowacka et al. 2009, 2010b, c
	'Late hybrid'	Immature inflorescences	Cultivar	Genetic manipulation	Engler and Chen 2011
	'Pure seed'	Seeds	Cultivar	Genetic manipulation	Engler and Chen 2011
	18 accessions JM0001–JM0138, Japan	Seeds	Wild collections,	Genetic manipulation	Wang et al. 2011
	China	Seedling explants	Wild collections	Propagation	Zhang et al. 2012
	Unstated	Seeds	–	Genetic manipulation	Yamada et al. 2010; Trieu 2010
	'Goliath', MS-88-110, S1-20	In vitro shoot apices	Cultivar and breeders lines	Micro-propagation, GM	Dalton (unpublished)

Table 11.2 Methods used for the surface-sterilisation of *Miscanthus* tissues for in vitro culture

Plant tissue	Explant	Treatment	Reference
Node sections	Node segment	NaOCl (1.5 %) 20 min	Nielsen et al. 1993
Node sections	Nodal bud	CaOCl (30 %) 20 min	Schwarz and Meyer (pers comm)
Node sections	Node segment	Bleach (25 %)	Rothrock 2010
Node sections	Node segment	Ethanol 1 min then NaOCl (4 %) 10 min	Gubisova et al. 2013
Tillers, node sections	Shoot apices, leaf and root sections, immature inflorescences	NaOCl (0.5 %) 2–3 min	Holme et al. 1997; Petersen et al. 1999 ;Yu et al. 2009; Kim et al. 2010
Tillers, node sections rhizome buds	Shoot apices, rhizome buds, leaf primordia, immature inflorescences	Ethanol (70 %) 3 s then NaOCl (15 %) 5 min	Lewandowski and Kahnt 1993, Lewandowski 1997
Tillers	Axillary buds	Ethanol (80 %)	Pepó and Tóth 2005
Rhizome shoots	Rhizome shoot	NaOCl (5 %) 10 min, tillers trimmed, then NaOCl (2 %) 5 min then explant in NaOCl (1 %) 1 min	Petersen et al. 2003
Rhizome sections, young tillers	Shoot apices, lateral buds	Bleach (50 % = 1 % chlorine) 20–30 min, then 100 mg L^{-1} cefataxime and vancomycin in culture medium	Dalton (unpublished)
Tillers	Immature inflorescences	Ethanol (70 %) 1 min then Bleach (10 %) 10 min	Płażek and Dubert 2010
Tillers	Immature inflorescences	Bleach (10 %) = 0.45 % NaOCl 20 min	Głowacka et al. 2010a, b, c
Tillers, seeds	Immature inflorescences	Bleach (20 %) +0.1 % liquinox 20 min	Trieu 2010
Tillers	Immature inflorescences	Bleach (20 %) +0.1 % Triton 5 min	Engler and Chen 2011
Tillers	Immature inflorescences	Ethanol 1 min then NaOCl (4 %) 20 min	Gubisova et al. 2013
Seeds	Seeds	NaOCl (1 % chlorine) 90 min then imbibe o/n H_2O 4 °C, NaOCl (1 % chlorine) 30 min	Wang et al. 2011
Seeds	Seeds	Bleach (20 %) +0.1 % Triton 20 min	Engler and Chen 2011
Seeds	Seeds	Bleach (100 %) 60 min, imbibe o/n H_2O 4 °C, Bleach (10 %) 10 min	Dalton (unpublished)
Seeds	Epicotyl, radical, young leaf	H_2O 2 h, 70 % ethanol 30 s, NaHOCl (5.25 %) 20 min	Zhang et al. 2012

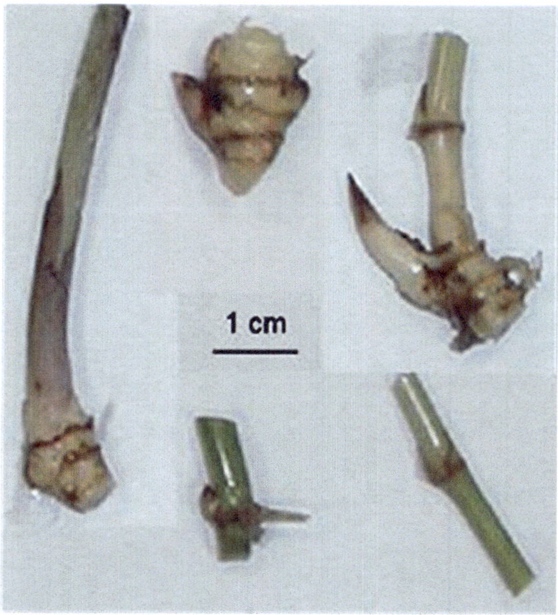

Fig. 11.2 Rhizomes and node sections used to isolate apical shoot tips, lateral buds, and nodal buds

Apical shoot tips and lateral buds from them were then cultured in medium containing the antibiotics that are normally used for eliminating *Agrobacterium*. Without antibiotics, contamination often affected more than 50 % of explants and in vitro plants could only be established from 10 % of explants and 75 % of genotypes. The addition of antibiotics (100 mg L^{-1} cefataxime, and vancomycin) markedly reduced contamination and plantlets could be established from 40 to 45 % of suitable explants. While an overnight pre-wash of the explants in the antibiotic solution again halved the contamination rate, it also reduced survival of the explants, and was not considered helpful. However, by using antibiotics it was possible to culture large apical meristem–containing shoot tip explants up to 4 cm long, which could be very quickly established. This method has the advantages that fewer, larger explants can be cultured in order to successfully establish a genotype in culture and that it can be used at any time of the year and with the hybrid and both parent species.

11.2.2 Shoot Initiation Media, Conditions, Vessels and Control of Browning

Murashige and Skoog (1962) (MS) medium with sucrose and up to 5 mg L^{-1} 6-benzylaminopurine (BAP) has been used as the initiation medium in all *Miscanthus* reports to date. Sometimes naphthalene acetic acid (NAA) and magnesium chloride have also been included (Table 11.3). Lower light levels of 30–50 $\mu E\ m^{-2}\ s^{-1}$ have been used for shoot initiation than the

Table 11.3 Culture media used for shoot initiation, maintenance and micro-propagation of *Miscanthus* cultures (all values in mg L^{-1} unless otherwise stated)

Basal Medium	Carbon source	Auxin	BAP	Other varied ingredients	Other ingredients	Best result	Reference
MS	2 % sucrose	0.24 NAA	0.002–22.5	0.002–22.5 KIN, 2iP, TDZ	750 $MgCl_2.6H_2O$, 0.3 % gelrite	Initiation: 4.5 BAP Propagation: 2.25–6.75 BAP	Nielsen et al. 1993, 1995
MS	3 % sucrose	0–0.45 IAA	1–5	0.8 % agar, liquid	50 cysteine	Initiation: 1-3 BAP, 0.8 % agar Propagation : 0.45 IAA, 5 BAP, liquid	Lewandowski 1997, 1998
MS	3 % sucrose	–	1.2	–	0.25 % gelrite	Maintenance	Hansen and Kristiansen 1997
MS	2 % sucrose	–	0.3	–	Liquid	Maintenance	Pepó and Tóth 2005
MS	2 % sucrose	0.24 NAA	2.5–5.0	–	0.2 % gelrite	Initiation: 0.24 NAA, 5 BAP Propagation : 2.5 BAP	Petersen et al. 2003
MS	3 % sucrose	–	2	–	0.8 % agar	Propagation	Głowacka et al. 2010aa
MS, MODMS	2 or 3 % sucrose or maltose	–	0.15–2.0	700–3,500 proline 100 cysteine, CH, MES	100 cefataxime, 100 vancomycin liquid	Initiation: MODMS medium, 3 % sucrose, 0.3 BAP, (+antibiotics) Maintenance and propagation : MODMS medium, 3 % sucrose, 1,500 proline, 0.3–1 BAP	Dalton (unpublished)
MS	3 % sucrose	0.25 NAA	5	0.25 TDZ, 1 PBZ pH 5.6+0.6 % agar pH 3.3+ liquid	750 $MgCl_2.6H_2O$,	Initiation: 0.24 NAA, 5 BAP Propagation: 5 BAP, 0.25 TDZ, 1 PBZ, solid or liquid	Rothrock 2010
MS	3 % sucrose	0.05–0.5 NAA	0.5–2	0.1–2 KIN	0.8 % agar	1 BAP, 2 KIN, 0.05 NAA	Zhang et al. 2012
MS	3 % sucrose	–	0.4–4	50 cysteine, PVP 150 citric, ascorbic acid 3000 ACT-charcoal	0.8 % agar	Initiation 0.4 BAP, 50 cysteine Propagation 2 BAP, 50 cysteine, liquid	Gubisova et al. 2013

70–100 μE m^{-2} s^{-1} normally used for maintenance and propagation. Incubation temperatures of 25–27 °C have been used most often, although Pepó and Tóth (2005) used 21 °C and Rothrock (2010) has used 31 °C.

Phenolic browning has been considered a major problem in *Miscanthus x giganteus* cultures and to counter this Lewandowski (1997) pre-treated apical meristem explants in 100 mg L^{-1} cysteine for 8 h and cultured them in medium containing 50 mg L^{-1} cysteine. Gubisova et al. (2013) tried citric and ascorbic acids, activated charcoal, polypyrrolidone (PVP) and cysteine and found only cysteine to be effective. We have found in our work that phenolic browning varied with genotype and was particularly present in hybrids and in *Miscanthus sacchariflorus*. However, in our experience it was still possible to obtain plantlets even from badly affected genotypes as generally one or two explants would establish even if the vast majority died. Browning could also be dispersed by using liquid medium for shoot initiation, maintenance and propagation, and although these cultures quickly became very acidic (pH 2.5) they did continue to grow. Rothrock (2010) deliberately used a very low pH of only 3.3 in liquid shoot initiation medium.

We have also successfully cultured small explants in 1 ml modified MS (MODMS) liquid medium containing extra thiamine (1 mg L^{-1}) and antibiotics in 25 compartment square Petri-dishes (Sterilin) and very large explants in Magenta vessels sealed with Nescofilm with 5–10 ml MODMS medium. Only 0.3 mg L^{-1} BAP was used in this work as root growth is easily inhibited in many genotypes (Kai Schwarz, pers. comm.). The explants were cultured in low light for 7–10 days and responding tissues sub-cultured to fresh medium without antibiotics in Magenta vessels and at a higher light intensity until plantlets were fully established. Maltose and additional proline were also tested, but were not considered to be helpful for shoot initiation (Fig. 11.3). Over one hundred genotypes of

Fig. 11.3 The effect of proline on growth of shoot tip cultures from rhizomes of *Miscanthus sinensis* 'Goliath' (**a**), and S21 (**b**)

M. x giganteus, M. sinensis and *M. sacchariflorus* have been established in this way. Most plantlets developed from apical shoot meristems rather than from lateral buds and larger explants were much more successfully established. *Miscanthus* is an unusual grass in that shoot cultures grow much more vigorously in liquid than in solid medium. However, it is more practical to establish sterile plants in solid medium for movement or export to other sites (Dalton unpublished).

11.3 In Vitro Micro-Propagation, Maintenance and Rooting

11.3.1 Culture Media and Conditions

To date, the propagation of *Miscanthus* species by tissue culture has been mainly of the original *Miscanthus x giganteus* genotype through the use of callus cultures (Lewandowski 1997; Holme and Petersen 1996; Christian and Haase 2001) and far less has been published on direct micro-propagation or with other genotypes. Nielsen et al. (1993, 1995) investigated the use of several cytokinins and found BAP to be most promising for micro-propagation, although there was some synergy with thidiazuron (TDZ) (Table 11.3). Hansen and Kristiansen (1997) used 1.2 mg L^{-1} BAP for stock plant maintenance and Lewandowski (1997, 1998) used 3–5 mg L^{-1} BAP for micro-propagation. Rothrock (2010) also used 5 mg L^{-1} BAP, but combined this with TDZ and paclobutrazol (PBZ) as well as a very low pH of only 3.3 in liquid medium and with a high culture temperature of 31 °C.

Miscanthus sinensis was found to be more sensitive to cytokinin than *M. x giganteus* and Petersen et al. (2003) and Głowacka et al. (2010b) used only 2.0–2.5 mg L^{-1} BAP for micro-propagation of colchicine treated plantlets. Zhang et al. (2012) used a combination of 1.0 mg L^{-1} BAP, 2 mg L^{-1} kinetin and 0.05 mg L^{-1} NAA in their best treatment. We have also found that tiller production was very genotype dependent and that growing *M. sinensis* with 1–2 mg L^{-1} BAP could sometimes reduce survival and tillering in subsequent growth passages. A lower concentration of 0.3 mg L^{-1} BAP was therefore routinely used in *M. sinensis* and *M. sacchariflorus* for both maintenance and propagation but increased to 1 mg L^{-1} BAP for slow tillering genotypes. Liquid medium was also found to be superior to solid medium and allowed much easier sub-culture.

Many authors have used a half strength rooting medium without BAP to induce roots in *M. x giganteus* (Table 11.4), but Rothrock (2010) used a liquid medium with high sucrose and NAA concentrations and a high pH of 6.5, stabilised with 2-(N-morpholino) ethanesulfonic acid (MES). Zili et al. (2004) added methionine (MET) to their rooting medium, while Lewandowski 1997 found that over 94 % of micro-propagated *M. x giganteus* plants were capable of rooting after transfer to soil directly from 3 mg L^{-1} BAP. However, in our work we found initially that survival of micro-propagated plants from 54 field and seed derived *Miscanthus* genotypes to be frequently very poor after transfer to soil, despite using only 0.3 mg L^{-1} BAP. Many genotypes were unable to grow roots despite numerous post-planting treatments including

Table 11.4 Culture media used for root induction in *Miscanthus* cultures (all concentrations mg L^{-1} unless otherwise stated)

Basal medium	Sucrose (%)	Auxin	Other ingredients	Reference
Half strength MS	2	1 NAA	5,000 activated charcoal, 750 $MgCl_2.6H_2O$, 0.3 % Gelrite	Nielsen et al. 1993
Half strength SH	2	0.2 NAA	1,000 activated charcoal 0.25 % Gelrite	Hansen and Kristiansen 1997
MS	2	0.5 IBA	Liquid	Pepó and Tóth 2005
Half strength MS	2	1 NAA	0.3 % Gelrite	Petersen et al. 2003
MS	3	0.5 NAA	0.25 MET	Zili et al. 2004
MS	3	1 NAA	0.8 % agar	Głowacka et al. 2009, 2010a, b, c
Half-strength MS	3	0.2 NAA	0.8 % agar	Zhang et al. 2012
MS	3	1 NAA	50 cysteine	Gubisova et al. 2013

indole acetic acid (IAA) treatment. Good existent roots were essential for establishment in most genotypes (Dalton unpublished).

We have also compared the use of BAP, IAA, proline and casein hydrolysate (CH) in liquid maintenance and propagation media with genotypes of *Miscanthus x giganteus, M. sinensis* and *M. sacchariflorus* (Table 11.1). All of the media tested became acidic (pH 3.4) after 2 weeks growth and the addition of MES had no effect on the decline of pH. Proline was very effective in promoting general vigour however, and for inducing root growth in propagated plants. A concentration of 1.5 g L^{-1} proline added to all maintenance, micro-propagation and rooting media improved growth in over forty of the *Miscanthus* genotypes that were tested. Propagating in liquid medium containing 0.3–1 mg L^{-1} BAP and a final subculture in medium without BAP ensured good root growth before transplanting to soil.

For micro-propagation some authors continued to grow the cultures in low light (Nielsen et al. 1993; Hansen and Kristiansen 1997; Rothrock 2010), while others increased the light intensity to 100 μE m^{-2} s^{-1} (Lewandowski 1997; Dalton and Donnison 2010a; Głowacka et al. 2010a). Most authors used a 16 h photoperiod and we have found that continuous light can induce viviparous development in at least one triploid *M. sinensis* genotype (MS19) (Fig. 11.4), where up to 78 small independent tillers were isolated from one viviparous inflorescence.

As we found that *Miscanthus x giganteus* genotypes 'Illinois' and 'Honda' were slower to tiller than most *M. sinensis* or *M. sacchariflorus* genotypes, we used 'Illinois' to further compare micro-propagation with different proline concentrations and 1 mg L^{-1} BAP (Fig. 11.5). Cytokinins generally cause increased tillering at the expense of shoot size and root growth (Nielsen et al. 1993), but proline appeared to be able to counteract these negative effects and particularly improved root growth and tiller survival. The highest proline concentration used (3.5 g L^{-1}) in conjunction with 1 mg L^{-1} BAP doubled the tillering rate in 'Illinois' and four surviving tillers were produced per passage. However, 1.5 g L^{-1} proline was chosen as a general level for maintenance and propagation. Holme (1998) considered that cell suspension cultures of *M. x giganteus* were using proline as a nitrogen source, so we also compared 1.5 g L^{-1} of a number

Fig. 11.4 Viviparous inflorescence of *Miscanthus sinensis* S19 induced in vitro by continuous light conditions

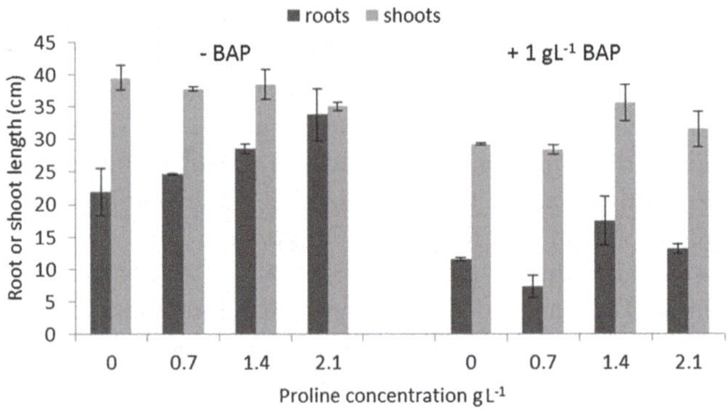

Fig. 11.5 Effect of adding proline to the medium in the presence and absence of BAP on growth of roots and shoots of *Miscanthus x giganteus* 'Illinois'

of nitrogenous compounds with 'Illinois' to test if the effect of proline was simply due to additional nitrogen. However, no other compound was as effective in promoting and maintaining root growth (Fig. 11.6). Proline accumulation has long been known to be a common metabolic response of higher plants to water deficits and salinity stress as well as having many other effects (Szabados and Savoure 2009). Possibly the effect of added proline is to enable *Miscanthus* plants to tolerate the rapid acidification of the culture medium.

Pepó and Tóth (2005) have examined the nutrient uptake of *Miscanthus* cultures growing in MS based maintenance medium containing 0.3 mg L^{-1} BAP. Potassium, calcium and magnesium concentrations in the culture medium declined slowly, but most of the nitrogen and nearly all of the phosphate was exhausted by 40 days, suggesting that phosphate was limiting.

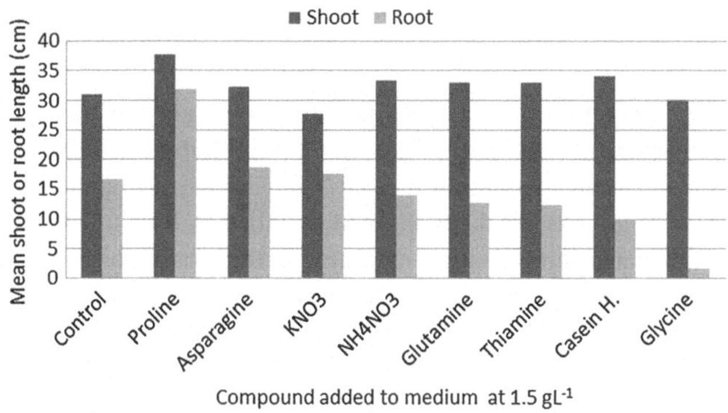

Fig. 11.6 The effect of nitrogen supplementation on in vitro root and shoot growth in *Miscanthus x giganteus* 'Illinois'

11.3.2 Micro-Propagation Rates

High rates of micro-propagation are essential if this method is to compete with traditional vegetative propagation methods via rhizomes. Nielsen et al. (1993, 1995) tested four different cytokinins and could produce ten tillers per 4 week passage, although half were less than 2 cm high. Pepó and Tóth (2005) reported that *M. x giganteus* multiplied at a rate of about three tillers over a 40 days cycle on a maintenance medium containing 0.3 mg L^{-1} BAP. Lewandowski 1997 also cultured in vitro shoots of *M. x giganteus* over four culture periods of 5 weeks with 3 or 5 mg L^{-1} BAP and obtained 3–4.5 tillers per passage. Thereafter the emphasis has been on using immature inflorescence calli for propagation, despite the possibility of auxin-induced somaclonal variation. However, Rothrock (2010) has described a method which more than doubled the propagation rate of *M. x giganteus* to eight tillers per three week passage using a combination of three cytokinins and growing the plantlets at a high temperature. This would theoretically give over sixty plants in only 6 weeks from one node explant, compared with the same number from an immature inflorescence explant via callus estimated to take 5 months. However, it is likely that subculturing calli is less laborious until the final regeneration stage is reached.

In our own work (Dalton unpublished) and that of Kai Schwarz (pers comm) we have found that propagation rates in over 100 *Miscanthus* genotypes, (mostly of *M. sinensis*) to be very variable, with some genotypes capable of producing up to 20 tillers in one passage. Longer passages (3–4 weeks) were considered more efficient than shorter passages because the plantlets were not so frequently disturbed by dividing and trimming. In their Patent Zili et al. (2010) also described a rapid propagation method for *M. sinensis* in which immature inflorescence callus was induced and immediately regenerated. The shoots were then micro-propagated in synchrony with root proliferation to produce numerous plantlets ready for planting. Zhang et al. (2012) describe micro-propagation of seedling shoot apices and obtained up to 6.7 plants per apex.

11.3.3 In Vitro Storage

In vitro storage is useful as a means to keep plants ready for future propagation and we find it particularly useful for keeping stock plants for callus culture from shoot apices. Hansen and Kristiansen (1997) developed an in vitro cold storage system for *M. x giganteus* stock plants and found that for storage, their rooting medium was superior to proliferation medium. In their best treatment, plants were stored at 16 °C for up to 27 weeks. We have also tested a system which has been successfully used for grasses over many years (Dale 1980) in which in vitro plantlets were stored on solid medium at 4 °C with an 8 h photoperiod. Storage plants of *Miscanthus* however, did not establish very well in solid medium and liquid medium was used instead. Forty *Miscanthus* genotypes of *M. x giganteus*, *M. sinensis* and *M. sacchariflorus* were stored for 10–12 months in Magenta vessels containing liquid maintenance medium before requiring subculture.

11.3.4 In Vitro Study of Stress Response

In an interesting study Grare (2010) used in vitro plantlets of two contrasting genotypes of *Miscanthus sinensis* (MS90-2 and MS88-110) which had been used as parents in a mapping family (Atienza et al. 2002) to examine salt tolerance. Rooted tillers were grown on media containing 0–200 mM salt and the fresh weight and other parameters were recorded. The initial results were inconclusive, except that the genotype MS90-2 produced a high level of anthocyanin in salt-stressed roots compared with MS 88–110. This approach may be applicable to other stresses or to pollutants in bio-remediation research.

11.3.5 Culture Vessels

Culture vessels used for *Miscanthus* plantlets have varied widely between laboratories, but most have been larger than the traditional universal tube. Lewandowski 1997 grew *Miscanthus x giganteus* plants in agar in 125 × 60 mm glass tubes with transparent covers and in 250 ml baby-food jars (Lewandowski 1998), while Głowacka et al. (2010a) used glass test-tubes with agar slants. We have found that plants grew very well in 100 ml medium in 1,400 ml transparent polypropylene containers with air-exchange filters (Combiness OS140). However, these were found to have no advantage over 50 ml medium in 380 ml Magenta GA-7 vessels, which were more robust and longer lasting, but require autoclaving for 45 min to ensure sterility. Magenta vessels containing 50 ml liquid medium and sealed with Nescofilm were ideal for in vitro maintenance, storage, propagation and regeneration and with solid medium were also ideal for the movement or export of sterile plants.

11.3.6 Transplanting to Soil

Efficient establishment of in vitro plants in soil is a key factor in developing reliable propagation systems. Lewandowski 1997 transplanted in vitro plantlets of *Miscanthus x giganteus* to soil and grew them in a glasshouse with a 16 h photoperiod at temperatures of 20 °C day/15 °C night. The plantlets were then grown under plastic film for one week and developed roots despite being transplanted directly from medium containing 3–5 mg L^{-1} BAP. Głowacka (2011) transplanted in vitro plantlets into a 1:3 mix of peat and sand and Kim et al. (2010) used an equal mixture of peat-moss, vermiculite and Perlite.

With other *Miscanthus* species and genotypes we found that existent roots were essential for establishment. However, the addition of 1.5 g L^{-1} proline to the maintenance and propagation media improved rooting to such an extent that all the genotypes that were subsequently propagated with proline could be successfully transplanted to soil. In this study eight transplanting protocols were compared using well-rooted plants of *M. x giganteus*, *M. sinensis* and *M. sacchariflorus* which had been propagated with 1.5 g L^{-1} proline and 0.3–1.0 mg L^{-1} BAP before a final subculture period without BAP. The plantlets were washed and half were pre-treated by growing for 12 days in 200 ml of half strength MODMS medium without sucrose in one litre beakers. Peat or a mixture of peat and soil were than added. Half were also planted directly

Fig. 11.7 *Miscanthus x giganteus* 'Honda' plants after transplanting to soil. Plants pre-treated in liquid medium, then grown in peat (**a**, **b**) or peat/soil mixture (**c**, **d**), or planted directly into peat (**e**, **f**) or directly into peat/soil mixture (**g**, **h**)

into beakers containing 200 ml of water 50:50 with either peat or the peat and soil mix. The plants were grown in a warm growth room (25 °C) or a cooler glasshouse (15 °C) and were placed under plastic film for one week to maintain humidity. The soil and water volume was kept to 200 ml.

Almost all of the plants successfully established as good root systems were already present after micro-propagation with proline. Algae grew in many of the beakers, but this was not detrimental to plant growth. The plants which had been transferred directly into the peat or peat/soil mix generally grew faster and had a larger root-ball than the pre-treated plants (Fig. 11.7). However, plants which were grown in peat became pale as they ran out of nutrients sooner than the plants grown in the peat/soil mix. Growing the plants in a warm growth room (25 °C) rather than a glasshouse had the greatest effect on mean tiller number and the production of new roots (Fig. 11.8 a, b). There was little difference in growth between normal *M. sinensis* plants and those derived from viviparous florets, or between frequently and infrequently sub-cultured *M. sacchariflorus* plants. The *M. x giganteus* plants produced more tillers after planting, but they had been micro-propagated at a higher concentration of BAP. Established, freshly rooting plants could then be transferred to normal pots of soil.

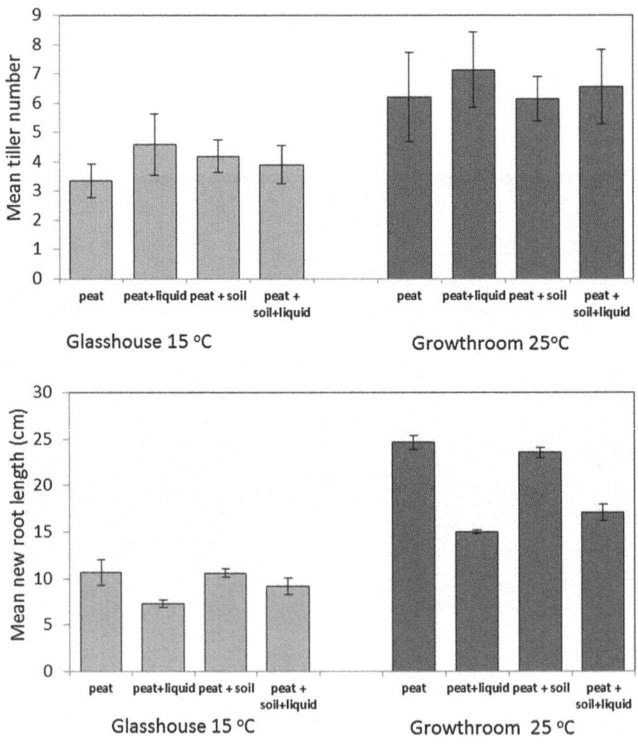

Fig. 11.8 Effect of rooting medium and growth environment on **a** mean in vivo tiller production and **b** new root growth in the three *Miscanthus* 'species' *M. sinensis*, *M. sacchariflorus* and *M. x giganteus*. Mean ± sem between species

11.4 Callus and Cell Suspension Cultures

11.4.1 Callus Induction

Many authors (Lewandowski and Kahnt 1993; Lewandowski 1997; Holme and Petersen 1996; Petersen 1997; Christian and Haase 2001; Zili et al. 2004; Głowacka et al. 2010a; Kim et al. 2010; Płażek and Dubert 2010; Wang et al. 2011) have described the emergence of different callus types in *Miscanthus x giganteus, M. sacchariflorus* and *M. sinensis* cultures (Fig. 11.9).

Type 1: Embryogenic callus described as white or light cream, compact, hard, smooth–surfaced, nodular, globular, solid, porous appearance, designated K2, slow to regenerate and regenerating via somatic embryogenesis.

Type 2: Shoot-forming callus described as compact, solid, friable, light yellow or white, glossy and with translucent, light green shoot-like structures or embryoids, designated K1 and regenerating via organogenesis.

Type 3: Rooty callus described as yellow, grey or pink, semi-soft to soft or watery, translucent, with anthocyanin-coloured spots, designated K1 or K3 and regenerating only roots.

Type 4: Non-morphogenic callus described as glassy, translucent, crystalline, rough, soft, watery, light brown, designated K4 and with no regeneration.

Embryogenic and shoot forming calli are the only useful callus types for propagation or genetic manipulation and it is surprising how many genotypes tested have proved to be embryogenic. We found that two-thirds of eighteen seed-derived

Fig. 11.9 Embryogenic callus of *M. x giganteus* 'Illinois' (**a**), *M. sacchariflorus* 'Robustus' (**b**) and *M. sinensis* S14 (with shoot forming callus also) (**c**) *top left* to *right*. Watery callus of *M. x giganteus* (**d**) and rooty callus of *M. sinensis* (**e**) *bottom left* and *right*

Miscanthus sinensis genotypes produced some embryogenic callus from in vitro shoot tips and that both *Miscanthus x giganteus* 'Honda' and 'Illinois', *M. sacchariflorus* 'Robustus' and *M. sinensis* 'Goliath', S19 and MS-88-110 were embryogenic. Only two field-derived genotypes tested have had no embryogenic capacity (Dalton unpublished). Wang et al. (2011) found that all eighteen seed accessions of *Miscanthus sinensis* that were collected wild in Japan, to be capable of producing embryogenic callus. Interestingly they discovered a significant correlation between the propensity to produce embryogenic callus and the mean 5 year temperature of the collection site, with the best accessions coming from southerly latitudes with mean temperatures of 17–25 °C.

11.4.2 Suitable Explants for Callus Induction

A large number of different explants have been used to initiate regenerable callus cultures, mostly of *Miscanthus x giganteus*. These include in vivo shoot apices, leaf primordia and immature inflorescences and in vitro leaf sections and shoot apices (Table 11.5). Isolated flowers, ovaries and microspores were also cultured by Płażek et al. (2007) but without success due to phenolic browning. Immature inflorescences have limited availability, but their low contamination rate and greater size compared to in vivo leaf primordia and shoot apices have made them much more useful (Lewandowski 1997). Inflorescences of 0.05–200 mm have been cultured and over two hundred 1 mm explants can be cultured from a single inflorescence. However, it has been shown several times that explants from smaller inflorescences tend to produce embryogenic and shoot forming calli more readily (Holme and Petersen 1996; Petersen et al. 1999; Głowacka et al. 2010a; Gubisova et al. 2013).

Immature inflorescences have been considered to be superior to shoot apices and shoot apices to be superior to leaf explants (Holme and Petersen 1996; Lewandowski 1997; Holme et al. 1997; Petersen et al. 1999). However, immature inflorescences are not always available. The most responsive leaf sections tended to be the youngest tissues and Lewandowski (1997) found that cutting leaf primordia in half nearly doubled the proportion of explants with embryogenic callus. However, leaf explants were generally still less successful than in vitro shoot apices.

Christian and Haase (2001) cultured very large shoot tip explants 6–9 cm long and obtained morphogenic callus from up to 100 % of explants depending on the *M. x giganteus* clone used. Dalton et al. (1999) used in vitro shoot tips and lateral buds from stock plants as a convenient and reproducible system in a number of grass species and has obtained embryogenic callus in this way from all three *Miscanthus* 'species'. In *Miscanthus x giganteus* 17 % of apical shoot apices responded, but only 10 % of lateral buds, while in the *M. sacchariflorus* genotype 'Robustus', 90 % of shoot apices were embryogenic compared to 31 % of lateral buds. In *M. sinensis* cultures embryogenic proliferation was more frequent if the stock plants had been grown in maintenance medium containing 0.3 mg L^{-1} BAP rather than no BAP. There was also faster growth, but more browning. In

Table 11.5 Explants used and the best results achieved (*Miscanthus x giganteus* unless stated)

Explant (and genotypes) used if different from *M. x giganteus*	Explant size (mm)	Highest proportion of explants producing callus	Highest proportion of explants producing embryogenic callus	Most responsive tissue	Reference
In vivo immature inflorescence (mm)	–	–	–	–	–
71–170	7	66 %	39 %	30–150 mm inflorescences	Lewandowski 1997
0.05–65	2	90 %	62 %	Younger inflorescences <25 mm	Holme and Petersen 1996
2–25	2	–	17 %	–	Holme et al. 1997
1–35	1–2, 7	65 %	79 %	2.5–8 mm inflorescences	Petersen et al. 1999
<90	1–2	–	–	–	Christian and Haase 2001
–	–	100 %	87 %	–	Zili et al. 2004
5–20	5–7	–	22 % (41 % shoot-forming)	–	Kim et al. 2010
5–25	1–5	100 %	–	5–15 mm inflorescences	Gubisova et al. 2013
100–200	5	–	84 %	Inflorescences over 80 mm	Płażek and Dubert 2010
1–50	1	–	0.1 % (25 % shoot-forming)	1–25 mm inflorescences	Głowacka et al. 2010a
1–50 (3 *M. sinensis*)	1	–	15 % (< 100 % shoot-forming)	1–25 mm inflorescences	Głowacka et al. 2010a
In vivo shoot apices, leaf primordia	3–5	25 % (48 % if cut in half)	–	Explants taken from 60–100 cm shoots	Lewandowski and Kahnt 1993,
‘’	–	–	100 %	60–90 mm explants	Christian and Haase 2001
In vivo leaf sections	2	79 %	5 %	Older leaf tissue	Holme and Petersen 1996
‘’	2	70 %	38 % (42 % shoot-forming)	Youngest leaf tissue	Petersen 1997
In vitro leaf sections	2	70 %	18 %	Youngest leaf tissue	Holme and Petersen 1996
‘’	2	51 %	41 % (93 % shoot-forming)	–	Petersen 1997
‘’	2	–	30 %	–	Holme et al. 1997
‘’	2	65 %	8 %	–	Petersen et al. 1999
In vitro shoot apices	1	100 %	50 %	–	Holme and Petersen 1996
‘’	–	90 %	42 % (81 % shoot-forming)	–	Petersen 1997

(continued)

Table 11.5 (continued)

Explant (and genotypes) used if different from *M. x giganteus*	Explant size (mm)	Highest proportion of explants producing callus	Highest proportion of explants producing embryogenic callus	Most responsive tissue	Reference
''	1	–	65 %	–	Holme et al. 1997
''	–	–	27 %	–	Petersen et al. 1999
'' (2 *M. x giganteus*)	2–3	–	17 %	Apical rather than lateral shoot apices	Dalton (unpublished)
'' (22 *M. sinensis*)	2–3	–	23 %	Apical rather than lateral shoot apices	Dalton (unpublished)
				Apices from stock plants grown with BAP	
'' (2 *M. sacchariflorus*)	2–3	–	90 %	Apical rather than lateral shoot apices	Dalton (unpublished)
In vitro roots and rhizome sections	2	75 %	0 %	Root tips	Holme and Petersen 1996
''	–	–	50 %	Clone dependent	Christian and Haase 2001
Anthers (4 *M. sinensis*)	–	29 %	–	Collection from donor plants as late as possible	Głowacka and Ježowski 2009
Mature seeds (*M. sacchariflorus*)	–	98 %	46 %	Embryos cultured rather than seeds	Zili et al. 2004
'' (18 *M. sinensis*)	–	50 %	25 %	Seeds from warm regions	Wang et al. 2011
Seedlings (*M. sinensis*)	–		82 %	Epicotyls from 10 days old seedlings	Zhang et al. 2012

practise, the culture of shoot tips from in vitro stock plantlets is so convenient that a low embryogenic callus percentage is of little consequence as long as several, good, fast growing embryogenic calli can be obtained. The culture of seeds is similarly easy, although each seed is a different genotype and a normal non-callused control plant can never be obtained. Zhang et al. (2012) have recently compared epicotyl, radicle and young leaf explants from 10 day old seedlings and have found epicotyls to be most responsive with 40 % producing embryogenic callus.

11.4.3 Culture Media and Conditions

A wide range of embryogenic callus inducing media have been used with different basal culture media and varied auxin and cytokinin concentrations (Table 11.6). Petersen et al.(1999) and Płażek and Dubert (2010) compared different carbon sources, while Holme et al. (1997); Holme (1998) and our laboratory have investigated the effect of adding proline to culture media. Lewandowski and Kahnt (1993), Płażek et al. 2007, Płażek and Dubert (2010) and Gubisova et al. (2013) also tried to mitigate the phenolic browning which occurs around many cut explants. Lewandowski and Kahnt (1993) successfully used liquid medium and filter paper rafts to disperse the phenolic compounds and Gubisova et al. (2013) used cysteine. Płażek et al. (2007) tested the phenolic inhibitors α-aminoxyacetic acid (AOA) or polyvinylpyrrolidone (PVP) but without success. They went on (Płażek and Dubert 2010) to test four compounds previously used either to reduce phenolic browning in foods or cultures of other species (chitosan, cysteine) or were carbon sources which contained either antioxidants (honey) or natural growth hormones (banana pulp). They found that a combination of honey and banana pulp was superior to sucrose for inducing regenerable embryogenic callus. Petersen et al. (1999) also compared several metabolisable and non-metabolisable sugars and found that a filter-sterilised combination of glucose and fructose was superior to sucrose.

Zili et al. (2004) compared MS, CC (Potrykus et al. 1979) and NB (Li et al. 1993) basal media. NB medium consisted of N6 medium (Chu et al. 1975) with Gamborgs B5 vitamins (Gamborg et al. 1968). Calli on CC medium grew more slowly because the osmotic potential of the medium was higher, but the calli retained their regeneration potential for a longer period. We have also compared five variants of MS medium and found that modified MS medium (MODMS) with increased thiamine was generally most helpful, because good callus growth was combined with good regeneration potential (Table 11.7). Trieu (2010) however, used N6 medium, but added some MS micro-elements and B5 vitamins, while Engler and Chen (2011) used MS medium with B5 vitamins. However, in our work we considered that this combination encouraged friable callus growth (Dalton unpublished). Zhang et al. (2012) compared MS, N6 and HB medium (Sun et al. 1999) and found HB medium consistently improved embryogenic response.

Table 11.6 Media used for callus or cell suspension culture of *Miscanthus* species (all concentrations mg L^{-1})

Basal Medium	Carbon source	2,4-D	BAP	Other varied ingredients	Other ingredients	Best result	Reference
MS, J25-8, Anderson, B5	3 % sucrose	2–10	0.25–2	50 cysteine, 4–8 2,4,5-T 0.25–2 KIN, 1 Zeatin, liquid + FP, 0.8 % agar	–	Liquid MS or JK25-8 medium 4-8 2,4-D or 2,4,5-T, 0.5-2 BAP or KIN	Lewandowski and Kahnt 1993, Lewandowski 1997
MS	2–3 % sucrose	1–7	0.1–0.8		750 $MgCl_2.6H_2O$ 0.2–0.3 % gelrite	3 2,4-D, 0.1 BAP, 3 % sucrose, 0.2 % Gelrite	Holme and Petersen 1996, Petersen 1997
MS/Chu N6 + MS vitamins	3 % sucrose	5–7	–	1,440–34,500 proline	750 $MgCl_2.6H_2O$ 0.2 % gelrite	MS medium, 1,440 proline 5 2,4-D (callus), 7 2,4-D (cell suspension)	Holme et al. 1997
MS	3 % sucrose or maltose 1.6 % glucose, fructose sorbitol, mannitol 0.8 % glucose+ fructose	3	–	Filter-sterilised or autoclaved	500 CH, 300 glutamine 750 $MgCl_2.6H_2O$, 2,880 proline, 0.2 % gelrite	Filter sterilised 0.8 % glucose+fructose	Petersen et al. 1999
MS	3 % sucrose	2.5	0.5	–	0.8 % agar	–	Christian and Haase 2001
MS	2 % sucrose	5	0.1	1,440 proline	0.2 % gelrite	1,440 proline (embryogenic callus)0.1 BAP (shoot-forming callus)	Petersen et al. 2002, 2003[a]
MS, CC, NB	3 % sucrose (MS,CC) +3.6 % mannitol CC) 3 % maltose (NB)	2	–	500 proline, 300 CH, 500 glutamine (All with NB)	Solid	CC medium, 3 % sucrose, 3.6 % mannitol	Zili et al. 2004[a]
190-2, C17, PG-96, R2 M	3 % sucrose 9 % maltose	0.5–2	0.5	0.5 NAA, 0.5 KIN 200–400 KH_2PO_4 0.8 % agar, 0.6 % agarose	–	C17 medium (Wang and Chen 1983), 2 2,4-D, 0.5 KIN, 9 % maltose, 200 KH_2PO_4, 0.6 % agarose	Głowacka and Jeżowski 2009[a]

(continued)

Table 11.6 (continued)

Basal Medium	Carbon source	2,4-D	BAP	Other varied ingredients	Other ingredients	Best result	Reference
MS, B5, SH,	3 % sucrose	0.75–5	0.1–1	0.5 KIN, 1430 proline 50 ACT-charcoal	0.8 % agar	MS medium, 5 2,4-D, 1,430 proline or 0.1 BAP	Głowacka et al. 2010a, b, c[a]
MS	3 % sucrose	3	0.1	–	750 $MgCl_2.6H_2O$, 2,880 proline, 0.2 % Phytogel	–	Yu et al. 2009; Kim et al. 2010
MS	10–50 % coconut milk	2–5	0.5–2	–	Solid	–	Zili et al. 2010[a]
MS	3 % sucrose	5	–	AOA or PVP	–	MS medium, 5 2,4-D	Płażek et al. 2007
MS	3 % sucrose or honey 6.5 % banana pulp	6.5	0.25	500 CH, 200 chitosan 100 cysteine	0.8 % agar	500 CH, 3 % honey +6.5 % banana pulp	Płażek and Dubert 2010
MS	3 % sucrose	1–10	0.01–1		0.2 % gelrite	5 2,4-D, 0.01 BAP	Wang et al. 2011[a], Yamada et al. 2010[a]
MS, LS, MODMS, MS+B5 or N6 vitamins	3 % sucrose, maltose	3	0.15–0.3	700–6,100 proline	0.3 % gelrite	Induction : MODMS medium, 2,100 proline, 3 % maltose, 0–0.15 BAP Selection: MODMS medium, 1,400 proline, 3 % maltose, 0–0.15 BAP	Dalton 2010b[a]
N6+B5 vitamins and micro-elements	3 % maltose	2–4	1	0.5 KIN, 1,000 asparagine 2,000 proline, 0.35 % gelrite, 0.7 % agar	1,000 CH 5 thiamine	Induction: 2 2,4-D, 0.5 KIN, 0.35 % gelrite Selection: 4 2,4-D, 1 BAP, 1,000 asparagine, 2,000 proline, 0.7 % agar	Trieu 2010[a]
MS+B5 vitamins	3 % maltose	2–5	1	–	Solid	–	Engler and Chen 2011[a]
HB, N6, MS	3 % sucrose	2–4	0.5–1	0–0.2 thiamine	0.8 % agar	HB, 2 2,4-D, 0.05 BAP, 0.1 thiamine	Zhang et al. 2012
MS	3 % sucrose	5	0.1	–	0.8 % agar 50 cysteine	–	Gubisova et al. 2013

[a] Includes genotypes other than *M. x giganteus*

Table 11.7 The effect of different vitamin combinations on regeneration from *Miscanthus sinensis* calli

Basal medium	Vitamins	Calli with regeneration (%)	Calli producing green shoots (%)	Mean tiller number	Mean root number
Murashige and Skoog (MS)	MS	36	36	6.5	8.3
MS	Gamborg's B5	58	75	16.3	6.0
Linsmaier and Skoog	0.4 mg L^{-1} thiamine only	75	75	15.8	8.6
MS	Nitsch	82	91	22.6	7.4
MS (MODMS)	MS+ 1 mg L^{-1} thiamine	83	75	20.1	6.9

With respect to growth regulators *M. x giganteus* appeared to be relatively insensitive to the auxin 2,4-dichlorophenoxyacetic acid (2,4-D) and could produce callus at any concentration from 1 to 7 mg L^{-1} (Holme and Petersen 1996). The related auxin 2,4,5-trichlorophenoxyacetic acid (2,4,5- T) has also been found to be effective for callus induction (Lewandowski and Kahnt 1993). Many authors have used 5 mg L^{-1} 2,4-D for callus induction, but Christian and Haase (2001); Zili et al. (2004); Trieu (2010) and in our own work we have found that 2–3 mg L^{-1} was enough to induce embryogenic callus in genotypes of all three *Miscanthus* 'species'. These concentrations are less likely to inhibit regeneration or to cause somaclonal variation, although the induction of somaclonal variation can be an objective in itself (Płażek et al. 2007).

BAP has been the most commonly used cytokinin at concentrations of 0.1–2 mg L^{-1}, although Wang et al. (2011) reported that very low concentrations of 0.01 mg L^{-1} BAP were beneficial in *M. sinensis* seed cultures. We have also confirmed that this concentration is helpful in other *Miscanthus* genotypes which hitherto did not grow very well with higher concentrations of BAP. The optimum BAP concentration for embryogenesis appears to be one variable which is very genotype dependent. In all reports, callus cultures were grown in darkness at temperatures of 23–29 °C. However, we have found that there are also genotypic differences in the preference for 25 or 28 °C.

Significant improvements in *Miscanthus* culture have been made by the inclusion of the amino acid proline. Holme et al. 1997 compared proline concentrations of 1.44–34.5 g L^{-1} and concluded that 1.44 g L^{-1} was the optimum concentration for growth of both calli and cell suspension cultures. We have also tested proline concentrations from 0.7 to 2.1 g L^{-1} and found that embryogenic calli grew faster as the proline concentration increased, but also became more friable (Dalton et al. 2010b). More shoot tip explants initially responded with 0.7 g L^{-1} proline, but 2.1 g L^{-1} proline was subsequently used because fast growth was such a pre-requisite for successful genetic transformation. After callus bombardment 1.4 g L^{-1} proline was used during subsequent growth under selection. Higher proline concentrations were tested with some recalcitrant genotypes, but over 3 g L^{-1} proline was never beneficial. Trieu (2010) used 1 g L^{-1} casamino acids (casein hydrolysate) for all callus media and also an additional 2 g L^{-1} proline and 1 g L^{-1} asparagine for the recovery and selection of *Agrobacterium* treated calli.

11.4.4 Plant Regeneration

Reliable regeneration is a pre-requisite for any propagation, chromosome doubling, double haploid production, or genetic manipulation protocol using callus. Most authors have reduced or omitted auxin and increased cytokinin in their callus induction medium to induce regeneration from embryogenic or shoot-forming callus (Table 11.8). Zili et al. (2004) included 5 mg L^{-1} abscisic acid (ABA) in a pre-differentiation medium and Engler and Chen (2011) included giberellic acid (GA3). Regenerating calli on cotton wool or filter paper in liquid medium also improved the regeneration rate compared to agar-solidified medium (Lewandowski and Kahnt 1993).

Petersen et al. (1999) investigated various carbon sources as well as filter-sterilised and autoclaved media. They discovered that immature inflorescence derived calli grew and regenerated best on a filter-sterilised medium with a combination of glucose and fructose, but that shoot apex and leaf explant-derived calli that had been grown on filter-sterilised maltose containing medium had much higher regeneration rates than those from the glucose and fructose combination. We have successfully used a maltose based regeneration medium developed for *Dicanthium annulatum* (Dalton et al. 2003) while Płażek and Dubert (2010) compared honey with sucrose and found that calli that had been grown with banana pulp to be particularly regenerative on honey based medium.

With regard to amino acids, Trieu (2010) included a small amount of glutamine in the regeneration medium, while Dalton et al. (2010b) compared the effect of four proline concentrations on the regeneration of *Miscanthus sinensis* calli which had been induced from the shoot apices of stock plants grown with and without 0.3 mg L^{-1} BAP. The calli induced with each proline concentration were divided and regenerated on four media containing the same series of proline concentrations to give sixteen treatments. The overall regenerability of calli derived from plants grown with and without BAP was similar, but the growth rate of calli from BAP-grown plants was higher and more friable in nature with high concentrations of proline. In both cases the highest regeneration rates were from calli grown with 1.4 g L^{-1} proline and regenerated with 0 or 0.7 g L^{-1} proline (Figs. 11.10, 11.11) although 1.4 g L^{-1} proline helped to maintain what regenerability there was in the more friable calli obtained with 2.1 g L^{-1}proline. The least regeneration was from calli grown without proline, but regenerated with 2.1 g L^{-1} proline, particularly the calli from plants grown without BAP. We have also assessed proline for direct regeneration from in vitro shoot tips (Fig. 11.12) and 0.7 g L^{-1} was most successful. This observation that high concentrations of proline are not helpful for plant regeneration is consistent with shoot initiation medium in which the addition of proline was not helpful either.

Table 11.8 Culture media used for regeneration from calli and cells suspensions of *Miscanthus* species (all concentrations mg L^{-1}unless otherwise stated)

Basal medium	Carbon source	Auxin	Cytokinin	Other variables	Other ingredients	Best result	Reference
MS	3 % sucrose	–	0.1–1.0 BAP, KIN, Zeatin	0.8 % agar, liquid+FP	50 cysteine	0.1 BAP, liquid + FP	Lewandowski and Kahnt 1993; Lewandowski 1997
MS	2–3 % sucrose	0–3 2,4-D	1–3 BAP	–	750 $MgCl_2.6H_2O$ 0.2 % Gelrite	2 BAP, 3 % sucrose	Holme and Petersen 1996; Petersen 1997, Petersen et al. 2002, 2003[a]
MS	3 % sucrose	1 2,4-D	5 BAP	–	750 $MgCl_2.6H_2O$ 0.2 % Gelrite	1 2,4-D, 5 BAP (callus) 5 BAP(cell suspension)	Holme et al. 1997
MS	3 % sucrose or maltose 1.6 % glucose or fructose 0.8 % glucose+ fructose	–	2 BAP	Filter-sterilised or autoclaved	750 $MgCl_2.6H_2O$ 0.2 % Gelrite	Filter sterilised maltose	Petersen et al. 1999
MS, NB, CC (Potrykus et al. 1979)	3 % sucrose (MS, CC) +3.6 % mannitol (CC) 3 % maltose (NB)	1 NAA	2 BAP	–	5 ABA solid	CC medium, 3 % sucrose, 3.6 % mannitol	Zili et al. 2004[a]
MS, FHG, 190-2, B5 (Gamborg et al. 1968)	3 % sucrose	0.5 NAA	0.1–1 BAP 0.5 KIN	0.8 % agar, 0.6 % agarose, liquid+CW	–	B5 medium, 0 .8 % agar, no growth regulators	Głowacka et al. 2009[a]
MS	3 % sucrose	–	2 BAP	–	750 $MgCl_2.6H_2O$ 0.8 % agar	–	Głowacka et al. 2010a, b, c[a]

(continued)

Table 11.8 (continued)

Basal medium	Carbon source	Auxin	Cytokinin	Other variables	Other ingredients	Best result	Reference
MS	2 % sucrose	0.24 NAA 1 2,4-D	5 BAP	–	750 $MgCl_2.6H_2O$ 0.3 % Phytogel	0.24 NAA, 5 BAP	Yu et al. 2009; Kim et al. 2010
MS	3 % sucrose or honey	–	0.2 BAP 0.05 KIN	–	0.8 % agar	0.05 KIN, 3 % honey	Plażek and Dubert 2010
MS	1–5 % sucrose	–	0–2 BAP	–	–	–	Zili et al. 2010[a]
MS	3 % sucrose	0.5 IAA	1 TDZ	–	–	–	Yamada et al. 2010[a]
MS	3 % sucrose	–	0.1–4 BAP	–	0.2 % Gelrite	2 BAP	Wang et al. 2011 [a]
MS, LS, MODMS, MS+B5 or N6 vitamins	3 % sucrose or maltose	1 NAA	0.3–1 BAP	700–2,100 proline	0.3 % Gelrite	MODMS medium, 3 % maltose, 1 BAP, 700 proline, 1 NAA,	Dalton 2010b[a]
MS+B5 vitamins	1 % maltose+ 2 % sucrose	0.1 NAA	2 BAP	–	50 glutamine 0.7 % agar	–	Trieu 2010[a]
MS+B5 vitamins	3 % maltose	–	–	–	0.5 GA3, solid	–	Engler and Chen 2011[a]
HB (Sun et al. 1999)	3 % sucrose	0.05–0.1 NAA	2–4 BAP 0.05–0.5 IBA	–	0.8 % agar	2 BAP, 0.05 IBA, 0.05 NAA	Zhang et al. 2012

[a]Includes genotypes other than *Miscanthus x giganteus*. (*CW* cotton wool, *FP* filter paper)

Fig. 11.10 *Miscanthus sinensis* calli, grown on callus induction medium containing 1.4 g L^{-1} proline, regenerating on regeneration medium containing 0.7 g L^{-1} proline

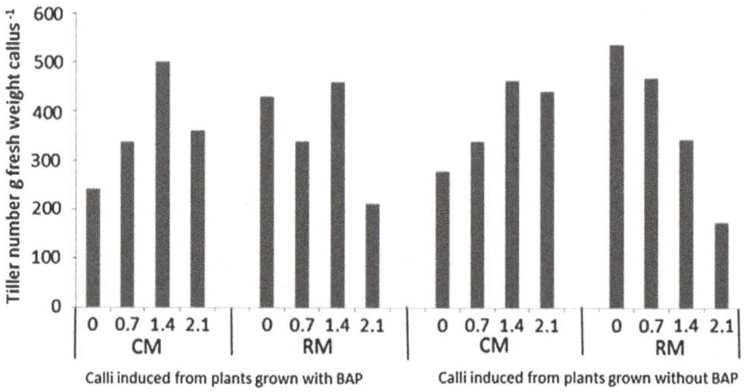

Fig. 11.11 The effect of proline (0–2.1 g L^{-1}) in callus medium (CM) or regeneration medium (RM) on mean shoot regeneration from calli of *Miscanthus sinensis* derived from shoot apices from plants grown with or without BAP (Means of 4 treatments)

11.4.5 Vegetative Propagation via Callus

In vitro propagation has been the main purpose of many callus induction experiments despite the risk of auxin-induced somaclonal variation. Most work has concentrated on *Miscanthus x giganteus* because it would be clearly economically beneficial to be able to propagate this cultivar more efficiently. Several authors have therefore defined the callus induction rate as the number or proportion of embryogenic and shoot-forming callus units produced per number of explants cultured. The callus regeneration rate has been defined as the number or proportion of

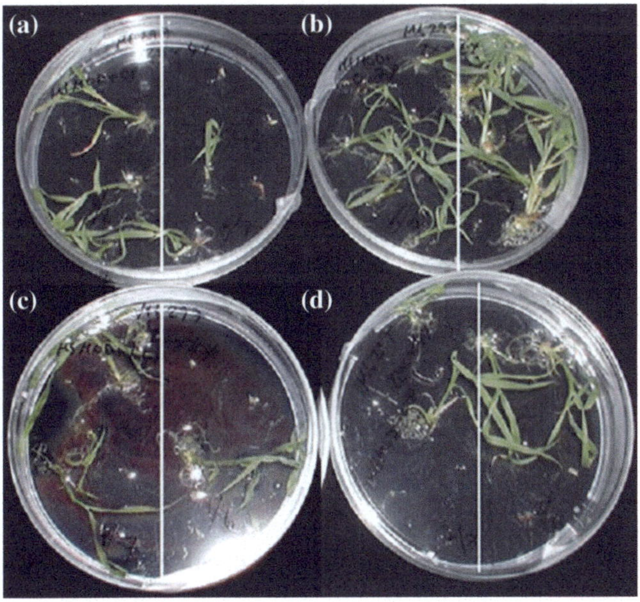

Fig. 11.12 *Miscanthus sacchariflorus* 'Robusta' plantlets developing from in vitro shoot apices cultured on medium containing 0 g L^{-1} (**a**), 0.7 g L^{-1} (**b**), 1.4 g L^{-1} (**c**) and 2.1 g L^{-1} proline (**d**). The *left* half of each dish contains diploid shoot apices and the *right* half contains tetraploid shoot apices

callus units which regenerate and the shoot regeneration rate has been defined as the number of shoots regenerated per callus unit.

Using shoot apices of *M. x giganteus* Petersen (1997) obtained a callus induction rate of 90 % of which 42 % of calli were embryogenic and produced 4 plants per 5 mm piece of callus. A further 26 % of calli were shoot-forming and regenerated 9 plants per 5 mm of callus. An estimated 360 plants at least could thus be theoretically regenerated from 100 shoot apices and Shin et al. (2011) have also reported similar figures.

Using young inflorescences of *M. x giganteus* Lewandowski and Kahnt (1993) calculated that a 100 mm inflorescence would yield 286 explants producing about 6 plants per explant (1,830 plants per inflorescence). This equates to 640 plants per 100 explants over a five month period and compares to a theoretical 800 plants produced from 100 shoot apices directly micro-propagated with BAP (Lewandowski 1997). Kim et al. (2010) produced immature inflorescence calli at 78 % frequency with 41 and 22 % shoot-forming and embryogenic callus respectively. From 100 explants callused for one month and immediately regenerated, approximately 32 shoot-forming calli could be induced which would produce 43 shoots (1.33 per callus). A further 17 embryogenic calli would produce 4 shoots in total, giving at most 50 plants per 100 explants.

Głowacka et al. (2010a) however, obtained a callus induction rate from immature *M. x giganteus* inflorescences of 8 %, defined as 8 × 1 mm callus pieces per explant, of which 33 % were embryogenic, with a regeneration rate of 45 % and a subsequent shoot regeneration rate of 1.85. Thus, 100 explants would yield 800 × 1 mm pieces of callus of which about 260 would be embryogenic. About 120 of these would regenerate 1.85 tillers each, thus producing 220 tillers from 100 explants. These were very good figures for plant regeneration until Rothrock (2010) greatly improved the direct micropropagation method. However, even more recently Gubisova et al. (2013) reported obtaining 64 plants per 5–15 mm inflorescence equating to over 1200 plants from 100 explants. This number could be further more than doubled by regenerating the calli on propagation medium. With other species, Głowacka et al. (2010a) found that two *M. sinensis* genotypes had similar callus induction rates to *M. x giganteus* (5.5–7 %), but that more of the calli were embryogenic (51–58 %) and able to regenerate (58–77 %) with shooting rates of about 6 tillers per callus. Thus, 100 explants of *M. sinensis* would produce about 625 × 1 mm calli of which 340 would be embryogenic and 230 calli able to regenerate about 6 tillers each, yielding about 1,420 tillers from 100 explants. Another *M. sinensis* genotype was similar to *M. x giganteus* in callus induction rate and embryogenic callus proportion, but had such poor regeneration potential that by these calculations only 41 plants would be produced per 100 explants.

In relation to the stability of plant regeneration, Lewandowski and Kahnt (1993) reported that with *M. x giganteus* callus plant regeneration decreased with age so that nearly 16 plants per gram of callus regenerated after 7 weeks callus growth, but only 8 plants per gram regenerated from 14 week old calli. Christian and Haase (2001) described experiments by the company Piccoplant in which calli produced 500–1,500 shoots per gram depending on the *M. x giganteus* 'clone' and continued to regenerate for 44 months. Kim et al. (2010) selected and grew embryogenic and shoot forming calli over a 12 month period and found that the regeneration frequency of embryogenic callus actually increased from one to two months, but that no regeneration was possible in either type of callus after 5 months. However, with careful selection cell suspensions were capable of some regeneration for 12 months. It has also been reported that compact callus can turn soft and watery if it is not sub-cultured at least once a month (Wang et al. 2011).

11.4.6 Cell Suspensions

Although callus culture has dominated work on *Miscanthus* to date, some work with cell suspension cultures has been reported. Holme et al. (1997) initiated cell suspensions of *M. x giganteus* from 6 month old embryogenic callus and grew them with MS and N6 medium and various concentrations of proline. Surprisingly some MS-grown cell suspensions were capable of shoot regeneration (mostly of albino shoots) for up to 18 months after initiation. However, these were also the slowest growing cell suspensions that had been grown without proline or conversely had been grown with such a high concentration of proline that the resulting osmotic stress inhibited their growth. Holme (1998) went on to determine the nutrient uptake of media components during the 15 day culture cycle. N6 medium (which contains less nitrogen than MS) required more proline to supply nitrogen for equivalent suspension growth, although the need for nitrogen was saturated by the addition of 5.75 g L^{-1} proline. Ammonium was more rapidly broken down than nitrate and sucrose was also rapidly broken down and the resultant glucose was metabolised before the fructose. Both 1.44 and 2.88 g L^{-1} proline were completely metabolised within 6 days and cultures with 1.44 g L^{-1} proline had the highest dry and fresh weight yields. Surprisingly, the pH remained within the region of pH 4.7–6.0.

Kim et al. (2010) also induced embryogenic cell suspensions from *M. x giganteus* and regenerated plants for up to 12 months, although the frequency declined after 6 months. Albino shoots only occurred during one period at 8 months. Fast growing regenerative cell suspensions of *M. x giganteus* capable of producing nearly 500 shoots per gram of tissue have also been described (Christian and Haase 2001).

11.4.7 Phenotypic Differences in Propagated Plants Transferred to Soil

Most in vitro propagation has been of the *Miscanthus x giganteus* genotype. Auxin-induced somaclonal variation is a consideration when cultures have been propagated through callus culture and cytokinins can have carry-over effects such as delayed flowering after the plants are transferred to soil. Lewandowski (1997, 1998) and Kim et al. (2010) in particular, have examined the characteristics of in vitro propagated plants after transplanting. A possible carry-over effect of BAP in directly micro-propagated *M. x giganteus* plants has been described in which many, thin shoots were produced giving them a bushy appearance. The thin stems made them more prone to lodging and they did not grow as fast as plantlets propagated through somatic embryogenesis. In vivo propagated plants were generally more robust, although the phenotypic differences between plants declined over 3 years (Lewandowski 1997, 1998).

Clifton-Brown et al. (2007) found no significant difference in the yield between plants established from rhizome cuttings or micro-propagated plants. However, reports from the Danish Institute of Agricultural Science showed that in vitro propagated plants of *M. x giganteus* produced only 16 % as much dry matter as rhizome derived plants and none survived the first winter, while propagated plantlets of *M. sinensis* 'Goliath' produced only 8 % as much dry-matter, but had an 86 % winter survival (Christian and Haase 2001). Kim et al. (2010) have also reported that plants propagated from immature inflorescence calli had poor winter survival compared to normally propagated plants, but otherwise had a normal phenotype. Conversely Płażek et al. (2011) compared compared survival of in vivo and in vitro propagated plants and concluded that in vitro derived plants were actually more cold tolerant in the first vegetative season compared to plants obtained in vivo. This was primarily because the commercially propagated in vivo plants were divided so finely that the rhizome reserves were very small. The differences again disappeared in the following years. At present there seems to be evidence for some carry-over effects of tissue culture, but no evidence of beneficial or detrimental somaclonal variation.

11.5 Applications of Cell Culture

11.5.1 Genetic Manipulation

Miscanthus is a commercial crop and many companies have their own genotypes, aims and methods, much of which are not published. The only detailed accounts so far of stable plant transformation whereby new genes are incorporated into the genome are by Zili et al. (2004) and Wang et al. 2011 and in two Patent applications, (Trieu 2010, Engler and Chen 2011), and in preliminary reports by Dalton et al. (2010b, 2011a, b). Two transformation methods have been successful; the bombardment of calli with DNA-coated gold particles and the preferred route of *Agrobacterium tumefaciens* infection. *Agrobacterium*-mediated transformation has been shown to result in lower copy number and fewer transgene re-arrangements than particle bombardment (Travella et al. 2005). So far the most successful production of transformed plants has been with *M. sinensis* (Dalton et al. 2010b, Wang et al. 2011, Engler and Chen 2011) with a single report of transformation of *M. sacchariflorus* (Zili et al. 2004). Widholm (2010) has also described transformed callus and plants of *M. x giganteus* whereas Trieu (2010) showed marker-transformed calli of *M. x giganteus* and the two parental species, although it is unclear whether plants of all three 'species' were regenerated. Seeds were collected implying that at least *M. sinensis* had regenerated and was fertile.

Using particle bombardment Zili et al. (2004) transformed *M. sacchariflorus* with a potato proteinase II gene and hygromycin resistance gene, Wang et al. (2011) produced *M. sinensis* plants expressing hygromycin resistance and a green fluorescent protein (GFP) and Yamada et al. (2010) regenerated *M. sinensis* plants containing a fructosyltransferase gene from *Lolium perenne*. We have also produced plants of two *M. sinensis* genotypes expressing Basta herbicide resistance (*bar*) and a fungal ferulic acid esterase gene (FAEA) gene (Dalton et al. 2010b, 2011a, b). By using *Agrobacterium*-mediated transformation Trieu (2010) produced plants expressing GFP and Engler and Chen (2011) produced plants expressing resistance to the herbicide Asulam. Widholm (2010) described plants of *M. x giganteus* transformed with the *bar* gene using both particle bombardment and *Agrobacterium*-mediated transformation methods. All these authors have used embryogenic or shoot-forming callus either from immature inflorescences (Engler and Chen 2011; Trieu 2010), seeds (Wang et al. 2011; Engler and Chen 2011; Trieu 2010) or shoot apices (Dalton et al. 2010b; Trieu 2010).

11.5.1.1 Transformation by Particle Bombardment

Particle bombardment involves bombarding suitable calli with DNA coated gold particles under a brief vacuum of ~700 mm Hg (Table 11.9). Transformation frequency is increased greatly by placing the calli as target tissue onto a medium with a high osmotic potential to plasmolyse the cells slightly before bombardment

Table 11.9 Special conditions and media used for particle bombardments of *Miscanthus*

Species	Additional osmoticum	Time pre and post bombardment (hrs)	DNA gun	Gold particle size	Distance	Pressure	Other	Reference
M. sacchariflorus	4.5 % mannitol	12/72	PDS 100/He(Bio-Rad)	1 μm	6 cm	1,300 psi (90 bar)	Bombarded twice	Zili et al. 2004
M. sinensis	9 % sorbitol	6/18	Particle inflow gun (Finer et al. 1992)	1 μm (Inbio Gold)	12 cm	7 bar	100 μm mesh	Dalton (unpublished)
M. sinensis	5.5 % mannitol+ 5.5 % sorbitol	8/24	IDERA GIE-III (Tanaka Co.)	–	5 cm	5 bar	–	Wang et al. 2011

(Vain et al. 1993). Normally 0.2 M mannitol and sorbitol is used (3.64 %). Zili et al. (2004) successfully bombarded *Miscanthus sacchariflorus* calli using a PDS 100/He gun (Bio-Rad) with 1 μm particles and compared a number of rupture disc pressures from 900 to 1,550 psi (62–107 bar) and found that the relatively high pressure of 1,300 psi was optimal. They also examined the effect of different hyper-osmotic media containing an additional 2–9 % mannitol for 4–48 h prior to bombardment and for 16–96 h after bombardment. A control treatment of an additional 2 % mannitol used throughout the culture period produced a transformation frequency of 13 %, but an additional 4.5 % mannitol applied for 12 h before bombardment and 72 h afterwards increased the transformation frequency to nearly 23 %. However, the calli had already been grown in CC medium, which contained 3 % sucrose and an extra 3.6 % mannitol, so that this osmotic treatment actually contained 8.1 % mannitol. They also examined the effect of multiple bombardments and found that the best results were obtained by bombarding the calli twice from different directions. Bombardment quantity was also examined and 250 μg gold coated with 0.5 μg DNA was optimal.

Zili et al. (2004) also screened a number of selection agents for common selection genes. The antibiotics geneticin (G418) and kanamycin can both be used to select for expression of the neomycin phosphotransferase gene (*nptII*). G418 was found to be more lethal than kanamycin, but sensitivity to the antibiotic hygromycin (*hpt* gene) and the herbicide Basta (*bar* gene) was even greater and these were considered to be the most useful selectable markers. They also considered a prolonged selection period (40 days) with a low selection pressure (30 mg L^{-1} hygromycin or 200 mg L^{-1} G418 or 20 mg L^{-1} Basta) to be most effective. They used a construct, (pIP101) that contained the hygromycin resistance gene under the CaMV35S promoter and a potato proteinase II (*pin II*) gene (Sanchez-Serrano et al. 1987) under an actin promoter to produce transformed plants at a transformation rate of 2.3 %. Surprisingly the commonly used β-glucuronidase marker gene (*gus or UidA)*, under the CaMV35S, actin1, ubiquitin or adh1 promoters was never expressed and they concluded that there might be an endogenous β-glucuronidase (GUS) inhibitor.

In our laboratory we have used a helium-powered Particle Inflow Gun to bombard *Miscanthus sinensis* calli. Young calli were transferred to high osmolarity medium 5–11 days after subculture and bombarded at a distance of 12 cm with two constructs, pAHC27 (Christensen and Quail 1991) and pRESQ48 (Sivamani and Qu 2006). These constructs encoded the *gus* gene under the maize and rice ubiquitin promoters respectively and were used to check transient GUS expression with different bombardment parameters such as gold particle size (0.6–1.6 μm). Calli were also bombarded at a shorter distance (8 cm), but although this greatly increased transient GUS expression, no stable transformants were obtained. As found in maize (Dalton et al. 2007) the transient expression of GUS under the rice ubiquitin promoter was much stronger than expression under the maize ubiquitin promoter.

Embryogenic calli were also bombarded with the constructs pROB5 encoding the *hpt* gene under the CaMV35S promoter (Bilang et al. 1991) or pUBA encoding the *bar* gene under the maize ubiquitin promoter (Toki et al. 1992). The selection genes were co-bombarded with an FAEA gene (pINH1d) controlled by the rice actin promoter and a vacuole targeting signal sequence (Buanafina et al. 2008). No plants were regenerated after selection on 25–50 mg L^{-1} hygromycin. However, plants were regenerated from bombardments with the *bar* gene after selection with 3–4 mg L^{-1} bialophos (Fig. 11.13). Transformation efficiency was very low with only one or two plants regenerating per ten bombardments. Unusually *Miscanthus* could regenerate plants in the presence of 3 mg L^{-1} bialophos in the light without photo-bleaching, thus allowing the selection to continue throughout regeneration. Bialophos resistant plants have been obtained from two *M. sinensis* genotypes—Genotype S14 (7 plants) and MS-88-110 (3 plants). Leaf sections from three in vivo S14 plants tested were resistant

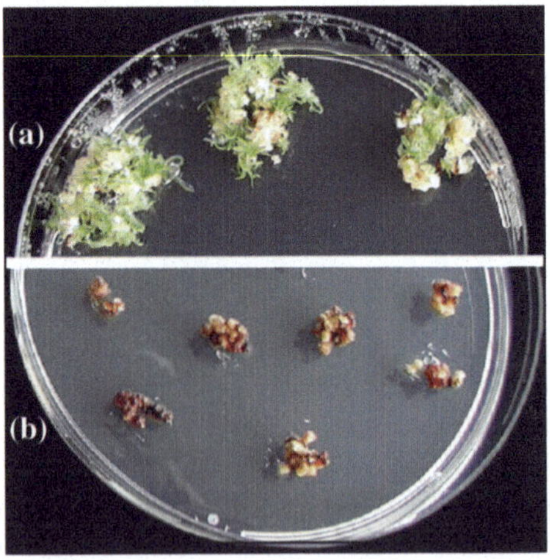

Fig. 11.13 Transformed (**a**) and control (**b**) calli growing on selective culture medium containing bialophos

in medium containing chlorophenol red and phosphinothricin (PPT) which is the active ingredient in Basta herbicide (Fig. 11.14). This test was also applicable to in vitro plantlets, leaves and calli. We found that two plants also showed ferulic acid esterase (FAE) activity in vitro and after transplanting to soil, (Fig. 11.15), and although the activities were not very high they were comparable to previous levels of expression reported for the same construct in *Festuca* and *Lolium* leaves (Buanafina et al. 2006, 2008) and were considerably higher in stems. We have re-transformed an FAE expressing plant with a senescence induced xylanase (*xyn2*) construct (pIOM6) co-bombarded with the *hpt* gene under the rice actin promoter in pAct1HPT-4 (Cho et al. 1998) for selection.

Wang et al. (2011) bombarded 10–14 day old seed-derived calli of *Miscanthus sinensis*. The calli were placed on a highly hyper-osmotic medium

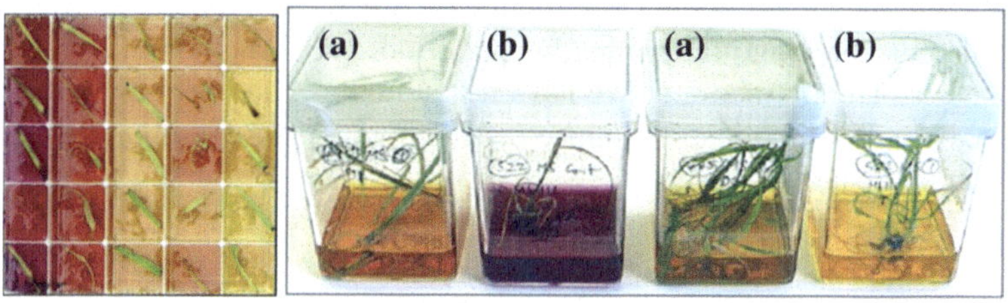

Fig. 11.14 *Miscanthus sinensis* in medium containing 5 mg L^{-1} PPT and chlorophenol *red*. *Bar* gene activity reduces the pH to turn the medium *yellow*. In vivo leaf material (*left*) and in vitro plants (*right*) Basta resistant transgenic (**a**) and control (**b**)

Fig. 11.15 *Miscanthus sinensis* plants expressing a vacuolar targeted fungal ferulic acid esterase gene

containing 0.3 M mannitol and sorbitol (5.46 %) before bombardment. A construct pAhcH1 (Spangenberg et al. 1995) containing the *hpt* gene under the rice actin promoter was co-bombarded with a construct pUC19-GFP, which contained the *gfp* gene under the CaMV35S promoter. Calli were selected on 50–150 mg L^{-1} hygromycin and over 100 co-expressing calli were recovered from 120 bombarded calli and four hygromycin resistant plants were regenerated, two of which also expressed GFP. Wang et al. (2011) also compared different types of calli and a higher frequency of transient expression of GFP was found in bombarded compact and friable (shoot-forming) calli than in soft callus. However, it is likely that the slow growth and hard, almost nut-like texture of *Miscanthus* embryogenic callus relative to other species or to other types of regenerable callus such as maize type II is a physical barrier to penetration by gold particles and to *Agrobacterium* infection, resulting in low transformation efficiency. Zili et al. (2004) used a long pre-bombardment period on hyper-osmotic medium to soften the callus and bombarded the calli twice. Wang et al. (2011) also used a long pre-bombardment period, but combined this with a very strong hyper-osmotic medium. We have tested maltose and proline in the callus medium, both of which tend to induce faster-growing, softer embryogenic callus (Dalton unpublished).

11.5.1.2 Transformation via *Agrobacterium tumefaciens*

Trieu (2010) and Engler and Chen (2011) have described the genetic transformation of *Miscanthus* via *Agrobacterium*-mediated transformation in their respective Patents (Table 11.10) and both used maltose based media and a long co-cultivation period. Interestingly Engler and Chen (2011) exposed the embryogenic *M. sinensis* calli to light to initiate shoot-growing callus prior to infection. They used the

Table 11.10 Special media used for *Agrobacterium*-mediated transformation of *Miscanthus* species (mg L^{-1} unless otherwise stated)

Purpose	Basal medium	Carbon source	2,4-D	BAP	Other ingredients	Author
Co-cultivation	1/10 MS+1/10 B5 vitamins	3 % maltose + 1 % glucose	2	0.5	200 cysteine, 50 glutamine Liquid medium, pH5.4, 300 μM acetosyringone	Trieu 2010
Recovery	N6+B5 vitamins and micro-elements	3 % maltose	2	1	1,000 CH, 1,000 asparagine 2,000 proline, 5 thiamine, 0.7 % agar	Trieu 2010
Co-cultivation	MS+B5 vitamins	3 % maltose	2, 5	1	Liquid medium, pH5.7 100 μM acetosyringone	Engler and Chen 2011

Agrobacterium tumefaciens strain GV3101 (pMP90) harbouring the PBI121 binary vector with the *gus* gene under the CaMV35S promoter and *nptII* gene under the nos promoter (Chen et al. 2003). Calli were incubated for 5–10 min in *Agrobacterium* solution and co-cultivated for 5 days at 25 °C. Surprisingly the nos promoter was active enough to enable selection with 100 ppm G418 and the CaMV35S promoter was strong enough to enable GUS expression in the leaves of transformed shoots. Subsequently the *nptII* gene was substituted with a sulfonamide herbicide resistance gene (DHPS) and the transgenic calli were selected with Asulam herbicide.

On the other hand, Trieu (2010) used the *Agrobacterium tumefaciens* strain EHA105 harbouring a NB4KAN:p326-EGFP construct with the *nptII* gene under the CaMV35S promoter and an enhanced *gfp* gene under a 326 FL promoter from *Arabidopsis* as markers. They incubated calli of *M. sacchariflorus* 'Robustus', *M. x giganteus* 'Illinois' and *M. sinensis* for 30 min at 22 °C in *Agrobacterium* solution containing a very high concentration of acetosyringone, before co-culture for 5–7 days at 25 °C on wetted filter paper. After resting for a further 5–7 days the calli were selected with 100 mg paromomycin and the visual selection of GFP expression. GFP was expressed in calli of *M.x giganteus* and both parental species and the transformation frequency was 5–15 %. When the calli were exposed to a 3 min 43 °C heat-shock before incubation with *Agrobacterium* the frequency rose to 20–30 %.

11.5.1.3 Targets for Genetic Manipulation

The list of potential uses for *Miscanthus* genetic transformation covers all the breeders objectives mentioned below and more besides (Trieu 2010; Engler and Chen 2011). However, in the short term there appears to be some interest in both herbicide resistance and particularly in manipulating genes with the potential for improving ethanol production from lignocellulose. Transgenic down-regulation of the enzymes involved in lignin biosynthesis using antisense and RNAi gene silencing could increase the cellulose to lignin ratio, as at present *Miscanthus* straw contains 20–30 % lignin which is higher than many other grasses (Visser and Pignatelli 2001). This would reduce the need for enzymatic pre-treatments and would potentially increase ethanol production (Heaton et al. 2010). We have produced *Miscanthus sinensis* plants expressing a fungal ferulic acid esterase gene (Dalton et al. 2010b, 2011a) and also a plant which expresses both the ferulic acid esterase gene and a senescence-induced xylanase gene.

There is also preliminary research into transcription factors involved in cell wall lignification. Jorgensen (2008) suggested using a xylanase gene from a heat tolerant bacterium, which would only be activated during the bioethanol production process. Yamada et al. (2010) have also reported the production of fructan-accumulating transgenic *Miscanthus sinensis* plants expressing a fructosyltransferase (FT) gene from a C3 grass, *Lolium perenne* which may have applications for fermentation. Other current targets include prolonging vegetative growth by reducing the expression of inflorescence induction genes (Juvik et al. 2007).

11.5.2 Production of Haploids via Anther Culture

In vitro anther culture is an established technique used in many crop species to induce haploid calli from pollen, which can then regenerate haploid or more commonly 'doubled haploid' (DH) plants. This enables a variety of homozygous plants to be obtained from a heterozygous parent in a single generation. There is only one report of anther culture in *Miscanthus sinensis* (Głowacka and Jeżowski 2009) in which anthers were cultured from *M. sinensis* inflorescences with pollen at the uni-nucleate stage. They compared four media and obtained most calli from C17 medium which had been devised for wheat (Wang and Chen 1983), but was modified by using maltose and agarose. There were genotypic differences in response and a 7–14 day period of cold treatment of the spikes containing the young inflorescence was found to be helpful. They regenerated calli on five media of which a Gamborg's B5 medium with 0.1 mg L^{-1} BAP was considered best and finally obtained nine plants of varying ploidy. Most significantly, they found that the anthers from spikes at the right stage in October produced 17 times as many calli as anthers from spikes taken in August (29 and 1.7 calli per 100 anthers respectively). It was considered that cooler, shorter days had a positive effect on the donor plant and subsequent androgenesis.

11.5.3 Chromosome Doubling

Chromosome doubling is of potential benefit because polyploid cells and plants are generally larger in size and there is also the possibility of being able to restore the fertility, or to create new hybrids with triploid plants such as *Miscanthus x giganteus*. There have been a several approaches to using in vitro cultures to induce polyploidy in *Miscanthus* species (Table 11.11). Petersen et al. (2002, 2003) tested a wide range of parameters, including the concentration of colchicine, oryzalin and dimethylsulphoxide (DMSO), the timing and repeat treatments in twelve *Miscanthus sinensis* genotypes (Table 11.1). They compared in vivo and in vitro plantlets, shoot apices and leaf explants as well as single in vitro shoots and embryogenic callus derived from immature inflorescences. Immature inflorescences, shoot apices and leaf explants were also induced to form calli in the presence of colchicine or oryzalin and then regenerated. Treating tissues during callus induction produced more tetraploid plants than treating plantlets and they concluded that the shoot apex was the most suitable explant because the production of shoot forming callus was less genotype dependent than the production of embryogenic callus. Fewer than 3 % of 538 plantlets regenerated from shoot-forming calli were chimaeras, triploid or more than tetraploid. Their best treatment appeared to be to culture in vitro shoot apices in callus medium containing 2.5 g L^{-1} colchicine for 4 days. Of 112 treated apices, nearly 80 % regenerated tetraploid shoots.

Table 11.11 Treatments used for chromosome doubling in *Miscanthus* species (g L^{-1} unless otherwise stated)

Genotype	Explants	Colchicine	Exposure	Other variables	Best result	Reference
M.sinensis one genotype	In vivo and in vitro shoots and plantlets, immature inflorescence and leaf explants, shoot apices, embryogenic calli derived from shoot apices and immature inflorescences	1.25–10	18 h–7 day,	5.2–10.4 oryzalin, 1.5 % DMSO	Unclear	Petersen et al. 2002
M.sinensis twelve genotypes	In vivo and in vitro shoots and plantlets, in vitro leaf explants and shoot apices, immature inflorescence derived embryogenic calli	0.6–10	5 h–7 day Low light(42 μE m^{-2} s^{-1})	0.4–41.5 oryzalin, 1.5 % DMSO	Shoot apices on callus medium, 2.5 colchicine, 4 days	Petersen et al. 2003
M.x giganteus	Immature inflorescence derived calli and cell suspensions	1.25–3.75	2–7 day	1.7–10.4 oryzalin Liquid, 0.3 % gelrite	Liquid medium 3.4 oryzalin, 4 days	Yu et al. 2009
M.sinensis (five genotypes), *M.x giganteus*	Four-week-old in vivo plants from in vitro cultures	2.5 and 5	6–24 h	1 and 2 % DMSO Tween (10drops/L)	5 colchicine+ 2 % DMSO	Głowacka et al. 2009
M.sinensis (four genotypes), *M.x giganteus*	In vitro shoots regenerated immature inflorescence derived calli	0.6–5	6–24 h Low light (40–50 μE m^{-2} s^{-1})	–	1.25 colchicine, 18 h	Głowacka et al. 2010b
M.sinensis (five genotypes), *M.x giganteus*	Immature inflorescence derived calli	1.25–2.5	2–7 day	Embryogenic calli, Regenerating calli	Regenerating calli, 1.25 colchicine,4 days	Głowacka et al. 2010c
M.sacchariflorus 'Robustus'	In vitro shoot apices	2	16 h darkness, shaking in liquid	–	–	Dalton (unpublished)

Yu et al. (2009) also tested various colchicine and oryzalin concentrations, timings and liquid and solid medium with embryogenic calli from immature inflorescences of *M. x giganteus*. They found that survival was higher with colchicine, but that oryzalin was more effective. With oryzalin in liquid medium, survival was further reduced, but more of the calli contained hexaploid sectors and more hexaploid plants were recovered. Seven out of 33 plants recovered from 426 regenerating calli were hexaploid. Five of the plants were from oryzalin treatments and two had been treated with 3.4 g L^{-1} oryzalin in liquid medium for 4 days. Preliminary results indicated that the hexaploid *M. x giganteus* had thicker stems and larger stomata, but otherwise had a similar growth rate to the triploid genotype. More pollen was viable in the hexaploid (Heaton et al. 2010) and hybrid progeny have recently been produced (Touchell and Ranney 2012).

Głowacka et al. (2009) exposed in vivo tillers regenerated from immature inflorescence derived calli to a range of colchicine and DMSO treatments and obtained polyploid plants from five *Miscanthus sinensis* genotypes, but not from *Miscanthus x giganteus*. They went on to treat in vitro tillers regenerated from calli with colchicine and then micro-propagated them. Polyploids of three diploid *M. sinensis* genotypes, the triploid *M. sinensis* 'Goliath' and *M. x giganteus* were recovered. The greatest efficiency of polyploidisation was 55 % in one genotype after 18 h of treatment with 1.25 g L^{-1} colchicine (Głowacka et al. 2010b). Regardless of the genotype tested, this treatment appeared to be the most effective combination for generating polyploidy and was optimal for shoot survival. Over 400 plants were evaluated and stomatal length and pollen grain diameter were shown to be good parameters to identify putative *Miscanthus* polyploids. Surprisingly flow cytometry showed no sectoring or aneuploidy at all. The beneficial effect of increased ploidy appeared to depend upon genotype as the stem diameter was significantly increased in four genotypes, the dry matter increased in two genotypes and the stem height increased in one genotype.

In subsequent experiments exposing immature inflorescence derived calli to colchicine during regeneration was found to increase polyploidy production threefold, compared to exposure during callus induction, when the best treatment of 1.25 g L^{-1} colchicine for 4 days was used (Głowacka et al. 2010c). The resulting tetraploid *M. sinensis* and hexaploid *M. x giganteus* plants were evaluated and two of the *M. sinensis* polyploids had significantly wider leaves, while all genotypes had higher pollen stainability.

In our laboratory we have used a method which has worked well in *Lolium multiflorum,* to produce tetraploid *M. sacchariflorus* 'Robustus' plants without any callus phase. In vitro shoot tips were cultured directly in colchicine to reduce the 'sectoring' found in colchicine treated whole tillers, whereby subsequent tillers could be tetraploid, aneuploid or diploid depending on which sector of a meristem they arose from. Small shoot apices <3 mm were dissected from in vitro stock plantlets and placed directly into liquid regeneration medium. After culturing, the medium was replaced with 2 g L^{-1} colchicine solution and the apices were shaken gently (30 rpm) for 16 h. The shoot apices were then washed and floated on fresh regeneration medium. The surviving shoots (29 from 93 shoot apices) were subsequently micro-propagated and bulk leaf samples from each multi-tillered plantlet derived from a single treated shoot apex were tested for ploidy. Seven plantlets (7.5 %) contained tetraploid tissue and were then divided into separate tillers and propagated once more. The ploidies of plantlets derived from the shoot apex with the most abundant tetraploid tissue were tested again. Three plantlets were tetraploid, three were diploid and another was aneuploid. In this way tetraploid plants were obtained after two rounds of selection, with no danger of somaclonal variation or any necessity for embryogenic capacity.

11.6 Plant Breeding

11.6.1 Genomic and Transcriptomic Characterisation of Genetic Resources

There has been a revolution in plant breeding with the development of molecular techniques over the last 30 years, enabling the deeper analysis of the genetic basis for traits of interest. These techniques have been applied to *Miscanthus* relatively recently to characterise the genetic resources available (Głowacka 2011). Initially generic approaches were used. Greef et al. (1997) and Hodkinson et al. (2002b) used amplified fragment length polymorphism (AFLP) fingerprinting to distinguish successfully between *Miscanthus* species, sub-species and hybrids. Hernandez et al. (2001) found that there was a high transferability of microsatellite simple sequence repeat (SSR) markers and restriction fragment length polymorphisms (RFLP) from maize to *Miscanthus* and more recently Yook et al. (2011) found SSR markers developed for maize and sugarcane were transferable to *Miscanthus*. Small numbers of SSR markers have also been developed from within *M. sinensis* and have shown that there is extensive genetic diversity within the species (Hung et al. 2009). Many of these markers are also applicable to related species such as *M. sacchariflorus*, *M. floridulus*, and *M. lutarioparius* (Zhou et al. 2011; Ho et al. 2011).

Jakob et al. (2009) considered SSRs to be the most useful markers to develop in *Miscanthus* as they allow differentiation between many alleles, can be easily automated and are transferable across genotypes and related species. Surprisingly Zhao et al. (2011) also discovered a high transferability of SSR markers from the model grass *Brachypodium distachyon* to *Miscanthus*. However, single nucleotide polymorphisms (SNPs) are the most numerous variations within genomes and as high throughput genomic sequencing has developed, SNPs have started to replace microsatellites as a way of compiling genetic maps (Hamblin et al. 2007). The *Miscanthus* genome is calculated to be very large at about 7 Gbp (Rayburn et al. 2009), although Swaminathan et al. (2010) considered that up to 98 % of this might be repeat sequences. They provided a 'snapshot' of the *Miscanthus x giganteus* genome using genomic and small RNA sequencing, with reference to sorghum and showed that 80 % of *Miscanthus* contigs matched sorghum genes.

Lee et al. (2011), using whole-genome transcriptome profiling, showed that a selected Korean *Miscanthus* genotype had EST (expressed sequence tag) sequences with considerable homologies to sorghum (43 %), maize (37 %), and rice (15 %). Ma et al. (2012) using genotyping-by-sequencing (GBS) methods have also shown that sorghum has the closest syntenic relationship to *M. sinensis* compared with maize, rice, and *Brachypodium distachyon*. They have also completed the first high-resolution, comprehensive genetic map of *M. sinensis* using more than 3,700 SNPs to map all 19 chromosomes.

11.6.2 Identification of Trait-Linked Markers and Marker Assisted Selection

Marker assisted selection (MAS) uses molecular markers to identify quantitative trait loci (QTL) associated with particular characteristics. These are usually found by analysing a mapping family of plants which are the progeny of two widely differing parents. Atienza et al. (2002) constructed a mapping population using MS-90-2 and MS-88-110 *M. sinensis* and created a genetic map using random amplification of polymorphic DNA (RAPDs). Biomass yield and combustion quality are the most important areas for *Miscanthus* breeders and from this material Atienza et al. (2003a, b, c, d, e) identified twenty potential QTLs associated with agronomic traits, such as biomass yield, stem-yield, leaf-yield and total top-yield. They also identified a further five potential QTLs associated with flowering date and fifteen associated with the minerals calcium, phosphorus, potassium, sodium and sulphur. These may be important for nutrient recycling and ash content.

This was a major advance, although RAPDs have been superseded by newer techniques. SNP markers and association mapping may be more precise than SSRs in identifying relevant genes and alleles because they are so abundant, but require more comprehensive genomic sequence information (Lister et al. 2009). This approach is being used at the University of Illinois where SNP arrays for *M. sinensis* and for hybrids with *M. sacchariflorus* are being developed (Heaton et al. 2010). At IBERS, Aberystwyth, Jensen et al. (2008) have generated mapping populations from early and late flowering *M. sinensis* and other populations with different leaf morphologies and stem traits in order to identify QTLs associated with flowering time, senescence and biomass yield. These can be tested, and the heritability of each trait measured, in a second larger synthetic population using SNPs (Clifton-Brown et al. 2013). Restriction site Associated DNA (RAD) markers are also being developed and have the advantage that the DNA sequencing is done at the screening stage allowing the identification of large numbers of SNPs (Baird et al. 2008). In addition several candidate genes from other species have been isolated from *Miscanthus* using a Bacterial Artificial Chromosome (BAC) library of genomic fragments. A number of these loci have been sequenced in a range of genotypes and compiled to produce haplotypes which it is hoped will be consistently present in plants with certain phenotypes (Farrar et al. 2011). Chloroplast CpSSRs and SNPs have also been identified (Cesare et al. 2010).

Many more QTLs remain to be discovered for key traits such as for biotic and abiotic stress tolerances, tillering, flowering, canopy architecture, seed quality, senescence, cell wall composition and potential invasiveness characters such as aggressive rhizome growth, seed shatter, or growth from nodes (Jakob et al. 2009).

11.6.3 Production of New Hybrids

There are reports of the successful hybridisation of *Miscanthus sinensis* x *Saccharum officinarum* × *S. spontaneum* (commercial sugarcane) (Park et al. 2011) to produce 'Miscane'. Several of these 'miscane' genotypes have shown potential for being more cold resistant than sugarcane. These two species hybridise naturally, but according to a Patent (Rooney et al. 2010) there have also been successful hybridisations between sorghum and both *Miscanthus floridulus* and *Miscanthus sinensis* and also between sorghum and 'miscane'. Interestingly there was a high incidence of vivipary and thus low seed set. The seeds required in vitro embryo rescue to germinate, which involved surface sterilising the seeds and germinating them on an MS medium with additional glycine, argenine and tyrosine. The intergeneric hybrids have the potential to be used directly, either as seed or as clonally propagated biofuel crops. They could also be used as a means of introgressing genes and traits from one species to another. As an example transgenic sugarcane can already produce isomaltulose, sorbitol, poly-β-hydroxybutyric acid (PHB) and p-hydroxybenzoic acid and could be crossed with *Miscanthus* (Poirier et al. 2007). Interspecific compatibility within the genus *Miscanthus* has also produced some interesting hybrids (John Clifton-Brown pers. comm.).

11.6.4 Endophyte Exploitation

Novel endophytic nitrogen-fixing bacteria of three different genera have been discovered in *Miscanthus* species (Kirchhof et al. 2001; Miyamoto et al. 2004; Eckert et al. 2005) and there has been considerable interest in looking at the effect and possible exploitation or genetic manipulation of endophytes. We have found that the use of in vitro plants simplifies screening for endophytes, because many confounding bacteria are eliminated (Cope-Selby et al. 2011).

11.7 Conclusions

Miscanthus biotechnology in terms of conventional in vitro initiation, storage, propagation and chromosome doubling methods are well developed and phenolic browning is no longer considered a barrier to successful culture. It may also be possible to induce and to exploit in vivo or in vitro vivipary as a means of producing 'propagules' which can be 'sown'. Genetic transformation of *Miscanthus* species is still at a very early stage compared to the sophisticated applications currently being developed with the globally important monocot crops of maize, rice and wheat, such as the manipulation of transgenic host plants to allow recombinase-mediated gene stacking (Ow 2011) and transformations using very large inserts (He et al. 2010). Genetic manipulation in sugarcane is much more advanced than in *Miscanthus* and it may be that transformed 'miscane' hybrids will be more exploitable than transformed *Miscanthus* in the short-term. The hard, compact nature of *Miscanthus* embryogenic callus is a problem and reminiscent of similar problems in maize until type II callus was identified and a means to induce it developed. Even now very specific protocols must be followed to have any chance of success (Frame et al. 2006). It is extremely heartening however, that a high proportion of *Miscanthus* genotypes are at least capable of embryogenic development, in contrast to many other species of grass. It is also very encouraging that three different selection systems with the *nptII*, *hpt* and *bar* genes have been successfully applied. This enables the re-transformation of interesting transgenic plants. The promoters which have enabled genes to be expressed have also ranged from the relatively weak nos and CaMV35S promoters through to the much stronger rice actin and maize and rice ubiquitin promoters. As transformation becomes more efficient and genomic and transcriptomic analysis accumulates ever more information concerning potential traits of interest, this is an exciting time to be involved with *Miscanthus* biotechnological research.

Acknowledgments I am extremely grateful to Phil Morris for critically reading the manuscript, helping with the figures and for such useful advice, discussion and editing. Thanks also to Ray Bilang for pROB5, Peggy Lemaux for pAct1HPT-4, Peter Quail for pAHC27, Rongda Qu and Elumalai Sivamani for pRESQ48, and Seiichi Toki for pUBA. In addition many thanks to colleagues at IBERS including Tim Langdon for pINH1D, pIOM6 and useful discussion, John Clifton-Brown for photographs and useful discussion, Emma Timms-Taravella for expert molecular analysis, Cathy Morris and Charlotte Hayes for expert cytometry, Ana Winters for collaboration over FAE expression, Samantha Gill and Sue Youell for assistance and to Ian Thomas, Elaine Jensen, Maurice Bosch, Joe Gallagher, Paul Robson, Kerrie Farrar and Iain Donnison for useful discussion and for which I also thank Kai Schwarz and Heike Meyer of the Julius Kühn-Institute.

References

AEBIOM (2011) Annual Statistical Report p 36 European Biomass Association

Allen D (2008) Genetic improvement of *Miscanthus* at mendel biotechnology. The 5th annual bioenergy feedstocks symposium. Urbana-Champaign, IL, 10 Jan 2008

Atienza SG, Satovic Z, Petersen KK, Dolstra O, Martín A (2002) Preliminary genetic linkage map of *Miscanthus sinensis* with RAPD markers. Theor Appl Genet 105:946–952

Atienza SG, Satovic Z, Petersen KK, Dolstra O, Martín A (2003a) Identification of QTLs influencing agronomic traits in *Miscanthus sinensis* Anderss. I. Total height, flag-leaf, height and stem diameter. Theor Appl Genet 107:123–129

Atienza SG, Satovic Z, Petersen KK, Dolstra O, Martín A (2003b) Identification of QTLs associated with yield and its components in *Miscanthus sinensis* Anderss. Euphytica 132:353–361

Atienza SG, Satovic Z, Petersen KK, Dolstra O, Martín A (2003c) Identification of QTLs influencing combustion quality in *Miscanthus sinensis* Anderss. II. chlorine and potassium content. Theor Appl Genet 107:857–863

Atienza SG, Satovic Z, Petersen KK, Dolstra O, Martín A (2003d) Influencing combustion quality of *Miscanthus sinensis* Anderss: identification of QTLs for calcium, phosphorus and sulphur content. Plant Breed 122:141–145

Atienza SG, Ramirez MC, Martin A (2003e) Mapping QTLs controlling flowering date in *Miscanthus sinensis* Anderss. Cereal Res Comm 31:3–4

Baird NA, Etter PD, Atwood TS, Currey MC, Shiver AL, Lewis ZA, Selker EU, Cresko WA, Johnson EA (2008) Rapid SNP discovery and genetic mapping using sequenced RAD markers. PLoS ONE 3(10):e3376. doi:10.1371/journal.pone.0003376

Bilang R, Iida S, Peterhans A, Potrykus I, Paszkowski J (1991) The 3′-terminal region of the hygromycin-B-resistance gene is important for its activity in *Eschiricia coli* and *Nicotiana tabacum*. Gene 100:247–250

de O Buanafina MM, Langdon T, Hauck B, Dalton SJ, Morris P (2006) Manipulating the phenolic acid content and digestibility of Italian ryegrass (*Lolium multiflorum*) by vacuolar targeted expression of a fungal ferulic acid esterase. Appl Bioch Biotech 129(132):416–426

de O Buanafina MM, Langdon T, Hauck B, Dalton SJ, Morris P (2008) Expression of a fungal ferulic acid esterase increases cell wall digestibility of tall fescue (*Festuca arundinacea).* Plant Biotechnol J 6:264–280

Cesare M, Hodkinson T, Barth S (2010) Chloroplast DNA markers (cpSSRs, SNPs) for *Miscanthus, Saccharum* and related grasses (*Panicoideae, Poaceae*) Mol Breeding 26: 539–544

Chen P-Y, Wang C-K, Soong S-C, To K-Y (2003) Complete sequence of the binary vector pBI121 and its application in cloning T-DNA insertion from transgenic plants Mol Breeding 11: 287–293

Christensen AH, Quail P (1991) Ubiquitin promoter-based vectors for high-level expression of selectable and/or screenable marker genes in monocotyledonous plants. Transgen Res 5:213–218

Christian DG, Haase E (2001) Agronomy of *Miscanthus* in *Miscanthus* for energy and fibre. James and James (ed) Mary Walsh pp 46–68

Cho M-J, Jiang W, Lemaux PG (1998) Transformation of recalcitrant barley cultivars through improvement of regenerability and decreased albinism. Plant Sci 138:229–244

Chu C-C, Wang C-C, Sun C-S, Hsu C, Yin K-C, Chu C-Y (1975) Establishment of an efficient medium for anther culture in rice through comparative experiments on the nitrogen sources. Sci Sinica 18:659–668

Clifton-Brown JC, Breuer J, Jones MB (2007) Carbon mitigation by the energy crop *Miscanthus*. Glob Change Biol 13:2296–2307

Clifton-Brown J, Robson P, Davey C, Farrar K, Hayes C, Huang L, Jensen E, Jones L, Hinton-Jones M, Maddison A, Meyer H, Norris J, Purdy S, Rodgers C, Schwarz K, Salvatore C, Slavov G, Valentine J, Webster R, Youell S, Donnison I (2013) Breeding *Miscanthus* for bioenergy. In: Saha M (ed) Biomass crops: breeding and genetics, Wiley

Cope-Selby N, Donnison IS, Farrar K (2011) Bacterial endophytes in the bioenergy grass *Miscanthus*. Soc Exp Biol Annual Meeting, Glasgow

Dale PJ (1980) A method for in vitro storage of *Lolium multiflorum* Lam. Ann Bot 45:497–502

Dalton SJ, Bettany AJE, Timms E, Morris P (1999) Co-transformed, diploid *Lolium perenne* (perennial ryegrass), *Lolium multiflorum* (Italian ryegrass) and *Lolium temulentum* (darnel) plants produced by microprojectile bombardment. Plant Cell Rep 18:721–726

Dalton SJ, Bettany AJE, Bhat V, Gupta MG, Bailey E, Timms E, Morris P (2003) Genetic transformation of *Dichanthium annulatum* (Forssk)—an apomictic tropical forage grass. Plant Cell Rep 21:974–980

Dalton SJ, Heywood E, Timms EJ, Morris P (2007) A comparison of maize and rice ubiquitin promoter activity using the *uidA* (*gus*) gene in maize: an enhanced rice ubiquitin promoter increases transgene expression in maize compared with the native maize ubiquitin promoter. Plant transformation technologies conference, Vienna, 4–7 Feb 2007

Dalton SJ, Donnison I (2010a) Improved in vitro propagation and establishment in soil of *Miscanthus* species. 18th European biomass conference, Lyon, 4–7 May 2010

Dalton SJ, Timms-Taravella E, Donnison I (2010b) Improved genetic transformation of *Miscanthus sinensis*. 18th European Biomass Conference, Lyon, 4–7 May 2010

Dalton SJ, Winters A, Langdon T, Timms-Taravella E, Donnison I (2011a) Genetic transformation of *Miscanthus sinensis* with FAE. Plant transformation technologies II conference, Vienna, 19–22 Feb 2011

Dalton SJ, Gallagher JA, Winters AL, Langdon T, Timms-Taravella E, Donnison IS (2011b) Genetic transformation of *Miscanthus sinensis* with a ferulic acid esterase gene. 33rd symposium on biotechnology for fuels and chemicals, Seattle, 2–5 May 2011

Deuter M (2000) Breeding approaches to improvement of yield and quality in *Miscanthus* grown in Europe, EMI Project, Final report, pp 28–52

Deuter M (2011a) *Miscanthus* plant named `MBS 7001' Mendel biotechnology, Inc. World intellectual property organization, United States Patent PP22033 ('Nagara')

Deuter M (2011b) *Miscanthus* plant named 'MBS 7002' Mendel biotechnology, Inc. World intellectual property organization, USA Patent PP22047 ('Lake Erie')

Deuter M (2011c) *Miscanthus* plant named 'MBS 1002' Mendel biotechnology, Inc. World intellectual property organization, USA Patent PP22127

Eckert B, Weber OB, Kirchhof G, Halbritter A, Stoffels M, Hartmann A (2005) *Azospirillum doebereinerae* sp. nov., a nitrogen-fixing bacterium associated with the C4-grass *Miscanthus*. Int J Syst Evol Microbiol 51:17–26

Engler D, Chen J (2011) Transformation and engineered trait modification in *Miscanthus* species. Mendel Biotechnology, Inc. World intellectual property organization, USA Patent US20110047651

Farrar K, Jensen E, Allison G (2011) Defining ideotypes in the biomass crop *Miscanthus*. Problems and reports from the fourth MPSSG, centre for plant integrative biology, Nottingham, 25 July 2011

Finer JJ, Vain P, Jones MW, McMullen MD (1992) Development of the particle inflow gun for DNA delivery to plant cells. Plant Cell Rep 11:323–328

Fowler PA, McLauchlin AR, Hall LM (2003) The potential industrial uses of forage grasses including *Miscanthus*. BioComposites Centre, University of Wales, Bangor

Frame BR, Paque T, Wang K (2006) Maize (*Zea mays* L.) In: Wang K (ed) Methods in molecular biology. Humana Press Inc., Totowa, NJ, pp185–199

Gamborg OL, Miller RA, Ojima K (1968) Nutrient requirements of suspension cultures of soybean root cells. Exp Cell Res 50:151–158

Głowacka K, Jeżowski S, Kaczmarek Z (2009) Polyploidization of *Miscanthus sinensis* and *Miscanthus x giganteus* by plant colchicine treatment. Ind Crop Prod 30:444–446

Głowacka K, Jeżowski S (2009) Genetic and nongenetic factors influencing callus induction in *Miscanthus sinensis* (Anderss.) anther cultures. J Appl Genet 50:341–345

Głowacka K, Jeżowski S, Kaczmarek Z (2010a) The effects of genotype, inflorescence developmental stage and induction medium on callus induction and plant regeneration in two *Miscanthus* species. Plant Cell Tiss Org Cult 10:79–86

Głowacka K, Jeżowski S, Kaczmarek Z (2010b) In vitro induction of polyploidy by colchicine treatment of shoots and preliminary characterisation of induced polyploids in two *Miscanthus* species. Ind Crop Prod 32:88–96

Głowacka K, Jeżowski S, Kaczmarek Z (2010c) Impact of colchicine application during callus induction and shoot regeneration on micropropagation and polyploidisation rates in two *Miscanthus* species. In vitro Cell Dev Biol Plant 46: 161–171

Głowacka K (2011) A review of the genetic study of the energy crop *Miscanthus*. Biomass Bioenerg 35:2445–2454

Grare MAR (2010) Variability of salinity response in *Miscanthus sinensis*, MSC thesis, edepot.wur.nl/155120

Greef JM, Deuter M (1993) Syntaxonomy of *Miscanthus x giganteus* GREEF and DEU. Angewandte Botanik 67:87–90

Greef JM, Deuter M, JungC Schondelmaier J (1997) Genetic diversity of European *Miscanthus* species revealed by AFLP fingerprinting. Genet Resour Crop Ev 44:185–195

Gubisova M, Gubis J, Zofajova A, Mihalik D, Kraic J (2013) Enhanced *in vitro* propagation of *Miscanthus x giganteus*. J Ind Crop Prod 41:279–282

Hamblin MT, Warburton ML, Buckler ES (2007) Empirical comparison of simple sequence repeats and single nucleotide polymorphisms in assessment of maize diversity and relatedness. PLoS ONE 2(12): e1367, doi:10.1371journal.pone.0001 0001367

Hansen J, Kristiansen K (1997) Short-term in vitro storage of *Miscanthus* × *ogiformis* Honda 'Giganteus' as affected by medium composition, temperature, and photon flux density. Plant Cell Tiss Org Cult 49:161–169

He R, Pan J, Zhu L, He G (2010) *Agrobacterium*-mediated transformation of large DNA fragments using a BIBAC vector system in rice. Plant Mol Biol Rep 28:613–619

Heaton EA, Dohleman FG, Miguez AF, Juvik JA, Lozovaya V, Widholm J, Zabotina OA, McIsaac F, David MB, Voight TB, Boersma NN, Long SP (2010) *Miscanthus*: a promising biomass crop. Adv Bot Res 56:75–135

Hernández P, Dorado G, Laurie DA, Martín A, Snape JW (2001) Microsatellites and RFLP probes from maize are efficient sources of molecular markers for the biomass energy crop *Miscanthus*. Theor Appl Genet 102:616–622

Ho C-W, Wu T-H, Hsu T-W, Huang J-C, Huang C-C, Chiang T-Y (2011) Development of 12 genic microsatellite loci for a biofuel grass, *Miscanthus sinensis* (Poaceae). Am J Bot 98:201–203

Hodkinson TR, Chase MW, Lledo MD, Salamin N, Renvoize SA (2002a) Phylogenetics of *Miscanthus, Saccharum* and related genera (*Saccharinae, Andropogoneae, Poaceae*) based on DNA sequences from ITS nuclear ribosomal DNA and plastid trnL intron and trnL-F intergenic spacers. J Plant Res 115:381–392

Hodkinson TR, Chase MW, Renvoize SA (2002b) Characterization of a genetic resource collection for *Miscanthus* (Saccharinae, Andropogoneae, Poaceae) using AFLP and ISSR PCR. Ann Bot 89:627–636

Holme IB, Petersen KK (1996) Callus induction and plant regeneration from different explant types of *Miscanthus x ogiformis* Honda 'Giganteus'. Plant Cell Tiss Org Cult 45:43–52

Holme IB, Krogstrup P, Hansen J (1997) Embryogenic callus formation, growth and regeneration in callus and suspension cultures of *Miscanthus x ogiformis* Honda 'Giganteus' as affected by proline. Plant Cell Tiss Org Cult 50:203–210

Holme IB (1998) Growth characteristics and nutrient depletion of *Miscanthus x ogiformis* Honda 'Giganteus' suspension cultures. Plant Cell Tiss Org Cult 53:143–151

Hung K-H, Chiang T-Y, Chiu C-T, Hsu T-W, Ho C-W (2009) Isolation and characterization of microsatellite loci from a potential biofuel plant *Miscanthus sinensis* (Poaceae). Conserv Genet 10:1377–1380

Jakob K, Zhou F, Paterson AH (2009) Genetic improvement of C4 grasses as cellulosic biofuel feedstocks. In vitro Cell Dev Biol Plant 45: 291–305

Jennings P (2011) Commercial scale Giant *Miscanthus*. Southeast bioenergy conference, Tifton, Georgia, 9 Aug 2011

Jensen E, Thomas-Jones S, Farrar K, Clifton-Brown J, Donnison I (2008) Unravelling the genetic control of flowering time in the bioenergy grass *Miscanthus*. Soc Exp Biol, Marseille, 6–10 July 2008

Jørgensen U (2008) Breeding and biotechnology perspectives in *Miscanthus*. EC-US task force on biotechnology research: workshop on biotechnology for sustainable bioenergy. San Francisco, 21–22 Feb 2008

Juvik J, Kim H-S, Ibrahim K (2007) *Miscanthus* breeding and improvement. Symposium on biomass feedstocks for energy production in Illinois. Urbana, 1 Nov 2007

Kim H-S, Zhang G, Juvik JA, Widholm JA (2010) *Miscanthus x giganteus* plant regeneration: effect of callus types, ages and culture methods on regeneration competence. Glob Change Biol Bioenerg 2:192–200

Kirchhof G, Eckert B, Stoffels M, Baldani JI, Reis VM, Hartmann A (2001) *Herbaspirillum frisingense* sp. nov., a new nitrogen-fixing bacterial species that occurs in C4-fibre plants. Int J Syst Evol Micr 51:157–168

Lee G-J, Jeon YJ, Ma K-Y, Kim I-K, Kim D-S (2011) A whole-genome transcriptome profiling of *Miscanthus* species in Korea plant and animal genomes XIX conference. San Diego, 15–19 Jan 2011

Lewandowski I, Kahnt G (1993) Development of a tissue culture system with unemerged inflorescences of *Miscanthus* 'Giganteus' for the induction and regeneration of somatic embryoids. Beitr Biol Pflanzen 67:439–451

Lewandowski I (1997) Micropropagation of *Miscanthus x giganteus*. Biotech Agr Forest 39:241–255

Lewandowski I (1998) Propagation method is an important factor in the growth and development of *Miscanthus x giganteus*. Ind Crop Prod 8:229–245

Li L, Qu R, Kochko A, Fauquet C, Beachy RN (1993) An improved rice transformation system using the biolistic method. Plant Cell Rep 12:250–255

Lister R, Gregory BD, Ecker JR (2009) Next is now: new technologies for sequencing of genomes, transcriptomes, and beyond. Curr Opin Plant Biol 12:1–12

Ma X-F, Jensen E, Alexandrov N, Troukhan M, Zhang L, Thomas-Jones S, Farrar K, Clifton-Brown J, Donnison I, Swaller T (2012) Flavell R (2012) High resolution genetic mapping by genome sequencing reveals genome duplication and tetraploid genetic structure of the diploid *Miscanthus sinensis*. PLoS ONE 7(3):e33821. doi:10.1371/journal.pone.0033821

Matumura M, Hasegawa T, Saijoh Y (1985) Ecological aspects of *Miscanthus sinensis* var. *condensatus, M. x sacchariflorus*, and their 3x-, 4x-hybrids. 1. Process of vegetative spread. Research Bulletin of the Faculty of Agriculture, Gifu University vol 50 pp 423–433

Mendel Biotechnology Inc. (2012) Mendel Biotechnology Inc, and BP biofuels to conduct demonstration field trial of PowerCane™ *Miscanthus*. http://www.mendelbio.com/newsevents/index.php#ai

Miyamoto T, Kawahara M, Minamisawa K (2004) Novel endophytic nitrogen-fixing *Clostridia* from the grass *Miscanthus sinensis* as revealed by terminal restriction fragment length polymorphism analysis. Appl Environ Microb 70:6580–6586

Murashige T, Skoog F (1962) A revised medium for rapid growth and bioassays with tobacco tissue culture. Physiol Plant 15:473–497

Nielsen JM, Brandt K, Hansen J (1993) Long-term effects of thidiazuron are intermediate between benzyladenine, kinetin or isopentenyladenine in *Miscanthus sinensis*. Plant Cell Tiss Org Cult 35:173–179

Nielsen JM, Hansen J, Brandt K (1995) Synergism of thidiazuron and benzyladenine in axillary shoot formation depends on sequence of application in *Miscanthus x ogiformis* 'Giganteus'. Plant Cell Tiss Org Cult 41:165–170

Nishiwaki A, Mizuguti A, Kuwabara S, Toma Y, Ishigaki G, Miyashita T, Yamada T, Matuura H, Yamaguchi S, Rayburn AL, Akashi R, Stewart JR (2011) Discovery of natural *Miscanthus* (Poaceae) triploid plants in sympatric populations of *Miscanthus sacchariflorus* and *Miscanthus sinensis* in southern Japan. Am J Bot 98:154–159

Ow DW (2011) Recombinase-mediated gene stacking as a transformation operating system. J Integrative Plant Biol 53:512–519

Park JW, Yu Q, Gracia NS, Acuna GM, da Silva JA (2011) Development of new intergeneric cane hybrids, Miscanes, as a source of biomass feedstock for biofuel production. Plant and animal genomes XIX conference, San Diego, 15–19 Jan 2011

Pepó P, Tóth S (2005) The role of nitrogen and phosphorus source in *Miscanthus* in vitro cultures. Cereal Res Comm 33:549–552

Petersen KK (1997) Callus induction and plant regeneration in *Miscanthus x ogiformis* Honda 'Giganteus' as influenced by benzyladenine. Plant Cell Tiss Org Cult 49: 137–140

Petersen KK, Hansen J, Krogstrup P (1999) Significance of different carbon sources and sterilization methods on callus induction and plant regeneration of *Miscanthus x ogiformis* Honda 'Giganteus'. Plant Cell Tiss Org Cult 58:189–197

Petersen KK, Hagberg P, Kristiansen K (2002) In vitro chromosome doubling of *Miscanthus sinensis*. Plant Breed 121:445–450

Petersen KK, Hagberg P, Kristiansen K (2003) Colchicine and oryzalin mediated chromosome doubling in different genotypes of *Miscanthus sinensis*. Plant Cell Tiss Org Cult 73:137–146

Płażek A, Dubert F, Zur I, Waligorski P (2007) In vitro culture of *Miscanthus x giganteus*. Zeszyty Problemowe Postepow Nauk Rolniczych 523:175–184
Płażek A, Dubert F (2010) Improvement of medium for *Miscanthus x giganteus* callus induction and plant regeneration. Acta Biologica Cracoviensia Series Botanica 52(1):105–110
Płażek A, Dubert F, Janowiak F, Krępski T, Tatrzańska M (2011) Plant age and in vitro or in vivo propagation considerably affect cold tolerance of *Miscanthus* × *giganteus*. Eur J Agron 34:163–171
Poirier Y, van Beilen J, Orts B (2007) Biopolymers: crops for biopolymer and platform chemicals. Epobio Workshop, Athens, May 2007
Potrykus I, Saul MW, Petruska J, Paszkowski J, Shillito RD (1979) Direct gene transfer to cells of a graminaceous monocot. Mol Gen Genet 199:183–188
Rayburn AL, Crawford J, Rayburn CM, Juvik JA (2009) Genome size of three *Miscanthus* species. Plant Mol Biol Rep 27:184–188
Rooney WL, Hodnett GL, Kuhlman LC, Stelly DM, Price HJ, Price PK (2010) Intergeneric hybrid plants and methods for production thereof. The Texas A&M University. World intellectual property organization, USA. US20100050501
Rothrock RE (2010) Propagation of switchgrass and *Miscanthus*. Ceres Inc. World intellectual property organization, USA. Patent WO2010011717A2
Sanchez-Serrano JJ, Keil M, O'Connor A, Schell J, Willmitzer L (1987) Wound-induced expression of a potato proteinase inhibitor II gene in transgenic tobacco plants. EMBO J 6:303–306
Shin S-B, Abdel-Ghany SE, Reddy ASN (2011) High efficiency regeneration of *Miscanthus x giganteus* plants. ASPB Plant Biology, Minneapolis, 6–10 Aug 2011
Sivamani E, Qu R (2006) Expression enhancement of a rice polyubiquitin gene promoter. Plant Mol Biol 60:225–239
Spangenberg G, Wang ZY, Wu XL, Nagel J, Potrykus I (1995) Transgenic perennial ryegrass (*Lolium perenne*) plants from microprojectile bombardment of embryogenic suspension cells. Plant Sci 108:209–217
Stewart R, Toma Y, Fernández FG, Nishiwaki A, Yamada T, Bollero G (2009) The ecology and agronomy of "*Miscanthus sinensis*", a species important to bioenergy crop development, in its native range in Japan: a review. Glob Change Biol Bioenerg 1–2:126–153
Sun GZ, Ma MQ, Zhang YQ, Xian XL, Cai XL, Li XP (1999) A medium adapted to embryogenic callus induction and subcultures of wheat. J Hebei Agric Sci 3:24–26
Swaminathan S, Alabady MS, Varala K, de Paoli E, Ho I, Rokhsar DS, Arumuganathan AK, Ming R, Green PJ, Meyers BC, Moose SP, Hudson ME (2010) Genomic and small RNA sequencing of *Miscanthus giganteus* shows the utility of sorghum as a reference genome sequence for Andropogoneae grasses. Genome Biol 11:R12
Szabados L, Savoure A (2009) Proline: a multifunctional amino acid. Trends Plant Sci 15:89–97
Toki S, Takamatsu S, Nojiri C, Ooba S, Anzai H, Iwata M, Christensen A, Quail PH, Uchimiya H (1992) Expression of maize ubiquitin gene promoter-bar chimeric gene in transgenic rice. Plant Physiol 100:1503–1507
Touchell DH, Ranney TG (2012) Chromosome Doubling and Fertility Restoration in *Miscanthus*× *giganteus*. ASHS Annual Conference, Miami, Florida
Travella S, Ross SM, Harden J, Everett C, Snape JW, Harwood WA (2005) A comparison of transgenic barley lines produced by particle bombardment and *Agrobacterium*-mediated techniques. Plant Cell Rep 23:780–789
Trieu ATN, (2010) *Miscanthus* transformation methods. Ceres Inc. World intellectual property organization, USA. Patent WO2010065534A3
USDA (2011) Proposed biomass crop assistance program (BCAP) giant Miscanthus (*Miscanthus x giganteus*) establishment and production in Arkansas, Missouri, Ohio, and Pennsylvania. Environmental assessment, USDA farm service agency. http://fsa.usda.gov/Internet/FSA_File/finaleagiantmcanthus.pdf
Vain P, McMullen MD, Finer JJ (1993) Osmotic treatment enhances particle bombardment-mediated transient and stable transformation of maize. Plant Cell Rep 12:84–88

Visser P, Pignatelli V (2001) Utilization of *Miscanthus*. In: Jones MB, Walsh M (eds) *Miscanthus* for energy and fibre. James and James, London, pp 109–154

Wang P, Chen Y (1983) Preliminary study on production of height of pollen H_2 generation in winter wheat grown in the field. Acta Agron Sin 9:283–284

Wang X, Yamada T, Kong F-J, Abe Y, Hoshino Y, Sato H, Takamizo T, Kanazawa A, Yamada T (2011) Establishment of an efficient in vitro culture and particle bombardment-mediated transformation systems in *Miscanthus sinensis* Anderss., a potential bioenergy crop. GCB Bioenergy 3:322–332

Widholm JM (2010) Improvement of bioenergy crops via transformation. Energy biosciences institute annual report

Yamada T, Wang X, Kanazawa A, Yamada T, Hoshino Y, Ukaji N (2010) Genetic improvement through a transgenic approach in *Miscanthus* ssp., a new bioenergy crop. Green revolution 2.0: Food + Energy and Environmental Security, Long Beach, 31 Oct–4 Nov 2010

Yook M-J, Li S-H, Kim D-S (2011) SSRs analysis for genetic diversity in *Miscanthus* species. Plant and animal genomes XIX conference, San Diego, 15–19 Jan 2011

Yu C-Y, Kim H-S, Rayburn AL, Widholm JM, Juvik JA (2009) Chromosome doubling of the bioenergy crop, *Miscanthus* × *giganteus*. GCB Bioenergy 1:404–412

Zhao H, Yu J, You FM, Luo M, Peng JH (2011) Transferability of microsatellite markers from *Brachypodium distachyon* to *Miscanthus sinensis*, a potential biomass crop. J Integr Plant Biol 53:232–245

Zili Y, Puhua Z, Chengcai C, Xiang L, Wenzhong T, Li W, Shouyun C, Zuoshun T (2004) Establishment of genetic transformation system for *Miscanthus sacchariflorus* and obtaining of its transgenic plants. High Tech Lett 10:27–31

Zili Y, Liang X, Jianxiong J, Zhiyong C, Jingping T, Lifang H (2010) Hunan agricultural university rapid breeding method of *Miscanthus sinensis*. Chinese Patent Application No. CN 201010259544 (24 Nov 2010)

Zhang QX, SunY, Hu HK, Chen B, Hong CT, Guo HP, PanYH, Zheng BS (2012). Micropropagation and plant regeneration from embryogenic callus of *Miscanthus sinensis*. In Vitro Cell Dev Biol Plant 48:50–57

Zhou H-F, Li S–S, Ge S (2011) Development of microsatellite markers for *Miscanthus sinensis* (Poaceae) and cross-amplification in other related species. Am J Bot 98:195–197

Zub H, Brancourt-Hulmel M (2010) Agronomic and physiological performances of different species of *Miscanthus*, a major energy crop. A review. Agron Sustain Dev 30:201–214

Part IV
Section 4

Chapter 12
Tissue Culture and Genetic Transformation of Safflower (*Carthamus tinctorius* L.)

M. Sujatha and S. Dutta Gupta

Abstract Safflower belonging to the family Asteraceae is an underutilized industrial crop of semi-arid regions. The oil is valued therapeutically due to its high degree of poly-unsaturation apart from its use in cosmetics, in the manufacture of urethane, resins, linoleum, and high quality emulsion paints. In vitro culture techniques, such as direct shoot organogenesis and somatic embryogenesis are successful from cotyledonary leaf explants. Direct and callus- mediated shoot regeneration from seedling tissues and anthers are also described. Cotyledon explant is highly responsive in adventitious shoot regeneration on medium supplemented with BA + NAA or TDZ + NAA. Somatic embryos were induced directly on adaxial surface of the cotyledonary leaves on Murashige and Skoog (MS) medium containing 5.37–10.74 μM NAA and 2.22 μM BA within 8–10 days of culture. Somatic embryo development was asynchronous and strongly influenced by auxin type and concentration. Various factors such as genotype, explant age, carbon source, and ethylene affected somatic embryogenesis frequency, number of somatic embryos per responding explant and somatic embryo maturation and germination. Medium containing 8.87 μM BA and 2.69 μM NAA promoted shoot regeneration from anthers and microspores. Genetic engineering has the potential to accelerate crop improvement programmes and has yielded encouraging results in several crop plants. In safflower, *Agrobacterium*-mediated transformation protocols have been reported for both Indian and American cultivars. Transformation studies were limited to constructs harbouring commonly used reporter (*uidA*, GFP) and selectable marker genes (*npt*II). The studies indicated influence of co-cultivation conditions, *Agrobacterium tumefaciens* strain, type of explant, genotype, etc. Transgene expression was confirmed with transient assays by GUS expression and molecular analysis through PCR and Southern hybridization assays

M. Sujatha (✉)
Directorate of Oilseeds Research, Rajendranagar, Hyderabad 500 030, Andhra Pradesh, India
e-mail: mulpurisujata@yahoo.com

S. Dutta Gupta
Department of Agricultural and Food Engineering, Indian Institute of Technology, Kharagpur 721 302, India
e-mail: sdg@agfe.iitkgp.ernet.in

S. M. Jain and S. Dutta Gupta (eds.), *Biotechnology of Neglected and Underutilized Crops*, DOI: 10.1007/978-94-007-5500-0_12,

of the primary transformants. Despite the high frequency of shoot regeneration from transformed tissues (15–34.3 %), rooting of transformed shoots is yet to overcome. So far, safflower has been used for the production of valuable plant made pharmaceuticals through agro-infection. However, the tissue culture based transformation protocols are yet to be exploited for the development of transgenics with agronomically desirable traits.

Abbreviations

NAA	α- napthaleneacetic acid
BA	Benzyl adenine
MS	Murashige and Skoog
TDZ	Thidiazuron
2-ip	N^6-[2-Isopentenyl] adenine
KN	6-furfurylaminopurine
p-CPA	*p*-chlorophnoxyacetic acid
IBA	Indole-3-butyric acid
2, 4-D	2, 4- dichlorophenoxyacetic acid
ACC	1 aminocyclopropane-1-carboxylate
$AgNO_3$	Silver nitrate
AS	Acetosyringone
CEPA	2 chloroethylphosphonic acid

12.1 Introduction

Safflower (*Carthamus tinctorius* L.), belongs to the family Asteraceae, is a minor oil seed crop of semi-arid regions. In the past, its flowers were used as a source of orange dye carthamin and later it was developed as an edible oil crop. The seed oil is valued therapeutically for its high degree of poly-unsaturation in the form of linoleic acid (Knowles 1989; Velasco and Fernandez-Martinez 2001). Safflower oil is also nutritionally desirable for human consumption because of elevated levels of α-tocopherol (Furuya et al. 1987), an antioxidant that improves the utilization of vitamin A in human body. This is particularly important since vitamin A deficiency is common in many tropical countries. The oil is also important for its industrial use like in cosmetics, in manufacture of urethane resins, linoleum, and high quality emulsion paints. The meal or flour (with or without hull) is used as poultry feed. The decorticated oil-free meal which is rich in protein and trace elements is used as human food in protein deficient diets (Weiss 1971). Safflower florets are also reported to have medicinal properties to cure hypertension, arthritis, and spondylosis (Li and Muendel 1996). In spite of its various potential uses, the plant has remained under-exploited and enough attention was not paid to its improvement using biotechnological approach.

Genetic improvement of safflower through conventional breeding is limited to the development of cultivars with high yield, stability and varying oil content and quality (Knowles 1989). Although conventional breeding has developed few elite cultivars with promise, in vitro technology could further help to improve the plant's genetic make-up. The most important requirement for application of in vitro or transgenic techniques for genetic improvement is the availability of a reliable and reproducible plant regeneration system through adventitious shoot organogenesis and somatic embryogenesis. Compared to shoot organogenesis, plant regeneration through somatic embryogenesis is preferred as it gives high rate of multiplication and often results true-to-type plants (Van Arnold et al. 2002). Moreover, the possibility of single cell origin of somatic embryos may make the regenerable target tissue amenable for genetic transformation. Plants regenerated from transformed tissues via organogenesis are often chimeras while regeneration through somatic embryos reduces the formation of chimeras. Regardless of the mode of regeneration, it is essential to optimize various factors which would help in developing a reproducible system of plant regeneration in this crop. The present chapter summarizes the work done on shoot organogenesis and direct somatic embryogenesis in safflower considering various factors influencing caulogenesis, somatic embryogenesis and ontogeny of somatic embryo induction and development. The current status of genetic transformation is also reviewed.

12.2 Tissue Culture

Tissue culture studies in safflower include callus-mediated regeneration (Rani et al. 1996; Rani and Rao 1998; Radhika et al. 2006; Sujatha and Dinesh Kumar 2007), direct organogenesis (George and Rao 1982; Tejovathi and Anwar 1987, 1993; Sujatha and Suganya 1996; Tejovathi and Das 1997; Nikam and Shitole 1999; Mandal and Dutta Gupta 2001; Radhika et al. 2006; Sujatha and Dinesh Kumar 2007) and somatic embryogenesis (Mandal et al. 1995; Mandal et al. 2001; Mandal and Dutta Gupta 2003; Walia et al. 2007; Kumar et al. 2008) (Table 12.1). The major constraints in safflower regeneration are low shoot regeneration rate, sensitivity of regenerated shoots to high relative humidity in culture vessel and differential rooting responses among cultivars.

12.2.1 *Shoot Organogenesis*

Direct adventitious shoot regeneration is achieved through culture of seedling tissues of both spiny and non-spiny genotypes. Juvenile tissues including roots expressed great propensity for de novo shoot bud induction. induction. Murashige and Skoog (1962) medium medium supplemented with plant growth regulators

Table 12.1 Tissue culture studies in safflower

Explant	Type of morphogenetic response	Medium used (μM)	Reference
Hypocotyl, cotyledons	Multiple shoots	MS + 0.88–8.87 BA + 2.69 NAA	George and Rao (1982)
Cotyledons	Induction of capitula	MS + 2.22 BA + 0.53 NAA MS + 2.32 KN + 0.53 NAA	Tejovathi and Anwar (1987)
Leaf	Oil accumulation	MS + 4.44 BA + 1.34 NAA + 5 % sucrose + 1 g/l CH + 10 % coconut water	Singh (1991)
Primary seedling explants of *C. tinctorius* shoot apices of *C. oxyacantha*	Shoot regeneration from leaf shoot proliferation	MS + 35.5 BA + 2.69 NAA + 37 adenine sulphate MS + 2.69 NAA + 57.74 GA_3 + 5.0 ascorbic acid	Singh (1991)
Anther	Shoot regeneration	MS + 8.87 BA + 2.69 NAA	Prasad and Anwar (1991)
Cotyledons	In vitro rooting	MS + 1.0–2.0 2,4,5-Cl_3POP	Tejovathi and Anwar (1993)
Leaf	Shoot buds	MS + 22.19 BA + 1.34 NAA	Chatterjee and Singh (1993)
Primary seedling explants	Adventitious shoot regeneration	MS + 4.44 BA + 0.53 NAA	Orlikowska and Dyer (1993)
Immature embryos	Shoot regeneration	MS +0.45 TDZ + 0.53 NAA MS + 0.045 TDZ + 0.53–53.71 NAA	
Cotyledons	Somatic embryos	MS + 2.22 BA + 5.37–10.74 NAA	Mandal et al. (1995)
Primary seedling explants	Adventitious shoots histological analysis	MS + 2.22–22.19 BA + 0.53–2.69 NAA	Zhanming and Biwen (1993)
Cotyledons, hypocotyl	In vitro rooting	MS + 49 IBA for 7 days followed by MS + 1.5 % sucrose + 1 g/l activated charcoal for 21 days	Baker and Dyer (1996)
Hypocotyl, cotyledons	Shoot regeneration	MS + 1.11 BA + 0.53 NAA	Jhansi Rani et al. (1996)
Primary seedling explants	Selection of calli and shoots resistant to *Fusarium oxysporum*	MS + 4.44 BA + 5.37 NAA for callus MS + 4.44 BA + 0.53 NAA for shoots	Suganya et al. (1997)
Cotyledons	Selection of calli resistant to sodium chloride	MS + 2.22 BA + 8.06 NAA	Nikam and Shitole (1997)

(continued)

Table 12.1 (continued)

Explant	Type of morphogenetic response	Medium used (μM)	Reference
Primary seedling explants	Direct shoot regeneration	MS + 4.44 BA + 10.0 CH	Nikam and Shitole (1999)
Cotyledons	Adventitious shoots	MS + 2.22–8.87 BA	Mandal and Dutta Gupta (2001)
Cotyledons	Somatic embryos	MS + 2.22 BA + 5.37 NAA	Mandal and Dutta Gupta (2003)
Cotyledonary node, stem node	Shoot buds	MS + B5 vitamins + 19.96 BA + 6.97 Kn	Kumar et al. (2008)
Primary seedling explants including roots	Shoot regeneration	MS + 2.27–22.71 TDZ + 0.53-2.69 NAA	Radhika et al. (2006)
Endosperm	Somtic embryos	MS + 2.22–8.87 BAP/2.32–9.29 KN/0.91–2.27 TDZ	Walia et al. (2007)
Leaf	Shoot regeneration	MS + 0.91–4.54 TDZ + 1.07–5.37 NAA MS + 4.44 BA + 2.69–5.37 NAA	Sujatha and Dinesh Kumar (2007)
Cotyledons, leaf	Somatic embryogenesis	MS + 27.25 TDZ + 6.78 2iP + 12.26 IBA	Kumar et al. (2008)
Cotyledonary leaf	Adventitious shoots	MS + 2.27 TDZ + 1.23 IBA	Basalma et al. (2008)

BA and NAA proved to be superior for morphogenic response (Prasad and Anwar 1991; Chatterji and Singh 1993). Shoot organogenesis and plant regeneration from hypocotyl segments are depicted in Fig. 12.1.

12.2.1.1 Direct Organogenesis

Tejovathi and Anwar (1993) observed that low concentrations of BAP and NAA induced 10–12 shoot buds per explants from seedling cotyledons of safflower cultivars, A-1 and Manjira. Although A-1 showed a wider plasticity of growth regulator requirement, it showed a slightly lesser degree of shoot bud regeneration than Manjira.

Orlikowska and Dyer (1993) reported direct shoot regeneration from primary seedling explants and immature embryos of the American safflower cv. Centennial. Direct shoot regeneration from primary explants was obtained on MS medium supplemented with 2.22 μM BAP or 0.45 μM TDZ and 0.53 μM NAA. Shoot regeneration from immature embryos was observed on MS medium containing TDZ and

Fig. 12.1 Shoot organogenesis and plant regeneration from hypocotyls in safflower. **a** Regeneration of shoots after 10 days of culture of hypocotyls on MS medium supplemented with 0.91 μM TDZ and 1.07 μM NAA. **b** Shoot multiplication on MS medium supplemented with 2.22 μM BAP. **c** Shoot proliferation on MS medium supplemented with 2.32 μM KN. **d** Elongated shoots on 2.32 μM KN, 5.29 μM 2iP. **e** Rooted shoots on half-strength MS medium supplemented with 5.37 μM NAA. **f** Acclimatization and establishment of a regenerated shoot

NAA. The shoots were healthy and elongated on MS supplemented with 4.92 or 9.84 μM 2-iP and rooted on half-strength MS with 5.37 μM NAA.

In studies of Mandal and Dutta Gupta (2001), adventitious shoot buds were induced from adaxial surface of cotyledons of eight safflower cultivars on MS medium supplemented with BAP. The variety S144 showed maximum shoot organogenesis (54.4 %) with 8.87 μM BA. Highest number of normal shoots was obtained with lower concentration of BA (2.22 μM).

Adventitious shoot regeneration from different explants of safflower including roots was described by Radhika et al. (2006). MS medium supplemented with TDZ (2.27–22.71 μM) and NAA (0.53–2.69 μM) showed high frequency of shoot regeneration. Shoot elongation was achieved on medium with 2.32 μM KN. Rooting was obtained on half-strength MS medium with 2.69 μM NAA. The study for the first time demonstrated the amenability of roots for in vitro regeneration and also the use of TDZ in induction of adventitious shoot buds.

Basalma et al. (2008) reported adventitious shoot regeneration from safflower cv, Dincer (Turkish cultivar) using cotyledonary leaf explants. Highest shoot regeneration was obtained on MS medium with 2.27 μM TDZ and 1.0 μM IBA. Shoot regeneration from cotyledonary nodes and meristem tips was obtained on MS medium supplemented with various concentrations of BA alone or in combination with NAA.

12.2.1.2 Callus-Mediated Shoot Regeneration

George and Rao (1982) have reported callus formation and differentiation of 2–4 shoot buds per explant on medium with MS salts + 0.88–8.87 μM BAP + 2.69 μM NAA in 40 % hypocotyl cultures of NP-9 black and Th-10 black. They also observed that cotyledons were more responsive than hypocotyl tissues. The shoot buds developed on 8.87 μM BA + 2.69 μM NAA medium were very small and most of them failed to develop into normal plantlets. In the variety partial hull black, good callus induction occurred, but shoot differentiation was observed only in 10 % of the cotyledon and hypocotyl explants.

Anwar et al. (1989) cultured cotyledonary explants on seven different media (LS, MS, B5, Blaydes, Chaleffs, Wood and Hsienmiaos). Callus initiation was observed within 7–10 days of inoculation and the frequency of callusing was maximum (91.1 %) on MS medium followed by Chaleff's and B5 media. Poor callusing was observed on Wood's medium. The ratio of dry weight, fresh weight was highest (38.9 %) on MS medium followed by B5 (34.1 %) whereas, it varied from 10.2 to 19.8 % on other media. In seven different explants tested (cotyledons, root, apical bud, hypocotyl, stem, leaf and seedling base), the frequency of callus induction was maximum with cotyledonary explants (90–95 %) followed by root (85–90 %), apical bud (80–90 %), hypocotyl (80–90 %), stem (60–68 %) and leaf (55–60 %). Minimum callusing (30–40 %) was noticed in seedling base explant. Similar results were reported by Padmaja et al. (1990) who found that MS medium supplemented with 9.05 μM 2,4-D was effective in callusing among several media tested. The study also showed that among the different explants tested, morphogenic ability was high in callus derived from cotyledons. Likewise, studies of Rani et al. (1996) indicated the efficiency of cotyledonary leaves for callus induction, growth and shoot differentiation.

Reddy and Devi (1991) cultured cotyledon explants on MS medium supplemented with 3 % sucrose and varying levels of 2,4-D (4.52, 9.05 and 22.62 μM) for 2–3 weeks. Frequency of callus induction was high on medium with 4.52 μM 2,4-D in Manjira and HUS-305 genotypes. Shoot differentiation was achieved with varying frequencies from cotyledon-derived callus on MS medium containing BA. Shoots were rooted on hormone-free MS medium. These studies suggested that shoot morphogenesis could be induced from cotyledonary callus culture.

Morphogenetic studies in callus and organ cultures of cultivated (*C. tinctorius* L. cv. CO-1) and wild safflower (*C. oxyacantha*) was carried out by Singh (1991). A higher cytokinin (4.44–11.09 μM BA) to auxin (1.34–5.37 μM NAA) ratio was found to be optimal for callus induction. Callus proliferation rate and greening was increased with addition of adenine sulphate (37.01–74.02 μM). Morphogenetic response through organogenesis was observed after 5 weeks when cultures were placed on medium containing enhanced levels of BAP (35.5–44.38 μM).

Zhanming and Biwen (1993) reported profuse callus growth of different seedling explants of safflower cultured on MS basal medium supplemented with 2,4-D (1.13 μM) singly or in combination with BA (2.22 μM). Adventitious bud differentiation was induced on medium supplemented with BA (2.22–22.19 μM) + NAA (0.53–2.69 μM) from the young leaf and shoot apex.

Chatterji and Singh (1993) studied the induction of morphogenetic calli from leaf explants of an elite safflower cultivar, CO-1. Leaf explants from 3-week-old in vitro grown seedlings exhibited callusing potential on MS medium with NAA (5.37–8.06 μM) and BA (2.22–4.44 μM). Shoot bud primordia surfaced when the level of BA was raised to 22.19 μM in combination with 1.34 μM NAA after 4 weeks of sub-culturing. For leafy shoot development, shoot buds were transferred to half-strength basal MS with adenine sulphate (148.04 μM).

Nikam and Shitole (1999) reported callus induction and in vitro plantlet regeneration systems for safflower cv. Bhima using root, hypocotyl, cotyledon and leaf explants. Supplementation of the medium with an auxin: cytokinin ratio >1 enhanced the growth rate of callus cultures, however for 2,4-D the ratio was <1.34. The growth regulators, IAA, NAA, BA and KN alone or in combinations were found effective for callus induction and regeneration in all the explants. The calli could be maintained for over 32 months. BAP (4.44 μM) combined with casein hydrolysate (CH 10 mg/l) yielded the highest rate of shoot production on hypocotyls (3–6) and cotyledon (5–7) explants and cotyledonary derived callus (4–8).

Genotypic differences were observed for both direct and callus-mediated regeneration. Among the various Indian cultivars tested, Manjira proved to be the most responsive for dedifferentiation and shoot differentiation (Anwar et al. 1989; Prasad et al. 1990; Tejovathi and Anwar 1993).

12.2.2 Somatic Embryogenesis

12.2.2.1 Somatic Embryo Induction and Development and Influence of Auxin on Somatic Embryogenesis

Direct somatic embryos were detected on the adaxial surface of explant within 8–10 days, different stages of somatic embryos within 15 days of culture initiation (Fig. 12.2a). Among the different NAA/BA treatments, somatic embryogenesis was evident with 2.69, 5.37, 10.74, and 21.48 μM NAA plus 2.22 μM BA. Highest response was obtained with 10.74 μM NAA plus 2.22 μM BA. Somatic embryogenesis from endosperm, cotyledon and leaf derived callus have also been reported in safflower (Walia et al. 2007; Kumar et al. 2008). Kumar et al. (2008) was able to induce cyclic somatic embryogenesis from cotyledon-derived embryogenic callus. High frequency (94.3 %) of embryogenic callus was developed on MS basal medium containing TDZ, 2-ip and IBA. Primary, secondary and cyclic somatic embryos were reported from embryogenic calli in media free of plant growth regulators.

Substitution of NAA with other auxins, viz., IAA and p-CPA (0.54–22.84 μM) along with 2.22 μM BA were also induced somatic embryos at varying frequencies (Table 12.2). Among the auxins tested, the highest number of somatic

Fig. 12.2 Direct somatic embryogenesis in cotyledonary explants of safflower. **a** Group of embryos at different developmental stages. **b** Germinated embryo with shoot axis (*arrow*). **c** Plantlet regenerated from somatic embryo. **d** Regenerated plant after transfer to soil. Adapted from Mandal (1996)

Table 12.2 Effect of auxins in presence of 2.22 μM BA on somatic embryogenesis in cotyledonary explants of safflower

Type of auxin	Concentration (μM)	% Explants responded	Mean no. of SE/responding explant	Developmental stages (%)			
				G	H	T	C
NAA	1.07	0.0^{a}	0.0^{a}	0	0	0	0
	2.69	27.6^{efgh}	8.3^{fgh}	46.3	19.5	18.7	15.4
	5.37	30.0^{ghi}	9.0^{ghi}	28.4	21.0	23.1	27.4
	10.74	55.0^{n}	14.2^{j}	27.0	24.9	24.9	23.2
	21.48	46.7^{m}	10.3^{i}	35.2	26.9	16.9	21.0
IAA	1.14	20.2^{cd}	6.3^{de}	47.6	24.4	13.4	14.9
	2.85	23.3^{cdef}	6.7^{def}	46.8	20.2	18.1	14.6
	5.70	25.0^{defg}	17.3^{j}	38.1	23.8	18.1	20.0
	11.42	36.8^{jkl}	20.7^{k}	34.9	25.0	19.9	20.2
	22.84	31.7^{hij}	15.7^{ij}	39.9	21.8	19.8	18.4
p-CPA	0.54	38.3^{kl}	17.3^{ij}	87.9	12.1	0	0
	1.07	28.3^{fghi}	9.4^{hi}	84.9	15.1	0	0
	2.68	21.8^{cde}	6.2^{de}	89.6	10.3	0	0
	10.72	18.3^{bc}	5.0^{cd}	100	0	0	0
	21.44	13.3^{b}	3.5^{bc}	100	0	0	0

Percent response, number of somatic embryos (SE) and developmental stages were recorded after 3 weeks and percent germination after 4 weeks of culture initiation. Means within columns followed by the same letter are not significantly different (LSD at P = 0.05). G: Globular embryo, H: Heart-shaped embryo, T: Torpedo-shaped embryo, C: Cotyledonary embryo. Reproduced from Mandal and Dutta Gupta (2003) with kind permission from Springer Science + Business Media B.V

embryos per responding explant was recorded at 11.42 μM IAA. There were significant differences among the auxin types, concentration and their interactions with respect to percent somatic embryogenesis and number of somatic embryos per responding explant. Contrary to this observation, George and Rao (1982); Tejovathi and Anwar (1987) reported shoot bud formation in safflower in NAA or IAA plus BA treatment. This difference in response may be due to different genotypes and the explants tried (Santarem et al. 1997). In all the treatments, continuous and asynchronous differentiation of somatic embryos were noticed which were affected by type and concentration of auxin. Higher concentrations of auxins favoured somatic embryo maturation except p-CPA where development ceased at globular or heart-shaped stage. Maximum percentage of somatic embryos (27.4 %) at cotyledonary stage was obtained with 5.37 μM NAA with 2.22 μM BA. Strong influence of type and concentration of auxin on somatic embryo development and morphology has been reported in many plants (Wetzstein and Baker 1993; Suhasini et al. 1996; Chengalrayan et al. 1997; Hofmann et al. 2004; Venkatesh et al. 2009).

Germination of somatic embryos was evident within 4 weeks with the emergence of shoots under the same cultural conditions used for embryo induction and development (Fig. 12.2b). Complete regeneration of plantlets was achieved after transferring the germinated somatic embryos with shoot axes onto half-strength MS medium containing 3 % sucrose, 0.8 % bacto agar and 1.07 μM NAA where 70 % germinated embryos were converted into plantlets (Fig. 12.2c). The regenerated plantlets were transferred to soil (Fig. 12.2d). The regenerated plantlets appeared morphologically normal and exhibited normal diploid chromosome complement of $2n = 24$ (Fig. 12.3).

12.2.2.2 Role of Genotype

Embryogenic response of eight cultivars indicates a strong influence of genotype (Table 12.3). The cv. A-300 failed to illicit embryogenic response, while the highest frequency of somatic embryogenesis was observed in cv. Girna (51.7 %). The cv A-1 was the least responsive with 16.7 % embryogenesis. In vitro plant regeneration often seems to be under genetic control. Influence of genotype on somatic embryogenesis is well documented in several oil seed crops (Hofmann et al. 2004; Zeynali et al. 2010). Transformation efficiency of safflower has also been regulated by the genotype (Shilpa et al. 2010). The differences in transformation efficiency among genotypes have been attributed to the differences in regeneration potential of genotypes.

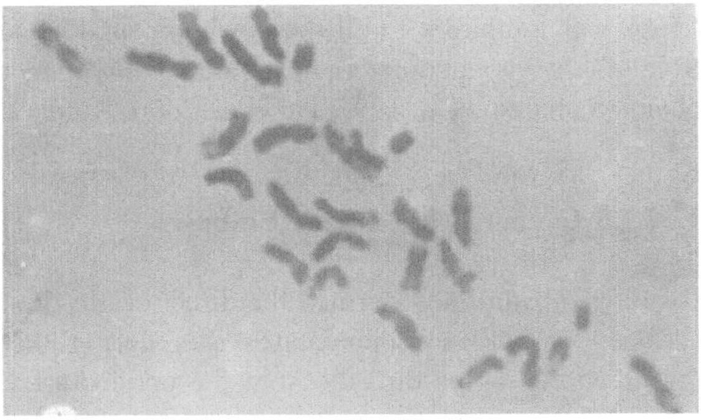

Fig. 12.3 Photomicrograph of somatic metaphase chromosomes ($2n = 24$) of safflower plantlet regenerated from somatic embryo. Adapted from Mandal (1996)

Table 12.3 Effect of genotype on somatic embryogenesis percentage and number of somatic embryos at the end of 3 weeks of culture on MS medium containing 2 mg/l NAA + 0.5 mg/l BA

Genotype	Embryogenesis (%)	No. of somatic embryos/explant
JLSF-1	40.0 ± 2.2^{cd}	12.0 ± 0.5^{fg}
APRR-3	33.3 ± 2.0^{c}	6.3 ± 0.2^{bc}
A-1	16.7 ± 2.1^{b}	7.2 ± 0.2^{cd}
Bhima	18.3 ± 2.4^{b}	4.5 ± 0.2^{b}
Girna	51.7 ± 2.0^{d}	11.2 ± 0.3^{ef}
Tara	20.7 ± 2.1^{b}	9.6 ± 0.3^{de}
S-144	35.0 ± 1.9^{c}	14.2 ± 0.4^{g}
A-300	0^{a}	0^{a}

Data are mean ± SE. Means in each column followed by the same letters are not significantly different at $P = 0.05$ according to Fisher's Least Significant Difference (LSD) test. Adapted from Mandal et al. (2001) with kind permission from Springer Science + Business Media B.V

12.2.2.3 Role of Explant Age and Position

Age and position of the explant significantly affected somatic embryogenesis. It has been observed that 10 d-old cotyledons were more responsive (56.7 %) than 5 (33.3 %) or 15 d (26.7 %) old seedlings. The influence of seedling age has been well demonstrated in safflower transformation following shoot organogenesis (Shilpa et al. 2010). Among the different sizes, whole cotyledons were more embryogenic than segments of cotyledons. Similarly, the frequency of shoot organogenesis form intact cotyledons was found to be more than of cotyledon segments (Basalma et al. 2008). However, the present findings contradict the role of wounding of explant in somatic embryo induction as noted in *Brassica napus* (Koh and Low 2000).

12.2.2.4 Regulation of Somatic Embryogenesis with Ethylene Inhibitors and Promoters

Among the three ethylene inhibitors used, only $AgNO_3$ has significant stimulatory effect on somatic embryogenesis of safflower. A maximum of 83.3 % embryogenic response with 22.1 numbers of somatic embryos per responding explant was obtained at 50 μM $AgNO_3$ treatment. Addition of ethylene promoters, ACC and CEPA was inhibitory to all the developmental stages of somatic embryogenesis. Role of ethylene modulators in somatic embryogenesis was demonstrated in many plants (Santarem et al. 1997; Fei et al. 2000; Kumar et al. 2009).

12.2.2.5 Ontogeny of Somatic Embryos

Histological studies confirmed the direct origin of somatic embryos from explant surface. Various developmental stages during the induction and development of somatic embryos directly from the cotyledonary leaf surface are depicted in Fig. 12.4. Meristematic centres were developed by anticlinal divisions of epidermal cells within 6 d of culture initiation within 8 days multicellular proembryonal complex was formed and developed further to form globular embryos. Various stages of somatic embryos were evident within 15 d of culture and indicative of asynchronous development of embryos. The ontogenic pattern suggests unicellular origin of somatic embryos.

12.2.3 Cell Cultures

In safflower, cell cultures have been established for production of tocopherol (Wang et al. 1999), antioxidant kinobeon (Wakayama et al. 1994) and pigments (Hanagata and Karube 1994; Gao et al. 2000). Callus and cell culture studies for selection of resistant *Fusarium oxysporum* f. sp *carthami* cell lines (Suganya et al. 1997) and for tolerance to sodium chloride (Nikam and Shitole 1997) were also reported. Studies

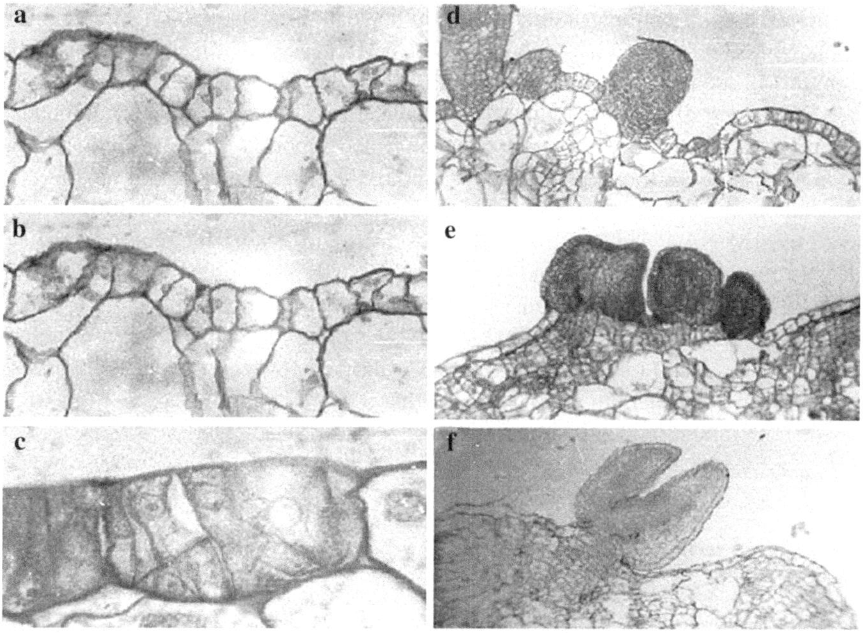

Fig. 12.4 Light micrographs illustrating somatic embryo origin directly from the cotyledonary explants of safflower. **a** Meristematically active adaxial epidermal cell with prominent nucleus at 2 days of culture initiation. **b** First anticlinal division in an adaxial epidermal cell at 4 days of culture initiation. **c** Multicellular proembryonal complex at 8 days of culture initiation. Note a group of small cells within a common boundary of thick cell wall. **d** Longitudinal section of a globular embryo and a globular proembryo developing from adaxial epidermal layer at 10 days of culture initiation. **e** Longitudinal section of one-heart shaped and two globular embryos at 15 days of culture initial. **f** Longitudinal section of a cotyledonary embryo at 15 days of culture initiation. Note the broad base attachment of globular embryo and absence of prominent suspensor. Reproduced from Mandal and Dutta Gupta (2003) with kind permission from Springer Science + Business Media B.V

of Nikam and Shitole (1997) revealed tolerance of callus cultures to sodium chloride, which was retained even after eight passages (30 d each). The tissue culture technique has been extended for production of plants resistant to *Alternaria carthami* (Kumar et al. 2008). In this protocol, embryogenic and organogenic calli were subjected to selection of shoots on medium supplemented with 40 % fungal culture filtrate of *A. carthami* (5×10^5 conidia/ml). Resistance in selected plants was increased to 100, 97.6 and 84 % over that of the control in R_0, R_1 and R_2 generations, respectively.

12.2.4 Anther Culture

Anthers of safflower were cultured in vitro and haploids were recovered successfully (Prasad and Anwar 1991). Prasad and Anwar (1991) obtained shoot regeneration from cultured anthers with uninucleate microspores on MS medium supplemented with 8.87 μM BA and 2.69 μM NAA and rhizogenesis on half-strength MS medium supplemented with 0.53 μM NAA and low level of sucrose (1 %). Prasad et al. (1990) studied the influence of culture medium, genotype and cold pre-treatment on anther culture response and reported that MS medium was the most responsive for callus induction among the five different basal media (N6, B5, Chaleff's, LS and MS) tested. The MS medium supplemented with 8.87 μM BA, 2.69 μM NAA and 2 % sucrose was selected to assess the genotypic differences in the induction of haploid calli in 10 genotypes of safflower including the local varieties. The most responsive genotype was Manjira, with 48.6 % callus induction. Prasad and Anwar (1991) studied the influence of plant growth factors such as NAA, 2,4-D and IAA (auxins), BA and KN (cytokinins) and gibberellin singly and in combination which revealed that a low cytokinin to auxin ratio stimulated shoot induction from gametic tissues.

12.2.5 In Vitro Rooting

Despite the research expanded over the past three decades, rooting of regenerated shoots and post acclimatization survival remained a problem in safflower tissue culture. Improvement in rooting was attempted (Tejovathi and Anwar 1987; Nikam and Shitole 1999; Orlikowska and Dyer 1993; Baker and Dyer 1996) which include high sucrose levels up to 9 % in the rooting medium (George and Rao 1982; Tejovathi and Anwar 1987; Nikam and Shitole 1999), increased sucrose (9 %) along with 5.37 μM of NAA and 1 mg/l riboflavin (Orlikowska and Dyer 1993), incorporation of 2,4,5,-trichlorophenoxypropionic acid (Tejovathi and Anwar 1993; Baker and Dyer 1996) and root induction with *Agrobacterium rhizogenes* (Baker and Dyer 1996). Baker and Dyer (1996) reported that a 7 days exposure to 49 μM IBA in root induction media, followed by incubation in media with 15 g/l sucrose and 1 g/l activated charcoal for 21 days enhances rooting frequency and reduces shoot hyperhydricity. Roots could be induced in different treatments at frequencies ranging from 10 to 95 % in various experiments but post acclimatization survival was successful only with shoots exhibiting less hyperhydricity and those that developed good tap roots with branches in vitro. Despite these improvements, the problem of rooting still persisted and frequency of rhizogenesis varied with genotype, shoot quality, the medium to which the shoots were habituated and the period of culture. The problem was further aggravated in genetic transformation experiments where the regenerated shoots were subjected to exposure to the bacteriostats and selective agents (antibiotics/herbicides) for selection of putative transformants. Extensive studies on improvements in rooting in the author's laboratory resulted in the identification of a hormonal combination involving IBA and the compound phloroglucinol, which gave >70 % rooting with high rate of survival following acclimatization.

12.2.6 *In Vitro Capitula Production*

In vitro capitula induction on media with appropriate combination of growth regulators is an interesting aspect in safflower tissue cultures (Tejovathi and Anwar 1987; Nikam and Shitole 1999; Radhika et al. 2006). Tejovathi and Anwar (1987) reported in vitro capitula induction in two varieties (A-1, Manjira) of safflower from the inner surface of the cotyledons on MS medium with 2.22 μM BAP or 2.32 μM KN and 0.53 μM NAA. Complete blooming of florets in a capitulum was observed within 55–90 days after culture initiation. In vitro produced flowers were normal with normal pollen production and seed set. The pollen fertility of in vitro produced flowers ranged from 90.0 to 95.0 %. Growth regulators type and concentration and genotype were found to have strong influence on flower formation in vitro. Among the cytokinins tested, capitula were induced frequently on media supplemented with BA + NAA and at a low frequency on media fortified with kinetin (Tejovathi and Anwar 1987). Concentration of BA + NAA should be optimum for production of flowers (Seeta et al. 1999). Under continuous illumination, many of the rooted plantlets as well as the rootless micro-shoots produced flowers (George and Rao 1982). Studies of Seeta et al. (1999) revealed that in vitro produced pollen(s) serve as an useful source for creation of genetic variation in safflower. Thus, in most of the tissue culture studies of safflower, MS medium supplemented with BA or TDZ and NAA at varying concentrations was used successfully for stimulating different organogenic responses from seedling tissues. For somatic embryogenesis, medium with 2,4-D, IAA and NAA was found favourable.

12.3 Genetic Transformation

In safflower, genetic engineering studies so far are confined to *Agrobacterium*-mediated transformation developed for the American (Centennial) and Indian cultivars (Table 12.4). Transformation through callus-mediated regeneration (Ying et al. 1992; Orlikowska et al. 1995; Rao and Rohini 1999) and zygotic embryo transformation through *in planta* (Rohini and Rao 2000) have been reported. Seedling explants were transformed which regenerated shoots on MS medium supplemented with BA and NAA (Ying et al. 1992; Rao and Rohini 1999); Orlikowska et al. (1995) succeeded in shoot induction on medium containing TDZ and NAA. In all these studies, the gene constructs harboured *npt* II as plant selection marker and *Uid A* or GFP as reporter gene. The introduced transgene was initially confirmed by GUS histochemical assay. Further, confirmation by molecular studies was done through PCR analysis for the genes of interest and the vector genes (*npt II, Uid A, virC*), Southern analysis, non-denaturing PAGE for neomycin phosphotransferase activity, Western blot and Dot blot analysis (Ying et al. 1992; Orlikowska et al. 1995; Rao and Rohini 1999; Rohini and Rao 2000; Shilpa et al. 2010; Srinivas et al. 2011).

Table 12.4 Genetic transformation studies in safflower

Explant	Stage of transformants characterized	Medium used (μM)	Reference
Primary seedling explants	Transformed shoots	MS + 4.44 BAP + 5.37 NAA	Ying et al. (1992)
Primary seedling explants	Transformed shoots	MS salts +B5 vitamins + 0.053 TDZ + 0.53NAA + 2.5 $AgNO_3$ + 500 carbenicilin	Orlikowska et al. (1995)
Primary seedling explants	Transformed shoots	MS + 4.44 BA +0.53 NAA	Rao and Rohini (1999)
Embryos	In planta transformed shoots	–	Rohini and Rao (2000)
Cotyledons	T_0 plants	0.91 TDZ + 1.07 NAA	Shilpa et al. (2010)
Cotyledons	T_0 and T_1 plants	4.54 TDZ + 0.53 NAA 4.44 BA + 0.53 NAA	Srinivas et al. (2011)

Most of the studies were confined to the characterization of primary transformants with the exception Srinivas et al. (2011) characterized T_1 progeny.

The major constraints in genetic transformation studies of safflower are: lack of genotype-dependent regeneration system, low frequency of transformation, growth retardation of shoots following long time exposure to selection, poor rooting and low survival following acclimatization of selected shoots.

Ying et al. (1992) studied *Agrobacterium tumefaciens*-mediated transformation of safflower cv. Centennial. Efficient callus formation and shoot regeneration were obtained from cotyledon, stem and leaf explants on MS medium amended with 4.44 μM of BA and 5.37 μM NAA. They further reported that substitution of 2,4-D for NAA did not improve efficiency of callus formation. Callus growth was visible on over 80.0 % explants after fourteen days. Shoot buds regenerated from 26.0 % leaf-derived calli on callus induction medium. However, attempts to induce roots in the regenerated shoots were not successful. Transformation and stable integration of transgene was confirmed by GUS assay and DNA hybridization.

Factors (such as bacterial strain, acetosyringone, co-cultivation period) influencing *A. tumefaciens*-mediated transformation and regeneration of safflower cultivar 'Centennial' were examined by Orlikowska et al. (1995). Shoots were formed on medium containing MS salts and B5 vitamins, TDZ (0.045 μM) + NAA (0.53 μM) + $AgNO_3$ (2.5 mg/l) + carbenicillin (500 mg/l). Shoot elongation was optimal on MS medium with KN (4.65 μM) + NAA (0.265 μM) + $AgNO_3$ (2.5 mg/l) + carbenicillin (500 mg/l). Rooting was induced on half-strength MS medium supplemented with NAA (5.37 μM) + $AgNO_3$ (2.5 mg/l) + vitamin B2 (1 mg/l) and incubated in dark for 5 days and then transferred to light. The *A. tumefaciens* strain EHA 105 with p35SGUSInt was more infective than LBA 4404 with pBI121 which was determined in terms of β-glucuronidase activity. The presence of the transgene was confirmed in the putative transformants by PCR and Southern hybridization assay.

Rao and Rohini (1999) studied gene transfer into two Indian cultivars (A-1 and A-300) of safflower using *A. tumefaciens*-mediated transformation method. The binary vector pKIWI 105 lacking bacterial ribosome binding site and harbouring *Uid A* gene was used. Shoot induction was obtained on MS medium supplemented with 0.44–22.19 μM BAP and 0.05–5.37 μM NAA. Putative transformants were confirmed for the presence of the introduced gene by GUS histochemical assay and western blot analysis for *npt* II expression assay. Transgene integration was examined by PCR and dot blot hybridization. Extended periods of callus-mediated regeneration led to hyperhydricity and vitrification of the shoots. The axillary shoots of A-1 did not root on medium containing NAA or IBA while the shoots of A-300 formed roots only on the medium added with 0.53 μM NAA.

In planta transformation using embryos, bypassing tissue culture regeneration in safflower was reported by Rohini and Rao (2000). Embryo axes of germinating seeds were used for *A. tumefaciens*-mediated transformation. After 24 h of co-cultivation, explants were decontaminated with cefotaxime and allowed to germinate and transferred to green house. Putative transformants were confirmed by GUS histochemical assay, PCR amplification of *Uid A* and *npt* II marker genes and Southern analysis. The frequency of transformation was 5.3 % in A-1 and 1.3 % in A-300.

Shilpa et al. (2010) evaluated the efficacy of different parameters such as, genotype, seedling age, co-cultivation period, bacterial titer, enzymatic pre-treatment of target tissues, use of acetosyringone (AS) that induce *vir*-gene enhancer, explant type and explant injury that enhance the transformation efficiency of cultured cotyledons. Transformation of safflower hypocotyls via *Agrobacterium*-mediated transformation and confirmation of putative transformants is presented in Fig. 12.5. Transformation frequency was high when root and hypocotyl explants of 8 day-old seedlings of safflower cv. HUS-305 were co-cultivated with a bacterial cell density of 0.5 OD_{600}, a co-cultivation period of 2 days followed by selection regime of 10-15-15 mg/l hygromycin. The frequency of rooting of the primary transformants was low (18.0 %) when compared to the regenerated shoots (70.0 %), and 7 shoots survived on transfer to soil. The putative transformants were confirmed by β-glucuronidase (GUS) histochemical assay, polymerase chain reaction (PCR), reverse transcriptase PCR (RT-PCR) and Southern blot analysis. With the optimized transformation protocol, putative transformed shoots were obtained with a frequency of 51.0 % within 8–10 weeks of culture initiation.

Srinivas et al. (2011) reported an efficient and reproducible protocol of vector-mediated transformation of cotyledons of high oleic (S-137) and high linoleic acid (WT) types with a transformation efficiency of 4.8 and 3.1 %, respectively. Use of compounds such as iota-carrageenan (1.5 g/l), cysteine (50 mg/l), ascorbic acid (1.5 mg/l) and increased concentration of agar (9 %) has controlled hyperhydration and necrosis effectively. The problem of poor rooting was overcome through grafting with ~50 % success. The integration and stable inheritance of the introduced genes was confirmed by PCR, Southern and Western blot analysis of GFP and hygromycin genes. Southern analysis showed different integration patterns of T-DNA and number of insertions ranged from 1 to 7 in T_0 plants indicating independent insertion events.

Fig. 12.5 Transformation in safflower. **a**. *i*. Regeneration control (without selection), *ii*. transformation control (15 mg/l hygromycin) and *iii*. transformation plates after 10 days of culture of hypocotyl from 8-day-old seedlings on MS medium supplemented with 0.91 μM TDZ and 1.07 μM NAA. **b** Putative transformed shoots obtained from different explants of 8-day-old seedlings of HUS- 305 after the third cycle of selection. Observations were taken after 4 days of culture on MS medium supplemented with 2.32 μM KN. **c** Elongated putative transformed shoots on 2.32 μM KN, 5.37 μM 2iP, 1.0 mg/l $AgNO_3$ after three cycles of selection. **d** Acclimatization and establishment of a transformed shoot. **e** GUS staining of putative transformants recovered after selection following transformation observed after 12 days of second selection from roots. **f** Molecular analysis of the putative transformants. A PCR analysis of putative transformants showing amplification of 1,200-bp fragment of the *UidA* gene using the primers 5′ ggt ggg aaa gcg cgt tac aag 3′ and 5′ ggt tac gcg ttg ctt ccg cca 3′. Lanes: M represents λ DNA double digest with *Eco*RI/HindIII, 1: pCAMBIA 1391Z with CaMV 35S DNA, 2: no DNA control, 3: DNA from untransformed safflower plant (control), lanes marked as 4–12: DNA from 9 independent putative transformants

12.4 Safflower for Plant Made Pharmaceuticals

In the last decade, plants have been exploited for production of large quantities of cost-effective recombinant proteins as these provide high capacity and low cost option when compared to animal cell cultures. The criteria for selecting a plant host include the knowledge of the plant (native constituents, toxicants, ability to produce large quantities of protein, breeding behavior), its agronomy ability for transformability and stable expression of the introduced protein and appropriateness as raw material for pharmaceutical products. Crops like spinach, tobacco, corn and potato have been exploited for production of PMPs. Safflower being a less common crop and owing to its predominantly self-pollinated breeding behavior, lower production and capital costs, easy transformability, ability for accumulation of high levels of recombinant proteins in seeds has gained importance as a protein production factory for therapeutic proteins for metabolic and cardiovascular diseases and non-pharmaceutical products addressing human topical, nutritional oils and agricultural biotechnology markets (www.sembiosys.com, www.arcadia.com). Most of the proteins are targeted for accumulation in the seeds to enable easy extraction and for utilization of long term storage capabilities. The oilbody-oleosin technology is based on two proprietary capabilities of expression of recombinant proteins in seed oil bodies (Stratosome^{TM} Biologics System) and ability to extract oilbodies inexpensively from seeds (Affinity Capture System) (Markley et al. 2006). Safflower has been successfully modified to express and produce apolipoprotein A1 and its variant apolipoprotein AI (Milano) collectively referred to Apo AI constituting the next generation cardiovascular drug that targets the removal of atherosclerotic plaque from arteries (Nykiforuk et al. 2010). SemBioSys is also engaged in production of insulin with accumulation levels of 1.2 % of total seed protein. The other therapeutic product of interest in safflower is the γ-linolenic acid (GLA). The high GLA safflower contains 35 % GLA which is an omega-6 fatty acid with health benefits that are similar and complementary to the benefits of fish oil derived omega-3 fatty acids and has been shown to have significant anti-inflammatory effects with benefits for cardiac, joint, skin and neurological health.

12.5 Conclusions

Safflower exhibits great propensity for plant regeneration through organogeneic and embryogenic pathways. The major limitations are hyperhydricity of the shoots and poor rooting which need to be circumvented. Robust protocols of anther culture are to be developed for production of double haploid lines as a prelude for accelerating the marker-assisted breeding programmes. Genetic transformation protocols are in place for Indian and American cultivars which need to be exploited for development of transgenics with agronomically desirable genes and seed quality traits.

References

Anwar SY, Khadeer MA, Tejovathi G (1989) Genetic and tissue culture studies in safflower. In: Farook SA, Khan IA (eds) Recent advances in genetics and cytogenetics, Premier Publishing House, Hyderabad, pp 457–463

Baker CM, Dyer WE (1996) Improvements in rooting regenerated safflower (*Carthamus tinctorius* L.) shoots. Plant Cell Rep 16:106–110

Basalma D, Uranbey S, Mirici S, Kolsarici O (2008) TDZ X IBA induced shoot regeneration from cotyledonary leaves and in vitro multiplication in safflower (*Carthamus tinctorius* L.). African J Plant Sci 7:960–966

Chatterjee AK, Singh HP (1993) Plant regeneration from leaf calli of safflower. In: Li Dajue and Han Yuanzhou (eds), Proceedings of third international safflower conference Beijing, China (June 14–18) pp 139–143

Chengalrayan K, Sathaye SS, Hazra S (1997) High frequency conversion of abnormal peanut somatic embryos. Plant Cell Rep 16:783–786

Fei S, Read PE, Riordan TP (2000) Improvement of embryogenic callus induction and shoot regeneration of buffalo grass by $AgNO_3$. Plant Cell Tiss Org Cult 60:197–203

Furuya T, Yoshikawa T, Kimura T, Kaneko H (1987) Production of tocopherols by cell culture of safflower. Phytochem 26:2741–2747

Gao WY, Fan L, Paek KY (2000) Yellow and red pigment production by cell cultures of *Carthamus tinctorius* in a bioreactor. Plant Cell Tiss Org Cult 60:95–100

George L, Rao PS (1982) In vitro multiplication of safflower (*Carthamus tinctorius* L.) through tissue culture. Proc Indian Natl Sci Acad 99:293–296

Hanagata N, Karube I (1994) Red pigment production by *Carthamus tinctorius* cells in two stage culture system. J Biotechnol 30:259–269

Hofmann N, Nelson RL, Korban SS (2004) Influence of media components and pH on somatic embryo induction in three genotypes of soybean. Plant Cell Tiss Org Cult 77:157–163

Knowles PF (1989) Safflower. In: Downey RK, Robellen G, Ashri A (eds) Oil crops of the world. Mc-Graw Hill, New York, pp 363–374

Koh WL, Loh CS (2000) Direct somatic embryogenesis and plant regeneration and in vitro rapid-cycling *Brassica napus*. Plant Cell Rep 19:1177–1183

Kumar V, Parvatam G, Ravishankar GA (2009) $AgNO_3$: a potential regulator of ethylene activity and plant growth modulator. Electron J Biotechnol 12:1–15

Kumar VJ, Ranjitha Kumari BD, Castano E (2008) Cyclic somatic embryogenesis and efficient plant regeneration from callus of safflower. Biol Plant 52:429–432

Li D, Mundel HH (1996) Safflower: (*Carthamus tinctorious* L.) promoting the conservation and use of underutilized and neglected crops. International Plant Genetic Resources Institute, Rome

Mandal AKA, Chatterjee AK, Dutta Gupta S (1995) Direct somatic embryogenesis and plantlet regeneration from cotyledonary leaves of safflower. Plant Cell Tiss Org Cult 43:287–289

Mandal AKA (1996) Shoot organogenesis and somatic embryogenesis in safflower (*Carthamus tinctorious* L.) and their improvement by manipulating various factors. Ph. D Thesis, Indian Institute of Technology, Kharagpur

Mandal AKA, Chatterjee AK, Dutta Gupta S (2001) Factors affecting somatic embryogenesis from cotyledonary explants of safflower. Biol Plant 44(4):503–507

Mandal AKA, Dutta Gupta S (2001) Direct shoot organogenesis and plant regeneration in safflower. In vitro Cell Dev Biol-Plant 37:50–54

Mandal AKA, Dutta Gupta S (2003) Somatic embryogenesis of safflower: influence of auxin and ontogeny of somatic embryos. Plant Cell Tiss Org Cult 72:27–31

Markley N, Nykiforuk C, Boothe J, Moloney M (2006) Producing proteins using transgenic oil-body-oleosin technology. Bhiopharm Inter 9:6–12

Murashige T, Skoog F (1962) A revised medium for rapid growth and bioassays with tobacco tissue cultures. Physiol Plant 15:473–497

Nikam TD, Shitole MG (1997) Sodium chloride tolerance in *Carthamus tinctorius* L. cv. A-1 callus culture. In: Carleto A, Mundel HH (eds) Proceedings of the IVth international safflower conference, Adriatica Eitrice, Bari, Italy, pp 175–178

Nikam TD, Shitole MG (1999) In vitro culture of Safflower L. cv. Bhima: initiation, growth, optimization and organogenesis. Plant Cell Tiss Org Cult 55:15–22

Nykiforuk CL, Shen Y, Murray E, Booth J, Busseuil D, Rheaume E, Tardif JC, Reid A, Moloney M (2010) Expression and recovery of biologically active recombinant apolipoprotein AI_{Milano} from transgenic safflower (*Carthamus tinctorius*) seeds. Plant Biotechnol J 9:250–263

Orlikowska TK, Dyer WE (1993) In vitro regeneration and multiplication of safflower (*Carthamus tinctorius* L.). Plant Sci 93:151–157

Orlikowska TK, Cranston HJ, Dyer WE (1995) Factors influencing *Agrobacterium tumefaciens* mediated transformation and regeneration of the safflower cultivar centennial. Plant Cell Tiss Org Cult 40:85–91

Padmaja G, Tejovathi G, Anwar SY (1990) Anatomical studies on certain in vitro induced abnormal variants in safflower (*Carthamus tictorius*). Phytomorphology 40 (3, 4): 233–241

Prasad BR, Khadeer MA, Seeta P, Anwar SY (1990) Influence of genotype and cold pretreatment on anther culture response in safflower (*Carthamus tictorius*). Indian J Exp Biol 28(10):924–927

Prasad BR, Anwar SY (1991) Anther culture studies in safflower (*Carthamus tictorius*). In vitro 27 (3): Pt.2, 69A

Rao SK, Rohini VK (1999) Gene transfer into Indian cultivars of safflower (*Carthamus tinctorius* L.) using *Agrobacterium tumefaciens*. Plant Biotechnol 16:201–206

Radhika K, Sujatha M, Nageshwar Rao T (2006) Thidiazuron stimulates adventitious shoot regeneration in different safflower explants. Biol Plant 50:174–179

Rani KJ, Rao TN, Raghunathnam G, Rao PV (1996) Studies on callus growth and differentiation in safflower. Indian J Genet 56:458–461

Rani KJ, Rao TN (1998) Callus differentiation in tissue cultures of safflower. J Oilseeds Res 15:258–260

Reddy MM, Devi P (1991) Plantlet regeneration from cotyledon cultures of safflower (*Carthamus tinctorius*). In vitro 27(3): Pt.2, 142A

Rohini VK, Rao KS (2000) Embryo transformation, a practical approach for realizing transgenic plants of safflower (*Carthamus tinctorius* L.). Ann Bot 86:1043–1049

Santarem ER, Pelissier B, Finner JJ (1997) Effect of explant orientation, pH, solidifying agent and wounding on initiation of soybean somatic embryos. In Vitro Cell Dev Biol Plant 33:13–19

Seeta P, Talat K, Anwar SY (1999) In vitro pollen(s): novel source of genetic variability in safflower (*Carthamus tinctorius*, L). Indian J Exp Biol 37:491–495

Shilpa KS, Dinesh Kumar V, Sujatha M (2010) *Agrobacterium*-mediated genetic transformation of safflower (*Carthamus tinctorious* L.). Plant Cell Tiss Org Cult 103:387–401

Singh HP (1991) Morphogenetic potential of callus and organ cultures of safflower. Narendra Deva J Agril Res 6(1):163–167

Srinivas B, Luch H, Green AG, Singh SP, Wood CC (2011) *Agrobacterium*-mediated transformation of safflower and the efficient recovery of transgenic plants via grafting. Plant Methods 7:12. doi:10.1186/1746-4811-7-12

Suganya A, Sujatha M, Kalpana Sastry R (1997) In vitro selection for resistance to *Fusarium oxysporum* Schelt. *Carthami* Klisiewicz and Houston in safflower. In: Carleto A, Mundel HH (eds) Proceedings of the IVth international safflower conference, Adriatica Eitrice, Bari, Italy, pp 305–308

Sujatha M, Dinesh Kumar V (2007) In vitro bud regeneration of *Carthamus tinctorius* and wild *Carthamus* species from leaf explants and axillary buds. Biol Plant 51:782–786

Sujatha M, Suganya A (1996) In vitro organogenic comparison of different seeding tissues of safflower (*Carthamus tinctorius* L.). Sesame and Safflower Newsletter 11:85–90

Suhasini K, Sagare AP, Krishnamurthy KV (1996) Study of aberrant morphogenesis and lack of conversion of somatic embryos of chickpea (*Cicer arietinum* L.). In Vitro Cell Dev Biol. Plant 32:6–10

Tejovathi G, Anwar SY (1987) Plantlet regeneration from cotyledonary cultures of safflower (*Carthamus tinctorius* L.). In: Reddy GM (ed) Plant cell and tissue culture of economically important plants. Osmania University, Hyderabad

Tejovathi G, Anwar SY (1993) 2,4,5-Trichloro phenoxy propionic acid induced rhizigenesis in *Carthamus tinctorius* L. Proc Indian Natl Sci Acad B59(6):633–636

Tejovathi G, Das RR (1997) In vitro multiplication of *Carthamus tictorius* L. Adv Plant Sci 10(2):149–152

Van Arnold S, Sabala I, Bozhkov P, Dyachok J, Filonova L (2002) Development pathways of somatic embryogenesis. Plant Cell Tiss Org Cult 69:233–249

Velasco L and Fernandez-Martinez J (2001) Breeding for oil quality in safflower. In: Proceeding 5th international safflower conference. Montana, USA 23–27 July. pp 133–137

Venkatesh K, Roja Rani A, Baburao N, Padmaja G (2009) Effect of auxins and auxin polar transport inhibitor (TIBA) on somatic embryogenesis in groundnut (*Arachis hypogaea* L.). African J Plant Sci 3:288–293

Wakayama S, Kusaka K, Kanehira T, Yamada Y, Kawazu K, Kobayashi A (1994) Kinobeon A, a novel red pigment produced in safflower tissue culture systems. Zeitshrift fur Naturforsch 49c: 1–5

Walia N, Kaur A, Babbar SB (2007) Proliferation and differentiation from endosperm of *Carthamus tinctorius*. Biol Plant 51:749–753

Wang P, Ou S, Peilin C (1999) Optimization of conditions for safflower cell culture and accumulation of cellicolous product tocopherols. Chinese J Biotechnol 15(4):231–237

Weiss EA (1971) Castor, sesame and safflower. Leonard Hill, London

Wetzstein HY, Baker CM (1993) The influence of somatic embryo morphology and histology on germination and conversion in peanut. Plant Sci 92:81–89

Ying MC, Dyer WE, Bergman JW (1992) *Agrobacterium tumefaciens* mediated transformation of safflower (*Carthamus tinctorius* L.) cv centennial. Plant Cell Rep 11:581–585

Zeylani M, Zanjani BM, Amiri ME, Noruzizn M, Aghajari SM (2010) Influence of genotype and plant growth regulator on somatic embryogenesis in rapeseed (*Brassica napus* L.). African J Plant Sci 9:4050–4055

Zhanming H, Biwen H (1993) The tissue culture of safflower and its histological and cytological study. In: Dajue L, Yunzhou H (eds) Proceedings of the 3rd international safflower conference, Beijing, China, pp 184–194

Chapter 13
Plant Regeneration and Genetic Transformation in *Jatropha*

M. Sujatha, S. Nithianantham and M. P. Reddy

Abstract *Jatropha curcas*, a non-edible oil bearing species with multiple uses, and considerable economic potential is emerging as a potential biofuel plant. The limited knowledge of this species, low and inconsistent yields, the narrow genetic variability, and vulnerability to insects and diseases are major constraints in successful cultivation of *Jatropha* as a biofuel crop. Hence, genetic improvement of *Jatropha* is essential by conventional and modern biotechnological tools to use as a viable alternative source of bio-diesel. Realising its potential as a bio-energy crop, in vitro regeneratio n methods have been established to meet the demand of large scale supply of superior clones, and also as a prelude for genetic improvement of the species through transgenic approaches. In this chapter, an overview of in vitro tissue culture and genetic transformation of *Jatropha* is discussed.

Abbreviations

NAA	α-napthaleneacetic acid
BA	Benzyl adenine
MS	Murashige and Skoog
Kn	6-furfurylaminopurine
IBA	Indole-3-butyric acid
TDZ	Thidiazuron
AFLP	Amplified fragment length polymorphism

M. Sujatha (✉)
Directorate of Oilseeds Research, Rajendranagar, Hyderabad, Andhra Pradesh 500 030, India
e-mail: mulpurisujata@yahoo.com

S. Nityanandam
Department of Environmental Sciences, Bharathiar University, Coimbatore, Tamilnadu 641 046, India

M. P. Reddy
Plant Stress Genomics Research Centre, King Abdullah University of Science and Technology, Thuwal 23955-6900, Kingdom of Saudi Arabia

S. M. Jain and S. Dutta Gupta (eds.), *Biotechnology of Neglected and Underutilized Crops*, DOI: 10.1007/978-94-007-5500-0_13,

ISSRs Inter simple sequence repeats
MSAP Methylation sensitive amplified polymorphism

13.1. Introduction

Fast depleting natural resources, increasing demand for energy, and environmental concerns compel to develop alternative fuel strategies globally. Further, political unrest in oil rich countries is another cause of concern. Bio-energy encompass a wide range of alternative sources of energy of biological origin, and offer the prospect of increasing energy supplies, and also counteract the increasing levels of greenhouse gases (Fairless 2007). There are many non-edible oleaginous plants such as neem (*Azadirachta indica* A.), karanja (*Millettia pinnata* L.), mahua (*Madhuca* sp), castor bean (*Ricinus communis* L.), simarouba (*Simarouba glauca* DC.), wild apricot (*Prunus armeniaca*), jojoba (*Simmondsia chinensis*), kokum (*Garcinia indica*), mahua (*Madhuca indica*), *Calophyllum ionophyllum*, Jatropha (*Jatropha curcas* L.) etc., which can provide oil for bio-diesel production (Carels 2009; http://www.novodboard.com/Jatropha-english.pdf). *Jatropha curcas* (Family: Euphorbiaceae), is a multipurpose, drought resistant, perennial plant of Latin American origin, but it is now widespread throughout the tropical regions of the world (Openshaw 2000). Besides the use of oil as a bio-diesel, most of the plant parts have medicinal importance (Gupta 1985; Openshaw 2000) and insecticidal properties (Adebowale and Adedire 2006). The recognition that *Jatropha* oil can yield high quality bio-diesel has led to a surge of interest in *Jatropha* across the globe, in view of its potential for avoiding the dilemma of "food versus fuel", hardiness, rapid growth, easy propagation, short gestation period, wide adaptation, relatively high oil content, ease of conversion to bio-diesel, and optimum plant size thus, making this species suitable for sustainable cultivation on wastelands (Gupta 1985; Openshaw 2000; Francis et al. 2005; Adebowale and Adedire 2006; Ghosh et al. 2007; Jha et al. 2007; Deore and Johnson 2008; Kumar et al. 2011).

Seed oil content of *J. curcas* is about 40 % higher than the typical oil crops such as soybean and rape and thus, has been considered as a strategic plant resource in many countries (Gubitz et al. 1999; Deore and Johnson 2008; Carvalho et al. 2008). The oil can be used in diesel engines after simple processing because it is similar to diesel oil in characteristics, being a potential substitute for fossil fuel and a renewable energy (Berchmans and Hirata 2008; Deore and Johnson 2008). Environmental benefits of its use (in place of diesel) include lower exhaust emissions of particulate matter and greenhouse gases, such as CO, CO_2, and SOx. Large scale production of bio-diesel from 25 kg of *Jatropha* oil has resulted in 24 kg of bio-diesel (96 % of yield), which is only reduced by 2 % as compared to lab scale (Chitra et al. 2005). Genetic engineering can be used to modify the metabolic pathways involved in fatty acid bio-synthesis and enhance oil accumulation to provide desirable and renewable raw materials for bio-diesel production.

Large scale cultivation of elite material remains the single most important factor that will ultimately determine the success of *Jatropha* as a source of biofuel. The limited knowledge of the species, low and inconsistent yields, narrow genetic variability, and vulnerability to insects and diseases are major constraints in successful cultivation of *Jatropha* as a biofuel crop (Jones and Miller 1991; Ginwal et al. 2004; Francis et al. 2005; Ghosh et al. 2007; Jongschaap et al. 2007; Kumar and Sharma 2008).

J. curcas is primarily propagated through seeds, and significant variations in seed yield and oil content has been observed in plants raised through seeds (Ghosh et al. 2007; Jha et al. 2007). Seed viability and the rate of germination are low (Heller 1996; Kochhar et al. 2005), and seed screening for quality is a laborious task, thus propagation through seed may not provide enough high quality material for sustainable agriculture. It has also been observed that a large amount of seed is required for raising the planting material. Alternative vegetative propagation techniques would not only help in raising high quality planting material, but also allow improving the species and use of seeds for bio-diesel preparation. Propagation can be carried out without the loss of traits by stem cuttings. However, the limitation in generation of large scale planting material is (a) availability of sufficient quantity of material, and (b) propagation is seasonal (Kumar and Reddy 2010). Trees propagated by cuttings show a lower longevity and possess a lower drought and disease resistance and poor seed set (Sujatha et al. 2005). Large scale cultivation of *J. curcas* remains the single most important issue that will ultimately decide the success of the crop. Thus, conventional propagation through seeds is not reliable and vegetative propagation by stem cuttings is inadequate to meet the demand (Heller 1996; Openshaw 2000). The crop has to be genetically upgraded for resistance to biotic and abiotic stresses and also the seed oil quality traits for successful cultivation and to meet the industry demand for oil. Therefore, improvement programmes of *J. curcas* by modern methods of biotechnology are of interest worldwide.

In vitro techniques have facilitated to produce rapidly millions of true-to-type plants, continuous supply of planting material, germplasm preservation, and genetic improvement of the species. Tissue culture-derived plants of *J. curcas* have advantage over seed propagated plants in yield (Sujatha et al. 2005). Also in vitro plant regeneration protocols have also been developed with the aim to genetically transform *J. curcas* with desirable agronomic traits. Several laboratories are working in parallel on various aspects to contribute to better understanding of the crop worldwide. This chapter provides up-to-date information on tissue culture methods and genetic engineering for improving *J. curcas*, highlighting the advances and limitations of the existing studies and future perspectives thereon, for making this crop economically sustainable.

13.2 Tissue Culture

Research efforts expanded over the past one and half decades in *J. curcas* tissue culture, resulted in development of reliable protocols of tissue culture and transformation (Sujatha and Mukta 1996; Wei et al. 2004; Sujatha et al. 2005; Rajore and Batra

2007; Jha et al. 2007; Deore and Johnson 2008; Kumar and Reddy 2010; Sharma et al. 2011a, b; Reddy and Pamidimarri 2010). In *J. curcas*, tissue culture protocols for shoot regeneration from seedling tissues and leaves from mature plants through both organogenic and embryogenic pathways are optimised. The media requirements are simple and medium supplemented with the cytokinin, benzyl adenine (BA) or thidiazuron (TDZ) in combination with the auxin indole-butyric acid facilitates shoot regeneration although the concentrations vary with the genotype and the explanted tissue. Somatic embryogenesis is promoted on medium supplemented with kinetin (Kn) and indole-butyric acid (IBA). The major concern is mass propagation of elite genotypes for obtaining large quantity of quality planting material. *J. curcas* plants can be propagated through vegetative cuttings but the plants have lower longevity, low clonal multiplication rates, are easily uprooted with strong wind, less resistant to drought to biotic stresses when compared to plants propagated by seeds. For propagation through tissue culture, a reproducible protocol of bud proliferation with acceptable multiplication index, good shoot elongation and rooting is a prerequisite. Reports of shoot proliferation using axillary and apical buds are available but commercial feasibility of these techniques has yet to be demonstrated. A wealth of information has been generated on biotechnological aspects for improvement of the crop through tissue culture and genetic transformation.

13.2.1 Direct Organogenesis

Shoot regeneration from different explants of *J. curcas* viz., cotyledon, petiole, hypocotyls, epicotyls, leaf tissues from seedling and mature plants has been reported (Table 13.1). Attempts have been made to regenerate toxic and non-toxic varieties of *J. curcas* (Sujatha and Mukta 1996; Wei et al. 2004; Sujatha et al. 2005; Rajore and Batra 2007; Jha et al. 2007; Deore and Johnson 2008; Singh et al. 2010; Kumar et al. 2010a; Kumar and Reddy 2010) (Table 13.1). The most preferred medium for direct and indirect organogenesis is Murashige and Skoog salt media while Warakagoda and Subasinghe (2009) used B5 medium for facilitating organogenesis. It has been reported that regeneration in *J. curcas* is highly genotype dependent (da Camara Machado et al. 1997; Kumar and Reddy 2010; Singh et al. 2010). Genotypic differences were observed and hypocotyls explants of toxic genotypes possessed higher regeneration frequency and in vitro rooting as compared to non-toxic genotypes (Sharma et al. 2011a, b). The main factors that affect morphogenesis were the type and concentration of cytokinin, source and orientation of explants. In general, in vitro explants had a high rate of regeneration and number of buds per responding explants as compared with in vitro explants (Kumar and Reddy 2010). Initial studies employed medium supplemented with BA + IBA for direct adventitious shoot regeneration but subsequent studies reported the efficacy of TDZ singly or in combination with BA or Kn in promoting shoot bud induction (Deore and Johnson 2008). The percentage response of explants forming shoot buds increased with increase in the concentration of TDZ (Kumar et al. 2010a;

Table 13.1 *In vitro* responses in *Jatropha curcas*

S. No	Type of explant	Mode of regeneration	Medium + plant growth regulators	References
1.	Hypocotyl, petiole and leaf disc	Adventitious shoot regeneration	BA (0.44–2.22 μM) + IBA (0.49–4.9 μM)	Sujatha and Mukta (1996)
2.	Shoot tip	Direct shoot regeneration	GA_3 (3 mg/l) + IAA (3 mg/l)	Sardana et al. (1998)
3.	Leaf explants	Somatic embryogenesis	MS salts and B5 vitamins + BA (3 mg/l) + IAA (1 mg/l)	Sardana et al. (2000)
4.	Nodal segments	Multiple shoot proliferation	Kn (2 mg/l) + IBA (1.5 mg/l) + Ascorbic acid (10 mg/l) + Citric acid (50 mg/l) + Adenine sulphate (25 mg/l) + Glutamine (100 mg/l)	Rajore et al. (2002)
5.		Callus induction	–	Lin et al. (2002a, b)
6.	Hypocotyl and petiole	Callus induction	BA (0.1–0.5 mg/l) + IBA (0.1–1 mg/l)	Lu et al. (2003)
7.		Callus induction	–	Weida et al. (2003)
8.	Epicotyl	Direct shoot regeneration	BA (0.2–0.7 mg/l) + IBA (0.1 mg/l)	Wei et al. (2004)
9.	Epicotyl	Direct organogenesis and shoot differentiation from callus	BA (1.0–10.0 mg/l) + IBA (1.0 mg/l)	Jin-xia et al. (2005)
10.	Axillary node and leaf sections	Direct adventitious bud regeneration Axillary bud proliferation	TDZ (0.5–1.0 mg/l)	Sujatha et al. (2005)
11.	Shoot tip	Multiple shoot proliferation	BAP (2.0 mg/l) + IAA (0.5 mg/l) + Adenine sulphate (25 mg/l) + Glutamine (100 mg/l)	Rajore and Batra (2005)
12.		Callus mediated regeneration	–	Sharma et al. (2006)
13.	Leaf segments	Somatic embryogenesis	Kn (0.5–2.0 mg/l) + IBA (0.1–1.0 mg/l) + Adenine sulphate (2.5 mg/l)	Jha et al. (2007)
14.	Node	Multiple shoot bud induction	BA (5.0 mg/l), Kn (0.5 mg/l) + IBA (0.1 mg/l) + Adenine sulphate (5–10 mg/l)	Datta et al. (2007
15.	Nodal segments	Multiple shoot proliferation, Somatic embryogenesis	BAP (1.5 mg/l) + Kn (0.5 mg/l) + IAA (0.1 mg/l)	Kalimuthu et al. (2007)
16.	Leaf	Callus mediated regeneration	BA (0.5-5 mg/l) + NAA (1 mg/l)	Rajore and Batra (2007)

(continued)

Table 13.1 (continued)

S. No	Type of explant	Mode of regeneration	Medium + plant growth regulators	References
17.	Leaf and hypocotyls	Callus and suspension culture	2,4 D (0.5 mg/l) + 2 % coconut milk	Soomro and Memon (2007)
18.	Leaf disc	Direct shoot regeneration	TDZ (0.5 mg/l) + BAP (0.5 mg/l) + IBA (0.1 mg/l)	Deore and Johnson (2008)
19.	Axillary node	Multiple shoot bud induction	BA (3 mg/l) + IBA (1 mg/l) + Adenine sulphate (25 mg/l) + Glutamine (50 mg/l) + L-arginine (15 mg/l) + Citric acid (25 mg/l)	Shrivastava and Banerjee (2008)
20.	Petiole	Multiple shoot proliferation	BA (0.5–1.0 mg/l) + IBA (0.01 mg/l)	Thepsamran et al. (2008)
21.	Petiole	Direct induction of shoot buds	TDZ (0.5 mg/l) + Kn (2.0 mg/l) + BAP (1.0 mg/l) + NAA (1.0 mg/l)	Kumar and Reddy (2010)
22.	Immature embryo	Indirect organogenesis (Callus mediated)	BA (1 mg/l) + Kn (0.5 mg/l) + IBA (0.25 mg/l) + PVP (500 mg/l) + Citric acid (30 mg/l)	Varshney and Johnson (2010)
23.	Cotyledonary leaf	Direct shoot regeneration	TDZ (2.0 mg/l) + Kn (2.0 mg/l) + BAP (1.0 mg/l) + NAA (1.0 mg/l)	Kumar et al. (2010a, b)
24.	Shoot apices	Shoot bud induction	BAP (0.5 mg/l) + GA_3 (0.2 mg/l)	Purkayastha et al. (2010)
25.	Epicotyls and hypocotyls	Direct and indirect organogenesis	TDZ (0.25 mg/l) + Kn (0.5 mg/l) + IBA (0.25 mg/l)	Kaewpoo and Te-chato (2010)
26.	Anther	Maturation and germination of microspores	MS salts + White vitamin + 22 % maltose, pH 6.5	Li et al. (2010)
27.	Leaf segments from seedlings	Direct oragnogenesis	BA (2.0 mg/l)	Krishna et al. (2010)
28.	Leaf explants from 2-year old plants	Adventitious shoot buds	TDZ (0.2 mg/l) + IBA (0.2 mg/l)	Khurana-Kaul et al. (2010)
29.	Stem	Direct shoot bud regeneration	BA (1.0 mg/l) + Kn (1.0 mg/l)	Singh et al. (2010)
30.	Petioles from non-toxic genotypes	Direct organogenesis	TDZ (0.5 mg/l) + Kn (2.0 mg/l) + BA (1.0 mg/l) + NAA (1.0 mg/l)	Kumar et al. (2011)
31.	Hypocotyl	Direct shoot bud regeneration	TDZ (0.5 mg/l)	Sharma et al. (2011a, b)

Kumar and Reddy 2010; Sharma et al. 2011a; Khemkladngoen et al. 2011). Kumar et al. (2011), reported that low concentration of TDZ induced relatively fewer shoot buds, and developed rapidly into shoots in subsequent culture. In contrast, media containing high concentration of TDZ had more visible primordial but, only a few were able to develop into shoots. Earlier reports on *J. curcas* revealed that TDZ was more effective than BA as cytokinin and the response of explants forming shoot buds decreased with the addition of IBA to TDZ or BA containing medium due to formation of callus (Sujatha et al. 2005; Reddy et al. 2008; Misra et al. 2010a). The inhibitory effect of high concentration of TDZ on shoot elongation has been reported and it is recommended that the TDZ concentration should be reduced and/or other cytokinins or combinations of cytokinins and auxins must be used for further shoot elongation (Preece and Imel 1991; Huetteman and Preece 1993). Enhanced concentration of copper (10-fold MS medium) improved adventitious shoot regeneration and reduced the time for shoot differentiation (Khurana-Kaul et al. 2010). The maximum shoot elongation was obtained using a combination of BA with IAA as compared to BA + IBA or BA + NAA. Elongation was reduced in the medium containing BA and IBA due to proliferation of axillary buds. The poor elongation observed in a medium containing BA and NAA is due to the profuse callusing at the basal end of proliferated shoots (Kumar 2008).

Several reports on micropropagation are also available which could be applied for mass propagation of the species with an average multiplication index of around 6 per subculture cycle (Table 13.1). A report on shoot tip culture (Rajore and Batra 2005) demonstrated that MS medium supplemented with BAP and IAA along with 100 mg/l glutamine and 25.0 mg/l adenine sulphate gave the best response for shoot proliferation. Datta et al. (2007) found that MS medium containing 5.0 mg/l BAP and 20.0 mg/l adenine sulphate would be most suitable for the highest (30.8 ± 5.48) axillary shoot bud proliferation from nodal explants. However, Shrivastava and Banerjee (2008), reported that when a combination of auxin (1.0 mg/l IBA) and cytokinin (3.0 mg/l BAP) along with growth additives (adenine sulphate, glutamine and L-arginine) was used, shoot regeneration from axillary nodes was observed. In conclusion, it is obvious that BAP with or without IBA is playing a very important role for direct shoot bud induction. TDZ had a higher impact on the induction of direct shoot buds in comparison to BAP. For further shoot bud multiplication a reduced concentration of auxin (IBA) was found to be effective. The high concentration of auxins is generally inhibitory to morphogenesis, and the use of an appropriate auxin-cytokinin ratio is essential to obtain proper shoots and root primordials (Kalimuthu et al. 2007).

13.2.2 Indirect organogenesis

Callus mediated regeneration is also reported and callus induction has been achieved from all the parts of the plant viz., hypocotyls, cotyledon, leaf, petiole and stem (Rajore and Batra 2007; Sujatha and Mukta 1996). Varshney and Johnson

(2010) reported plant regeneration from immature embryos and cotyledon-derived cultures and found that morphogenic callus induction and subsequent plant regeneration was dependent on several factors like age of explants, size of the explants, and combinations of various auxins and cytokinins and growth additives like L-proline, silver nitrate, copper sulphate, etc. Histological techniques are widely used in plant tissue culture to understand the process of morphogenesis (Kothari and Varshney 1998; Yeung 1999). Varshney et al. (2011) reported the histological events leading to shoot organogenesis in immature embryonal axis and cotyledon cultures of *J. curcas*. Shoot organogenesis in cotyledons was adventitious while shoot regeneration from immature embryonal axis followed both organogenesis and through multiplication of pre-existing meristems.

13.2.3 Somatic Embryogenesis

Somatic embryogenesis is the developmental process by which somatic cells other than a gamete under suitable induction conditions undergo restructuring to generate embryogenic cells (somatic or non zygotic embryos) capable of regenerating into plants, and serves as a powerful tool of plant biotechnology for faster and quality plants generation. These cells then undergo a series of morphological and biochemical changes that result in the formation of somatic or non-zygotic embryos capable of regenerating into plants. In *J. curcas*, Sardana et al. (2000) described a two-step protocol of somatic embryogenesis on MS salts + B5 vitamins with 3.0 mg/l BA + 1.0 mg/l IAA. Kalimuthu et al. (2007) reported the induction of somatic embryos from cotyledons on medium supplemented with 2.0 mg/l BA. According to Jha et al. (2007), the type and concentration of the plant growth regulators were the strong determining factors for somatic embryogenesis in *J. curcas.* Highest frequency (80 %) of globular somatic embryos (58.5 ± 12.7) from leaf-derived callus was recorded on medium supplemented with 0.5 mg/l Kn and 0.2 mg/l IBA after 4–6 weeks of culture. The synergistic combination of kinetin, IBA and adenine sulphate favoured secondary embryogenesis. Addition of 2.5 mg/l adenine sulphate stimulated the process of development of somatic embryos. The whole process of induction of somatic embryos to whole plant regeneration is accomplished in 12–16 weeks. Plantlets derived through somatic embryogenesis flowered early and produced flowers almost throughout the year.

13.2.4 Microsporogenesis

Pollen as a vector of genetic material is critical for successful pollination and fertilisation, seed set and yield of *J. curcas*. In vitro maturation and germination of *J. curcas* microspores can provide a novel system for transformation and thus, have great potential for genetic improvement. However, information on *J. curcas* microsporogenesis, pollination ecology, physiological and biochemical basis, molecular mechanism and factors influencing pollen developmental process is rather limited (Liu et al. 2007). In maturation culture of microspores of *J. curcas*, MS salts produced large amount of mature pollen than BK salts which indicated that inorganic salts had an important impact on male gametophyte of *J. curcas*. Further, optimising the combination of inorganic salts will more remarkably improve mature frequency of *J. curcas* microspores. Li et al. (2010) studied in vitro maturation and germination of *J. curcas* microspores, and reported that the medium with MS salts + White vitamins + 22 % maltose, pH 6.5 and hanging culture to be optimal for microspore maturation; while medium with 20 % sucrose + 10 % PEG-4,000 + BK ($Ca(NO_3)_2$-400 mg/L, $MgSO_4$-200 mg/L, boric acid-100 mg/L, KNO_3-100 mg/L) + 2.5 % coconut water, pH 6.5 with a pre-cold store time was the most appropriate for pollen germination. Microspore culture is better than anther culture in terms of recovery of haploids and could be used for production of doubled haploids which are invaluable in transgenic breeding and marker assisted breeding.

13.2.5 Factors Affecting Micropropagation

Regeneration from the various explants of *J. curcas* is reported (Sujatha and Mukta 1996; Wei et al. 2004; Sujatha et al. 2005; Rajore and Batra 2005; Jha et al. 2007; Misra et al. 2010b; Sharma et al. 2011a, b), however, very few studies have been made on assessment of factors affecting the in vitro response that are known to influence organogenesis (Josephina and van Staden 1990; Wu et al. 2009; Kumar and Reddy 2010; Kumar et al. 2010a; Singh et al. 2010).

13.2.5.1 Effect of Source and Age of Explant on Shoot Buds Induction

Significant differences in percentage response and the number of shoot buds per explant was observed between in vitro and ex vitro generated explants and a better response was observed when in vitro explants were used (Kumar and Reddy 2010; Sharma et al. 2011a, b). Explants collected from 30-days-old in vitro raised seedlings were best responsive as compared to 15 or 45-days-old, and percentage of responding explants significantly decreased with increase in the age of explants to 45 days (Sharma et al. 2011a, b), which could be due to differences in endogenous growth regulators concentration and their metabolism which changes with age.

13.2.5.2 Effect of Orientation and Position of Explant on Shoot Bud Induction

The orientation of the explant (horizontal or vertical) on the medium significantly influenced the regeneration and shoot buds induction in *J. curcas* (Kumar and

Reddy 2010; Singh et al. 2010). The position of the explant on the seedling axis also affected the response of the explants for number of shoot buds formed. Regeneration potential of explants increased with an increase in distance from the root and explants adjacent to cotyledonary node showed the highest response (Sharma et al. 2011a, b). Misra et al. (2010a) reported that segments from second and third leaves (from the top) responded well, as compared to other leaves. This differential behaviour can be related to different mechanisms for control of the endogenous PGRs metabolism and/or contents as cells within the same plant can have different endogenous levels of plant growth regulators, and variation in receptor affinity (Minocha 1987).

13.2.5.3 Establishment of Long-Term Proliferating Cultures

Shoot multiplication and their growth is not difficult in *J. curcas*, but the regenerating shoots turn brown/necrose within 15–20 d of culture (He et al. 2009a, b), leading to difficulties in maintaining cultures for a long time which is a major limitation for large scale commercial propagation. It has been observed that *J. curcas* impregnates endophytic bacterial contamination which expresses after 2–3 subcultures in the medium (Misra et al. 2010a). Initially, these bacteria do not hinder the formation and development of new shoots, but later they affect the growth of newly developed shoots and these shoots turn yellowish brown and become necrotic making it impossible to maintain cultures beyond three subcultures (Misra et al. 2010a). Tissue browning generally inhibits the growth and formation of adventitious shoots (Wei and Newton 2004) and subsequently endophytic bacteria destroy the cultures (Thomas 2004). Supplementing antioxidants, namely, reduced glutathione, ascorbic acid, tocopherol and cysteine individually and in combination in the medium could help in overcoming the problems of tissue browning and necrosis (Misra et al. 2010a).

13.2.5.4 Genetic Fidelity of Micropropagated Plants

Tissue culture has high potential for the commercialised production of economic crops. However, the occurrence of somaclonal variation in micropropagated plants has brought into question the validity of the micropropagation protocols (Karp 1994; Rani and Raina 2000). This thus, warrants the need for assessment of the micropropagated plants for their clonal status. Use of differences in morphological characters and isozyme patterns (Zhao et al. 2005; Shen et al. 2007) has limitations in identifying the somaclonal variation due to (1) limited number and low heritability of phenotypic characters; (2) late expression of phenotypic characters; (3) difficulty in morphological identification of stable genetic somaclonal variations in the species having narrow genetic diversity and (4) their unstable patterns and expression levels due to the environmental changes.

Molecular markers offer numerous advantages over conventional phenotype based alternatives as they are stable and detectable in all tissues regardless of growth, differentiation, development as they are not confounded by environment, pleiotropic and epistatic effects (Agarwal et al. 2008). The use of molecular markers such as random amplified polymorphic DNA (RAPD), amplified fragment length polymorphism (AFLP), methylation sensitive amplified polymorphism (MSAP), restriction fragment length polymorphism (RFLP) and inter simple sequence repeats (ISSRs) are being rapidly integrated as routine laboratory tools available for quick assessment of the genetic stability of micropropagated plants (Rani and Raina 2000; Palombi and Damiano 2002; Martins et al. 2004; Saker et al. 2006; Smykal et al. 2007). Such studies are limited in tissue cultured plants of *J. curcas* (Sharma et al. 2011b). Shoot tip cultures from different genotypes of *J. curcas* subjected to RAPD and AFLP analysis showed no significant differences as compared to their mother plants up to 16th generation (sub culture). Leela et al. (2011) using nodal explants of selected accessions of *J. curcas* reported similar results. Micropropagated plants from the cultures of preformed structures, such as shoot tips and axillary buds from the hardwood shoot cuttings are reported to maintain clonal fidelity (Ahuja 1987; Wang and Charles 1991; Ostry et al. 1994).These results indicate that the axillary shoot bud proliferation can safely be used as an efficient micropropagation method for mass propagation of selected elite adult trees of *J. curcas*. Flow cytometric analysis revealed that there is no variation in ploidy levels of micropropagated plants irrespective of the explants used (Kaewpoo and Te-chato 2010).

13.2.6 Micropropagation of Non-Toxic J. curcas

A non-toxic *J. curcas* variety has been reported from Mexico, and its seeds can be used for human consumption after roasting (Makkar et al. 1998). Cultivation of non-toxic varieties could provide oil for bio-diesel and de-oiled cake as livestock feed, and thus add value to the crop (Becker and Makkar 1998). Micropropagation techniques were successfully developed for non-toxic *J.curcas* (Sujatha et al. 2005; Kumar et al. 2010b, 2011).

13.2.7 In Vitro Plant Regeneration in Jatropha Species and Interspecific Hybrids

Jatropha species are amenable to tissue culture manipulations, which indicate scope for widening the genetic base through parasexual hybridization and biotechnological tools (Sujatha 2006). Sujatha and Dhingra (1993) reported a simple, rapid and reproducible protocol for direct shoot regeneration from different explants of *J. integerrima* (Table 13.2). Prolific adventitious shoot bud initiation was obtained using a combination of 0.5 or 1.0 mg/l benzyladenine and 1.0 mg/l

Table 13.2 In vitro responses in *Jatropha* species

S. No	*Jatropha* species	Type of explant	Mode of regeneration	Medium + plant growth regulators	References
1.	*J. panduraefolia*	Endosperm	Proliferation and differentiation from endosperm	BM + 2,4-D, KN + YE	Johri and Bhojwani (1965)
2.	*J. panduraefolia*	Endosperm	Proliferation and differentiation from endosperm	BM + 2,4-D, KN + YE	Srivastava (1971)
3.	*J. panduraefolia*	Endosperm	Proliferation and differentiation from endosperm	BM + NAA + KN + Casein hydrolysate	Srivastava and Johri (1974)
4.	*J. integerrima*	Stem, leaf, hypocotyls, peduncle	Adventitious shoot regeneration	BA (0.5–1.0 mg/l) + IBA (1.0 mg/l)	Sujatha and Dhingra (1993)
5.	*J. tanjorensis*	Leaf discs	Adventitious shoot regeneration	BA (0.5–5.0 mg/l) + IBA (0.5–1.0 mg/l)	Prabakaran and Sujatha (1999)
6.	*J. integerrima*	Hypocotyl, Stem, peduncle, leaf	Adventitious shoot regeneration	BA (0.1–2 mg/l) + IBA (1 mg/l)	Sujatha and Reddy (2000)
7.	*J. curcas* × *J. integerrima*	Leaf explants	Adventitious shoot regeneration	BA (1.0–10.0 mg/l) + IBA (1.0 mg/l)	Sujatha and Prabakaran (2003)

indole-3-butyric acid (IBA). Reduction of IBA concentration (0.5 mg/l) promoted further development of shoots. Regenerated shoots rooted readily on Murashige and Skoog (MS) medium lacking growth regulators. Plantlets were acclimatised and successfully transferred to pots. Interspecific hybridization has been undertaken with an objective of combining the desirable traits such as high oil, oil quality, resistance to insect pests and shattering of these two economically important species. Successful artificial hybridization between the two species was reported earlier (Rupert et al. 1970; Dehgan 1984).

13.3 Genetic Transformation

Advances in genetic transformation and the availability of characterised genes with many advantages have made it possible to transfer chimeric gene(s) of academic/agronomic importance to the genome of recipient species to produce transgenic progeny with desired characteristics. This technology may help bypass some of the limitations of classical breeding programmes and reduce the time required to produce improved varieties. Genetic transformation of any crop species through genetic engineering techniques requires an efficient in vitro regeneration system which is rapid, reproducible and applicable to a broad range of genotypes. The most widely used method of genetic transformation is *Agrobacterium*-mediated gene transfer for reasons like simplicity, cost effectiveness, fewer re-arrangements of transgene, ability to transfer relatively larger DNA segments (Hamilton et al. 1997), and preferential integration of foreign genes into transcriptionally active regions (Konez et al. 1989; Ingelbrecht et al. 1991), thereby ensuring long-term stability of expression of transgenes in plants (Hernandez et al. 1999).

13.3.1 Agrobacterium-Mediated Transformation

Agrobacterium-mediated transformation of *J. curcas* cotyledons and leaf explants is reported. Li et al. (2006) were the first to report *Agrobacterium*-mediated transformation of callus cultures of *J. curcas* and the transformation efficiency was restricted to transient GUS assay. Li et al. (2008) used cotyledon discs as explants for *Agrobacterium tumefaciens* mediated transformation and reported that the *Agrobacterium* strain, explant type, marker gene, and bacterial incubation time play an important role in transformation efficiency. The transformation efficiency following selection on phosphinothricin was 13 %. Cotyledon age and orientation were critical and cotyledonary leaves from germinated seed with abaxial side touching the medium surface resulted in higher transformation efficiency (Mazumdar et al. 2010). Kumar et al. (2010a, c) described the fundamental key parameters affecting the efficiency of *Agrobacterium*-mediated transformation of leaf, stem and petiole explants (Figs. 13.1, 13.2, 13.3). Efficient transformation was achieved using 4 days pre-cultured non-wounded leaf explants infected with *Agrobacterium* culture of optical density $A_{600} = 0.6$ for 20 min followed by co-cultivation for 4 days. The results of these studies consistently demonstrated that stable genetic transformation can be achieved using the *Agrobacterium*-mediated method, and the transformation efficiency was 29 %. Naruemon et al. (2011) reported that highest stable transformation rate (53 %) was achieved when explants were subjected to 1 min of sonication followed by 9 min of shaking in *Agrobacterium* suspension. The transformation efficiency obtained using this method was four-fold and two-fold higher than those reported using cotyledons (Li et al. 2008) and young leaves (Kumar et al. 2010c), respectively. A wounding treatment is crucial for efficient transformation and has been successfully employed to improve transformation efficiency in many plants. Using different explants, He et al. (2009a, b), Mazumdar et al. (2010), Pan et al. (2010), Zong et al. (2010), Misra et al. (2011) also reported transformation in *J. curcas* via *Agrobacterium*-mediated method. Depending on the construct, different selection agents based on herbicides and antibiotics viz., phosphinothricin (PPT), hygromycin and kanamycin were used. Li et al. (2006) tested the sensitivity of cotyledons to the commonly used selection

Fig. 13.1 Direct shoot bud induction from petiole explants of non-toxic *J. curcas*. Direct shoot bud induction from **a** in vitro petiole in *horizontal position*, **b** *in vivo* petiole in *horizontal position*, **c** in vitro petiole in *vertical position*, **d** *in vivo* petiole in *vertical position* on MS medium with 0.5 mg/l TDZ after 6 weeks, **e** shoot proliferation on MS medium with 2.0 mg/l Kn + 1.0 mg/l BAP + 1.0 mg/l NAA after 4 weeks, **f** elongation of shoots on MS medium with 0.5 mg/l BAP and 1.5 mg/l IAA after 6 weeks, **g** development of roots on half strength of MS medium with 3.0 mg/l IBA + 2.0 mg/l IAA +1.0 mg/l NAA + 0.25 mg/l activated charcoal after 4 weeks, **h** regenerated plant in polybag (*Source* Kumar et al. 2010c)

Fig. 13.2 Regeneration from stem explants of *J. curcas* placed **a** *horizontally* and **b** *vertically* on MS medium supplemented with 1.0 mg/l BAP and 1.0 mg/l Kn, **c** 15-days-old and **d** 28-days-old explant in medium supplemented with 1.0 mg/l IAA and 0.5 mg/l BAP, **e** shoots rooted in agar medium, **f** encapsulated shoots, **g** and **h** direct rooting of encapsulated shoot in sterile soil, **i** Well established plant in pot. *Arrows* pointed to the specified tissue

agents and the results showed inhibition of shoot regeneration from untransformed cells with 50 mg/l kanamycin, 0.5 mg/l PPT or 5 mg/l hygromycin and complete kill of untransformed shoots with 100 mg/l kanamycin, 1.0 mg/l PPT or 7.5 mg/l hygromycin. Regardless of the explants used, studies on

Fig. 13.3 Shoot regeneration from cotyledonary leaf explants of *J.curcas*. Direct organogenesis from **a** in vitro cotyledonary leaf explant (bar 5 mm), **b** *in vivo* cotyledonary leaf explant (bar 5 mm) on MS medium with 2.27 0.5 mg/l thidiazuron (TDZ) after 6 weeks, **c** shoot proliferation of induced shoot buds on MS medium with 2.0 mg/l kinetin (Kn) + 1.0 mg/l 6-benzyl aminopurine (BAP) + 1.0 mg/l α-naphthaleneacetic acid (NAA) after 4 weeks (bar 100 mm), **d** elongation of proliferated shoot on MS medium with 0.5 mg/l BAP + 1.5 mg/l indole-3-acetic acid (IAA) after 6 weeks (bar 5 mm), **e** elongated shoot cultured on half strength basal MS liquid medium supplemented with 3.0 mg/l indole-3-butyric acid (IBA) + 1.0 mg/l + 1.0 mg/l NAA for root induction (bar 5 mm), **f** development of roots at the base of auxins treated elongated shoot on half strength basal MS medium with 0.25 mg/L activated charcoal after 4 weeks (bar 1 mm), **g** regenerated plant in polybag after 4 weeks (bar 150 mm), **h** regenerated plant in pot soil after 6 month under natural condition (100 mm) (*Source* Kumar et al. 2010a)

Agrobacterium-mediated transformation were limited to characterisation of T_0 plants. Hence, information of inheritance and stability of the introduced gene is virtually lacking.

13.3.2 Particle Bombardment

Microprojectile bombardment, also called the biolistic method or the particle gun method has been used for genetic transformation in many laboratories. The ability to deliver foreign DNA into regenerable cells, tissues or organs appears to provide the best method for achieving genotype independent transformation bypassing *Agrobacterium* host specificity. There are only two reports on genetic transformation in *J. curcas* using microprojectile bombardment (Purkayastha et al. 2010; Joshi et al. 2011) and are confined to the optimisation of physical parameters and characterisation of primary transformants. In studies of Purkayastha et al. (2010) using shoot apices as target tissues for bombardment, maximum transformation frequency was obtained when the explants were bombarded twice with 1.0 μ gold particles at a pressure of 1,100 psi and target distance of 9 cm. Similar conditions with minor modifications were found suitable for bombardment of embryonic axis with an overall transformation efficiency of 44.7 % (Joshi et al. 2011). Genetic engineering in combination with a conventional breeding programme would be a potential tool for the improvement of *J. curcas*. However, research pertaining to the genetic transformation of *J. curcas* is limited to few reports on the use of cotyledons and leaf explants and has not made much headway with regard to development of transgenics with desirable traits.

13.4 Functional Proteins in *J. curcas*

Jatropha proteins having specific biological roles with respect to metabolic, physiological or defensive activities are identified (Table 13.3). The plant expresses proteins (aquaporins and betaine aldehyde) necessary to sustain environmental pressures such as, drought and arid conditions. The abundance of *JcPIP2* transcripts was increased under serious drought stress indicating its role in the drought resistance of *J. curcas* (Zhang et al. 2007). In the biological systems (plants, animals, bacteria, etc.) many organic osmolytes (e.g. betaine) or other substances are synthesised for protection against osmotic stresses, drought, high salinity or high temperature (Kumar et al. 2004). Many of the applications in industry have a demand for enzymes that are highly stable against high temperature and pressure and have broad specificity for substrates and organic solvents (Alain et al. 2004). The bioactive proteins, particularly, curcin has potential to be used as a successful immunoconjugate in chemotherapy. Furthermore, several cyclic peptides present in *Jatropha* seeds are of clinical significance and show their potential in pharmacy. Gressel (2008) has provided an exhaustive list of genes that could be deployed into *J. curcas* for developing genotypes with dwarf plant type, suppressed branching, anti-shattering, reduced levels of curcin and suppressed phorbol ester production.

13.5 Conclusion and Future Perspectives

Genetic engineering appears to be an effective approach to reduce the levels of toxic substances in seeds, and increase resistance to biotic stresses, and furthermore offers opportunity to modify seed oil quality for higher engine efficiency. The reported protocols for genetic transformation need to be refined to make them sufficiently robust to take advantage of the great potential. Micropropagation for large scale multiplication needs further refinements to allow commercial application by improving the rooting and also including more diverse accessions and

Table 13.3 Biological activities of *J. curcas* proteins

S. No	Functional proteins	Biological activity	Reference
1.	Aquaporins	Drought resistance	Zhang et al. (2007)
2.	Betaine aldehyde dehydrogenase	Drought resistance	Zhang et al. (2008)
3.	Esterase and lipase	Hydrolysis of triglycerides	Staubmann et al. (1999), Abigor et al. (2002)
4.	Curcain	Wound healing property	Nath and Dutta (1992)
5.	Curcin	Inhibits protein synthesis, immunotoxins	Stripe et al. (1976), Lin et al. (2002a, b), Weike et al. (2006)
6.	β-glucanase	Antifungal activity	Wei et al. (2005), Jin-xia et al. (2005)
7.	Jatrophidin	Antifungal activity	Altei et al. (2008)
8.	Curcacydine	Antimalarial activity, inhibits cell proliferation and inhibits classical pathway of human complement	Auvin et al. (1997), Van den berg et al. (1995), Baraguey et al. (2001)
9.	DGAT-1, oleosins, GPD, acetyl CoA Carboxylase, ACP-acyltranferase, ACP thioesterase A	Fatty acid metabolism, oil accumulation and TAG synthesis	Sato et al. (2011), Xu et al. (2011)

low-cost tissue culture techniques to reduce the unit cost plant production without compromising the quality (Singh et al. 2010, Fig. 13.2). Secondary somatic embryogenesis is also desirable for large scale micropropagation, which is especially important for woody plants having long generation cycles, and with low frequencies of somatic embryogenesis. So, far there are no reports on proliferation and germination of secondary somatic embryos in *Jatropha*.

Reducing phorbol esters through a transgenic approach would be a protracted balancing act. Molecular genetic tools may allow genetic improvement however, we got to be cautious regarding the stability of integration and expression of foreign gene/genes when taking a transgenic approach in long-lived plants like *Jatropha*. Plant transformation methods and enhanced gene silencing technology can effectively be used to evaluate and authenticate newly discovered endogenous genes to validate their function in plants as well as to genetically manipulate trait quality and productivity. Further, the technology helps in better understanding and subsequent improvement of the species for higher yield and reducing the world's dependence on fossil reserves. Metabolic engineering and RNAi technology can be considered as the most effective tools for further improvement to achieve the goal. These will be possible through an efficient and reproducible in vitro plant regeneration and genetic transformation system.

References

Abigor RD, Uadia PO, Foglia TA, Haas MJ, Scott K, Savary BJ (2002) Partial purification and properties of lipase from germinating seeds of *Jatropha curcas* L. J Amer Oil Chem Soc 79:1123–1126

Adebowale KO, Adedire CO (2006) Chemical composition and insecticidal properties of the underutilized *Jatropha curcas* seed oil. Afr J Biotechnol 10:901–906

Agarwal M, Shrivastava N, Padh H (2008) Advances in molecular marker techniques and their applications in plant sciences. Plant Cell Rep 27:617–631

Ahuja MR (1987) Gene transfer in forest trees. In: Hanover JW, Keathley DE (eds) Genetic manipulation of woody plants. Plenum Press, New York, pp 25–41

Alain H, Ali K, Danielle L (2004) Lipases and their industrial applications: an overview. Appl Biochem Biotechnol 118:155–170

Altei WF, Picchi DG, Barbosa SC et al (2008) NMR studies, solid phase synthesis and MD/SA simulation as a tool for structural elucidation of new bioactive peptides from the latex of *Jatropha curcas* L. Planta Med 74:1–338 (SL65)

Auvin C, Baraguey C, Blond A et al (1997) Curcacycline B, a cyclic nonapeptide from *Jatropha curcas* enhancing rotamase activity of cyclophilin. Tetrahedron Lett 38:2845–2848

Baraguey C, Blond A, Cavelier F et al (2001) Isolation, structure and synthesis of mahafacyclin B, a cyclic heptapeptide from the latex of *Jatropha mahafalensis*. J Chem Soc Perkin Trans 1:2098–2103

Becker K, Makkar HPS (1998) Toxic effects of phorbol esters in carp (*Cyprinus carpio* L). Vet Hum Toxicol 40:82–86

Berchmans HJ, Hirata S (2008) Biodiesel production from crude *Jatropha curcas* L. seed oil with a high content of free fatty acids. Bioresour Technol 99:1716–1721

Carels N (2009) *Jatropha curcas*: a review. Adv Bot Res 50:39–86

Carvalho CR, Clarindo WR, Praca MM et al (2008) Genome size, base composition and karyotype of *Jatropha curcas* L., an important biofuel plant. Plant Sci 174:613–617

Chitra P, Venkatachalam P, Sampathrajan A (2005) Optimization of experimental conditions for biodiesel production from alkali-catalysed transesterification of *Jatropha curcas* oil. Energy Sustain Dev 9:13–18

da Camara Machado A, Frick NS, Kremen R et al (1997) Biotechnological approaches to the improvement of *Jatropha curcas*. In: Proceedings of the International Symposium on *Jatropha* pp. 15 Managua, Nicaragua

Datta MM, Mukherjee P, Ghosh B et al (2007) *In vitro* clonal propagation of biodiesel plant (*Jatropha curcas* L). Curr Sci 93:1438–1442

Dehgan B (1984) Phylogenetic significance of interspecific hybridization in *Jatropha* (Euphorbiaceae). Syst Bot 9:467–478

Deore AC, Johnson TS (2008) High-frequency plant regeneration from leaf-disc cultures of *Jatropha curcas* L.: an important biodiesel plant. Plant Biotechnol Rep 2:7–11

Fairless D (2007) Biofuel: the little shrub that could-maybe. Nature 449:652–655

Francis G, Edinger R, Becker K (2005) A concept for simultaneous waste land reclamation, fuel production, and socio-economic development in degraded areas in India: need, potential and perspectives of *Jatropha* plantations. Nat Resourc Forum 29:12–24

Ghosh A, Chaudhary DR, Reddy MP et al (2007) Prospects for Jatropha methyl ester (biodiesel) in India. Int J Environ Stud 64:659–674

Ginwal US, Rawat PS, Srivastava RL (2004) Seed source variation in growth performance and oil yield of *Jatropha curcas* L in Central India. Silvae Genet 53:186–192

Gressel J (2008) Transgenics are imperative for biofuel crops. Plant Sci 174:246–263

Gubitz GM, Mittelbach M, Trabi M (1999) Exploitation of the tropical oil seed plant *Jatropha curcas* L. Bioresour Technol 67:73–82

Gupta RC (1985) Pharmacognostic studies on 'Dravanti'. Part I. *Jatropha curcas* Linn. Indian Acad Sci 1:65–81

Hamilton C, Frary A, Lewis C et al (1997) Stable transfer of intact high molecular weight DNA into plant chromosomes. Proc Natl Acad Sci USA 93:9975–9979

He Y, Pasapula V, Li X et al (2009a) *Agrobacterium tumefaciens*-mediated transformation of *Jatropha curcas*: factors affecting transient transformation efficiency and morphology analysis of transgenic calli. Silvae Genet 58:123–128

He Y, Guo X, Lu R et al (2009b) Changes in morphology and biochemical indices in browning callus derived from *Jatropha curcas* hypocotyls. Plant Cell Tiss Org Cult 98:11–17

Heller J (1996) Physic nut *Jatropha curcas* L In: Promoting the Conservation and use of Underutilized and Neglected Crops. 1. Institute of Plant Genetics and Crop Plant Research. Gatersleben/International Plant Genetic Resources Institute, Rome

Hernandez JBP, Remy S, Sauco VG et al (1999) Chemotactic movement and attachment of *Agrobacterium tumefaciens* to banana cells and tissues. J Plant Physiol 155:245–250

Huetteman CA, Preece JE (1993) Thidiazuron: a potent cytokinin for woody plant tissue culture. Plant Cell Tiss Org Cult 33:105–119

Ingelbrecht I, Breyne P, Vancomperonolle A et al (1991) Transcriptional interferences in transgenic plants. Gene 109:239–242

Jha TB, Mukherjee P, Datta MM (2007) Somatic embryogenesis in *Jatropha curcas* Linn., an important biofuel plant. Plant Biotechnol Rep 1:135–140

Jin-xia Z, Qin W, Li-jun Z et al (2005) The extraction of β-1,3-glucanase and analysis of partial characteristics in *Jatropha curcas*. Southwest China J Agric Sci 18:328–333

Johri BM, Bhojwani SS (1965) Growth response of mature endosperm in cultures. Nature 298:1345–1347

Jones N, Miller JH (1991) *Jatropha curcas* a multipurpose species for problematic sites. Land Resour Ser 1:1–12

Jongschaap REE, Corré WJ, Bindraban OS et al (2007) Claims and facts on *Jatropha curcas* L.: Wageningen, The Netherlands: Plant Research International. http;//factfuels.org/media_en/Climes_and_Facts_on_Jatropha_WUR? Session = isgsklbna58j7grrfst888n5r7

Josephina GN, van Staden J (1990) The relationship between genotype, tissue age and endogenous cytokinin levels on adventitious bud formation on leaves of Lachenalia. Plant Cell Tiss Org Cult 22:223–228

Joshi M, Mishra A, Jha B (2011) Efficient genetic transformation of *Jatropha curcas* L. by microprojectile bombardment using embryo axes. Ind Crops Prod 33:67–77

Kaewpoo M, Te-chato S (2010) Study on ploidy level of micropropagated *Jatropha curcas* L. via flow cytometry. J Agri Tech 6:391–400

Kalimuthu K, Paulsamy S, Senthil Kumar R et al (2007) *In vitro* propagation of bio-diesel plant of Jatropha curcas L. Plant Tiss Cult Biotech 17:137–147

Karp A (1994) Origins, causes and uses of variation in plant tissue cultures. In: Vasil IK, Thorpe TA (eds) Plant Cell and Tissue Culture. Kluwer Acad Publ Dordrecht, The Netherlands, pp 139–151

Khemkladngoen N, Caragena J, Shibagaki N et al (2011) Adventitious shoot regeneration from juvenile cotyledons of a biodiesel producing plant, *Jatropha curcas* L. J Biosci Bioeng 111:67–70

Khurana-Kaul V, Kachhwaha S, Kothari SL (2010) Direct shoot regeneration from leaf explants of *Jatropha curcas* in response to thidiazuron and high copper contents in the medium. Biol Plant 54:369–372

Kochhar S, Kochhar VK, Singh SP et al (2005) Differential rooting and sprouting behaviour of two *Jatropha* species and associated physiological and biochemical changes. Curr Sci 89:936–939

Konez C, Martini N, Mayerhofer R et al (1989) High frequency T-DNA mediated tagging in plants. Proc Nat Acad Sci USA 86:8467–8471

Kothari SL, Varshney A (1998) Morphogenesis in long term maintained immature embryo derived callus of wheat (*Triticum aestivum* L.): histological evidence for both somatic embryogenesis and organogenesis. Plant Biochem Biotech 7:93–98

Krishna G, Sairam Reddy P, Bhattacharya PS et al (2010) Shoot organogenesis and plantlet regeneration from *in vitro* raised leaf segments of *Jatropha curcas* L. Trends Biosci 3:45–48

Kumar N (2008) Studies on regeneration and genetic transformation in *J. curcas*. PhD thesis, Bhavngar University, Bhavngar, India

Kumar N, Reddy MP (2010) Plant regeneration through the direct induction of shoot buds from petiole explants of *Jatropha curcas*: a biofuel plant. Ann Appl Biol 156:367–375

Kumar A, Sharma S (2008) An evaluation of multipurpose oil seed crop for industrial uses: a review. Ind Crops Prod 28:1–10

Kumar S, Dhingra A, Daniell H (2004) Plastid-expressed betaine aldehyde dehydrogenase gene in carrot cultured cells, roots, and leaves confers enhanced salt tolerance. Plant Physiol 136:2843–2854

Kumar N, Vijayanand KG, Reddy MP (2010a) Shoot regeneration from cotyledonary leaf explants of *Jatropha curcas*: a biodesel plant. Acta Physiol Plant 32:917–924

Kumar N, Vijayanand KG, Reddy MP (2010b) *In vitro* plant regeneration of non-toxic *Jatropha curcas* L: direct shoot organogenesis from cotyledonary petiole explants. J Crop Sci Biotech 13:189–194

Kumar N, Vijayanand KG, Pamidimarri DVNS et al (2010c) Stable genetic transformation of *Jatropha curcas* via *Agrobacterium tumefaciens*-mediated gene transfer using leaf explants. Ind Crops Prod 32:41–47

Kumar N, Vijayanand KG, Reddy MP (2011) *In vitro* regeneration from petiole explants of non-toxic *Jatropha curcas*. Ind Crops Prod 33:146–151

Leela T, Naresh B, Srikanth Reddy M et al (2011) Morphological, physico-chemical and micropropagation studies in *Jatropha curcas* L. and RAPD analysis of the regenerants. Appl Energy 88:2071–2079

Li MR, Li HQ, Wu GJ (2006) Study on factors influencing Agrobacterium-mediated transformation of *Jatropha curcas*. J Mol Cell Biol 39:83–87

Li M, Li H, Jiang H et al (2008) Establishment of an *Agrobacterium*-mediated cotyledon disc transformation method for *Jatropha curcas*. Plant Cell Tiss Organ Cult 92:173–181

Li C, Yu M, Chen F et al (2010) *In vitro* maturation and germination of *Jatropha curcas* microspores. Intl J Agric Biol 12:541–546

Lin J, Tang L, Chen F (2002a) Tissue culture and plantlet regeneration of *Jatropha curcas*. Plant Physiol Commun 38:252

Lin J, Yan F, Tang L et al (2002b) Isolation, purification and functional investigation on the N-glycosidase activity of curcin from the seeds of *Jatropha curcas*. High Technol Lett 11:36–40

Liu HF, Kirchoff BK, Wu GJ et al (2007) Microsporogenesis and male gametogenesis in *Jatropha curcas* L. (Euphorbiaceae). J Torrey Bot cL 134:335–343

Lu WD, Wei Q, Tang L et al (2003) Induction of callus from *Jatropha curcas* and rapid propagation. Chin J Appl Environ Biol 9:127–130

Makkar HPS, Becker K, Schmook B (1998) Edible provenances of *Jatropha curcas* from Quintana Roo state of Mexico and effect of roasting on antinutrient and toxic factors in seeds. Plant Foods Human Nutr 52:1–36

Martins M, Sarmento D, Oliveira MM (2004) Genetic stability of micropropagated almond plantlets as assessed by RAPD and ISSR markers. Plant Cell Rep 23:492–496

Mazumdar P, Basu A, Paul A et al (2010) Age and orientation of the cotyledonary leaf explants determine the efficiency of de novo plant regeneration and *Agrobacterium tumefaciens* mediated transformation in *Jatropha curcas* L. South Afr J Bot 76:337–344

Minocha SC (1987) Plant growth regulators and morphogenesis in cell and tissue cultures of forest trees. In: Bonga JM, Durzan DJ (eds) Cell and Tissue Culture in Forestry, Vol I, pp 50–66. Martinus Nijhoff Publisher, Dordrecht

Misra P, Gupta N, Toppo DD et al (2010a) Establishment of long-term proliferating shoot cultures of elite *Jatropha curcas* L. by controlling endophytic bacterial contamination. Plant Cell Tiss Org Cult 100:189–197

Misra P, Toppo DD, Gupta N et al (2010b) Effect of antioxidants and associate changes in antioxidant enzymes in controlling browning and necrosis of proliferating shoots of elite *Jatropha curcas* L. Biomass Bioenergy 34:1861–1869

Misra P, Toppo DD, Mishra MK et al (2011) *Agrobacterium tumefaciens*-mediated transformation protocol of *Jatropha curcas* L. using leaf and hypocotyl segments. Plant Biochem Biotechnol. doi:101007/s13562-011-0072-3

Naruemon K, Cartagena JA, Kiichi F (2011) Physical wounding assisted *Agrobacterium* mediated transformation of juvenile cotyledons of a biodiesel producing plant, *Jatropha curcas* L. Plant Biotechnol Rep 5:235–243

Nath LK, Dutta SK (1992) Wound healing response of the proteolytic enzyme curcain. Indian J Pharmacol 24:114–115

Openshaw K (2000) A review of *Jatropha curcas*: an oil plant unfulfilled promise. Biomass Bioenergy 19:1–15

Ostry ME, Hackett W, Michler C et al (1994) Influence of regeneration method and tissue source on the frequency of somatic variation in Populus to infection by *Septoria musiva*. Plant Sci 97:5222–5226

Palombi MA, Damiano C (2002) Comparison between RAPD and SSR molecular markers in detecting genetic variation in kiwifruit (*Actinidia deliciosa* A. Chev). Plant Cell Rep 20:1061–1066

Pan J, Fu Q, Xu ZF (2010) *Agrobacterium tumefaciens*-mediated transformation of biofuel plant *Jatropha curcas* using kanamycin selection. Afr J Biotech 9:6477–6481

Prabakaran AJ, Sujatha M (1999) *Jatropha tanjorensis* Ellis and Saroja, a natural interspecific hybrid occurring in Tamil Nadu, India. Genet Resour Crop Evol 46:213–218

Preece JE, Imel MR (1991) Plant regeneration from leaf explants of Rhododendron 'P.J.M. Hybrids'. Sci Hortic 48:159–170

Purkayastha J, Sugla T, Paul A et al (2010) Efficient *in vitro* plant regeneration from shoot apices and gene transfer by particle bombardment in *Jatropha curcas*. Biol Plant 54:13–20

Rajore S, Batra A (2005) Efficient plant regeneration via shoot tip explant in *Jatropha curcas*. J Plant Biochem Biotechnol 14:73–75

Rajore S, Batra A (2007) An alternative source for regenerable organogenic callus induction in *Jatropha curcas*. Indian J Biotech 6:545–548

Rajore S, Sardana J, Batra A (2002) *In vitro* cloning of *Jatropha curcas* L. J Plant Biol 29:195–198

Rani V, Raina SN (2000) Genetic fidelity of organized meristem derived micropropagated plants: a critical reappraisal. In Vitro Cell Dev Biol Plant 36:319–330

Reddy MP and Pamidimarri DVNS (2010) Biology and biotechnological advances in *Jatropha curcas*—a biodiesel plant. Ramawat KG (ed) Desert Plants. Springer, Berlin Heidelberg

Reddy MP, Kumar N, Vijay Anand KG et al (2008) Method for micropropagation of *Jatropha curcas* plants from leaf explants. Patent filed US and PCT, Application No. 2537de2008

Rupert EA, Dehgan B, Webster GL (1970) Experimental studies of relationships in the genus *Jatropha* L. *J. curcas* × *J. integerrima*. Bull Torrey Bot Cl 97:321–325

Saker MM, Adawy SS, Mohamed AA et al (2006) Monitoring of cultivar identity in tissue culture-derived date palms using RAPD and AFLP analysis. Biol Plant 50:198–204

Sardana J, Batra A, Sharma R (1998) *In vitro* plantlet formation and micropropagation of *Jatropha curcas* L. Adv Plant Sci 11:167–169

Sardana J, Batra A, Ali DJ (2000) An expeditious method for regeneration of somatic embryos in *Jatropha curcas* L. Phytomorphology 50:239–242

Sato S, Hirakawa H, Isobe S et al (2011) Sequence analysis of the genome of an oil bearing tree, *Jatropha curcas* L. DNA Res 18:65–76

Sharma A, Kansal N, Shekhawat GS (2006) *In vitro* culture and plant regeneration of economically potent plant species *Jatropha curcas*. Biochem Cell Arch 6:323–327

Sharma S, Kumar N, Reddy MP (2011a) Regeneration in *Jatropha curcas*: Factors affecting the efficiency of *in vitro* regeneration. Ind Crops Prod 34:943–951

Sharma S, Sudheer PDVN, Vijay Anand KG et al (2011b) Assessment of genetic stability in micropropagules of *Jatropha curcas* genotypes by RAPD and AFLP analysis. Ind Crops Prod 34:1003–1009

Shen X, Chen J, Kane ME et al (2007) Assessment of somaclonal variation in Dieffenbachia plants regenerated through indirect shoot organogenesis. Plant Cell Tiss Org Cult 91:21–27

Shrivastava S, Banerjee M (2008) *In vitro* clonal propagation of physic nut (*Jatropha curcas* L): Influence of additives. Intl J Integrative Biol 3:73–79

Singh A, Reddy MP, Chikara J et al (2010) A simple regeneration protocol from stem explants of *Jatropha curcas*—a biodiesel plant. Ind Crops Prod 31:209–213

Smykal P, Valledor L, Rodriguez R et al (2007) Assessment of genetic and epigenetic stability in long-term *in vitro* shoot culture of pea (*Pisum sativum* L.). Plant Cell Rep 26:1985–1998

Soomro R, Memon RA (2007) Establishment of callus and suspension culture in *Jatropha curcas*. Pak J Bot 39:2431–2441

Srivastava PS (1971) *In vitro* induction of triploid roots and shoots from mature endosperm of *Jatropha panduraefolia*. Z Pflanzenphysiol 66:93–96

Srivastava PS, Johri BM (1974) Morphogensis in mature endosperm cultures of *Jatropha panduraefolia*. Beitr Biol Pflanz 50:255–268

Staubmann R, Ncube I, Gubitz GM et al (1999) Esterase and lipase activity in *Jatropha curcas* L. seeds. J Biotechnol 75:117–126

Stripe F, Pession-Brizzi A, Lorenzoni E et al (1976) Studies on the proteins from the seeds of *Croton tiglium* and of *Jatropha curcas*. Biochem J 156:1–6

Sujatha M (2006) Genetic improvement of *Jatropha curcas* (L.) possibilities and prospects. Indian J Agrofor 8:58–65

Sujatha M, Dhingra M (1993) Rapid plant regeneration from various explants of *Jatropha integerrima*. Plant Cell Tiss Org Cult 35:293–296

Sujatha M, Mukta N (1996) Morphogenesis and plant regeneration from tissue cultures of *Jatropha curcas*. Plant Cell Tiss Org Cult 44:135–141

Sujatha M, Prabakaran AJ (2003) New ornamental *Jatropha* hybrids through interspecific hybridization. Genet Resour Crop Evol 50:75–82

Sujatha M, Reddy TP (2000) Morphogenic responses of *Jatropha integerrima* explants to cytokinins. Biologia (Bratisl) 55:99–104

Sujatha M, Makkar HPS, Becker K (2005) Shoot bud proliferation from axillary nodes and leaf sections of non-toxic *Jatropha curcas* L. Plant Growth Regul 47:83–90

Thepsamran N, Thepsithar C, Thongpukdee A (2008) *In vitro* induction of shoots and roots from *Jatropha curcas* L. explants. J Hortic Sci Biotechnol 83:106–112

Thomas P (2004) *In vitro* decline in plant cultures: detection of a legion of covert bacteria as the cause for degeneration of long-term micropropagated triploid watermelon cultures. Plant Cell Tiss Org Cult 77:173–179

Van den Berg AJJ, Horsten SFAJ, van den Bosch Kettenes JJ et al (1995) Curcacycline A—a novel cyclic octapeptide isolated from the latex of *Jatropha curcas* L. FEBS Lett 358:215–218

Varshney A, Johnson TS (2010) Efficient plant regeneration from immature embryo cultures of *Jatropha curcas*, a biodiesel plant. Plant Biotech Rep 4:139–148

Varshney A, Sangapillai R, Patil M et al (2011) Histological evidence of morphogenesis from various explants of *Jatropha curcas* L. Trees. doi:10.1007/s00468-011-0546-x

Wang PJ, Charles A (1991) Micropropagation through meristem culture. In: Bajaj YPS (ed), High tech and micropropagation biotechnology in agriculture and forestry vol. 17, pp 32–52. Springer, Berlin Heidelberg

Warakagoda PS, Subasinghe S (2009) *In vitro* culture establishment and shoot proliferation of *Jatropha curcas* L. Trop Agric Res Ext 12:77–80

Wei T, Newton RJ (2004) Increase of polyphenol oxidase and decrease of polyamines correlate with tissue browning in Virginia pine (*Pinus virginiana* Mill). Plant Sci 167:621–628

Wei Q, Lu WD, Liao Y et al (2004a) Plant regeneration from epicotyl explant of *Jatropha curcas*. J Plant Physiol Mol Biol 30:475–478

Wei Q, Lu WD, Liao Y et al (2004b) Plant regeneration from epicotyl explants of *Jatropha curcas*. J Plant Physiol Mol Biol 30:475–478

Wei Q, Liao Y, Chen Y et al (2005) Isolation, characterisation and antifungal activity of β-1,3-glucanase from seeds of *Jatropha curcas*. South Afr J Bot 71:95–99

Weida L, Qim W, Lin T et al (2003) Induction of callus from *Jatropha curcas* and its rapid propagation. Yingyong Yuhuan Jingshengwu Xuebao 9:127–130

Weike C, Qi Z, Xingchun C et al (2006) Chemical modification of *Jatropha curcas* RIPs (curcin) and effect of the modification on relative activity of curcin. Chin J Appl Environ Biol 12:329–333

Wu JH, Miller SA, Hall HK et al (2009) Factors affecting the efficiency of micropropagation from lateral buds and shoot tips of Rubus. Plant Cell Tiss Org Cult 99:17–25

Xu R, Wanga R, Liu A (2011) Expression profiles of genes involved in fatty acid and triacylglycerol synthesis in developing seeds of Jatropha (*Jatropha curcas* L.). Biomass Bioenerg. doi:10.1016/j.biombioe.2011.01.001

Yeung EC (1999) The use of histology in the study of plant tissue culture systems. Some practical comments. In Vitro Cell Dev Biol-Plant 35:137–143

Zhang Y, Wang Y, Jiang L (2007) Aquaporin JcPIP2 is involved in drought responses in *Jatropha curcas*. Acta Biochim Biophys Sin 39:787–794

Zhang FL, Niu B, Wang YC et al (2008) A novel betaine aldehyde dehydrogenase gene from *Jatropha curcas*, encoding an enzyme implicated in adaptation to environmental stress. Plant Sci 174:510–518

Zhao Y, Zhou Y, Brian WWG (2005) Variation in leaf structures of micropropagated rhubarb (*Rheum rhaponticum* L.) PC49. Plant Cell Tiss Org Cult 85:115–121

Zong H, Wang S, Ouyang C et al (2010) *Agrobacterium*-mediated transformation of *Jatropha curcas* young leaf explants with lateral shoot-inducing factor (LIF). Intl J Agric Biol 12:891–896

Chapter 14
Genetic Transformation of Bael (*Aegle marmelos* Corr.)

Rajesh Pati and M. Muthukumar

Abstract *Aegle marmelos* Corr. is better known for its values rather than its edible quality. Every part of the plant viz. leaves, fruits, roots, and the bark are used for treating various diseases related to heart, stomach, bacterial, fungal, viral, worms, fertility, fever and cosmetics. Marmelosin is one of the major compounds (coumarin), used to treat various diseases. Till date, there are no varieties developed for commercial orchards of Bael and hence, regarded as one among the underutilized fruits of India. Therefore, there is an urgent need for breeding new and cultivable varieties in *Aegle marmelos*. But the problem for the breeders would be the long gestation period for flowering and fruiting with seedling plants or vegetative propagules. Micro propagation seems to be a promising tool for large scale multiplication of this woody species and as its response to in vitro culture was quite successful, it offers possibilities for taping their potentials for extraction of secondary metabolites and medicinal compounds. This poses concerns on the yield, quality, stability and functional properties of the medicinal principles that are derived directly from the tree and from suspension cultures. An alternative and fruitful strategy that is possible is to develop a stable transformation system which can stack only potential genes of medicinal importance and silence inhibitory compounds so as to maintain the quality and their bioactive properties. This chapter deals with the micro propagation protocol, strategies for genetic transformation, challenges and prospects, biosafety, regulatory environment, intellectual property, and transgenic development

Abbreviations

NAA	α- napthaleneacetic acid
BA	benzyl adenine
MS	Murashige and Skoog

M. Muthukumar
Biotechnology Laboratory, Crop Improvement and Biotechnology Division, Central Institute for Subtropical Horticulture (ICAR), Rehmankhera P O. Kakori, Lucknow 227107, India

R. Pati (✉)
R&D Biotechnology (Plant Tissue Culture Lab), Jain Irrigation Systems Ltd., Jain Agri Park, Jain Hills, PO. Box: 72, Jalgaon 425 001, Maharashtra, India
e-mail: drrajeshpati@gmail.com

S. M. Jain and S. Dutta Gupta (eds.), *Biotechnology of Neglected and Underutilized Crops*, DOI: 10.1007/978-94-007-5500-0_14,

TDZ	thidiazuron
2-ip	N^6-[2-Isopentenyl] adenine
IBA	indole-3-butyric acid
2, 4-D	2, 4- dichlorophenoxyacetic acid

14.1 Introduction

According to World Health Organization (WHO), more than 21,000 plant species are being used for medicinal purposes worldwide. India ranks second for exporting medicinal plants in the world (Gupta et al. 2006). Despite our rich heritage and knowledge on use of medicinal plants in ayurveda, little attention has been paid to harness the inexpensive remedies to modern requirements (Setia et al. 2005) and only 40 plant species are currently used by the pharmaceutical industries. Similarly, cosmetic industries are currently using 42 plant species from a diverse genetic base of more than 1,300 plant species known for aromatic and pigment traits. Tendon and Thayil (1995) stated that India's medicinal heritage; one of the world's oldest living traditions, is in danger to extinction.

Bael (*Aegle marmelos* Corr.), belonging to family Rutaceae and growing wildly in India, is in high demand for pharmaceutical purposes (Pati et al. 2004). The leaves, fruits (Fig. 14.1), roots, and bark are used for medicinal purposes (Table 14.1). The *A. marmelos* tree grows throughout the dry hilly areas, reaches up to 1,300 m tall high in the Western Himalayas, Burma, Pakistan, Bangladesh, Sri Lanka, Northern Malaysia, Java, and Philippine Islands. Bael fruits were introduced in Europe in 1959 (Knight Jr 1980). In India, it is known with different names in (Dhiman 2003; Purohit and Vyas 2004) such as Bael, Beli, Belgiri, Vilwa, Bengal quince, Golden apple, Bel Kham Bel and Bilivaphal. The medium sized tree grows up to 8 m, with auxiliary spines. The bark is bluish–grey, soft, with irregular furrows on the younger branches. The leaves are alternate, ovate, trifoliate and aromatic. The flowers are stalked, sweet-scented, erect, axillary or terminal cymes. The fruits are round and size is usually 2–5 inches in diameter, woody berry with yellowish rind. Seeds are slightly compressed, slimy and embedded in sweet gummy pulp. Physico-chemical studies have revealed that bael fruit is rich in minerals and vitamins (Jauhari and Singh 1971; Shankar and Garg 1967; Paricha 2004). Major

Fig. 14.1 Fruits of *Aegle marlelos* (Corr.) (**a** and **b**)

Table 14.1 Medicinal use of various parts of *A. marmelos*

Plants parts	Biological effect	Plant part and preparation	References
Leaves	Diabetes	Leaf extract	Karunanayake et al. (1984), Kar et al. (2003), Akhtar et al. (2005)
	Cardiac depressant,	Leaf extract	Dhiman (2003), Purohit and Vyas (2004), Parmar and Kaushal (1982)
	Fertility control		Tuticorin and Manakkal (1983), Shankar and Garg (1967)
	Antibacterial, anti fungal, Ranikhet viral disease and intestinal parasites	Leaf oil/extract	Ansary (2005), Farooq (2005), Rana et al. (1997)
	Peptic ulcer	Leaf extract	Goel et al. (2000)
	Cold and respiratory infections	Leaf juice	Paricha (2004), Reddy et al. (2006)
	Jaundice, wounds, leucorrhea, conjunctivitis, and deafness		Devadi (2002), Kala et al. (2005)
Unripe fruits	Chronic diarrhea constipation and amoebic dysentery	Dried fruit powder	Shoba and Thomos (2001), Roy and Singh (1980)
	Anticancer activity	Bael pulp extract	Lambertine et al. (2004), Costa-Luatufo et al. (2005), Jagetia and Baliga (2004)
	Healing burn cases	Dried fruit powder with mustard oil	Parmar and Kaushal (1982), Oudhia (2005)
Ripe Fruits	Radio protective activity	Bael pulp extract	Jagetia et al. (2004), (2005)
	Constipation	Bael pulp extract	Tiwari and Joshi (1990), Roy and Singh (1980), Shailajan et al. (2011)
	Thyroid related disorders, snakebite,	Dried fruit powder	Kar et al. (2002)
	Anaemia, fractures, swollen joints, pregnancy troubles, typhoid, coma, colitis, bleeding sores and cramps	Dried fruit powder	Paricha (2004), Ahuja (1965), Agarwal (1990), (1997)
	Acute shigellosis, diuretic, gonorrhea, conjunctivitis	Dried fruit powder	Haider et al. (1991)
	Irritable bowel syndrome	Dried fruit powder	Yadav et al. (1989)
Roots and Bark	Vata diseases, insomnia, seizures, typhoid fever, hysteria	Decoction of root	Parichha (2004)
	Diabetes, stomatitis	Bark powder	Narayanan et al. (1999), International Diabetes Federation (2006). (www.idf.org)
Seeds	Antibacterial activity against different strains of vibrios and inhibits the growth of *Vibrio cholerae*, *Staphylococcus aureus* and *Escherichia coli* antifungal activity	Seed oil	Banerji and Kumar (1949), Jain (1977)

nutritional components of bael are listed in Table 14.2. Various parts of bael plant are being used in many Ayurveda and Unani patented drugs in India for treatment of a variety of diseases. The fruits and roots of *A. marmelos* Corr. possess antiamoebic and hypoglycaemic activity (Ponnachan et al. 1993). The alkaloid 'aegeline' present in the leaf is an effective antiasthamatic agent (Haravey 1968). The bark contains tannin, coumarin, and aegelinol; also furocourmarin, marmesin; umbelliferone, a hydroxy coumarin; and the alkaloids, fagarine and skimmianine. Coumarins are derived from 1,2-benzopyrones. These molecules are found in higher plants where they originate from the general phenylpropanoid pathway (Harborne 1999) and are subject to numerous modifications. Some natural coumarins have been used as human therapeutics, while 4-hydroxycoumarins are prominent examples of microbial modification which gave rise to the first generation molecules developed along with aspirin and heparin as anticoagulants (Mueller 2004). Marmelosin ($C_{13}H_{12}O_3$) derived from the pulp is laxative and diuretic. Shailajan et al. (2011) reported that higher concentration (Table 14.3) of marmelosin is found in ripe fruit pulp as compared to other parts of *Aegle marmelos* tree. The bael fruit is commonly multiplied by seed in nurseries and the seedlings show great variation in morphological and biochemical characters due to heterozygous nature of the plant. However, its commercial orchards are not expanding at a faster rate due to severe shortage of planting material. Many researchers developed micro propagation protocols for *A. marmelos* (Table 14.4). Clonal micro propagation technique of *Aegle marmelos* from mature tree using nodal explant has been developed for two improved cultivars i.e. CISH B1 and CISH B2 by Pati et al. (2008a, b). Since *A. marmelos* has pharmacological property it is a suitable candidate for genetic transformation. Till date until now there is no report on genetic transformation of *Aegle marmelos*.

Table 14.2 Nutritional value of bael (*A. marmelos*) fruit (100 gm)

Edible protein	64 %	Energy	137 K.cal
Moisture	61.5 gm	Calcium	85 mg
Protein	1.8 gm	Phosphorus	50 mg
Fat	0.3 %	Vitamin-C	8 mg
Mineral	1.7 gm	Potassium	600 mg
Fiber	2.9 gm	Vitamin B	Rich in B1 and B2
Carbohydrate	31.8 gm	Sodium	Nil

Table 14.3 Amount of marmelosin available in different parts of *A. marmelos*

Sample tested	Content of marmelosin[a] in mg/g
Rip fruit pulp	2.9 ± 0.00035
Unripe pulp	2.7 ± 0.0002
Seeds	1.3 ± 0.0001
Leaves	0.45 ± 0004
Rind	0.22 ± 0.0002
Inner stem	Nil
Outer stem	Nil

[a] Mean ± SD, *Source* Shailajan et al. (2011)

Table 14.4 List of developed micropropagation protocol for *A. marmelos*

Serial no.	Type of explants	Media composition	Remarks	References
1.	Cotyledons	MS + 2 mg/1 2,4-D + 1 mg/1 BAP- Callus formation	Callus/Somatic embryogenesis	Islam et al. (1993)
		MS + 1 mg/1 BA + 0.1 mg/1 IAA-shoot regeneration	Rooting 70 %	
		30 mg/1 IBA-rooting		
2.	Cotyledons	MS + 2 mg/l BAP + 0.2 mg/l IAA-shoot regeneration	The rooted plantlets were successfully transplanted to soil	Hossain et al.(1994)
		MS + 0.5 mg/l Kinetin + 0.1 mg/l GA_3-Shoot elongation		
		MS + 0.5 mg/l IBA-rooting		
3.	Nodal segments	MS + 0.5 mg/l BAP-regeneration	Organogenesis	Hazarika et al. (1996)
		MS + 0.5 mg/l IBA-rooting		
4.	Auxiliary bud	MS + 2.5 mg/l BAP + 1.0 mgl^{-1} IAA-bud induction	88 % plant survival during acclimatization	Ajithkumar and Seeni (1998)
		MS + 1 mg/l BAP-proliferation		
		1/2 MS + 0.5 mg/l IAA or 10.0 mg/l IBA-rooting		
5.	Callus culture	MS + 2.0 mg/l Kinetin + 0.5 mg/l NAA-Calli formation	This protocol is suitable to produce number of plants from cotyledon, hypocotyls and immature leaves of *A. marmelos*	Prematilake et al. (2006)
		MS + 1.0 mg/l NAA-rooting		
6.	Cotyledonary nodes from in vitro-seedlings	MS + 6.6 μM BA + 1.14 μM IAA -regeneration	80 % plant survival during acclimatization	Nayak et al. (2007)
		MS + 14.7 μM IBA-rooting		
7.	Nodal explants	MS + 0.5 mg/l BAP and 0.1 mg/l IAA-shoot proliferation	33 % plant survival during acclimatization	Sandhya et al. (2008)
		MS + 0.1 mg/l IAA-rooting		

(continued)

Table 14.4 (coninued)

Serial no.	Type of explants	Media composition	Remarks	References
8.	Nodal shoots	MS +3.0 mg/l kinetin-bud induction	83.33 % plant survival during acclimatization	Mishra et al. (2008)
		MS + 3.0 mg/l BAP-proliferation MS + 25 mg/l IBA-rooting	RAPD of regenerated plants	
9.	Nodal shoots	MS + 8.84 μM BAP + 5.7 μM IAA-bud induction	83.33 % Survival during acclimatization	Pati et al. (2008a)
		MS + 8.84 μM BAP + 5.7 μM IAA–proliferation	Biochenical and anatomical analysis	
		1/2 MS + 49.0 IBA μM + 5.7 μM IAA-rooting	Plants were tested for its genetic fidelity using 13 RAPD, 3 ISSR and 2 DAMD primers.	
10.	Nodal shoots	MS + 2.0 mg/l BAP + 1.0 mg/l IAA-bud induction	80.33 % Survival during acclimatization	Pati et al. (2008b)
		MS + 2.0 mg/l BAP + 1.0 mg/l IAA-proliferation	Biochemical and anatomical analysis	
		1/2 MS + 10 mg/l IBA + 1.0 mg/l IAA-rooting	Plants were tested for its genetic fidelity using 13 RAPD, 3 ISSR and 2 DAMD primers.	
11.	Leaf primordial	MS + 1.5 mg/l Kinetin + 0.5 mg/l 2,4-D + 1.0 mg/lNAA-production of organogenic calli and shoot	Callus/somatic embryogenesis	Ramanathan et al. (2010)
		MS + 1.5 mg/l NAA + 0.5 mg/l IBA-rooting	Rooted plants survived under acclimatization	
12.	Nodal segment	MS + 2.0 mg/l BAP + 1.0 mg/l IAA-proliferation	70 % Survival during acclimatization	Yadav and Singh (2011)
		1/2 MS + 1.0 mg/l IAA-rooting	Biochemical analysis	

14.2 Chemical Composition

Bael has various chemical compounds such as alkaloids, coumarins, steroids, polysaccharides, steroids and oil (Table 14.5) Besides this, minor constituents like ascorbic acid, sitosterol, crude fibres, tannins, α-amyrin, carotenoids, psoralen, xanthotoxin scopoletin and tembamide have also been isolated and identified from different parts of tree (Fig. 14.2).

Table 14.5 Chemical composition of *Aegle marmelos* Corr

Chemical composition		
Alkaloids: (Aegelin, aegelenine, marmeline, dictamine, fragrine, O-methylhalfordinine, O isopentenylhalfordinol, N-2-[4-(3′,3′-dimethylallyloxy) phenyl]ethyl cinnamide, N-2-hydroxy-2-[4-(3′,3′-dimethylallyloxy) phenyl] ethyl cinnamide, N-4 methoxystyryl cinnamide, N-2- hydroxy-2-(4 –hydroxy-phenyl) ethyl cinnamide, O-(3,3-dimethylallyl) halofordinol, N-2-ethoxy-2-(4-methoxy phenyl) ethyl cinnamide, N-2-methoxy-2-[4-(3′,3′-dimethylallyloxy)phenyl] ethylcinnamide, N-2-methoxy-2-(4-methoxyphenyl)-ethylcinnamide)	*Coumarins*: Marmelosin, marmesin, imperatorin, marmin, alloimperatorin, methyl ether, xanthotoxol, scoparone, scopoletin, umbelliferone, psoralen, marmelide and marmenol.	*Polysaccharides*: (Galactose, arabinose, uronic acid and L-rhamanose). *Seed oil*: Composed of palmitic, stearic, oleic, linoleic and linolenic acid *Tannins Carotenoids*

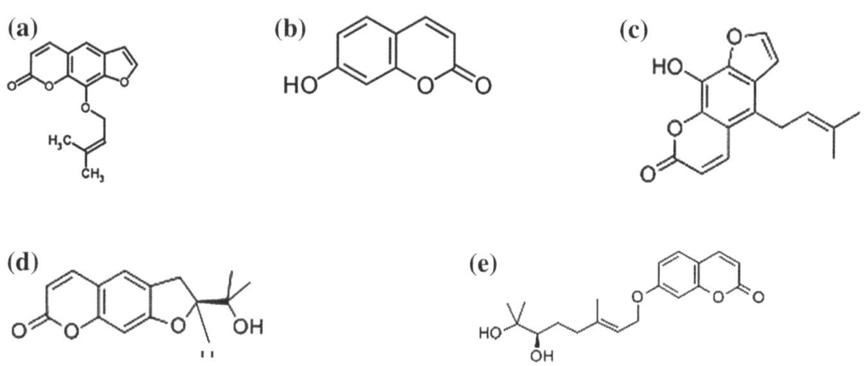

Fig. 14.2 Structure of some common coumarins. **a** Marmelosin, **b** Umbelliferone, **c** Alloimperatorin, **d** Marmesin, **e** Marmin

14.3 Varieties

A number of varieties have been identified and the following are among the best with regards to yield and fruit quality (Table 14.6). Other promising varieties of bael are Narendra Bael 1, Narendra Bael 2, Mirzapuri, Kagzi Gonda, Kagzi Etawah and Kagzi.

Table 14.6 Promising cultivars of *Aegle marmelos* Corr

Name	Characters
CISH-B1	Fruits are oval round (15–19 cm in length, 39–41 cm in circumference); weight varies from 0.8–1.2 kg, skull papery (0.10–0.12 cm). Pulp is orange yellow in color and having pleasant aroma
CISH-B2	Fruit is round (14.80 cm in length and 52.0–64.0 cm in circumference). Average weight 1.80–2.70 kg, orange yellow pulp, TSS 39^0 bricks with thin skull (0.24–0.26 cm), moderate amount of fiber and seed content. Plant can yield up to 40–50 kg fruit/tree
Narendra Bael 5	This variety has dwarf (3–5 m) and broad stature trees. The fruit in moderate size (21 × 25 cm) and sweet flavored (35–38°Brix) with flat ends and less number of seeds. Pulp is soft, tasty, low mucilage and less fibers. The average fruit weight is 900–1,000 g and average yield is 50–60 kg fruits/tree
Narendra Bael 6	Fruit is size medium sized, round with smooth surface, thin rind, few seeds, soft flesh, low mucilage and, mildly acidic
Narendra Bael 7	The average height of the plants is 5–7 m and canopy is of 3–5 m^2. Fruits are bigger 17.5 × 74 cm and weigh between 3 and 4.5 kg. They are of average sweetness 27–30°Brix and contains less fiber and seeds. The average yield of plant is 40–50 kg fruits/tree
Narendra Bael 9	The plants of this variety are of medium height 4–7 m and of broader stature the fruits are bigger in size (26–33 cm), round elliptical and very sweet 35–40°Brix. Average yield per plant is 50–60 kg fruits/tree. The fruits have good storage capacity
Pant Shivani	Mid season cultivar, shape ovoid oblong, size around 2 kg, color lemon yellow when ripe, fiber and mucilage content medium, rind medium thick, pulp light yellow with very good taste and pleasant flavor
Pant Aparna	Late cultivar, fruit size small (0.6–0.8 kg), globes, seed, mucilage, fiber and acidity low. Flesh yellow, sweet, tasty and having good flavor rind medium thick
Pant Urvashi	The plants are densely branched and taller in stature. The fruits are eleptical in shape (49.7 × 47.7) and weigh around 1.6 kg. The fruit has thin epicarp, sweet pulp (32°Brix) and good flavor. The fruit have 68.5 % pulp and less amounts of fibers. Average yield per plant is 25–30 kg fruits/tree
Pant Sujata	The plants are of medium height, dense and spreading nature. This is an early bearing variety and fruits have rectangular shape. Average weight of fruit is 1.14 kg and it contains sweet pulp (30°Brix) and less fibers. The fruit contains 77.8 % pulp. The average yield per plant is 45–50 kg fruits/tree

TSS Total soluble liquid, ^{0} bricks: unit

14.4 Disease

Accurate information on disease, diagnosis and management is essential for sustainable crop production. This section provides comprehensive coverage of important diseases affecting the commercial cultivation of *A. marmelos*. The common disease, diagnosis and their management are discussed in Table 14.7.

14.4.1 Strategies for Improving Disease Resistance

A. marmelos is a woody perennial tree were recalcitrance in the nature, thus it is very difficult to conduct breeding programs in this crop. So in order to overcome the recalcitrance, there is urgent need to transfer certain genes into the bael which is increase the regeneration potential. Genetic engineering and transgenic technology can play a major role in improvement of this crop. Most of the diseases are caused due to fungal infections. The tree contains a number of alkaloids and phenolics which are of great medicinal values. These secondary metabolites help the plant to fight the bacterial and fungal diseases. Marmelosin ($C_{13}H_{12}O_3$) derived from the pulp is a laxative and diuretic. This compound is widely used in various drugs related to stomach problems. This is produced in very less amount in the ripe fruits. By the use of genetic engineering we can increase its production level several folds. *Agrobacterium rhizogenes* mediated genetic transformation could be possible for increasing the amount of secondary metabolites production multifold, thus increasing the inborn resistance in the transformed plants. The fungal cell wall contains glucan and chitin which are readily hydrolyzed by enzymes glucanases and chitinases. The cDNA clones of these enzymes are available and can be used to develop transgenic plants expressing these enzymes. Somatic embryogenesis based regeneration system (Arumugam and Rao 2000) is available in bael, which can be used for efficient transformation. Plant regeneration using cotyledonary explants are also successful in bael and can be exploited for future genetic transformation. Another approach is tissue culture based cellular selection system to select the somaclonal variants resistant to fungal toxins.

14.5 Traditional Breeding versus Genetic Engineering

Genetic engineering is a modern form of crop modification system which is different from conventional plant breeding in basically two ways: (1) Plant breeding allows gene transfer only between closely related species, whereas in case of genetic engineering, genes are transferred from the same species or from any other species, even from animals, they can be introduced into a plant. Therefore genetic engineering creates a vast potential for crop alteration. (2) Plant breeding mixes large sets of genes of unknown function, whereas genetic engineering generally introduces only one to a few well characterized genes at the same time.

Table 14.7 List of disease and their management

Disease	Causal agent	Symptoms	Control measure
Leaf black spots	*Fungus*: *Ejiropsis spp.*	Black spots of 2–3 mm were formed on both side (upper and lower) of leaf	0.1 % bavestine (Carbendazime) or 0.2 % Difolaton are sprayed twice or thrice every 15 days, when new leaf comes
Die Back	*Fungus*: *Laciodiplodia spp.*	Plants start drying from top and gradually towards the base. Brown marks were seen on leaves and branches and later on the leaves fall	Dry branches are removed and copper oxychloride (0.3 %) should be sprayed every 15 days
Small fruit drop	*Fungus*: *Fusarium spp.*	Small fruits (5–8 cm) start fallen down. Small brown ring forms near the fruit-branch junction. As the disease spreads this junction become weak and fruits fall	0.1 % carbendazime should spray every 15 days when the fruits are small. Growth-regulators 2, 4-D, GA_3 and 2, 4, 5-T at various concentration check fruit drop to a reasonable extent
Bacterial shoot-hole and fruit canker	*Bacteria*: *Xanthomonas bilvae* Nov.	*Leaves*: Showing round, water-soaked spots (0.5 mm) surrounded by a clear halo. Gradually the spots increase in the size (3–5 mm) and form brown lesions with saucer-like depressions in the centre surrounded by an oily raised margin. Many times, few spots coalesce and the infected portions become chlorotic. Bacterial ooze in the form of shining beads and scales is appears (seen) on both sides of spots. The primary localized lesions all over the leaf are always followed by falling out of the dead tissues leaving circular or slightly irregular perforations or "shot-holes". Very often the dead portion of the lesion remains attached, through separated from the surrounding healthy tissue *Fruits*: the pathogen produces round, raised, water soaked spots measuring 4 mm to 7 mm with no halo. After some time, the crater-like depressions are noticeable in the centre of spots surrounded by irregular, oily, raised margins. The lesions then become very corky, irregular in shape and chocolate brown The pathogen infects twigs and thorns also. Later stage large lesions (150 × 4 mm) are commonly formed. The infected tissue ruptures in the centre giving a rough, corky appearance. Bacterial gummy exudates are commonly found on all infected parts	Use Streptocycline 200 PPM or Streptomycin sulphate 500 ppm for spraying after 15 days intervals
Bael canker	*Bacteria*: *Xanthomonas silvae*	Water soaked marks, which later on become brown black in color. As the disease advances the tissues of the affected parts fall off and form holes in the leaves	200 ppm of streptocycline is mixed in water and sprayed every 15 days

14.6 Developments of Transgenic

14.6.1 Donor Gene

For gene transfer, several opportunities are available as there are vast information of genes in the public domain databases and literature. Moreover, sequencing projects in crop plants have yielded huge information on genes. Manipulations of genes and developments of gene constructs with specific promoters are well established in many crop plants. Based on need, through gene specific PCR or isolation cDNA clone from a library, the genes of interest may be used for developing gene construct. The possible avenues are dwarfing the tree stature using dwarfing genes reported in crop plants especially perennials and decreasing shell hardiness in fruits by manipulation of genes responsible for fruit peel because the hard peel (shell) is the major drawback in bael. These are the areas where transgenics in bael is essential.

14.6.2 Methods of Transformation

Early experiments targeting the genetic transformation of medicinal plants were carried out using *Agrobacterium rhizogenes* to obtain hairy root cultures, or *Agrobacterium tumefaciens* to produce transformed cells to be maintained in culture or to regenerate whole plants. Transformation was successful in earlier stages of cell lines but the further regeneration was not achieved. Key limitations include the high cost of bioreactors and instability of cell lines, which often lose their capacity to produce target molecules over time. In plants, two methods viz. biolistics (Gene gun) and *Agrobacterium*-mediated are generally used. *Agrobacterium*-mediated transformation method is widely used because of stable transformation and less chances of gene silencing.

14.6.2.1 Explant to be Transformed

The purpose of most plant transformation experiments in plant biotechnology is to produce whole transgenic plants. The explants used in transformation must be capable of producing whole plants by regeneration and should contain a high number of cells that are competent for transformation.

14.6.2.2 The Vector Used for Transgene Integration into Plant Genome

Vector used in *Agrobacterium*-mediated plant transformation are derivatives of the naturally occurring Ti plasmid. Generally these are extensively modified so that the most features of a natural occurring Ti plasmid are removed, only the left and right border sequences being used to ensure transfer of the Ti–DNA region between them. The vector also contains a selectable marker (generally Kanamycin sulphate) on the T-DNA so that transformed plants can be separated from other non-transformed plants. Virulence genes required for transfer of the T-DNA are often located on a separate plasmid in the bacterium. There are two vector systems used for *Agrobacterium* mediated transformation, viz., binary and co-integrate vector systems. The gene constructs are tailor-made based on the need and inserted into the vector system.

14.6.2.3 *Agrobacterium* Strain to be Used

Several widely used *Agrobacterium* strains are available for plant transformation. For more recalcitrant plant species, the choice of strain is a major factor contributing to the success or failure of the experiment. A general protocol for *Agrobacterium*-mediated plant transformation is discussed below.

14.6.2.4 Sterilization and Explant Preparation

From transformation point of view immature zygotic embryo of *A. marmelos* will be the best explants as it was successful in many of the crop plants. The immature fruits were washed under running tap water and soaked in 0.1 % Carbendazime (Bavestine) containing 1 drop of Tween 20 for surface sterilization for one hour and then washed five times with sterile distilled water. Pre-surface sterilized explants were further surface sterilized with 0.1 % $HgCl_2$ for 5 min for effective control of contaminants, after that explants were washed five times with sterile distilled water. The immature zygotic embryo was scooped out from unripened fruits and transferred to culture medium.

14.6.2.5 Co-cultivation

Single colony of *Agrobacterium* containing gene of interest were inoculated in 50 ml of LB broth (Tryptone 10 gm/l, Yeast extract 5 gm/l and NaCl 10 gm/l) containing 50 mg/l kanamycin (Strain LBA4404). The culture is incubated in the dark at 28 °C for 24 h (0.8–1.0 OD at 600 nm) in incubator shaker at 100 rpm. Culture was centrifuged (10,000 rpm) and pellet was dissolved in 10 ml of liquid MS medium containing spermidine (1 μM) and acetosyringone (100 μM) (Mishra et al. 2010). The suspension was kept for 3 h prior to infection. The explants were sectioned into 2–3 mm length and placed into a culture of *A. tumefaciens* (which contains the vector) for 30 min. To eliminate any bacterial excess, tissues were washed in MS liquid medium followed by blotting drying on sterile filter paper and transfer on MS semisolid medium (no selective agent) for co-cultivation for 72 h to allow transfer of the T-DNA to the plant cells. After co-cultivation the explants were washed with antibiotic solution (cefotaxime 500 mg/l or carbenicillin 250 mg/l).

14.6.2.6 Selection and Regeneration

After co-cultivation, the explants were transferred to somatic embryo induction medium (1/2 strength MS medium supplemented with 400 mg/l glutamine, MS vitamins, 10 mg/l 2,4-D, 6 % sucrose, and 0.8 % agar having 5.8 pH) containing 150 mg/l kanamycin. Along with cefotaxime 500 mg/l or carbenicillin 250 mg/l to prevent the growth of non-transformed cells. The cultures were incubated at 25 ± 2 °C in the dark for 4 weeks. After 4 weeks the kanamycin concentration was increased up to 150 mg/l.

After 4 weeks, several white to light yellow globular embryos appeared (depends on plants species). The clumps of somatic embryos were transferred in regeneration medium (1/2 strength MS medium containing 400 mg/l glutamine, MS vitamins, 0.5 mg/l BAP and 0.1 mg/l NAA 3 % sucrose and 0.8 % agar) to convert into plants. Cultures were incubated in growth room at 25 ± 2 °C temperature for 16 h photoperiod with light at 40 μmolm^2 s^{-1}at 55 % relative humidity (Fig. 14.3).

Fig. 14.3 Steps in genetic transformation in *Aegle marmelos* Corr. **a** Transformed embryos, **b** selection of transformed embryos in Kanamycin 150 mg/l, **c** GUS assay, **d** regeneration in transformed callus, (**e** and **f**) proliferation in regenerated plantlets, **g** PCR analysis for *npt* II (480 bp), **h** Hardening/Acclimatization, **i** plant in transgenic glasshouse under controlled condition

14.6.2.7 Rooting and Acclimatization

Finally 3–4 cm long plantlets were transferred on kanamycin free rooting medium (IBA 3 mg/l + activated charcoal 500 mg/l) for roots induction. Rooted plantlets were transferred into culture bottles containing autoclaved vermiculite/coconut husk supplemented with 1/2 ms salt solutions and cultures were incubated in growth room at 25 ± 2^0c temperature for 16 h photoperiod with light at 40 μmolm^2 s^{-1}at 55 % relative humidity for 2–3 weeks. Once plant develops 4–6 leaves, transfer it in poly bags filled with soil + sand + field yard manure (1:1:1) in poly house provided with 50 % shade and misting.

14.6.2.8 Confirmation of Transformation/Testing

Transformation was confirmed by several means:

1. Transformed tissues were tested for their continued growth in the presence of specific antibiotics (Kanamycin 150 mg/l).
2. Transformed calli, somatic embryos, leaf sections, roots and fruits sections from transformed plants were tested for GUS expression by a histochemical GUS assay (Jefferson 1987).
3. PCR analysis was performed using *npt* II specific primers, which is given 480 bp band (Mishra et al. 2010) using DNA from transgenic and control plants prepared essentially as described (Dellaporta et al. 1983).
4. For DNA hybridization, genomic DNA was isolated from the plants essentially as reported (Murray and Kennard 1977). Fifty micrograms DNA was restricted with specific restriction enzyme and separated on a 0.8 % agarose gel. DNA was transferred onto a nylon membrane for hybridization with a cDNA fragment of specific promoter as a probe. The probe was designed to detect border sequences to give an estimation of T-DNA copies.
5. Double Antibody Sandwich Enzyme-linked Immunosorbent Assay (DAS-ELISA) technique (Clark and Adam 1977), employing polyclonal and monoclonal antibodies, used to assay for presence of desired proteins (e.g.—produced from PRV CP gene) and *npt II* in putative transgenic leaves (Fitch et al. 1992).

Before releasing the transgenic several steps are needed to strictly follow which is discussed in Table 14.8.

14.7 Prospects in Genetic Transformation

The present review article will enhance the existing knowledge of bael and creates awareness with the medicinal and therapeutic properties of bael which can further help in genetic transformation. Several genes have been characterized from chloroplast DNA of *A. marmelos* (Table 14.9).

Genetic transformation of bael with donor genes may explore the way to gain enhanced quality of marmelosin ($C_{13}H_{12}O_3$) production, secondary metabolites and

Table 14.8 Transgenic crop development and evaluation

Stage	Lab/Field	Analysis	Remark
T0	Lab testing or Green house/Phytotron	• Molecular analysis • Target trail data	Select a few promising plants
T1	Green house/Phytotron	• Molecular analysis of selected single plant progenies • Trial segregation analysis	Identify individual positive plants
T2	Green house/Phytotron	• Molecular analysis of single plant progenies • Target trial analysis	Identify non segregating progenies
T3	Limited field trial	• Agronomic performance including DUS(distinctiveness, Uniformity and Stability) features • Target trial expression • Toxicity data • Allergenicity data • Environmental impact analysis • Pollen flow study • Soil data	• Integration of transgenic into IARI breading programme • Identify promising progenies with desired level of target trait expression and advance to T4 generation
T4 and T5	Multilocation field evaluation	• Agronomic data • Target trait expression • Pollen flow data	Dialogue with project Directors/Coordinators of respective crops for VCU testing
T6	Value for cultivation and use (VCU) testing in all India coordinated trials	• Seed multiplication	Variety identification and release

Source A scheme of Indian Agricultural Research Institute, New Delhi, India

disease free quality planting material. Production of high level of marmelosin would make the bael (*Aegle mormelos* Corr.) commercially more important although no reports of genetic transformation in bael have been reported. Introduction of dwarfing gene for high density planting, shortening of juvenile period to enable early flowering, disease resistance, insect resistance and high marmelosin producing gene for enhancing the production of marmelosin and other secondary metabolites would further increase the commercial importance of this crop. Because of the problems associated with conventional breeding program, genetic transformation in bael (*Aegle mormelos*) holds promise for introducing donor genes for high value horticultural traits which could be potentially helpful in increasing the economic importance of the bael crop.

14.8 Intellectual Property Rights

The present and future status of genetically modified (GM) crops has been the subject of several recent reviews (Dunwell 2000, 2002, 2004, Gomez-Galera et al. 2007, 2010; Pati and Verma 2009; Srivastava et al. 2011). Although these reviews have included some information extracted from patent databases, this analysis has been necessarily limited in scope.

Table 14.9 List of genes sequence and characterized from *A. marmelos* retrieved from NCBI database

S. No.	Source	Locus	Genes	Base pair	References
1.	Chloroplast (CP DNA) Aegle marmelos	AF066839	ATP synthase beta subunit (atpB) gene	1,432 bp	Chase et al. (1999)
2.	Chloroplast *Aegle marmelos*	AF066811	Ribulose 1,5-bisphosphate carboxylase (rbcL) gene	1,342 bp	Chase et al. (1999)
3.	Chloroplast *Aegle marmelos*	AF025507	Chloroplast trnL-trnF intergenic spacer region	384 bp	Scott et al. (2000)
4.	Chloroplast *Aegle marmelos*	AF320882	Ribulose 1,5-bisphosphate carboxylase gene	954 bp	Samuel et al. (2001)
5.	Chloroplast *Aegle marmelos*	AY295294	Chloroplast trnL gene	972 bp	Morton et al. (2003)
6.	Chloroplast *Aegle marmelos*	AY295268	rps-16 gene	814 bp	Morton et al. (2003)
7.	Chloroplast *Aegle marmelos*	AY116508	Chloroplast trnT-trnL intergenic spacer region	304 bp	de Araujo et al. (2003)
8.	Chloroplast *Aegle marmelos*	AY115638	tRNA-Leu (trnL) gene	260 bp	de Araujo et al. (2003)
9.	Chloroplast *Aegle marmelos*	AY115615	Chloroplast trnL-trnF intergenic spacer	251 bp	de Araujo et al. (2003)
10.	Chloroplast *Aegle marmelos*	EF138836	SHA0574 maturase K gene	778 bp	Bayer et al. (2009)
11.	Chloroplast *Aegle marmelos*	EF134628	SHA0790 ribosomal protein S4 (rps4) gene	528 bp	Bayer et al. (2009)
12.	Chloroplast *Aegle marmelos*	EF164808	tRNA-Asp (trnD-GUC) gene	1206 bp	Bayer et al. (2009)
13.	Chloroplast *Aegle marmelos*	EF176492	tRNA-Gly (trnG) gene	1,516 bp	Bayer et al. (2009)
14.	Chloroplast *Aegle marmelos*	ABO28194	Maturase K	17 aa	Bayer et al. (2009)
15.	Chloroplast *Aegle marmelos*	ABO28111	Ribosomal protein S4	58 aa	Bayer et al. (2009)
16.	Chloroplast *Aegle marmelos*	FJ434197	Ribulose-1,5-bisphosphate carboxylase/ oxygenase large subunit (rbcL) gene	762 bp	Morton (2009)
17.	*Aegle marmelos*	FJ434169	5.8S ribosomal RNA gene	713 bp	Morton (2009)
18.	*Aegle marmelos*	HM627647	Miraculin-like protein 2 gene	522 bp	Gahloth et al. (2010)
19.	Chloroplast *Aegle marmelos*	AB505961	Chloroplast rbcL gene	1277 bp	Penjor et al. (2010)
20.	Chloroplast *Aegle marmelos*	HM163957	Voucher K:Chase 1340 maturase K (matK) gene	1,521 bp	Salvo et al. (2011)
21.	Chloroplast *Aegle marmelos*	HM163857	Voucher K:Chase 1340 ribosomal protein L16 (rpl16) gene	976 bp	Salvo et al. (2011)
22.	Chloroplast *Aegle marmelos*	HM163756	Voucher K:Chase 1340 tRNA-Leu (trnL) gene	483 bp	Salvo et al. (2011)
23.	Chloroplast *Aegle marmelos*	HM163653	Voucher K:Chase 1340 tRNA-Lys (trnK) gene	202 bp	Salvo et al. (2011)

Chloroplast DNA (*CP DNA*)

As technical limitations are overcome, it is possible that commercial limitations will become more serious barriers to exploitation of genetic transformation. New technologies developed in this area are effective inventions and are therefore eligible for patent protection (Kjeldgaard and Marsh 1994; Peet 1995). For example, patents have already been issued on most established or promising plant genetic transformation strategies (Schilperoort et al. 1990; Coffee and Dunwel 1995; Hiei and Komari 1994; Kryzyzek et al. 1995; Paszkowski et al. 1995; Lacroix et al. 2008) and on many isolated genes, promoters, and techniques for plant gene manipulation. The patent literature has become an important source of information in plant transformation research, albeit more difficult and expensive to search than the scientific literature (Electronic Data Systems Corporation 1995; http://igmoris.nic.in/; http://www.gmo-compass.org/eng/). A patent provides the inventor or assignee with a period of exclusive ownership, or formally a right to exclude others from making, using, or selling the invention. There is no statutory exclusion for infringement when patented products or methods are used for research purposes. Patents are intended to encourage and reward useful invention and technical innovation, and the new technology enters the public domain after a period of 17–20 years (Enayati 1995). In the interim period, commercial restrictions can appear quite ruthless as patent holders adopt commercialization strategies to capture the value of protected intellectual property (Hoyle 1996). In an era of tight public sector research funding and high research and development costs, the benefits of corporate investment to develop transformation technologies outweigh the inconvenience of patent restrictions. Debate continues on mechanisms to balance the competing interests.

14.9 Biosafety Regulations of GMOs

The countries participating on the Earth Summit in 1992 have agreed upon the fact that biotechnology can offer indubitable benefits to sustainable development, world food supplies and economic prosperity. The international rules for biosafety of transgenic crops are reflected both in the Cartagena protocol on biosafety and the WTO agreements. According to these rules built on scientific basis and promoted on the case-by-case approach, every transgenic event should undergo separate risk assessment and potential hazards, specific to this event should be identified and specific risk management measures assigned. The specific international agreements that treat different aspects of modern biotechnology (GMOs) products similar to any other international products, are results of negotiations and compromises. In the last years, there are some tendencies in the negotiations of a few international instruments, e.g. the Cartagena protocol on biosafety and the Aarhus convention on public participation that diverge from the initial idea of the Earth Summit, by taking into consideration only the eventual negative effects that might be associated with the deliberate release into environment of the products of modern biotechnology. Some countries and nongovernmental organizations have expressed their willingness for stricter liability regimes that are in the position to hinder the development of public research in the countries, particularly developing countries and countries with economies in transition. The policy makers in these countries should take into account the fact that public research is always oriented response to a specific problem in the country's agriculture or medicine and has a clear social benefit driven feature. Moreover, the scientific problems of the public research is not addressed by the multinational biotech companies, which products are mostly in commodity crops that are able to bring fast and considerable profits. In the past years several international instruments that consider different aspects of the trade, transboundary movement and potential adverse effects for the environment of GMOs have been agreed. In most of the cases, closer interaction and cooperation, as well as further harmonization among these agreements would be recommendable. Unified stricter regimes may lose on flexibility and would not be able to satisfy the needs and interests of every country, particularly developing countries. Other than relying on international instruments e.g. the Cartagena Protocol on Biosafety, the countries all over the world are highly encouraged to develop their own national biosafety frameworks (GEF project) that would better reflect the countries' needs in terms of import- export of the products of modern biotechnology. There are several models of such national regulations, overviewed in this paper that may be effective in building up a workable biosafety system. Which model to be chosen depends on the policy of the given country and should be in accordance with its international and regional obligations. It is commonly understood that international and regional harmonization, in addition to synchronizing the national regulatory frameworks, should focus on the issues of strengthening capacities and information sharing for biotechnological safety. Many countries, especially in the developing world, need to acquire the technology and the capacities necessary to sustainably handle the results of modern biotechnology. Therefore, public awareness, education and technology transfer play an important role. A number of international organizations such as FAO, WHO, UNEP, UNIDO, OECD, ICGEB and CGIAR, also show examples of regional cooperation, and are in the position to offer necessary assistance in capacity-building and dissemination of information on biosafety.

14.10 Potential Risks of GM Crops

In the first decades of the Green Revolution, risks to human health and to the environment from the accompanying pesticides, fertilizers, and irrigation were not a significant concern. Consequently, the Green Revolution had little problem achieving the level of public acceptance that was necessary for it to have a revolutionary impact. The introduction of genetically modified crops into the food supply has generated a number of concerns about potential risks associated with this new agricultural technology. Depending on the seriousness of these risks, they may limit the GM movement's impact. The risks fall under the broad categories of human health and the environment. When a new gene producing a novel protein is introduced into a crop, there is a chance that human subpopulations may have an allergic reaction to the protein. Also, GM crops may adversely affect non-target species; for example, a GM pest-protected plant targeting lepidopteran pests may spread toxin- containing tissues, such as pollen, that may contaminate the food of non-target lepidopterans. 2 Cross-pollination between GM crops and non-GM crops, or GM crops and wild plant species, may occur, with unknown consequences. A host of other risks surrounding GM crops may exist of which we are currently unaware (Wu 2002). Many of these risks, particularly those regarding food safety and impacts on non-target species, have undergone extensive scientific research (U.S. Environmental Protection Agency [EPA] 2001). Most of this research has found no evidence that such risks exist among current GM crops. Indeed, 81 scientific studies financed by the European Commission have all shown no evidence of risk to human and animal health or to the environment from genetically modified crops (Paarlberg 2003). However, it is impossible at this stage to fully investigate all potential risks, current and future, of GM crops, and the challenges posed for such investigations by the ongoing development of GM varieties are substantial. Both the benefits and the risk of transgenic plants may very spatially and temporarily on a case-by-case basis and to compare transgenic plants with traditional plants and other agricultural practices for elucidating the relative benefits and risk of the transgenic plants.

14.11 Conclusion

Regeneration and transformation systems using mature plant material of woody fruit species have to be achieved as a necessary requirement for the introduction of useful genes into specific cultivars and the rapid evaluation of resulting horticultural traits. Although the commercial production of transgenic annual crops is a reality, commercial genetically-engineered fruit trees are still far from common. In most woody fruit species, transformation and regeneration of commercial cultivars are not routine, generally being limited to a few genotypes or to seedlings. The future of genetic transformation as a tool for the breeding of fruit trees requires the development of genotype-independent procedures, based on the transformation of meristematic cells with high regeneration potential and/or the use of regeneration promoting genes. For some time, there was good reason to believe that *Agrobacterium tumefaciens* was the vector system with the capacity for gene transfer to any plant species and variety. However, some of the long-standing problems such as disease resistance, increased branching, dwarfism, early flowering, improved rooting for secondary metabolite production, short shelf life of fruit and abiotic stress sensitivity requires urgent attention of researchers.

Introduction of *rol* genes that confers dwarfing tendency of trees offers a pivotal role in paving the way to plant architectural traits an opportunity to explore for its medicinal trait values in the horticultural, agricultural and industries.

This endophytic fungus produced 187.6 μg/l of taxol which could be a potential source for the genetic engineering to improve the production of taxol as anticancer drug. Transgenic plants produced normal, fertile flowers that set fruits with seeds. These traits were transmitted to the progeny, resulting in trees with a generation time of 1 year from seed to seed. Whereas LFY lines showed alterations in growth and development, AP1 plants were adult and fully normal. There is need to exploitation of modern tools of biotechnology in improvement of bael. An increase in genetic transformation studies aimed at improving visual and growth characteristics of the plants has been hindered by low transformation efficiencies and genotype dependence of protocols. As a result, bael regeneration studies have once again emerged as an essential complement of transformation studies.

References

Agarwal VS (1990) Economic Plants of India. Kailash Parkashan, Calcutta, pp 3–9

Agarwal VS (1997) Rural economics of medicinal plants: vegetation in the forests. In: Drug plants of India, vol 1. Kalyani Publishers, New Delhi

Ahuja BS (1965) Aegle marmelos. In: Medicinal plants of Saharanpur, 1st edn. Dhrampal st Vidhyalankar Publisher, Gurukul Kangri Vishawvidyalaya, Hardwar, pp 10–11

Ajithkumar D, Seeni S (1998) Rapid clonal multiplication through in vitro axillary shoot proliferation of *Aegle marmelos* (L.) Corr. a medicinal tree. Plant Cell Rep 17:422–426

Akhtar J, Jamil S, Azhar MU (2005) Diabetes mellitus: prevention and management. Nat Prod Rad 4(5):413–415

Ansary PY (2005) A hand book on the plant sources of indigenous drugs. International Book Distributors, Dehra Dun 36

Arumugam S, Rao MV (2000) Somatic embryogenesis in *Aegle marmelos* (L.) Corr., a medicinal tree. In: Jain SM, Gupta PK, Newton RJ (eds) Somatic embryogenesis in woody plants, vol. 6. Kluwer Academic Publication, Dordrecht, pp 605–656

Banerji N, Kumar R (1949) Study of endophytic fungal community from different parts of *Aegle marmelos* Correae (Rutaceae) from Varanasi (India). J Inst Chem Soc 71:606

Bayer RJ, Mabberley DJ, Morton C, Miller CH, Sharma IK, Pfeil BE, Rich S, Hitchcock R, Sykes S (2009) A molecular phylogeny of the orange subfamily (Rutaceae: Aurantioideae) using nine cpDNA sequences. Am J Bot 96(3):668–685

Chase MW, Morton C, Kallunki JA (1999) Phylogenetic relationships of Rutaceae: a cladistic analysis of the subfamilies using evidence from RBC and ATP sequence variation. Am J Bot 86(8):1191

Clark M, Adams A (1977) Characteristics of the microplate methods of enzyme-linked immunosorbent assay (ELISA) for the detection of the plant viruses. J Gen Virol 34:475–483

Coffee RA, Dunwel JM (1995) Transformation of plant cells (to Zeneca Ltd.). US Patent No. 5,464,765

Costa-Luatufo LV, Khan MTH, Ather A, Wilke DV, Jimenez PC, Pessoa C, Moaraes MEA, Moreas MOD (2005) Studies of anticancer potential of plants used in Bangladeshi folk medicine. J Ethnopharmacol 99(1):21–30

de Araujo EF, de Queiroz LP, Machado MA (2003) What is Citrus? Taxonomic implications from a study of cp-DNA evolution in the tribe Citreae (Rutaceae subfamily Aurantioideae). Org Divers Evol 3(1):55–62

Dellaporta SL, Wood J, Hicks JB (1983) A plant DNA minipreparation: version II. Plant Mol Biol Reptr 1:19–21

Devadi G (2002) Importance of Bael and its use in health care. In: Kalyan Aarogaya Ank, vol 2. Geeta Press, Gorakhpur, pp 401–402

Dhiman AK (2003) Discussion of plants, sacred plants and their medicinal uses. Daya Publication House, New Delhi, pp 18–19

Dunwell JM (2000) Transgenic approaches to crop improvement. J Exp Bot 51:487–496

Dunwell JM (2002) Future prospects for transgenic crops. Phytochem Rev 1:1–12

Dunwell JM (2004) Transgenic crops: the current and next generations. Methods Mol Biol 286:377–398

Electronic data systems corporation (1995) Shadow Patent Off., http://www.spo.eds.com/spo

Enayati E (1995) Intellectual property under GATT. Biotechnology 13:460–462

Farooq S (2005) Medicinal plants: field and laboratory manual. International Book Distributors, Dehra Dun, pp 40–42

Fitch MM, Manshardt RM, Gonsalves D, Slinghtom JL, Sanford JC (1992) Transgenic papaya plants from *Agrobacterium*- mediated transformation of somatic embryos. Plant Cell Rep 12:245–249

Gahloth D, Selva Kumar P, Tomar PS, Kumar P, Sharma AK (2010) Miraculin-like proteins from Rutaceae family. Biotechnology http://www.uniport.org/uniport/GOWPSO

Goel RK, Maiti RN, Manickan M, Ray AB (2000) Antiulcer acitivity of naturally occurring pyrano-coumarin and isocoumarin and their effect on prostanoid synthesis using human colonic mucosa. Indian J Exp Biol 35(10):1080–1083

Gomez-Galera S, Pelacho AM, Gene A, Capell T, Christou P (2007) The genetic manipulation of medicinal and aromatic plants. Plant Cell Rep 26:1689–1715

Gomez-Galera S, Rojas E, Sudhakar D, Zhu C, Pelacho AM, Capell T, Christou P (2010) Critical evaluation of strategies for mineral forticication of staple food crops. Transgenic Res 19:65–80

Gupta M, Biswas TK, Saha S, Debnsth PK (2006) Therapeutic utilization of secretory products of some Indian medicinal plants: a review. Indian J Trad Knowledge 5(4):569–575

Haider R, Khan AK, Aziz KM, Chowdhury A, Kabir I (1991) Evaluation of indigenous plants in treatment of acute shizellosis. Trop Geogr Med 43(3):266–270

Harborne JB (1999) Classes and functions of secondary products from plants. In: Walton NJ, Brown DE (eds) Chemicals from plants. Imperial College Press, London, pp 1–25

Harvey SK (1968) A preliminary communication of action of *Aegle marmelos* (bael) on heart. Ind J Med Res 56:327

Hazarika BN, Nagaraju V, Parthasarathy VA (1996) Morphgenetic response of bale (*Aegle marmelos* L.) microshoots to benzyl amino purine. Ann Plant Physiol 10(1):40–44

Hiei Y, Komari T (1994) Transformation of monocotyledons using *Agrobacterium*. Int Patent WO 94/00977

Hossain M, Islam R, Karim MR, Joarder OI, Biswas BK (1994) Regeneration of plantlets from in vitro cultured cotyledons of *Aegle marmelos* Corr. (Rutaceae). Curr Sci 69(6):494–495

Hoyle R (1996) Another salvo in the patent wars. Nat Biotechnol 14:680–682

Islam R, Karim MR, Haque A, Hossain M, Joarder OI (1993) High frequency plant regeneration from Cotyledon cultures of *Aegle marmelos* Corr. Plant Tiss Cult 3(2):107–110

Jagetia GC, Baliga MSVP (2004) Fruit extract of *Aegle marmelos* protects mice against radiation-induced lethality. Integr Cancer Ther 3(4):323–332

Jagetia GC, Venkatesh P, Baliga MS (2004) Evaluation of the radioprotective effect of Bael leaf (*Aegle marmelos*) extract in mice. Int J Radiat Biol 80(4):281–290

Jagetia GC, Venkatesh P, Baliga MS (2005) *Aegle marmelos* Correa inhibits the proliferation of transplanted Ehrlich ascites carcinoma in mice. Biol Pharm Bull 28(1):58–64

Jain NK (1977) Antifungal activity of essential oil of *Aegle marmelos* Correa (Rutaceae). Ind Drugs Pharmaceut Ind 12:55

Jauhari OS, Singh RD (1971) Bael-a valuable fruit. Indian Hortic 16:9–10

Jefferson RA (1987) Assaying chimaric genes in plants, the GUS reporter gene fusion system. Plant Mol Biol Rep 5:387–405

Kala CP, Farooquee NA, Majila BS (2005) Indigenous knowledge and medicinal plants used by Vaidyas in Uttaranchal, India. Nat Prod Rad 4(3):195–204

Kar A, Choudhary BK, Bandyopadhyay NG (2003) Comparative evaluation of hypoglycaemic activity of some Indian medicinal plants in alloxan diabetic rats. J Ethnopharmacol 84(I):105–108

Kar A, Panda S, Bharti S (2002) Relative efficacy of three medicinal plant extracts in the alteration of thyroid hormone concentration in male mice. J Ethnopharmacol 81(2):281–285
Karunanayake EH, Welihinda J, Sirimanne SR, Sinnadoria H (1984) Oral hypoglycemic activity of some medicinal plants of Sri Lanka. J Ethnopharmacol 1(2):223–231
Kjeldgaard RH, Marsh DR (1994) Intellectual property rights for plants. Plant Cell 6:1524–1528
Knight RJ Jr (1980) Origin and world importance of tropical and subtropical fruit crops, In: Nagar S, Shah PE (eds) Tropical and subtropical fruits. AV, Westport, pp 1–120
Kryzyzek R, Laursen CRM, Anderson PC (1995) Stable transformation of maize cells by electroporation. US Patent 5, 472, 869
Lacroix B, Kozlovsky SV, Citovsky V (2008) Recent patents on *Agrobacterium*-mediated gene and protein transfer, for research and biotechnology. Recent Pat DNA Gene Seq 2(2):69–81
Lambertine E, Piva R, Khan MT, Lampronti I, Bianchi N, Borgatti M, Gambari R (2004) Effects of extract from Bangladeshi medicinal plants on in vitro proliferation of human breast cancer cell lines and expression of estrogen receptor alpha gene. Int J Oncol 24:419–423
Mishra M, Chandra R, Pati R (2008) In vitro regeneration and genetic fidelity testing of *Aegle marmelos* (L.) Corr. Plants. Ind J Hort 65(1): 6–11
Mishra M, Chandra R, Pati R, Jain RK, Agarawal Sulekha (2010) Shoot tip transformation of papaya (*Carica papaya* L.). Acta Hort 851:219–226
Morton CM (2009) Phylogenetic relationships of the Aurantioideae (Rutaceae) based on the nuclear ribosomal DNA ITS region and three noncoding chloroplast DNA regions, atpB-rbcL spacer, rps16, and trnL-trnF. Org Divers Evol 9(1):52–68
Morton CM, Grant M, Blackmore S (2003) Phylogenetic relationships of the Aurantioideae inferred from chloroplast DNA sequence data. Am J Bot 90(10):1463–1469
Mueller RL (2004) First-generation agents: aspirin, heparin and coumarins. Best Pract Res Clin Haematol 17:23–53
Murray MG, Kennard WC (1977) Altered chromatin conformation of the higher plant gene phaseolin. Biochemistry 23:4225–4423
Narayanan N, Thirugnanasambantham P, Viswanathan S, Vijayasekaran V, Sukumar E (1999) Antinociceptive, anti-inflammatory and antipyretic effects of ethanol extract of *Clerodendron serratum* roots in experimental animals. J Ethnopharmacol 65(3):237–241
Nayak P, Behera PR, Manikkannan T (2007) High frequency plantlet regeneration from cotyledonary node cultures of *Aegle marmelos* (L.) Corr. In Vitro Cell Dev Biol Plant 43(3):231–236
Oudhia P (2005) Bael (*Aegle marmelos* syn. *Crataeva marmelos*) as medicinal herb in Chhattisgarh, Research Notes, Article 107. www.botanical.com
Paarlberg R (2003) Reinvigorating genetically modified crops. Issues in ScI Techno 19(3):86–92
Paricha S (2004) Bael (*Aegle marmelos*) nature's most natural medicinal fruit, Orissa Rev 41:16–17
Parmar C, Kaushal MK (1982) *Aegle marmelos* Correa. In: Wild fruits of the sub-Himalayan region. Kalyani Publishers, New Delhi, pp 1–5
Paszkowski J, Potrykus I, Hohn B, Shillito RD, Hohn T et al. (1995) Transformation of hereditary material in plants. US Patent 5, 453, 367
Pati R, Verma J (2009) Genetic engineering: a new vistas to support feeding the global population. Plant Arch 9(2):531–536
Pati R, Chandra R., Chauhan UK, Mishra M (2008b) In vitro plant regeneration from mature explant of *Aegle marmelos* Corr.) cv. CISH-B2. Sci Cult 74(9–10): 359–367
Pati R, Chandra R, Chauhan UK, Mishra M, Srivastiva N (2008a) In vitro clonal propagation of bael (*Aegle marmelos* Corr.) cv. CISHB1 through enhanced axillary branching. Physiol Mol Biol Plants 14(4): 337–346
Pati R, Mishra M, Pandey D Chandra R (2004) Beal ka aushawadhiya mahatava. Fal Phooll 27(3):28–29
Peet RC (1995) Protection of plant-relatedinventions in the United States. http://biotechlaw.ari.net/plnt-ovr.html
Penjor T, Anai T, Nagano Y, Matsumoto R, Yamamoto M (2010) Phylogenetic relationships of citrus and its relatives based on rbcL gene sequences. Tree Genet Genomes 6:931–939

Ponnachan PTC, Paulose CS, Panikkar KR (1993) Effects of leaf extract of *Aegle marmelos* in diabetic rats. Ind J Exp Biol 31:345–347

Prematilake DP, Nilmini HAS, Kudagamage C (2006) Establishment of an in vitro plant regeneration system for *Aegle marmelos* (L.) Corr. via organogenic callus culture. Cey J Sci (Bio Sci) 35 (1): 87–90

Purohit SS, Vyas SP (2004) *Aegle marmelos* Correa ex Roxb. (Bael), medicinal plant cultivation-a scientific approach. Agrobios, Jodhpur, pp 280–285

Ramanathan T, Gurudeeban S, Satayavani k (2010) In vitro plant regeneration from leaf primordial of gum bearing tree *Aegle marmelos*. Res J Forestry 4(4):208–212

Rana BK, Singh UP, Taneja V (1997) Antifungal activity and kinetic of inhibition by essential oil isolated from leaves of *Aegle marmelos*. J Ethnopharmacol 57(1):29–34

Reddy KN, Reddy CS, Trimurthulu S (2006) Ethnobotanical survey on respiratory disorders in Eastern Ghats of Andhra Pradesh, India, www.sis.edu/~ebl/leaflets/reddy.htm

Roy SK, Singh RN (1980) Studies on changes during development and ripening of Bael fruit. Punjab Hortic J 20(3&4):190–197

Salvo G, Manafzadeh S, Ghahremaninejad F, Tojibaev K, Zeltner L, Conti E (2011) Phylogeny, morphology, and biogeography of Haplophyllum (Rutaceae), a species-rich genus of the Irano-Turanian floristic region. Taxon 60(2):513–527

Samuel R, Ehrendorfer F, Chase MW, Greger H (2001) Phylogenetic analyses of Aurantioideae (Rutaceae) based on non-coding plastid DNA sequences and phytochemical features. Plant Biol 3:77–87

Sandhya G, Devendra C, Manju B (2008) Micropropagation of Bael [*Aegle marmelos* (L.) Corr.]—an indigenous medicinal fruit tree of India. Indian J Plant Genet Resour 21(3):213–216

Schilperoort RA, Hoekema A, Hooykaas PJJ (1990) Process for the incorporation of foreign DNA into the genome of dicotyledonous plants. US Patent 4,940,838

Scott KD, McIntyre CL, Playford J (2000) Molecular analyses suggest a need for a significant rearrangement of Rutaceae sub-families and a minor reassessment of species relationships within Flindersia. Plant Syst Evol 223(1–2):15–27

Setia G, Luthra P, Sharma PC (2005) Siddha system: an ancient heritage of India. In: Proceedings of national seminar on role of medicinal and aromatic plants in ayurvedic, unani and siddha systems of medicine, Hisar, India, pp 11–14

Shailajan S, Menon S, Hande H (2011) Method validation of Marmelosin from fruit pulp of *Aegle marmelos* (L.) Correa using HPTLC Technique. J Pharm Res 4(5):1353–1355

Shankar G, Garg KL (1967) Nutritional value of some important fruits. In: Handbook of Horticulture, Kitabistan, Allahabad, pp 37–41

Shoba FD, Thomos M (2001) Study of antidiarrhoeal activity of four medicinal plants in castor-oil induced diarrhea. J Ethnopharmacol 76(1):73–76

Srivastava N, Gupta VK, Pati R, Gaur RK (2011) Genetically modified crops: an overview. Biotechnol 10:136–148

Tandon V, Thayil S (1995) Saving of medicinal plants in South India. Plants Talk 2:16–17

Tiwari NN, Joshi MP (1990) Medicinal plants of Nepal II. J Nep Med Assoc 28:181–190

Tuticorin RG, Manakkal SP (1983) Some alkaloids from *Aegle marmelos*. Phytochem 22(3):755–757

U.S. Environmental Protection Agency, Biopesticides Registration Action Document: *Bacillus thuringiensis* Plant-Incorporated Protectants, Environmental Protection Agency, 2001. Available at http://www.epa.gov/ pesticides/biopesticides/pips/bt_brad.htm as of May 2004

Wu F (2002) Tools for regulatory decisions concerning genetically modified corn, Ph.D. dissertation, Carnegie Mellon University, Pittsburgh

Yadav K, Singh N (2011) In vitro propagation and biochemical analysis of field established wood apple (*Aegle marmelos* L.). Analele Universităţii din Oradea—Fascicula Biologie Tom XVIII(1):23–28

Yadav SK, Jain AK, Tripathi SN, Gupta JP (1989) Irritable bowel syndrome: therapeutic evaluation of indigenous drugs. Ind J Med Res 90:496–503

Chapter 15
Genetic Transformation of Taro

Xiaoling He, Maureen M. M. Fitch, Yun J. Zhu and Susan C. Miyasaka

Abstract Taro (*Colocasia esculenta* (L.) Schott) is cultivated worldwide for its edible corms and leaves. It was the world's fourteenth most-consumed vegetable and the fifth most-produced root crop in the world during 2010. However, various pests and diseases, especially fungal and oomycete diseases, are major problems causing steep declines in taro production. Conventional breeding of disease resistant cultivars is ongoing, although it is a lengthy process. Tissue culture and genetic transformation of taro are alternative options to improve yields, quality, and disease resistance. Compared with conventional breeding, genetic engineering has unique advantages, such as a much broader gene pool for selection of genes of interest and the capability of transferring only a few transgenes, thus maintaining all other desirable crop characteristics. Only a few reports are available on the regeneration and genetic transformation of taro. The first report of taro transformation described insertion of a reporter *gus* gene and a selection gene *hpt* into a Japanese taro cultivar via particle bombardment with a very low transformation efficiency. More recently, particle bombardment and *Agrobacterium*-mediated transformation methods have been used to transform a Chinese taro cultivar with a disease resistance gene *chi*11 from rice. The *Agrobacterium*-mediated method had much higher transformation efficiency than particle bombardment. Insertion of this rice chitinase gene into taro resulted in moderately increased disease resistance against the fungal pathogen *Sclerotium rolfsii*. These results demonstrate the potential usefulness of genetic transformation to increase

X. He · M. M. M. Fitch · Y. J. Zhu
Hawaii Agriculture Research Center, 94-340 Kunia Rd, Waipahu HI 96797, USA
e-mail: xhe@harc-hspa.com

M. M. M. Fitch
e-mail: mfitch@harc-hspa.com

Y. J. Zhu
e-mail: jzhu@harc-hspa.com

S. C. Miyasaka (✉)
Department of Tropical Plant and Soil Sciences, University of Hawaii,
875 Komohana St, Hilo HI 96720, USA
e-mail: miyasaka@hawaii.edu

S. M. Jain and S. Dutta Gupta (eds.), *Biotechnology of Neglected and Underutilized Crops*, DOI: 10.1007/978-94-007-5500-0_15,

disease resistance of taro, particularly in instances where there are no naturally occurring resistances within the taro germplasm or the elite taro cultivars are difficult to breed conventionally.

15.1 Introduction

Taro [*Colocasia esculenta* (L.) Schott] is a member of the Araceae family and is one of the oldest food crops in the world. Archaeological evidence indicated that taro probably originated in India and adjacent areas of Southeast Asia approximately 50,000 years ago (White and O'Connell 1982). Today taro is widely cultivated throughout the South Pacific, South America, Asia, Africa, and the Caribbean (Caillon et al. 2006; Kreike et al. 2004; Quero-Garcia et al. 2006; Wang 1983). Taro was the world's fourteenth most-consumed vegetable and the fifth most-produced root crop in the world during 2010, with global production of 9.0 billion kg (FAO 2010).

The underground stem of taro, commonly known as a corm or cormel, is consumed for its easily digested starch. In comparison with potato and other starchy crops, taro corms or cormels have much smaller starch grains (one-tenth that of potato) that are fairly rich in the soluble starch amylose. The soluble amylose is excellent for people with digestive problems (Perez et al. 2005; Sefa-Dedeh and Sackey 2002). Further, the taro corm has a high proportion of protein (1.5–3.0 %) and is a good source of carbohydrates, calcium, potassium, iron, phosphorus, and fiber (Gibson 1999; Hussain et al. 1984; Shewry 2003). In addition, taro leaves serve as a vegetable, providing good sources of dietary fiber, vitamins A, B, C, and calcium (Ferguson et al. 1992; Pinto et al. 2000a, b).

Taro corms contain other healthful chemical constituents, such as anthocyanins reported to have antioxidant and anti-inflammatory properties (e.g., ancyanidin 3-glucoside, pelargonidin 3-glucoside and cyanidin 3-rhamnoside) (Cambie et al. 2003). In addition, taro extracts containing monogalactosyl diacylglycerol (MGDG) and digalactosyl diacylglycerol (DGDG) were reported to exhibit anti-hyperlipemia activity and anti-tumor-promoting activity (Tanaka et al. 2005). Recently, Kundu et al. (2012) reported that a water-soluble extracts of the taro corm could inhibit the proliferation of some breast and prostate cancer cell lines and completely blocked the migration of tumor cells.

Taro leaves also have been found to contain healthful constituents. Ferguson et al. (1992) found that the dietary fiber from taro leaves adsorbed a hydrophobic mutagen, suggesting one important reason for the much lower incidence of colorectal cancer in Pacific Islanders who eat taro as a staple food compared to European populations. Recently, Nair et al. (2005) evaluated the antibacterial activity of several plant extracts to determine their therapeutic potential, and found that taro leaf extracts showed the strongest antibacterial activity against a drug-resistant bacterial strain of *Klebsiella pneumonia*, the causal agent of several infections including pneumonia.

Taro corms and leaves are commonly used to make various dishes in many countries, especially in East Asia and the Pacific Islands. In China, taro corms are used to make ice cream products which are appreciated by many consumers due to their unique flavor, smooth texture, and special color (Lu et al. 2002). Also, the taro corm is an important base for industrial food products such as taro flour, chips and baby food (Gibson 1999; Huang 2000; Hollyer et al. 2000; Nip et al. 1995). Taro flour is used in infant formula and canned baby foods, and is good for people with allergies, such as lactose intolerance (Huang 2000; Sefa-Dedeh and Sackey 2002; Standal 1983). In Hawaii, indigenous food products include poi (cooked and mashed taro corm with water), kulolo (a pudding made with grated taro corm and shredded coconut), and laulau (a bundle of 5–6 taro leaves containing pork and fish). Recently, Brown et al. (2005a) demonstrated that poi significantly inhibited colon cancer cell growth and non-specifically activated lymphocytes, indicating that poi had anti-cancer effects. Moreover, Brown et al. (2005b) suggested that poi consumed as a non-dairy probiotic by cancer patients had greater therapeutic advantages than yogurt.

15.1.1 Major Diseases of Taro

Pathogens of taro include oomycetes, fungi, bacteria, nematodes, viruses, and mycoplasmas, and they are responsible partly for worldwide declines in taro production (Ooka 1994; Revill et al. 2005). A lethal virus of taro (i.e., Alomae or Chuaka) is found in the Solomon Islands (Kastom Gaden Association 2005). Due to global transportation and economic trade, taro pathogens have the potential of spreading rapidly throughout the world.

Oomycete and fungal diseases are the most economically significant among the taro disease classes (Ooka 1994; Philemon 1994). In Hawaii, there are four oomycete and fungal diseases that cause major losses of taro yield ranging from 25 to 50 %: Taro Leaf Blight (TLB); Taro Pocket Rot (TPR); Soft Rot; and Southern Blight (Takahashi 1953; Trujillo 1967; Miyasaka et al. 2001). Taro Leaf Blight, caused by the oomycete pathogen *Phytophthora colocasiae* Rac., usually occurs during rainy or overcast weather and reduces yields significantly (Miyasaka et al. 2012; Nelson et al. 2011; Ooka 1994). Taro Leaf Blight invaded Hawaii during the 1920s, resulting in extinction of many TLB-susceptible, traditional Hawaiian cultivars (Cho et al. 2007; CTAHR 2009). In 1993, this disease was introduced accidentally to American and Western Samoa, where it devastated yields of the traditional, highly susceptible taro cultivars during 1994–1998 (Nelson et al. 2011; Trujillo and Menezes 1995; Trujillo 1996). Taro production in these South Pacific Islands resumed only after the introduction of disease-resistant cultivars.

Recently, a new species of *Phytophthora* was found to cause Taro Pocket Rot (TPR) in wetland-grown taro in Hawaii (Uchida et al. 2002). This disease results in pockets of diseased tissues in the corm. Taro production in Hawaii during 2005 hit a record low, and two reasons given were increases of TPR and TLB fostered by unfavorable rainy weather conditions (NASS 2006).

Soft Rot is caused by several oomycete pathogens: *Pythium aphanidermatum* Fitzpatrick; *P. graminicola subramaniam*; *P. splendens* Braun; *P. irregulare* Buisman; *P. myriotylum* Drechsler; *P. carolinianum* Matthews; and *P. ultimum* Trow. This disease causes the taro corm to form a soft, often malodorous mass (Ooka 1994). Southern Blight, caused by the fungal pathogen *Sclerotium rolfsii* Sacc., is generally a problem of dryland-grown taro and results in corm rots and stunted plants (Ooka 1994).

15.2 Improvement by Conventional Breeding

Taro as a species is characterized by rare and erratic flowering (Ivancic et al. 2004). Flowers of taro consist of a spadix (i.e. small, individual flowers on a fleshy spike) with the female, pistillate flowers located at the base of the spadix and male, staminate flowers are located near the top (Strauss 1983). Typically, cross-pollination is required, because the female flowers become receptive before the pollen is shed (Ivancic and Lebot 2000). Modern conventional taro breeding programs started during the 1970s after the discovery that treatment with gibberellic acid (GA) induced flowering and allowed synchrony of flowering for hand-pollination (300–1000 mg GA L^{-1}) (Ivancic and Lebot 2000; Wilson 1979a). Hand-pollination is required, because specialized insect pollinators are either rare or non-existent outside of the Solomon Islands and Papua New Guinea (Plucknett 1970). Although taro can be self-pollinated by hand, loss of vigor occurs within the next generation (Wilson 1979b). To avoid inbreeding depression, Ivanic and Lebot (2000) recommended that when backcrossing, the recurrent parent should be alternated with another similar genotype.

Conventional breeding could improve yields, quality, and disease resistance of taro, provided that: (a) naturally occurring genes of interest are found within the taro gene pool; and (b) selected parents either flower naturally in synchrony or can be induced to flower in synchrony through GA application. For example, the commercial cultivar ‘Bun Long’ flowers only rarely under the environmental conditions of Hawaii, even with GA application, making it difficult to breed conventionally.

To prevent inbreeding depression, a better understanding of the genetic diversity in the taro germplasm is essential when selecting parents. Since taro was brought to Hawaii and many other South Pacific Islands as a canoe plant, it has a narrow genetic base (Lebot and Aradhya 1991). Greater genetic diversity of taro has been found in Asia and Southeast Asia as shown by random amplified polymorphic DNA (RAPD), isozymes (Ochiai et al. 2001; Xu et al. 2001), and amplified fragment length polymorphism (AFLP) (Kreike et al. 2004).

Resistance to TLB has been found within the taro germplasm and conventional breeding efforts to increase TLB resistance have been successful (Cho et al. 2007;

Trujillo and Menezes 1995; Trujillo 1996). However, preliminary observations of these new cultivars indicated that environmental conditions that are conducive to the disease could still break down the disease resistance and particular cultivars are more susceptible to other diseases such as Soft Rot (Trujillo et al. 2002). These issues over quality and susceptibility to other pests indicate the importance of maintaining or recovering desired traits of the elite cultivars (background selection) during the process of introgression of desired genes (foreground selection).

Conventional breeding of taro is a lengthy process, since it is a 6–13 month crop (Plucknett 1970). Genetic markers associated with the desired traits could be used to increase the efficiency of conventional breeding. Quero-Garcia et al. (2006) presented the first taro genetic map containing 161 amplified fragment length polymorphisms (AFLPs) and 8 simple-sequence repeats (SSRs). They identified several quantitative trait loci (QTLs) for corm yield and corm dimensions. Recently, Sharma et al. (2008) analyzed AFLP markers associated with TLB-resistance genes.

Only a few genes or their coding proteins that are related to yield, quality, or disease resistance have been isolated and identified from taro. Lin and Jeang (2005a, b) isolated, expressed and characterized a soluble starch synthase I gene and a novel soluble starch synthase II gene. Researchers have isolated genes for tarin, a family of storage proteins that accounts for approximately 40 % of the total soluble proteins in taro corms (Guimaraes et al. 2001; Monte-Neshich et al. 1995). In addition, a gene that encodes a tarin isoform was isolated and characterized in taro; it shared a high homology with lectins that are thought to play a role in defense (Bezerra et al. 1995; Guimaraes et al. 2001; Van Damme et al. 1995; Powell 2001). A novel, mannose-binding lectin gene from *Arisaema heterophyllum*, which belongs to the Araceae family, has been cloned and characterized (Zhao et al. 2003). A novel phytocystatin gene was cloned from the fungal resistant taro cultivar 'Kaosiung no. 1', and the expressed cystatin showed antifungal activity against the pathogen *Sclerotium rolfsii* (Yang and Yeh 2005).

15.3 Tissue Culture of Taro

Traditionally, taro is propagated vegetatively through corms or cormels, and approximately 10 % of the yield from a previous crop is used for propagation (de la Pena 1983). An efficient tissue culture system is important for the multiplication of desired cultivars. In addition, it is critical for obtaining virus-free plants. Dasheen Mosaic disease is caused by the dasheen mosaic virus (DMV), and it is known to decrease taro corm yields (Malamug et al. 1992; Ooka 1994). The DMV is carried by several aphid species (e.g., *Myzus persicae* Sulzer, *Aphis craccivora* Koch., and *A. gossypii* Glov.) and there are no known taro varieties that are immune to this virus (Ooka 1994; Philemon 1994). Many protocols of shoot tip culture have been found to successfully eliminate DMV from infected taro shoots (Hartman 1974; Hain 1991; Jackson et al. 1977; Mapes and Cable 1972;

Sabapathy and Nair 1995; Yam et al. 1990). He et al. (2008, 2010) modified earlier methods as follows: 0.5–1.5 mm shoot tips were excised from corms under the microscope, and then surface-sterilized for very short time (16s). Using this quick method, no visible contamination was observed and most explants remained healthy after surface-sterilization.

Different procedures for in vitro culture of various taro cultivars have been developed (Table 15.1; Chand et al. 1999; Deo et al. 2009, 2010; Hartman 1974; Hain 1991; Jackson et al. 1977; Malamug et al. 1992; Mapes and Cable 1972; Murakami et al. 1995; Sabapathy and Nair 1995; Yam et al. 1990). In addition, plantlet regeneration via callus (Malamug et al. 1992; Sabapathy and Nair 1995), via protoplasts (Murakami et al. 1995), and via somatic embryogenesis (Deo et al. 2009, 2010) has been established. However, certain taro cultivars are difficult to tissue culture.

In addition to multiplication of virus-free planting materials, tissue culture is essential for the genetic transformation of plants. He (2006) tested two commercially important taro cultivars 'Maui Lehua' and 'Bun Long' to develop a tissue culture system capable of producing highly regenerative calli. Medium M5 [MS (Murashige and Skoog 1962) + 2 mg L^{-1} benzyladenine (BA) + 1 mg L^{-1} napthaleneacetic acid (NAA)] resulted in highly regenerative calli of 'Bun Long,' and medium M15 (MS + 4 mg L^{-1} BA) resulted in induction of multiple shoots from calli. In contrast, none of the 40 media tested produced highly regenerative calli of 'Maui Lehua', illustrating the differences between cultivars in response to phytohormones.

15.4 Genetic Transformation

Compared with conventional breeding, genetic engineering has significant advantages. It can improve crop yields, quality or disease resistance via insertion of a single gene or a few genes of interest into target plants. In principle, the genes of interest could come from any organism: plants, animals, or microorganisms. Therefore, there is no species barrier and gene pools are enriched. Unlike conventional breeding, where half the genome is transferred from each parent, resulting in many undesirable traits being exhibited in progeny, genetic engineering enables insertion of only one or a few genes of interest, thus preserving all the desirable background traits of the elite cultivar. In practice, genetic engineering has achieved huge advances globally in recent decades. For example, the papaya 'Rainbow' was transformed with a resistance gene against Papaya Ringspot Virus, saving the Hawaiian papaya industry from a devastating epidemic caused by the virus (Goldstein 2004).

Various plant transformation methods have been developed successfully to improve the quality, yield, disease resistance, and other agronomic traits of a wide range plant species (Taylor and Fauquet 2002; Veluthambi et al. 2003). These transformation methods can be divided into two major methods: (a) direct DNA

Table 15.1 Various tissue culture methods to induce shoots (*SIM*), induce callus (*CIM*), culture protoplasts (*PCM*), or induce somatic embryogenesis (*SEM*) in taro

Cultivars	Explants	Media (mg L^{-1}) [a, b, c]	References
Ninu	Shoot tips	SIM: MS + coconut water + 1.13 BA + 0.63 TDZ	Chand et al. 1999
CPUK	Corm slices	CIM: ½ MS + 2 2,4-D + 1 TDZ	Deo et al. 2009, 2010
		SEM : ½ MS + 0.05 2,4-D + 0.1 TDZ + 1 biotin	
		SIM: MS + 0.05 BA + 0.1 IAA	
Eguimo	Shoot tips	CIM: LS + 2BA + 1NAA	Fukino et al. 2000
		CPM: liquid LS	
		SIM: LS	
Kaohsiung 1	Shoot tips	CIM: MS + 2BA + 0.02 NAA	Hain 1991
		SIM: MS + 2.5 kinetin + 0.02 NAA + 10 putrescine	
Unknown	Shoot tips	CIM: MS + 1 kinetin + 15 IAA	Hartman 1974
		SIM: MS + 1 kinetin + 15 IAA	
Bun long	Shoot tips	CIM: MS + 2BA + 1 NAA	He et al. 2008, 2010
		SIM: MS + 4 BA	
Dotare	Apical shoots	CIM: Nitsch + 1 2,4-D + 1 BA	Malamug et al. 1992
		SIM: Nitsch + 1 BA	
Eguimo	Etiolated stems	CIM: MS + 2 2,4-D + 2 2ip	Murakami et al. 1995
		PCM: ½ MS + KM + 2BA	
		SIM: MS + 0.2NAA + 2BA	
Keladi birah	Shoot apices	CIM: LS + 2 2,4,5-T + 0.2 kinetin	Sabapathy and Nair 1995
		SIM: LS + 2 kinetin	

(continued)

Table 15.1 (continued)

Cultivars	Explants	Media (mg L^{-1}) [a, b, c]	References
Akalomamale	Axillary buds	CIM: MS + 20 TE + 2 NAA + 0.2 BA	Yam et al. 1990
		SIM: MS + 20 TE	
Akalomamale	Axillary buds	CIM: ½ MS + 25TE + 2 2,4,5-T + 200 glutamine	Yam et al. 1991
		SIM: 1/2MS + 100 ml coconut water + 25 TE	
Unknown	Axillary buds	SIM: ½ MS + 2.2–4.4 BA	Zhou et al. 1999

[a] Abbreviations for types of medium as follows: CIM: Callus Inducing Medium; PCM: Protoplast culture medium; SEM: Somatic Embryogenesis Medium; SIM: Shoot Inducing Medium

[b] Abbreviations for specific medium: KM: organic substances of Kao and Michayluk (1975); LS: Linsmaier and Skoog (1965) medium; MS: Murashige and Skoog medium (Murashige and Skoog 1962); TE: taro extract (Yam et al. 1990)

[c] Abbreviations for phytohormones: BA: benzyladenine; IAA: indoleacetic acid; NAA: naphthaleneacetic acid; TDZ: thidiazuron; 2,4-D: 2,4-dichlorophenoxyacetic acid; 2,4,5-T: trichlorophenoxyacetic acid

delivery transformation; and (b) *Agrobacterium*-mediated transformation. Of the direct gene transformation methods, particle bombardment (also called micro-projectile bombardment or biolistic) transformation is the most widely used and developed tool to date (Taylor and Fauquet 2002; Veluthambi et al. 2003). The major advantage of particle bombardment transformation is that there is no incompatibility problem as occurs between *Agrobacterium* and certain plant species.

Agrobacterium transfers a segment of its plasmid DNA (T-DNA) and integrates it into the target plant genome (Veluthambi et al. 2003; Zupan et al. 2000). A range of *Agrobacterium* transformation strains and binary transformation vectors have been developed, including strains that harbour the so-called 'supervirulent' *vir* genes that can infect a broader range of host plants and result in a greater transformation efficiency (Veluthambi et al. 2003; Zupan et al. 2000). Earlier, a major limitation of *Agrobacterium*-mediated transformation was due to the fact that most monocotyledonous plants are not natural hosts of *Agrobacterium*, because they do not secret the molecular signal elicitors (e.g., acetosyringone) that are needed to induce *vir* genes. However, addition of acetosyringone and use of more efficient *Agrobacterium* strains have resulted in a wide range of transgenic monocotyledonous plants, including rice, maize, wheat and barley (Luo et al. 2004; Veluthambi et al. 2003). *Agrobacterium*-mediated transformation has the following advantages over particle bombardment: (a) it is simpler and less expensive; (b) it integrates a single copy or a low copy number of the transgene, resulting in lower incidences of gene rearrangement and gene silencing (Luo et al. 2004; Veluthambi et al. 2003); and (c) it is capable of transferring a larger sized transgene or several transgenes linked together. In contrast, particle bombardment transformation usually inserts multiple copies of the transgene into plant genome, resulting in greater risk of gene silencing and frequent truncation of the transgene DNA.

Very little research has been conducted on genetic engineering of taro. To date, only a few workers have reported genetic transformation of taro (Table 15.2). Fukino et al. (2000) was the first research group to transform taro, using particle bombardment to insert a reporter gene, ß-glucuronidase (*gus*) gene and a selection gene hygromycin phosphotransferase gene *hpt* into a Japanese triploid taro cultivar 'Eguimo'. However, the efficiency of transformation was very low, less than 1 % (Fukino et al. 2000).

A second transformation method was developed using *Agrobacterium* to transform a Chinese taro cultivar 'Bun Long' using a rice chitinase gene *chi*11 (He 2006; He et al. 2008). The plasmid pBI121/*chi*11 that was used for the transformation contained the *npt*II selection gene, the *gus* reporter gene and a rice chitinase gene *chi*11 driven by the *CaMV* 35S promoter (Fig. 15.1). Geneticin (G418) was used for the selection of transformed calli and plants, because the *npt*II gene confers resistance to this toxin. The transformation process is summarized in the flow chart (Fig. 15.2).

A comparison of both methods revealed the greater transformation efficiency of the *Agrobacterium*-mediated method for taro. Particle bombardment resulted in only one transgenic line (0.5 % efficiency) (He 2006; He et al. 2008). In contrast, the *Agrobacterium*-mediated method resulted in six transgenic lines (3 % transformation

Table 15.2 Genetic transformation of taro (*Colocasia esculenta*)

Cultivar and Transformation materials	Transformed gene	Transformation method [a]	Transformation efficiency (%)	Reference
Var. *esculenta* CPUK Somatic embryos	*gfp* reporter gene	PB and AMT	Unknown	Deo 2008
Triploid 'Eguimo' Regenerative calli	*gus* reporter gene *hpt* selection gene	PB	0.1	Fukino et al. 2000
'Bun Long' Regenerative calli	*gus* reporter gene *npt*II selection gene rice chitinase gene *chi* 11 wheat oxalate oxidase gene *gf*2.8	PB and AMT	0.05–3	He 2006
'Bun Long' Regenerative calli	*gus* reporter gene *npt*II selection gene rice chitinase gene *chi* 11	AMT	3	He et al. 2008
'Bun Long' Regenerative calli	*gus* reporter gene *npt*II selection gene rice chitinase gene *chi* 11	PB	0.5	He et al. 2010

[a] Abbreviations: PB = particle bombardment; AMT = *Agrobacterium*-mediated transformation

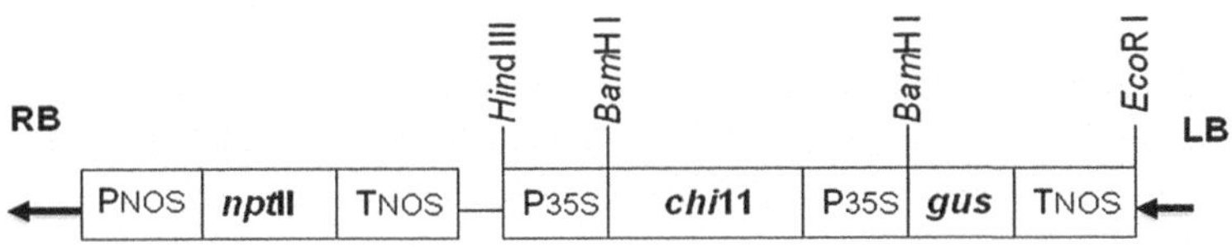

Fig. 15.1 Diagram of the transformation plasmid pBI121/*chi*11 consisting of the *npt*II selection gene, the *gus* reporter gene and a rice chitinase gene *chi*11 driven by the *CaMV* 35S promoter

Regenerative calli on medium M5

↓

Co-cultivated with *Agrobacterium* strain EHA 105/pBI121/*chi*11 (4 days)

↓

Selection on medium M5 with G418 (50 mg L^{-1}) (3 months)

↓

GUS test and PCR analyses for screening of transformed calli

↓

Shoots induced from transformed calli on medium M15 with G418 (50 mg L^{-1}) (3 months)

↓

PCR and Southern blot analyses for confirmation of independent transgenic lines

↓

Multiply shoots of independent transgenic lines on medium M15

Fig. 15.2 The flow chart of the *Agrobacterium*-mediated transformation of taro. Abbreviations: *G*418 = geneticin; Medium *M5* = *MS* (Murashige and Skoog 1962) +2 mg L^{-1} benzyladenine (*BA*) +1 mg L^{-1} napthaleneacetic acid (*NAA*); medium *M*15 = *MS* + 4 mg L^{-1} *BA*; *PCR* = polymerase chain reaction

efficiency) (He et al. 2010). In addition, the transformation via *Agrobacterium* probably will be more stable due to a single-copy or a low-copy insertion of the transgene (13 copies in the taro line transformed via particle bombardment *versus* 1 or 2 copies in the lines transformed via *Agrobacterium*) (He 2006; He et al. 2008, 2010). Recently, Deo (2008) in a Ph.D. thesis reported a transformation system with much higher transformation efficiency using a somatic embryogenesis method; however, the molecular characterization of these transgenic embryos has not been conducted yet to confirm the presence and integration of transgenes.

15.4.1 Transformation of Taro With Disease Resistance Genes

To transform taro for increased disease resistance, a rice chitinase gene was initially selected (He et al. 2008, 2010). Chitinase breaks down chitin, a major structural component in insect exoskeletons, crustacean shells, egg shells of nematodes, and the cell walls of many fungi such as Zygomycetes (e.g. *Mucor*, *Rhizopus*), Pyrenomycetes (e.g. the powdery mildews), Discomycetes (e.g. *Sclerotinia*), Hyphomycetes (e.g. *Botrytis*, *Penicillium*, and *Trichoderma*), and Agonomycetes (e.g. *Rhizoctonia*) (Gooday 1990; Riccardo and Muzzarelli 1999; Ruiz-Herrera 1992). In contrast, no chitin has been found in higher plants (Muzzarelli 1999; Ruiz-Herrera 1992).

Evidence indicates that plant chitinases are directly or indirectly associated with plant disease resistance against a wide range of fungal pathogens. Several transgenic plants with inserted chitinase genes have significantly increased resistance to fungal disease. For example, Kishimoto et al. (2002) developed a transgenic cucumber (*Cucumis sativus*) with a chitinase cDNA (RCC2) of rice, and demonstrated increased disease resistance of this transgenic plant to gray mold (*Botrytis cinerea*). Transgenic rice (*Oryza sativa*) with class I rice chitinase gene *chi11* showed increased resistance to *Rhizoctonia solani*, the rice sheath blight pathogen (Lin et al. 1995). Kumar et al. (2003) developed a simple and high throughput method for functional assessment of resistance against *Rhizoctonia solani*, and further confirmed that transgenic rice with the *chi11* gene had enhanced fungal disease resistance.

He et al. (2008, 2010) selected the rice chitinase gene *chi11* to transform taro 'Bun Long' through both particle bombardment and *Agrobacterium*-mediated transformation methods. All transgenic plants showed significantly increased resistance to the fungal pathogen *Sclerotium rolfsii* that causes Southern Blight. To our knowledge, this was the first report on transformation of taro with a transgene that conferred greater disease resistance.

Another disease resistance gene, a wheat oxalate oxidase gene (*oxo* gene), was successfully transformed into the taro cultivar Bun Long (He 2006). Transgenic plants showed significantly increased disease resistance to both the fungal pathogen *S. rolfsii* and the oomycete pathogen *Phytophthora colocasiae* that is the causal agent of Taro Leaf Blight.

15.4.2 Controversies Surrounding Genetic Engineering of Taro

Earlier, three promising new taro cultivars with greater resistance to Taro Leaf Blight were developed in Hawaii using conventional breeding and patented (Trujillo et al. 2002). However, a controversy developed over intellectual property rights of indigenous people, since one of the parents was a modern Hawaiian cultivar. In response, the University of Hawaii decided to release the patent rights into the public domain (Bhattacharjee 2006; CTAHR 2009).

Taro is highly valued culturally by native Hawaiians, and controversy erupted over the genetic engineering of taro, although only a Chinese taro cultivar had been transformed (CTAHR 2009). One concern was over the possible accidental movement of transgenes from transformed Chinese taro lines to Hawaiian cultivars. Based on scientific facts, this risk was practically nil in Hawaii, because: (a) the cultivar 'Bun Long' flowers rarely under the environmental conditions of Hawaii even with application of GA; and (b) taro is typically propagated vegetatively.

The State of Hawaii Legislature ranked first among states in the United States during 2005–2006 in the number of bills dealing with the regulation of agricultural biotechnology, including proposed moratoriums on transgenic taro (Pew Initiative on Food and Biotechnology, 2007). Although no bills that regulated genetic engineering of taro passed the State Legislature, bans on such research were passed by the County Councils of Hawaii (Armstrong 2008) and Maui (Tanji 2009). As a result of these recent controversies and bans, research on genetic transformation of taro is being discontinued in Hawaii.

15.5 Conclusion and Future Perspectives

Little research has been conducted on the genetic transformation of taro; only three peer-reviewed papers have been published to date on this subject. Transformation of a Chinese taro cultivar 'Bun Long' with a rice chitinase gene resulted in transgenic lines that all showed significantly increased disease resistance against the fungal pathogen *Sclerotium rolfsii*, the causal agent of Southern Blight. Taro plants transformed with another disease resistance gene, a wheat oxalate oxidase gene (*oxo* gene) showed significantly increased disease resistance to both the fungal pathogen *S. rolfsii* and the oomycete pathogen *Phytophthora colocasiae.* However, although the genetic transformation of taro has the potential to enhance resistances to diseases that are reducing production in many countries, such research in Hawaii is being discontinued due to recent controversies. Future research on genetic transformation of taro for improved yield, quality, and disease resistance likely will be conducted in other states or countries.

References

Armstrong J (2008) GMO taro, coffee banned: council overrides mayor's veto by unanimous vote. Hawaii Tribune Herald, p A-1–A-6

Bezerra IC, De Castro L, Neshich G, de Almeida E, de Sa MF, Mello LV, Monte-Neshich DC (1995) A corm-specific gene encodes tarin, a major globulin of taro (*Colocasia esculenta* L. Schott). Plant Mol Biol 28:137–144

Bhattacharjee Y (2006) Universal ownership. Newsmakers Sci 313:295

Brown AC, Reitzenstein JE, Liu J, Jadus MR (2005a) The anti-cancer effects of poi (*Colocasia esculenta*) on colonic adenocarcinoma cells in vitro. Phytotherapy Res 19:767–771

Brown AC, Shovic A, Ibrahim S, Holck P, Huang A (2005b) A non-dairy probiotic's (poi) influence on changing the gastrointestinal tract's microflora environment. Altern Ther Health Med 11(1):58–64

Caillon S, Quero-Garcia J, Lescure JP, Lebot V (2006) Nature of taro (*Colocasia esculenta* (L.) Schott) genetic diversity prevalent in a pacific ocean island, Vanua Lava. Vanuatu. Genet Res Crop Evol 53:1273–1289

Cambie RC, Ferguson LR (2003) Potential functional foods in the traditional maori diet. Mutation Res 523(524):109–118

Chand H, Pearson MN, Lovell PH (1999) Rapid vegetative multiplication in *Colocasia esculenta* (L) Schott (taro). Plant Cell Tiss Org Cult 55:223–226

Cho JJ, Yamakawa RA, Hollyer J (2007) Hawaiian kalo, past and future. University of Hawai'i, College of Tropical Agriculture and Human Resources, Honolulu, HI, Sustainable Agriculture, SA-1. Available at: http://www.ctahr.hawaii.edu/oc/freepubs/pdf/SA-1.pdf. Accessed 6 Apr 2012

College of tropical agriculture and human resources (CTAHR) (2009) CTAHR and Taro. University of Hawai'i, CTAHR, Honolulu, HI, USA. Available at: http://www.ctahr.hawaii.edu/oc/freepubs/pdf/CTAHR_and_taro.pdf. Accessed 6 Apr 2012

de la Pena RS (1983) Agronomy. In: Wang JK (ed) Taro, a review of *Colocasia esculenta* and its potentials. University of Hawaii Press, Honolulu

Deo PC (2008) Somatic embryogenesis and transformation in taro (*Colocasia esculenta* var. esculenta). Ph.D. thesis, The university of the south pacific, Suva, Fiji

Deo PC, Harding RM, Taylor M, Tyagi AP, Becker DK (2009) Somatic embryogenesis, organogenesis and plant regeneration in taro (*Colocasia esculenta* var. esculenta). Plant Cell Tiss Org Cult 99:61–71

Deo PC, Taylor M, Harding RM, Tyagi AP, Becker DK (2010) Initiation of embryogenic cell suspensions of taro (*Colocasia esculenta* var. esculenta) and plant regeneration. Plant Cell Tiss Org Cult 100:283–291

Ferguson LR, Roberton AM, Mckenzie RJ, Watson ME, Harris PJ (1992) Adsorption of a hydrophobic mutagen to dietary fiber from taro (*Colocasia esculenta*), an important food plant of the South Pacific. Nutr Cancer 17(1):85–95

Food and Agriculture Organization of the United Nations (FAO) (2010) http://faostat.fao.org/. Accessed 12 March 2012

Fukino N, Hanada K, Ajisaka H (2000) Transformation of taro (*Colocasia esculenta* Schott) using particle bombardment. JARQ 34(3):159–165

Gibson AC (1999) The potato of the humid tropics. http://www.botgard.ucla.edu/html/botanytextbooks/economicbotany/Colocasia/index.html. Accessed 6 Apr 2012

Goldstein C (2004) Cutting through biotech myths. Agric Hawaii 5(3):6–7

Gooday GW (1990) The ecology of chitin degradation. In: Marshall KC (ed) Advances in Microbial Ecology II. Plenum Press, New York

Guimaraes RL, Marcellino LH, de Grossi sa MF (2001) A storage protein gene from taro shows tuber-specific expression in transgenic potato. Physiol Plant 111:182–187

Hain C (1991) The studies of rapid propagation and protoplast culture of taro (Colocasia esculenta Schott). Master's thesis, Research Institute of Horticulture, National Chung-Hsing University, Taichung, Taiwan

Hartman RD (1974) Dasheen mosaic virus and other phytopathogens eliminated from caladium, taro, and cocoyam by culture of shoot tips. Phytopathology 64:237–240

He X (2006) Transformation and regeneration of taro with two plant disease resistance genes: a rice chitinase gene and a wheat oxalate oxidase gene. Ph.D. dissertation. University of Hawaii. ProQuest, UMI number: 3251049

He X, Miyasaka SC, Fitch MM, Zhu YJ, Moore P (2008) *Agrobacterium tumefaciens*-mediated transformation of taro (*Colocasia esculenta* (L.) Schott) with a rice chitinase gene for improved tolerance to a fungal pathgen *Sclerotium rolfsii*. Plant Cell Rep 27:903–909

He X, Miyasaka SC, Fitch MM, Zhu YJ (2010) Regeneration and transformation of taro (*Colocasia esculenta*) with a rice chitinase gene enhances resistance to *Sclerotium rolfsii*. HortSci 45:1014–1020

Hollyer J, Paull R, Huang A (2000) Processing taro chips. University of Hawai'i, College of Tropical Agriculture and Human Resources, Food Manufacturing and Technology, FMT-1

Huang AS (2000) Nutrient composition of taro corms and breadfruit. J Food Compos Anal 13(5):859–864

Hussain M, Norton G, Neale RJ (1984) Composition and nutritive value of cormels of *Colocasia esculenta* (L.) Schott. J Sci Food Agric 35:112–119

Ivancic A, Lebot V (2000) The genetics and breeding of taro. Centre de cooperation internationale en recherché agronomique pour le development (CIRAD). Montpellier, France
Ivancic A, Lebot V, Roupsard O, Quero-Garcia J, Okpul T (2004) Thermogenic flowering of taro (*Colocasia esculenta*, Araceae). Canadian J Bot 82:1557–1565
Jackson GV, Ball EA, Arditti J (1977) Tissue culture of taro, *Colocasia esculenta* (L.). Schott J Hort Sci 52:373–382
Kao KN, Michayluk MR (1975) Nutritional requirements for growth of *Vicia hajastana* cells and protoplasts at a very low population density in liquid media. Planta 126:105–110
Kastom Gaden Association. (2005) People on the edge: a report of the 2005 Kastom Gaden Association assessment of the food security, livelihoods potential and energy resources of the Guadalcanal Weather Cost, Solomon Islands. http://terracircle.org.au/publications/reports/. Accessed 11 March 2012
Kishimoto K, Nishizawa Y, Tabei Y, Hibi T, Nakajima M, Akutsu K (2002) Detailed analysis of rice chitinase gene expression in transgenic cucumber plants showing different levels of disease resistance to gray mold (*Botrytis cinerea*). Plant Sci 162:655–662
Kreike CM, Van Eck HJ, Lebot V (2004) Genetic diversity of taro, *Colocasia esculenta* (L.) Schott, in Southeast Asia and the Pacific. Theor Appl Genet 109:761–768
Kumar KK, Poovannan K, Nandakumar R, Thamilarasi K, Geetha C, Jayashree N, Kokiladevi E, Raja JAJ, Samiyappan R, Sudhakar D, Balasubramanian P (2003) A high throughput functional expression assay system for a defense gene conferring transgenic resistance on rice against the sheath blight pathogen, *Rhizoctonia solani*. Plant Sci 165:969–976
Kundu N, Campbell P, Hampton B, Lin CY, Ma X, Ambulos N, Zhao XF, Goloubeva O, Holt D, Fulton AM (2012) Antimetastatic activity isolated from *Colocasia esculenta* (taro). Anticancer Drugs 23:200–211
Lebot V, Aradhya KM (1991) Isozyme variation in taro *Colocasia esculenta* (L.) Schott from Asia and Oceania. Euphytica 56:55–66
Lin W, Anuratha CS, Datta K, Potrykus I, Muthukrishnan S, Datta SK (1995) Genetic engineering of rice for resistance to sheath blight. Bio/Technol 13:686–691
Lin DG, Jeang CL (2005a) Cloning, expression, and characterization of soluble starch synthase I cDNA from taro (*Colocasia esculenta* var. esculenta). J Agric Food Chem 53:7985–7990
Lin DG, Jeang CL (2005b) cDNA cloning, expression, and characterization of taro SSII: a novel member of starch synthase II family. J Agric Food Chem 53:7958–7964
Linsmaier EM, Skoog F (1965) Organic growth factor requirements of tobacco tissue cultures. Physiol Plant 18:100–127
Lu TJ, Chuang CW, Chang YH (2002) Sensory and physicochemical analyses on commercial taro ice products. J Food Drug Anal 10(1):55–63
Luo H, Hu Q, Nelson K, Longo C, Kausch AP, Chandlee JM, Wipff JK, Fricker CR (2004) *Agrobacterium tumefaciens*-mediated creeping bentgrass (*Agrostis stolonifera* L.) transformation using phosphinothricin selection results in a high frequency of single-copy transgene integration. Plant Cell Rep 22(9):645–652
Malamug JJF, Inden H, Yazawa S, Asahira T (1992) Plantlet regeneration from taro (*Colocasia esculenta* Schott) callus. J Japan Soc Hort Sci 60(4):935–940
Mapes MO, Cable MJ (1972) Mericloning of taro *Colocasia esculenta*. Hawaii Agricultural Station Journal Series No. 1694
Miyasaka SC, Hollyer JR, Kodani LS (2001) Mulch and compost effects on yield and corm rots of taro. Field Crops Res 71:101–112
Miyasaka SC, Lamour K, Shintaku M, Shreshta S, Uchida J (2012) Chapter 20. Taro leaf blight caused by *Phytophthora colocasiae*. In: *Phytophtora*: a global perspective Lamour K (ed.) CABI International, Wallingford UK (In Press)
Monte-Neshich DC, Rocha TL, Guimaraes RL, Santana EF, Loureiro ME, Valle M, de Grossi sa MF (1995) Characterization and spatial localization of the major globulin families of taro (*Colocasia esculenta* L. Schott) tubers. Plant Sci 112:149–159
Murakami K, Kimura M, Matsubara S (1995) Plant regeneration from protoplasts isolated from callus of taro. J Japan Soc Hort Sci 63(4):773–778

Murashige T, Skoog F (1962) A revised medium for rapid growth and bioassays with tobacco tissue cultures. Physiol Plant 15:473–497

Muzzarelli RAA (1999) Native, industrial and fossil chitins. In: Jolles P, Muzzarelli RAA (eds) Chitin and Chitinases., pp 1–6

Nair R, Kalariya T, Chanda S (2005) Antibacterial activity of some selected Indian medicinal flora. Turk J Biol. 29:41–47

National agricultural statistics service (NASS) and department of agriculture state of Hawaii. 2006. Hawaii Taro: taro production hits record low http://www.nass.usda.gov/Statistics_by_State/Hawaii/Publications/Archive/xtar05.pdf. Accessed 07 Apr 2012

Nelson S, Brooks F, Teves G (2011) Taro leaf blight in Hawai'i. University of Hawai'i, College of tropical agriculture and human resources, PD-71, pp 14, Available at: http://www.ctahr.hawaii.edu/oc/freepubs/pdf/PD-71.pdf. Accessed 07 April 2012

Nip WK, Whitaker CS, Vargo D (1995) Application of taro flour in cookie formulations. American Samoa Community College, Land Grant Program, American Samoa

Ochiai T, Nguyen VX, Tahara M, Yoshino H (2001) Geographical differentiation of Asian taro, *Colacasia esculenta* (L.) schott, detected by RAPD and isozyme analyses. Euphytica 122:219–234

Ooka JJ (1994) Taro diseases, a guide for field identification. University of Hawaii, Hawaii Inst Trop Agr Human Res, Res Ext Ser, 148 pp 13

Perez E, Schultz FS, de Delahaye EP (2005) Characterization of some properties of starches isolated from *xanthosoma sagittifolium* (tannia) and *Colocassia esculenta* (taro). Carbohyd Polym 60:139–145

Pew Iniative on Food and Biotechnology (2007) Factsheet: state legislative activity related to agricultural biotechnology in 2005–2006, http://www.pewtrusts.org/uploadedFiles/wwwpewtrustsorg/Reports/Food_and_Biotechnology/PIFB_State_Legislature_2005-2006Session.pdf

Philemon EC (1994) An overview of the pathology of genus *Colocasia*. Papua New Guinea J Agric Fish 37(2):53–61

Pinto NAV, de Carvalho VD, de Vava B, Moraes AR (2000a) Determination of the potential in dietary fibers of the leaves of taro (*Xanthosoma sagittifolium* Schott). Alimentaria 37(312):87–90

Pinto NAV, de Carvalho VD, da Conceicao A, Abreu CMP (2000b) Evolution of the contents of vitamin C and losses with the drying of the leaves of taro (*Xanthosoma sagittifolium* Schott). Alimentaria 37(312):83–85

Plucknett DL (1970) *Colocasia*, *Xanthosoma*, *Alocasia*, *Cyrtosperma*, and *Amorphophallus*. In: Plucknett DL (ed) Tropical Root and Tuber Crops Tomorrow, vol 1. Univ Hawaii, CTAHR, Honolulu, pp 127–135

Powell KS (2001) Antimetabolic effects of plant lectins towards nymphal stages of the planthoppers *Tarophagous proserpina* and *Nilaparvata lugens*. Entomol Exp Appl 99:71–77

Quero-Garcia J, Courtois B, Ivancic A, Letourmy P, Risterucci AM, Noyer JL, Feldmann Ph, Lebot V (2006) First genetic maps and QTL studies of yield traits of taro (*Colocasia esculenta* (L.) Schott). Euphytica 151:187–199

Revill PA, Jackson GVH, Hafner GJ, Yang I, Maino MK, Dowling ML, Devitt LC, Dale JL, Harding RM (2005) Incidence and distribution of viruses of taro (*Colocasia esculenta*) in Pacific Island countries. Australas Plant Path 34:327–331

Ruiz-Herrera J (1992) Fungal cell wall: structure, synthesis, and assembly. CRC Press, Boca Raton, p 248

Sabapathy S, Nair H (1995) In vitro propagation of taro, with spermine, arginine and ornithine II. Plant regeneration via callus. Plant Cell Rep 14(8):520–524

Sefa-Dedeh S, Sackey EK (2002) Starch structure and some properties of cocoyam (*Xanthosoma sagittifolium* and *Colocasia esculenta*) starch and raphides. Food Chem 79:435–444

Sharma K, Mishra AK, Misra RS (2008) Analysis of AFLP variation of taro population and markers associated with leaf blight resistance gene. Acad J Plant Sci 1(3):42–48

Shewry PR (2003) Tuber storage proteins. Ann Bot 91:755–769

Standal BR (1983) Nutritive value. In: Wang JK (ed) Taro, a review of *Colocasia esculenta* and its potentials. University of Hawaii Press, Honolulu, pp 141–147

Strauss M (1983) Anatomy and morphology of taro. In: Wang JK (ed) Taro: a review of *Colocasia esculenta* and its potentials. University of Hawaii Press, Honolulu, pp 20–33

Takahashi M (1953) Report of taro diseases in Hawaii. Hawaii Agricultural Experiment Station, Hawaii 63

Tanaka R, Sakano Y, Nagatsu A, Shibuya M, Ebizuka Y, Goda Y (2005) Synthesis of digalactosyl diacylglycerols and their structure-inhibitory activity on human lanosterol synthase. Bioorg Med Chem Lett 15:159–162

Tanji M. (2009) Council approves ban on GMO taro. Maui News. http://www.mauinews.com/page/content.detail/id/524344.html. Accessed 07 Apr 2012

Taylor NJ, Fauquet CM (2002) Microparticle bombardment as a tool in plant science and agricultural biotechnology. DNA Cell Biol 21(12):963–977

Trujillo EE (1967) Diseases of the genus *Colocasia* in the Pacific area and their control. Proc Int Symp Trop Root Crops 2:13–19

Trujillo EE, Menezes T (1995) Field resistance of Micronesian taros to *Phytophthora* blight. Phytopathology 85:1564

Trujillo EE (1996) Taro leaf blight research in the American Pacific. ADAP Bulletin 1:1–3

Trujillo EE, Menezes T, Cavaletto C (2002) Promising new taro cultivars with resistance to taro leaf blight: 'Pa 'lehua', 'Pa 'akala', and 'Pauakea'. Coll of Trop Agri and Human Res New plants for Hawaii. NPH-7, 4 Available at: http://www.ctahr.hawaii.edu/oc/freepubs/pdf/NPH-7.pdf. Accessed 07 Apr 2012

Uchida JY, Silva J, Kadooka CY (2002) Improvements in taro culture and reduction in disease levels. University of Hawaii, College of Tropical Agriculture and Human Resources. Plant Disease PD-22, pp 4. Available at: http://www.ctahr.hawaii.edu/oc/freepubs/pdf/PD-22.pdf. Accessed 07 Apr 2012

Van Damme EJ, Goossens K, Smeets K, Van Leuven F, Verhaert P, Peumans WJ (1995) The major tuber storage protein of Araceae species is a lectin. Characterization and molecular cloning of the lectin from *Arum maculatum* L. Plant Physiol 107:1147–1158

Veluthambi K, Gupta AK, Sharma A (2003) The current status of plant transformation technologies. Curr Sci 84:368–378

Wang JK (1983) Introduction. In: Wang JK (ed) Taro, a review of *Colocasia esculenta* and its potentials. University of Hawaii Press, Honolulu, pp 1–3

White JP, O'Connell JF (1982) A prehistory of Australia, New Guinea and Sahul. Academic Press, Sydney

Wilson JE (1979a) Promotion of flowering and production of seed in cocoyam (*Xanthosoma* and *Colocasia*). pp. 703–706. Int Symposium on Taro and Cocoyam, Visayas College of Agriculture, Baybay, Leyte, Philippines, 24–25 Sept

Wilson JE (1979b) Progress in the breeding of cocoyam (*Xanthosoma* and *Colocasia*), pp. 709–713. Int Symposium on Taro and Cocoyam, Visayas College of Agriculture, Baybay, Leyte, Philippines, Sept 24–25

Xu J, Yang Y, Pu Y, Ayad WG, Eyzaguirre PB (2001) Genetic diversity in taro (*Colocasia esculenta* Schott, Araceae) in China: an ethnobotanical and genetic approach. Econ Bot 55:14–31

Yang AH, Yeh KW (2005) Molecular cloning, recombinant gene expression, and antifungal activity of cystatin from taro (*Colocasia esculenta* cv. Kaosiung no.1). Planta 221:493–501

Yam TW, Webb EL, Arditti J (1990) Callus formation and plantlet development from axillary buds of taro. Planta 180:458–460

Yam TW, Chihashi S, Arditti J (1991) Callus growth and plantlet regeneration in taro, *Colocasia esculenta var. esculenta* (L) Schott (Araceae). Ann Bot 67:317–323

Zhao X, Yao J, Liao Z (2003) Molecular cloning of a novel mannose-binding lectin gene from *Arisaema heterophyllum*. Plant Sci 165:55–60

Zhou SP, He YK, Li SJ (1999) Inducing and characterization of in vitro corms of diploid-taro. Plant Cell Tiss Org Cult 57:173–178

Zupan J, Muth TR, Draper O, Zambryski P (2000) The transfer of DNA from *Agrobacterium tumefaciens* into plants: a feast of fundamental insight. Plant J 23:11–28

Chapter 16
Bioprospecting and Genetic Engineering of Mangrove Genes to Enhance Salinity Tolerance in Crop Plants

Anath Bandhu Das

Abstract Salinity in agricultural land is a major problem world wide, placing a severe constraint on crop growth and productivity in many regions and increased salinization of arable land is expected to have devastating global effects. Though plants vary in their sensitivity to salt stress, high salinity causes water deficit and ion toxicity in many plant species. Considerable efforts have therefore been made to investigate how genes respond to salt stress in various plants by using several approaches, including proteomics. Proteomic approaches for identifying proteins that are regulated in response to salt stress are becoming common in the post-genomics era of crop research. In this chapter, a detailed description of physiological, biochemical and antioxidative genes of salt stress in mangroves is given. Recent developments on salt stress genes of mangrove origin are also discussed and their identification applying bioinformatics approach as well as their validation in lower unicellular organisms. In addition, genomic biological changes in the proteomes of mangroves under salt stress condition are discussed. This chapter will provide a viewpoint into how proteomics and genomic based research is likely to develop in this field. Accumulation of compatible solutes and induction of antioxidative enzymes are other mechanisms of salt tolerance in mangroves. Mangroves also provide a reservoir for some of the best known, and at times, novel genes and proteins, involved in tolerance to salinity stress, that are likely at work in other plants. The salt tolerance genes listed in this review most likely represent only the tip of the iceberg, and continuous efforts to isolate and identify novel useful genes and promoters from mangroves are necessary; DNA microarray technology in particular is likely to become a powerful tool for this purpose. Eventually, the largest challenge will be to combine these genes and promoters in a systematic and logical way in order to maximize plant salinity tolerance. When realized, genetic engineering of crop and industrial plant for salinity tolerance using genes isolated from mangroves will be a vitally important tool in the quest to alleviate the earth's future problems concerning food, energy, and the environment.

A. B. Das (✉)
Molecular Genetics Laboratory, Department of Agricultural Biotechnology, Orissa University of Agriculture and Technology, Bhubaneswar 751003, Orissa, India
e-mail: a_b_das@hotmail.com

S. M. Jain and S. Dutta Gupta (eds.), *Biotechnology of Neglected and Underutilized Crops*, DOI: 10.1007/978-94-007-5500-0_16,

Abbreviations

ROS	Reactive oxygen species
EST	Expressed sequence tags
SOD	Superoxide dismutase
MDA	Malondialdehyde
CAT	Catalases
APX	Ascorbate peroxidases
AOC	Allene oxide cyclase
MT	Metallothioneins
ABA	Abscisic acid

16.1 Introduction

Mangroves are probably the most important single feature of tropical coastlines that occur in relatively sheltered areas, along estuaries, coastal lagoons and back-waters where the regular ebb and flow of the tides mix the fresh water from land drainage with marine coastal water. A variety of plants are seen putting up luxuriant growth in this intertidal habitat. These species are endemic to the area and are called mangroves (Figs. 16.1, 16.2, 16.3, 16.4, 16.5 and 16.6). Some species occurring outside the inter-tidal environment are called associate mangroves. As many as 110 species of mangroves and their associates belonging to 45 families are available in India. Mangrove species fall under the category of endangered/ threatened plants and occur in a highly fragile eco-system. The essential factors

Fig. 16.1 Vegetation of *Rhizophora apiculata* with pneumatahores

Fig. 16.2 *Bruguiera gymnorhiza* in natural mangrove habitat

Fig. 16.3 *Bruguiera parviflora* in viviverious propagules

for the maintenance of these eco-systems are fresh water influx from the tributaries and the tidal inflow from the sea. Changing the regime of either factor, whether in quality or quantity, is likely to affect a correspondence in the mangrove systems. Water salinity exhibits a gradient change from near fresh water to very saline conditions in mangrove ecosystem. So, vegetation has evolved and adopted accordingly to withstand wide range of salinity gradients. Some plants have specialized organs to thrive in narrow limiting high salinities. It is therefore likely that plant diversity would be very wide with higher potential for having a gene of interest for salt tolerance. The important mangrove families represented in

Fig. 16.4 *Rhizophra mucronata* in flowering condition

Fig. 16.5 *Avicennia marina* in flowering condition

Fig. 16.6 *Aegiceras corniculatum* in fruiting condition

India are Avicenniaceae, Arecaceae (Palmae), Combretaceae, Rhizophoraceae and Sonneratiaceae which are recognized as true mangroves. The mangrove flora of India includes trees, shrubs, herbs and grasses from 37 families (Banerjee et al. 1989; Kathiresan and Quasim 2005). *Rhizophora stylosa*, *Sonneratia griffithii*,

Avicennia marina, *Heritiera kanikensis* and *Xylocarpus moluccensis* are new records for India.

All the mangrove species have the same physiognomy, physiological characteristics and structural adaptations (Yenney -Esinsine 1980). The plant species in this particular ecosystem are constantly under varied environmental stress conditions including high salinity and extremes of temperatures and these plants have adapted themselves to these frequent and fluctuating changes (Lakshmi et al. 1997). Large areas of mangrove forests throughout the world are being converted for agriculture or being exploited for wood and other forest products. The Indian coastline covers about 7,500 km and it accounts for 8 % of the world's mangrove area and the eastern coast of India accounts for about 82 % of the mangrove forest cover in India (Parida et al. 1998a, b). In the absence of any national plan for conservation and sustainable utilization, mangroves along the Indian coast have reached an alarming stage of depletion. Reduction of 25 % mangrove forest cover has been reported along the Indian region during the last 25 years (Parida et al. 1998b). The coastal regions are also regions where there is intense agricultural activity. Increased soil erosion and water pollution caused by intensive farm practices in the inland area gets transported through the river and canal and adversely affect the coastal agro-ecosystem. The seawater intrusion has given rise to an increase in the level of abiotic stresses such as salinity, alkalinity, and drought in many coastal regions. Climate change and consequent rise in sea level even in case of Tsunami in the recent past of December 2005 is one of the major impending dangers affecting the coastal ecosystem. The rise in sea level has been expected to be in the order of 8–29 cms due to global warming by 2025 (Parida et al. 1998b). This could cause large scale inland flooding. Salinization is posing an increasing problem in coastal and agricultural areas reducing plant productivity and yield. Salinity is one of the major abiotic stresses decreasing the plant productivity. Tolerance to salt stress is a complex trait, which involves various aspects such as osmotic, ionic and oxidative stress. Salt stress leads to dehydration and osmotic stress with the reduced availability of water resulting in stomatal closure, reduced supply of carbon dioxide leading to a high production of reactive oxygen species (ROS) in the chloroplasts (Tanaka et al. 1999). This effect causes an irreversible cellular damage. Similar effects are also seen to occur during periods of high photosynthetic activity when the plants have a disturbance in photosynthetic activity and this causes photoinhibition (Bowler et al. 1991, 1992). Photoinhibition and salinity stress together cause severe damage to the cellular processes in the plant. Mangroves are plants that are capable of surviving in highly saline environments coupled with high capacity to maintain active leaves in conditions, which will be expected to reduce the photosynthesis severely through photoinhibition (Cheeseman et al. 1997). In order to combat such abiotic stress effects, a number of studies have been taken to conserve the mangrove genetic resources, characterize and harness the genes involved in salinity/abiotic stress tolerance from mangroves and transfer these genes to crop plants so as to generate crops with enhanced stress tolerance capacity. This review highlights various aspects of molecular markers and potential genes of interest from mangroves for future utilization in crop improvement programme.

Recent development of molecular methods has provided a wonderful tool to take mangrove research in new directions and to solve a number of unresolved problems. Molecular markers like protein, RAPD, RFLP, AFLP, ISSR and SSR markers have been used for lineage analysis and studying the population structure. Inferences concerning the histories of phylogenetic groups and geographical structure of populations are based on divergence of DNA sequences between lineages and on frequencies of allelic forms of genes within populations and differences in frequencies between populations. Mangrove habitats are being destroyed worldwide at an alarming rate (Farnsworth and Ellison 1997). Molecular methods are important tools for identification of appropriate population sources for reforestation of these unique and important habitats.

16.2 Isozyme Markers in Mangroves

Non-denatured proteins with different net charges with different allelic forms (allozymes) have been used to address phylogenetic relationships and estimation of population parameters such as gene flow, local differentiation and migration pathways. Allozyme markers have been utilized in studying genetic diversity in *Rhizophora stylosa*, *Kandelia candel*, *Lumnitzera racemosa*, *Avicennia marina*, and *A. germinas* (McMillan 1986; Baba et al. 1989). Difference in morphological characters like cold tolerance could not be correlated using allozyme markers (Markley et al. 1982). Recently, Duke et al. (1998) used allozyme markers to quantify gene flow and infer dispersal within and between the populations of *Avicennia* and they demonstrated that gaps in geographical distribution represent significant barriers to gene flow, suggesting limited long-distance dispersal capacity in *Avicennia*. While examining the genetic structure of *Ceriops*, Ballment et al. (1988) showed that *Ceriops tagal* var *tagal* and *C. tagal* var *australis* are completely reproductively isolated. These studies indicate that molecular markers might be extremely useful to solve the problems related to population genetics where normal taxonomical markers are not able to shed light.

16.3 Mangroves as a Model System for Studying Biochemical Adaptations During Salt Stress

Research on salt stress tolerance mechanisms was initially done using tobacco (*Nicotiana tabacum*), *Arabidopsis thaliana* and tomato (*Lycopersicon esculentum*) (Xiong and Zhu 2002). *A. thaliana* has many useful features that are favourable for its use as a genetic model for salt tolerance studies, including a small genome (100 Mb), rapid life cycle, and well-established transformation system. Moreover, after the advent of the whole-genome map, a large set of T-DNA insertional mutant stocks, and mutant screens, *A. thaliana* has been widely used for research on other abiotic stress factors as well. For instance, elucidation of the regulatory pathway for ion homeostasis and characterization of salt-overly-sensitive (SOS) genes were rendered possible by use of the *A. thaliana* model system. SOS gene mutations make *A. thaliana* more sensitive to Na+ stress (Zhu et al. 1997, 1998). These advantages have made *A. thaliana* the most widely used system in understanding the mechanism of salt tolerance. However, a big disadvantage is that *A. thaliana* shows limited plant growth in salt stress conditions and therefore cloning of novel genes and understanding of processes unique to salt-stress-tolerant plants might not be possible in this system (Zhang et al. 2001). The search for salt-tolerant genotypes has resulted in the use of halophytic models like *Mesembryanthemum crystallinum*, *Atriplex* spp., *Thellurgiella halophila*; mangrove models like *Avicennia marina*, *Acanthus ilicifolius*, *Aegiceras corniculatum*, *Bruguiera* spp., *Cressa cretica*, *Excoecaria agallocha*, *Kandelia candel*, *Lumnitzera racemosa* and *Rhizophora* spp.; and mangrove associates like *Sesuvium* spp., *Suaeda* spp. and *Porteresia coarctata*—a wild relative of rice.

Interest in mangrove models for salt stress research is because of their capacity to maintain active leaves under conditions that severely reduce photosynthetic capacity through photoinhibition. These are hardy plants capable of maintaining high cellular water potential (Cheeseman et al. 1997). The mangrove ecosystem is dynamic; these plants are able to adapt not only to stress conditions ranging from high salinity at one extreme, but also to a complete lack of salinity due to freshwater conditions. Thus, there may be no other group of plant communities with such adaptation for extreme conditions. The high salinity in these wetlands has been reported to be due to two important factors: low rainfall and freshwater runoff. Mangroves possess an efficient ion influx and efflux regulatory mechanism by means of which they regulate their cellular ionic conditions. Mangroves and their associates have been classified into three groups, (1) salt-excluding species (e.g. *Bruguiera* spp., *Excoecaria* spp., *Lumnitzera* spp., *Rhizophora* spp.): In these species, the root system possesses an ultrafiltration mechanism. The mechanism of ultrafiltration is still largely unknown but has been thought to occur at the root endodermis and is achieved by a physical process (2) salt secretors that regulate internal salt levels through foliar glands are represented by *Acanthus* spp., *Aegiceras corniculatum*, *Aegialitis* spp. and *Avicennia alba*, *A. marina*, *A. officinalis*, *A. marina*, for example, is a dominant mangrove tree species growing in Bhitarkanika and Pichavaram mangrove forests in India. These plants can tolerate and survive in a wide range of salinity. Salt is partially excreted through roots by an ultrafiltration mechanism. More importantly, these plants are characterized by presence of salt glands, by means of which they excrete salts (Tomlinson 1986; Meher-Homji 1988; Hogarth 1999; Parida et al. 2002, 2003b, 2004a, c, d, e, 2005; Parida and Das 2005; Das et al. 2006; Liangs et al. 2008; Parida and Jha 2010a, b) (3) Salt accumulators, represented by, *Sesuvium portulacastrum*, *Suaeda maritima* etc. These species accumulate high concentration of salts in their cells and tissues. These plants overcome salt toxicity by developing succulence (Tomlinson 1986).

Screening methods for the genes that are important for salt tolerance in mangroves and other halophytes have used large-scale transcript profiling/mRNA profiles and expressed sequence tags (ESTs). Among these approaches, large-scale transcript profiling has been used extensively because of the easy and cost-effectiveness of this approach (Ohlrogge and Benning 2000). A large collection of EST data is available for model halophytes like *M. crystallinum* and *T. halophila* (Koreeda et al. 2004; Wang et al. 2004b). However, in mangroves and their associates, EST collections are limited to *A. marina* (Mehta et al. 2005) and *Suaeda salsa* (Zhang et al. 2001). The screens were done using either hydroponically grown seedlings or using artificial sand culture for comparing untreated and salt-treated plants (300–500 mM NaCl concentration) for short-term-stress responses (48 h) or for longer duration (3 d). All of these studies have led to the identification of genes putatively involved in salt stress response. The genes thought to be important for salt stress responses in these studies are ones that are involved in osmotic balance, oxidative protection (antioxidants), ion homeostasis (salt antiporters) and signal transduction. For instance, the presence of a large collection of ROS-scavenging enzymes of the catalase and ascorbate peroxidase families suggests that there is increased ROS production in plants under salt stress conditions and, therefore, these enzymes are needed to maintain redox homeostasis (Zhang et al. 2001).

Although EST data might lead to an overestimation of the number of genes supposedly involved in salt tolerance, these data have also provided unique clones and low-abundance transcripts (Zhang et al. 2001). The significance of these data could be further validated by performing systematic analysis of gene expression using RNA analysis or large-scale transcript analysis using microarrays. A Transcriptomic approach using RNA expression data has been employed to identify genes that are expressed or repressed in response to saline stress. Halophytes exposed to salt stress cope with the deleterious effects of NaCl by alterations in gene expression. Indepth analysis of RNA expression profiles has resulted in characterization of salt-induced genes for antioxidant enzymes in mangroves (Jithesh et al. 2006) and salt antiporters in mangrove associates like *Porteresia coarctata* (Senthilkumar et al. 2005). Microarray analysis helps in differentiating two genotypes differing in salt sensitivity and provides a unique method of comparing their global gene expression profiles. A recent study distinguished key determinants of salt tolerance between *A. thaliana* and *T. halophila*. This study indicated that the genes that were upregulated only in *T. halophila* included genes that are either involved in protein folding or in post transcriptional modification. A systematic analysis of the encoded proteins might resolve the importance of genes upregulated only in *T. halophila*. *T. halophila* also exhibited higher prestress concentrations of several compounds involved in osmotic balance, such as proline and sugar-alcohols, and also in redox control (Gong et al. 2005). However, microarrays are at present not available for mangrove models though these have been recommended (Zhang et al. 2001). These genomic approaches would be useful for dissecting the salt-tolerance determinants in mangroves and in other halophyte models.

16.4 Enzymatic Pathways and Antioxidative Genes in Mangroves

16.4.1 Superoxide Dismutases

Superoxide dismutase (SOD) is an important antioxidant enzyme and is the first line of defence against oxidative stress in plants. SOD causes dismutation of superoxide radicals at almost diffusion-limited rates to produce H_2O_2 (Salin 1987). It plays an important part in determining the concentration of O_2 and H_2O_2 in plants and hence performs a key role in the defence mechanism against free-radical toxicity (Bowler et al. 1992). SODs are categorized into three main groups on the basis of the metal cofactor: (1) Cu/Zn SOD: these enzymes have copper and zinc as their cofactors and are localized in plants mainly in the cytosol and chloroplasts. Cytoplasmic Cu/Zn SODs are homodimeric and chloroplastic Cu/Zn SODs are homotetrameric in plants (2) Mn SOD: Mn SOD possesses manganese as its cofactor and is localized in mitochondria and also in the peroxisomes. (3) Fe SOD: Fe SODs are found predominantly in chloroplasts in plants and are absent in animals (Alscher et al. 2002). Each of these SOD forms can exist as different isoforms in plants (Bowler et al. 1992). For example, *A. thaliana* has three Cu/Zn SOD isoforms (CSD_1, CSD_2 and CSD_3), one Mn SOD (MSD_1) and three Fe SOD isoforms i.e. FSD1, FSD2 and FSD3 (Kliebenstein et al. 1998). Effects of salinity stress in halophytes are analysed by studying the total SOD protein activity and also by transcript profiles of individual mRNAs encoding different SOD isoforms before and after salt stress treatments. Total enzyme activity assays by spectrophotometry and individual SOD isozyme activity assays after electrophoresis in gels and mRNA profiles have been done in mangroves during salt stress. Steep increase in total SOD activity levels has been recorded in *Bruguiera gymnorhiza* and *B. parviflora* during salt stress (Takemura et al. 2000; Parida et al. 2003a, 2004a, c, d). When *B. gymnorhiza* seedlings were placed in 500 mM NaCl for 9 days, the total SOD activity increased by almost 8 times in comparison to unstressed controls (Takemura et al. 2000). Again, analysis of the total SOD activity in *B. parviflora* seedlings subjected to 45 d of NaCl stress at 200 and 400 mM NaCl revealed an increase of 128 % (Parida et al. 2004b). In *Avicennia marina* it was observed that the total SOD activity increased in leaf tissues after six weeks of application of salt stress (Cherian et al. 1999). Another interesting study showed that the SOD activities in field-grown mangrove *Rhizophora stylosa* were more than 40 times those of peas. This robust activity is considered to be effective in protecting mangroves from excess irradiance (which generates ROS) under field-grown conditions (Cheeseman et al. 1997). A few studies also suggest lack of a general rule with regard to SOD protein synthesis under salt stress. Variations in response were observed even across tissues that were studied. Surprisingly, while total SOD activity increased in response to salinity stress in leaf tissues of *A. marina*, a decrease in total SOD activity was observed in shoot and root tissues (Cherian et al. 1999). Again, total SOD activity was shown to increase in shoot tissues of a halophyte, *Crithmum maritimum*, in presence of 50 mM NaCl stress; however at higher

NaCl stress, at a concentration of 200 mM of NaCl, no significant difference in SOD activity was detected (Ben Amor et al. 2005). On the contrary, SOD activity in root tissues of *C. maritimum* decreased upon treatment with 50 mM as well as 200 mM of salt (Ben Amor et al. 2005). Decrease in total SOD activity was also observed in NaCl-treated callus cultures of *Suaeda nudiflora*. This result is not in agreement with studies conducted on whole plants, wherein SOD activity was found to be increased under salt stress (Cherian and Reddy 2003). Dey and Kar (1995) observed that callus tissues had lower activities of antioxidative enzymes like SOD when compared to whole plant seedlings. Reduced activities of SOD in *S. nudiflora* calli reveal the declining capability of these tissues to scavenge singlet oxygen in comparison to whole plant seedlings. According to Cherian and Reddy (2003), the presence of a strong antioxidative response mechanism combined with physiological specialization in the plant contributes to the difference in salt tolerance capacity between whole plants and calli in *S. nudiflora*.

Regulation of individual isoforms of SOD in response to salt stress in halophytes is examined by activity using polyacrylamide staining on gels after polyacrylamide gel electrophoresis. In this method, nitroblue tetrazolium is reduced by the superoxide radical into formazan dye by a photo catalysed reaction. The regions of migration of the SOD isozymes appear white on a blue-black background due to the scavenging of the superoxide radicals by SOD (Beauchamp and Fridovich 1971). Studies on the activity of individual SOD isoforms in many halophytes during salt stress have revealed increased specific activities of Mn SOD and Fe SOD isoforms localized in mitochondria and chloroplasts respectively. Experiments in *S. salsa*, showed that the total Fe SOD activity increased during NaCl and osmotic stress. However, while Fe SOD I isoform was induced by NaCl-mediated oxidative stress and not by osmotic stress, only Fe SOD II and III increased in response to osmotic stress and were completely inhibited by high NaCl stress (Wang et al. 2003, 2004a). This study also revealed the role of K^+ ion in tolerance to NaCl stress. Cu/Zn SOD I isoform activity decreased and the activity of Cu/Zn SOD II isoform remained unaltered in response to NaCl and in presence of K^+ ions. In absence of K^+ ions, activities of the both Cu/Zn SOD isoforms were unaltered under NaCl stress. On the other hand, Mn SOD and Fe SOD I specific activities increased in absence of K^+ ions during NaCl stress. Mn SOD isoform was also induced specifically in response to osmotic stress with PEG. However, in the presence of K^+ ions during NaCl stress, only FeSOD I activity increased while Mn SOD activity remained unaltered. This indicated that K^+ starvation, as in PEG treatment, resulted only in increased Mn SOD activity. It was suggested that K^+ starvation in *S. salsa* mimicked physiologically an osmotic deficiency caused by ionic imbalance. Therefore, this reveals that mangrove associates in which Na^- accumulation occurs, as in *S. salsa*, K^+ is used as a cheap solute for osmotic adjustment to compensate for changes in the osmotic potential. K^+ is an important macronutrient and is required in the plant cell in high concentrations. It plays essential roles in growth regulation, osmotic adjustment and regulation of stomatal opening (Wang et al. 2004a).

This finding is interesting because a role for K^+ in Na^- tolerance has been established. Recent evidence also suggests that maintenance of tissue K^+ levels may be an important strategy in halophytes during high-salt conditions (Wang et al. 2006). *A. thaliana* mutants that were sensitive to Na^- were also impaired in media containing low K^+. It was found that in *A. thaliana*, maintenance of high cellular K^+ content above threshold levels in presence of excess external Na^+ is critical for plant growth and a key element of salt tolerance (Zhu et al. 1998). This might explain why a strong antioxidant response mechanism (ARM) is important under conditions where physiological adaptations like low K^+ are not sufficient at cell level to cope with salinity stress. Hence, this observation that halophytes activate ARM to cope with salt stress under physiological conditions of low K^+ nutrition is significant. Wang et al. (2004a) also demonstrated in *S. salsa* that decrease in activity of the Cu/Zn SOD I was compensated by increase in the Fe SOD and Mn SOD activities in chloroplasts and mitochondria respectively and therefore no change was observed in total SOD activity in response to NaCl stress. Again, in *Bruguiera parviflora*, Mn SOD and Fe SOD II activities increased in response to salt stress but no increase was found in activities of other SOD isoforms (Parida et al. 2004a). However, in the halophyte *M. crystallinum*, only Fe SOD activity increased immediately in response to salt stress and increase in Mn SOD and Cu/Zn SOD was observed only at a later stage of salinity stress (Slesak et al. 2002). Certain other halophytes also exhibit high levels of SOD expression even in unstressed conditions. In *T. halophila*, Fe SOD gene was overexpressed even under non-salt stressed conditions in comparison to *A. thaliana*. The higher levels of expression of Fe SOD plays a key role in conferring tolerance to oxidative stress and salinity stress in *T. halophila* (Taji et al. 2004). Similarly in *M. crystallinum*, mRNA transcript levels of Fe SOD isoform (Slesak et al. 2002) along with one isoform of *Cu/Za* SOD (Hurst et al. 2004) increased in response to salt stress whereas the mRNA levels of another isoform of Cu/Zn SOD in *M. crystallinum* remained unaltered (Slesak et al. 2002). Similarly, mRNA transcript levels of cytosolic Cu/Zn SOD-I isoform were unaltered even at 500 mM of NaCl stress in the mangrove *A. marina* (Jithesh et al. 2006). The increase in the activities of specific isoforms and in the mRNA transcript levels of SODs particularly that of Fe SOD and Mn SOD, localized in the chloroplasts and mitochondria act as an ARM in most halophytes during salt stress. This is not surprising because the immediate targets of salt stress are the chloroplasts and mitochondria. Superoxide radicals and other ROS are formed in the chloroplasts during photosynthetic light reactions at the acceptor side of PSI, reducing site of PSII (Eltsner and Formmeyer 1979) and at the oxygen-evolving complex (Wydrzynski et al. 1989). Generation of superoxide radical due to reduction of oxygen could also take place at different points of the respiratory chain such as the flavor protein region of NADH dehydrogenase and the ubiquinone-cytochrome region in the mitochondria (Arora et al. 2002).

The superoxide radicals may not be limited to chloroplasts in the plant cell, but could also leak into the cytosol, resulting in induction of the cytosolic SOD isoforms. Accumulation of transcript mRNA of cytosolic Cu/Zn SOD in the mangrove species *Brugnira gymnorhiza* was increased after the 1st and 5th day of NaCl treatment and also in presence of mannitol and abscisic acid (Takemura et al. 2002). However, there are very few reports about the mechanisms of generation of superoxide radicals in the cytosol and role of cytosolic SOD in halophytes, and more studies are required in this regard (Takemura et al. 2002). On the other hand, studies using transgenic approaches amply demonstrate the importance of chloroplastic antioxidative enzymes during oxidative and salinity stress conditions. Overexpression of Fe SOD and of Mn SOD isoforms in chloroplasts of tobacco and rice plants respectively the genes has been shown to confer tolerance to oxidative stress and salinity stress in transgenic plants compared with untransformed control plants (Van Camp et al. 1996). Studies on differential activation of Fe SOD and Mn SOD isoforms from halophytes are in agreement with reports of transgenic approaches. In conclusion, increased activity of Fe SOD and Mn SOD isoforms is therefore an effective mechanism to prevent oxidative damage during salinity stress in halophytes. Physiological adaptability of three mangrove species *Rhizophora stylosa*, *Sonneratia apetala* and *S. caseolaris* to salt stress were investigated by Yan and Guizhu (2007). The response for adaption to salinity in three months old seedlings showed decreased net photosynthesis rate, stomata conductance and transpiration rate of leaves and soluble sugar content in leaves increased. The malondialdehyde (MDA) content in stems and leaves of *Sonneratia apetala*, *S. caseolaris* somewhat decreased when the salinity was lower but rapidly increased with increasing salt concentration. The MDA content in stems and leaves of *Rhizophora stylosa* increased only when salinity was greater. No changes were observed in the MDA content of roots in the three mangrove species. The adaptabilities of *S. apetala* and *S. caseolaris* to salt tolerance were limited. In the more salt tolerant mangrove *R. stylosa*, it seems more likely that the free oxygen radicals were eliminated through the increase in activity of superoxide dismutase (SOD).

16.4.2 Catalase

Catalases (CAT) are haem-containing tetrameric enzymes involved in the removal of H_2O_2 (Guan and Scandalios 1993). Plant catalases are involved in photorespiratory functions (Canvin 1990), and scavenging of H_2O_2 during oxidation of fatty acids in germinating seeds (Willekens et al. 1995) and also during salt stress and other abiotic stress conditions (Willekens et al. 1997). The multifarious functions of plant catalases require that plants evolve different isozymes of this important enzyme for the various roles it plays in plant systems. Plants have therefore evolved multiple CAT isozymes, which are encoded by members of a small gene family: two isozymes in castor bean (Ota et al. 1992), and three in tobacco *Nicotiana tabacum* (Havir and McHale 1987), *Nicotiana plumbaginifolia* (Willekens et al. 1994a), maize (Scandalios et al. 1980) and *Arabidopsis* (Frugoli et al. 1996).

In *Bruguiera*, catalase enzyme activity after salt stress varied with species. When *B. parviflora* plants were subjected to 400 mM NaCl stress conditions, a decrease in total catalase activity was observed. Activity staining on electrophoresis gels revealed four discrete catalase-active zones, CAT-1, CAT-2, CAT-3 and CAT-4. The extent of decrease in enzyme activity upon NaCl stress varied among

CAT-1, CAT-2, CAT-3 and CAT-4. While, CAT-2 decreased by 45 %, only a 29 % loss in activity was observed for CAT-3. Most importantly, total catalase activity decreased (35 % loss of activity) at all concentrations of NaCl stress. The inactivation of catalase was possibly due to enhancement of H_2O_2 levels. On the other hand, contradictory results were observed in a different species, *B. gymnorhiza*. In this study, catalase activity assays were done before and after salt stress treatments in 500 mM NaCl stress medium for 9 days. It was shown that catalase activity increased to about 4.9 times more that in unstressed control plants. This experiment revealed that catalase retained full activity at least up to seawater salt concentration. However, the enzyme activity levelled off later till the end of the experimental treatment time of 16 days. The differences could be species-specific biochemical response or the particular experimental conditions used and/or the plant growth conditions which need validation.

In a different study, NaCl tolerance in *Suaeda nudiflora* callus cultures was evaluated to investigate whether alterations in antioxidant activity occur at the cell level (Cherian and Reddy 2003). Plant growth was retarded with respect to both fresh and dry mass with increasing in NaCl concentrations. Maximum reductions of 92 and 94 % were found for fresh and dry mass at 200 mM NaCl. There was 60 % decrease in total catalase activity. These experiments showed the declining capacity of callus tissues to scavenge H_2O_2 with increasing salt stress. In another study, total catalase activity decreased with increasing NaCl concentration in *Crithmum maritimum* (Ben Amor et al. 2005). Though studies on catalases in mangroves are few, these clearly showed a decrease in activity. It is possible that catalases respond similarly to salt stress conditions in mangroves and halophytes as in other glycophytes (Streb et al. 1993). Most of the studies on the effect of salt stress on catalase activity in mangrove species were performed using spectrophotometric (total catalase activity) or activity staining methods. Total catalase activity measurements do not provide information on the role of individual isoforms of catalase. On the other hand, activity assays on electrophoresis gels are also not conclusive (Eising et al. 1990). An alternative method is to monitor mRNA level changes for the enzymes in each of the three catalase classes after salt stress treatment. Monitoring changes in mRNA levels using gene-specific probes might clarify the role of individual catalase forms (Jithesh et al. 2006). However, there are very few reports of isolation and characterization of catalase genes from mangrove species. Gene accessions of mangrove catalases are limited to *Avicennia marina* and *Bruguiera gymnorhiza*.

Catalase mRNA level has been shown to remain unaltered after NaCl treatment in *B. gymnorhiza* (Takemura et al. 2002). However, the catalase cDNA that was used for this study encodes a partial catalase protein and has not been fully characterized. It was suggested that this particular isoform might not have a contributory role in salt stress response (Takemura et al. 2002). In another study, a class II enzyme, CAT I from *A. marina*, was reported to be induced upon salt and other oxidative stress, such as exposure to H_2O_2 and light, in leaves (Jithesh et al. 2006). An analysis performed using RNA isolated from root tissues of *A. marina* and probed with CAT II revealed a similar trend. CAT I mRNA levels increased with 12 h of NaCl stress and declined mRNA analysis of *Avicennia marina* roots with CAT I in NaCl stress. One-month-old *A. marina* seedlings acclimatized in greenhouse conditions were treated with NaCl (500 mM) and RNA was isolated from root tissue. For comparison, a control (0 h) was also used and RNA isolated, (a) Isolated RNA was loaded in a denaturing formaldehyde gel and probed with 3′ UTR-specific CAT I probe (Jithesh et al. 2006). (b) To confirm equal loading of RNA, the same blot was stripped and probed with 18S rRNA gene. An abundance of a particular isoform of catalase over other isoforms was reported from a salt-stressed *Thellurgiella. halophila* EST library. Seventeen EST clones of CAT III gene were retrieved from this library. Catalase 3 from *T. halophila* shows highest identity in nucleotide sequence (90 %) and in encoded amino acid sequence (95 %) to catalase 3 from *A. thaliana*. The *A. thaliana* enzyme has been characterized as a class II isozyme (Klotz et al. 1997). At the time of writing, a study reported the differential induction of *A. thaliana* catalase 3 by oxidative stress caused due to senescence; expression of the gene for catalase 3 differed from that of catalase 1 and catalase 2 (Zimmermann et al. 2006). In the future, many more catalase genes may be isolated and characterized from halophytes and mangroves. It would then be possible to find if oxidative and stress responses of class II catalases are clearly differentiated from constitutively synthesized enzymes both in structure and in function.

16.4.3 Ascorbate Peroxidases

Ascorbate peroxidases (APX) are antioxidants performing the same general function as catalases. However, unlike catalase, they catalyse removal of H_2O_2 by using ascorbate as a reductant. APX play an important role in regulation of intracellular level of H_2O_2 in higher plants (Van Breusegem et al. 2001). APX are a family of isozymes with different characteristics. APX are localized in all the four major cellular compartments: stromal (sAPX) and thylakoid-membrane-bound APX in chloroplasts (tAPX), microbody (including glyoxysomes and peroxisomes) membrane-bound APX (mAPX), cytosolic APX (cAPX), and lastly a mitochondrial-membrane-bound form (mitAPX). Studies on the role of APX in tolerance to salt stress of mangroves and other halophytes have been restricted to few reports. Ascorbate-mediated H_2O_2 dismutation was assessed by in vitro spectrophotometric assays and quantifying the activity of APX before and after salt stress in *Bruguiera parviflora* (Parida et al. 2004a, b). APX activity increased significantly by 2.5 times when *B. parviflora* plants were exposes to two different levels (200 and 400 mM) of salt stress in the first week. Furthermore, there was a 4.5-fold enhancement in APX activity at 45 days of salt stress exposure, measured by densitometry analysis. It was hypothesized that the increase in activity of APX could be due to activation of preexisting APX or due to synthesis of new APX upon salt exposure (Parida et al. 2004a). In *B. parviflora*, superoxide radicals O_2^- formed owing to salt stress could be scavenged by increased activity of SOD. The end product of this reaction, H_2O_2, is partially dismutated by the increased activities of APX (Parida et al. 2004a, b, c). The increase in APX activity and the concomitant increase in SOD activity suggest that this is an adaptation to remove the excess H_2O_2 generated due to increase in SOD activity. It has been observed that cytosolic APX transcripts as well as cytosolic APX activities are increased after salt stress treatment in *Mesembryanthemum crystallinum*. The induction of APX clearly shows that this enzyme plays a critical role in controlling increase of H_2O_2 concentration in plant cells during the initial salt-induced oxidative stress. Presence of three APX was reported in a salt-treated EST cDNA library of *Thellungiella halophila*. The presence of coding sequences for ascorbate peroxidase and other ROS scavenging enzymes in a small *T. halophila* EST library suggest that these enzymes are responsible for effective ROS detoxification under salinity (Wang et al. 2004b). All these reports reveal that APX have a role in tolerance of NaCl-mediated oxidative stress in halophytes.

Therefore, in contrast to the hypothesis that APX is involved in H_2O_2 signalling in plants, an important role for APX against salt stress is suggested in halophytes (Mittler 2002). Plants contain abundant amounts of peroxidases (POX) that are associated with the cell wall where they generate phenoxy compounds from cinnamic acids. POX are also involved in other plant defence mechanisms, including responses to insects (Salin 1987), and, in a coordinated response known as the oxidative burst (Kawano 2003). Increased POX activity was observed in root and shoot tissues in *Avicennia marina* in comparison to control under NaCl stress conditions; however, no significant changes in POX activity were observed in leaf tissues (Cherian et al. 1999). Enhanced activity of POX was also observed in callus cultures of *Suaeda nudiflora* at higher salt concentrations. NaCl-induced enhancement of POX activity in salinized cells of *S. nudiflora* indicated that these cells had a higher capacity for decomposition of H_2O_2 and that POX perform a major role in this particular reaction (Cherian and Reddy 2003). A transient increase of POX activity was also observed in halophytes like *Crithmum maritimum* at different NaCl concentration studied (Ben Amor et al. 2005).

Changes in proteins and antioxidative enzymes in tree mangroves *Bruguiera parviflora* and *Bruguiera gymnorhiza* under high NaCl stress reported recently by Behera et al. (2009). In order to access the role of protein and antioxidative enzymes defense system during high salt shock, two *Bruguiera* species *B. parviflora* and *B. gymnorhiza* of the family Rhizophoraceae were exposed to high salt shock (500 mM NaCl) for a short period of 6 days. Total protein content in both the species decreased upon salt shock but the rate of degradation was more rapid in *B. parviflora* than in *B. gymnorhiza*. SDS-PAGE protein profiling revealed that the protein having apparent molecular mass 90, 49, 33, 23, 10 kDa reduced very first after 4 days of salt treatment of *B. parviflora* than in *B. gymnorhiza*. The specific activities of catalase increased 2.4 times and 2.1 times in *B. parviflora* and *B. gymnorhiza* upon exposure to high salt. Out of the four isoforms of catalase, CAT-2 activities enhanced 1.5 times and 1.2 times upon initial salt treatment for 4 days and 6 days. APX activity increased 1.3 times and 2.3 times in *B. parviflora* and *B. gymnorhiza* under salt shock. A single isoform of APX enhanced 1.3 times and 1.7 times in *B.*

parviflora and *B. gymnorhiza*. The GPX activity increased 1.5 times and 1.3 times in leaves of salt treated seedlings of *B. parviflora* and *B. gymnorhiza*.

It was suggested that accumulation of ferritin under salinity stress conditions provides chloroplasts with available iron for efficient synthesis of iron-containing enzymes, proteins and photosynthetic complexes, and also helps in protection of cells against toxic hydroxyl radicals. An active role for this protein in the oxidative stress network could be surmised because ferritin could potentially increase the availability of iron for enzymes that use iron as a cofactor (Paramonova et al. 2004). Considering the importance of this protein, research has focussed on ferritin mRNA regulation in response to different oxidative stresses like salt, iron H_2O_2, nitric oxide and abscisic acid in glycophytes like rice (Rabbani et al. 2003), *A. thaliana*, (Gaymard et al. 1996; Petit et al. 2001). However, the paucity of studies in mangroves and halophytes is evident in the number of genes that have been isolated from these plants. Currently, there is just one full-length ferritin gene from a mangrove (*A. marina*) was found in GeneBank. Again, an EST library from salt-stressed *M. crytallinum* revealed three EST accessions (AI861070, BE036876, BE036500) that showed identity to ferritin. The genes involved in iron homeostasis were shown to be abundant in a cDNA library of salt-treated *A. marina* (Mehta et al. 2005). Jithesh et al. (2006) assess the role of ferritin in response to salt-mediated oxidative stress in *A. marina*, increased *FerI* transcript levels were observed in leaf tissues within 12 h of salt stress treatment and then a decline with increased time of salt stress exposure. This suggests that *FerI* transcript accumulation was an immediate and short-term response to salt stress. Ferritin was shown to be an important constituent of the oxidative stress response in halophytes and may participate in the defence of chloroplasts against oxidative stress.

16.4.4 Regulatory Role of Reactive Oxygen Species in Mangroves

In addition to antioxidative enzymes, nonenzymatic antioxidants also play a key role in scavenging free radicals in plants during salinity stress. The nonenzymatic component comprises molecules such as ascorbic acid, glutathione, α-tocopherol and carotenoids that can scavenge ROS. Ascorbic acid and tocopherols have a poor ability to donate electrons and thereby transfer of single hydrogen atoms making them efficient antioxidants (Arora et al. 2002). Reduced glutathione is a powerful reductant and hence a very efficient scavenger of ROS. Carotenoids scavenge free radicals that are generated owing to excess excitation energy from chlorophyll during photosynthesis (Arora et al. 2002). Superoxide radicals generated in plants are scavenged nonenzymatically by reduced ascorbate and glutathione (Cheeseman et al. 1997). High concentration of ascorbate has been observed in whole leaves of *Rhizophora stylosa*. Therefore, an important role for ascorbate as a major nonenzymatic means of scavenging superoxide radicals in mangroves during oxidative stress conditions was suggested (Cheeseman et al.

1997). Ascorbate is the reductant necessary for the functioning of ascorbate peroxidase. On the other hand, glutathione is required for conversion of dehydroascorbate to ascorbic acid (Bowler et al. 1992). Parida and Das (2004) studied the effect of salt stress on concentrations of ascorbic acid and glutathione in the halophyte *Brugniera parviflora*. A reduction in ascorbic acid and glutathione levels were observed during salt stress in *B. parviflora*. However, this reduction was accompanied by an increase in the activities of APX and glutathione reductase. This suggested that a reduction in the levels of glutathione and ascorbic acid was due to consumption of these antioxidants in the ascorbate-glutathione oxidative pathway (Parida et al. 2004a, b, c).

16.4.4 Osmolytes and ROS Scavenging

The most important difference that separates halophytes (mangroves) from glycophytes with respect to satinity tolerance is the presence of 'physical' or 'structural' attributes in halophytes. These include salt-excreting glands like in *Avicennia marina* and root hairs in *Porteresia coarctata*. Apart from these, the 'genetic controls' for biochemical mechanisms are of considerable interest. These biochemical mechanisms, it appears, are regulated in a way that allows halophytes to grow and thrive in salt stress conditions by offering them a distinct advantage over other plants, such as glycophytes (Hasegawa et al. 2000a, b, Fig. 16.7). Recent evidence

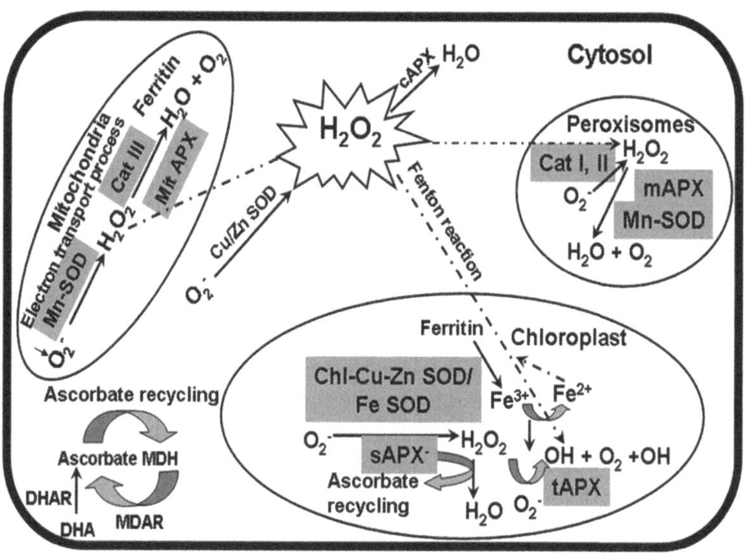

Fig. 16.7 Schematic representation of antioxidant enzyme activity in different organelles of the plant cell

on differential expression of genes encoding or concerned with antioxidants and osmolytes in halophytes supports this view and could be the reason for improved salt tolerance capacity in *Thellungiella halophila* in comparison to *Arabidopsis thaliana* (Kant et al. 2006). The use of halophytes as models for gene expression studies would, therefore provide vital clues in future. It is clear that salinity induces oxidative stress in plants (Xiong and Zhu 2002). In vitro detection of free radicals is difficult in tissues owing to their transient nature (Kennedy et al. 1990). Therefore, increases in malondialdehyde and extent of lipid peroxidation are often used as indicators for ROS production in planta during environmental stress conditions. Data from halophytes suggest maintenance of malondialdehyde levels after salt stress treatments and this, together with induction of antioxidants, confirms the role of antioxidants in conferring the salt tolerance trait in halophytes (Wang et al. 2004a; Fang et al. 2005).**Fig. 16.7** Schematic representation of antioxidant enzyme activity in different organelles of the plant cell

In these cases, induction of antioxidant enzymes was shown to protect halophytes against ROS, thus preventing lipid peroxidation during salt stress. This suggests that these antioxidant enzymes are essential components of an adaptive defence mechanism against salt stress in halophytes. Some of the major antioxidant enzymes involved in scavenging SOD, CAT, APX and POX. Considerable information emerging from halophytes suggests preponderance of sequences that code for certain specific isoforms like class II catalases in EST libraries from salt-stressed plants, and increased expression of Fe SOD, Mn SOD and APX during salt stress in halophytes. More emphasis, therefore, needs to be given to addressing the specificity of these isoforms. Also, more information on ROS scavenging and repair mechanism in plants by osmolytes is vital. Studies on ascorbate and glutathione and the role of ascorbate-regenerating enzymes should be carried out in halophytes, as emphasized by Fang et al. (2005). Most salt tolerance studies have explored antioxidant status in leaves. This is not surprising considering the fact that salt stress aggravates ROS mainly in the chloroplastic compartments of leaf cells. However, roots being the tissue in direct contact with saline water, studies on antioxidant enzyme status in roots might reveal rapid changes in gene expression. A few studies have found differences in root and leaf antioxidant responses that are contradictory (Cherian et al. 1999; Ben Amor et al. 2005). More indepth analyses are necessary for understanding the decrease in activity of certain antioxidant enzymes in roots. Alternatively, construction of root cDNA EST libraries and proteome analysis in root tissues could be done and comparisons drawn with analysis of leaf tissues. In the near future, EST libraries from salt-stressed plants and proteomic data would provide information on novel uncharacterized salt-responsive genes typical of mangroves. Recently, proteomic approaches have also been adopted in mangroves (Askari et al. 2006). Genetic information that would be obtained using these recent molecular tools from mangrove and halophytic models would eventually help in understanding and discovery of key genes involved in salt tolerance. It is possible that induction of antioxidants and osmolytes is part of an integrated strategy for stress defense in halophytes. Studies on superoxide dismutases during salt stress in mangroves are extensive.

Table 16.1 List of studies carried out on antioxidative enzymes during salt stress in mangroves and other halophytes

Enzyme/gene	Plant species	Method used	Enzyme expression	References
APX	*Bruguiera parviflora*	Activity staining on gel	Enhancement of APX activity in presence of NaCl stress in hydroponic culture	Parida et al. (2004a), Parida and Das (2005a)
APX	*Thellungiella halophila*	EST analysis	Presence of three APX in cDNA library from salt-treated plants	Wang et al. (2004a)
APX	*Aegiceras corniculatum*	Spectrophotometric assay	Activity decrease 58 % on 4th day in root	Mishra and Das (2003), (2004)
CAT	*Avicennia marina*	m RNA detection	Induction of AmCat I in salinity till 12 h and subsequent decreased	Jithesh et al. (2006)
CAT	*Bruguiera parviflora*	Spectrophotometric assay and activity staining on gel	Loss of total catalase activity with NaCl stress, CAT-2, and CAT-3 isoforms decreased	Parida et al. (2004a, b, c)
CAT	*Bruguiera gymnorhiza*	Spectrophotometric assay	Increase in total catalase up to sea water concentration	Takemura et al. (2000)
CAT	*Bruguiera gymnorhiza*	m RNA detection	A particular catalase transcript level did not change with salinity stress	Takemura et al. (2002)
CAT	*Crithmum maritimum*	Spectrophotometric and isozyme analysis	Decrease in catalase activity with increase NaCl	Ben Amor et al. (2005)
Ferritin	*Mesembryanthemum crystallinum*	Electron microscopic studies	Ferritin deposits present with NaCl stress in chloroplasts	Parmonova et al. (2004)
Ferritin	*Avicennia marina*	mRNA analysis	Early induction of AmFer1 in salinity	Jithesh et al. (2006)
CAT	*Aegiceras corniculatum*	Spectrophotometric assay	Activity decrease 72 % in root	Mishra and Das (2003)
CAT	*Aegiceras corniculatum*	Activity staining on gel	Two isoforms CAT 1,2 decreased	Mishra and Das (2003)
CAT	*Suaeda nudiflora*	Spectrophotometric analysis	Decrease in catalase activity in callus cultures with salinity stress	Cherian and Reddy (2003)

(continued)

Table 16.1 (continued)

Enzyme/gene	Plant species	Method used	Enzyme expression	References
POX	*Avicennia marina*	Spectrophotometric analysis	Enhanced peroxidase activity in root and shoot and no significant changes in leaf tissue	Cherian et al. (1999)
POX	*Crithmum maritimum*	Spectrophotometric analysis	Transient increase in peroxidase activity	Ben Amor et al. (2005)
POX	*Suaeda nudiflora*	Spectrophotometric analysis	Enhanced activity with NaCl stress	Cherian and Reddy (2003)
GPX	*Aegiceras corniculatum*	Spectrophotometric assay	Activity decrease 80 % on 4th day	Mishra and Das (2003)
GPX	*Aegiceras corniculatum*	Activity staining on gel	Four isoforms of GPX-1,2,3,4 decreased upon salt treatment of 6d	Mishra and Das (2003)
GPX	*Bruguiera parviflora*	Spectrophotometric assay	169 % of GPX enhancement at 45d exposure of NaCl at 400 mM	Parida et al. (2004a, b), Parida and Das 2005
GPX	*Bruguiera parviflora*	Activity staining on gel	GPX-6 show maximum 73 % increment in activity without any change in GPX-4,5 where as GPX-1 increase by 61 %,GPX-2 and 3 changed 34–38 %	Parida et al. (2004a, b),
SOD	*Avicennia marina*	Spectrophotometric analysis	Decrease in total activity in roots and shoots but increase in leaves with salinity stress	Cherian et al. (1999)
SOD	*Avicennia marina*	mRNA analysis	Constitutive expression of Cu/Zn SOD in salinity	Jithesh et al. (2006)
SOD	*Bruguiera parviflora*	Activity staining on gel	Increase in activity of Mn SOD and Fe SOD and no increase in activity of Cu/Zn SOD isoform with salinity stress	Parida et al. (2004a, b), Parida and Das (2005)
SOD	*Bruguiera gymnorhiza*	Spectrophotometric assay	Steep increase in SOD activity in salinity stress	Takemura et al. (2000)

(continued)

Table 16.1 (continued)

Enzyme/gene	Plant species	Method used	Enzyme expression	References
SOD	*Bruguiera gymnorhiza*	mRNA detection	Induction of Cu/Zn SOD activity in salinity stress	Takemura et al. (2002)
SOD	*Crithmum maritimum*	Spectrophotometric assay and isozyme analysis	Increase in total SOD activity in shoots and decrease in roots	Ben Amor et al. (2005)
SOD	*Suaeda nudiflora*	Spectrophotometric assay	Decrease in total SOD activity in callus culture with salinity stress	Cherian Reddy (2003)
SOD	*Suaeda salsa*	Isozyme assay and spectro-photometric assay	Increase in activity of thylakoid bound SOD	Fang et al. (2005)
SOD	*Suaeda salsa*	Staining on gel	Increase in Mn SOD and Fe SOD isoforms, Cu/Zn SOD I isoform activity decreased in presence of salinity stress, no change Cu/Zn SOD II isoform	Wang et al. (2004a)
SOD	*Thellungiella halophila*	mRNA detection	Higher mRNA transcript levels of Fe SOD	Taji et al. (2004)
SOD	*Mesembryanthemum crystallinum*	mRNA detection	Increase in Fe SOD and Cu/Zn SOD during salt stress	Slesak et al. (2002), Hurst et al. (2004)
SOD	*Mesembryanthemum crystallinum*	mRNA detection	Induction of all three SOD isoforms in salt stress	Slesak et al. (2002)
Acid phosphatase	*Bruguiera parviflora*	Activity staining on gel	Out of the three isoforms of ACP, ACP-3 increased by 37 % without much change in other isoforms	Parida et al. (2004a, b)
Alkaline phosphatase	*Bruguiera parviflora*	Spectrophotometric assay	Activity increased to 10, 20, 39 % dose dependent in 100, 200, 400 mM NaCl	Parida et al. (2004a, b), (2005)
Nitrate Reductase	*Bruguiera parviflora*	Spectrophotometric assay	Activity increased in 100 mM NaCl and gradually decline up to 63 % in 400 mM NaCl	Parida et al. (2004a, b), Parida and Das (2005)

Various antioxidative enzymes/genes are reported in a number of mangrove species (Table 16.1).

16.5 Genomic DNA Polymorphism

16.5.1 RAPD and AFLP Markers

Restriction Fragment Length Polymorphism (RFLP) is one of the indirect methods of studying genetic divergence. RFLPs have been utilized to study the genomic relations among 24 mangroves and its associates (Parani et al. 1998, 2000). Their study also provided molecular data favouring the separation of *Avicennia* species from the Verbenaceae to create a monotypic family the Avicenniaceae. The separation of *Viscum orientate* into the Viscaceae was also favoured. Lakshmi et al. (2000) demonstrated that considerable intra-population and inter-population genetic variations exist in *Excoecaria agallocha*, and that lack of genetic variation is not the reason for the morphological uniformity observed across the range of the species using RFLP markers along with RAPD based markers. They also did not find any marker differentiating male and female plants. Random Amplified Polymorphic DNA (RAPD) is another indirect approach extensively utilized to study the genetic differences among different mangrove species. This method is based on polymerase chain reaction (PCR) where short sequences of DNA fragments are randomly amplified using decamer oliginucleotide sequences. The RAPD markers alone or along with other markers have extensively been used to study genetic variations in different mangrove species like, *Acanthus*, *Avicennia*, *Bruguiera*, *Excoecaria*, *Suaeda*, etc. RAPD markers have been helpful to study the parentage analysis in *Rhizophora* hybrid (Parani et al. 1997). Mukherjee et al. (2004) successfully separated two non-mangrove species of Rhizophoraceae from nine mangrove species using RAPD markers along with AFLP (Amplified Fragment Length Polymorphism) markers. Similar result did not hold true when genomic relations among 31 species were studied using RAPD and AFLP markers. Jena and Das (2006) suggested the probability of ecotype adaptability in different saline habitat in *Suaeda nudiflora* using RAPD and cytology based markers. Sahoo et al. (2007) demonstrated that molecular markers together with cytological evidence provide an effective tool to access the existing interspecific genetic polymorphism in mangrove species, to solve the taxonomic problems and to design their conservation strategy. Other PCR based markers like AFLP and ISSR (Inter Simple Sequence Repeats) have also been used to study the genetic variations in mangroves. Das et al. (2002), (2004) used RAPD and chromosome markers and Mukherjee et al. (2003) used AFLP markers to study the genomic relations in three species of *Heritiera* Alton and they got clearcut segregation of true mangrove species *H. fomes* from *H. littoralis* (mangrove associate) and *H. macrophylla* (land race). Similar result was reported in the group Rhizophoraceae where two non-mangrove species were separated out from the rest nine mangrove species (Mukherjee et al. 2004). Das and Jena (2008) also distinguish eight ecotypes of *Xylocarpus granatum* from Bhitarkanika ecosystems using RAPD markers. Ge and Sun (1999) and, Li and Chen (2004) have used ISSR and allozyme markers for studying the reproductive biology and genetic diversity in *Aegiceras corniculatum* and *Sonneratia alba* respectively. They found that the genetic diversity was very low in comparison with other mixed mating or outcrossing trees. Genetic diversity studies were carried out for intra/interspecific variations in different populations of the genus *Avicennia* (Parani et al. 1997). The variation in two *Bruguiera* species in populations from different climatic and macrogeographic regions along the western to southeastern coastal area in Sri Lanka was studied by Abeysinghe et al. (2000) using RAPD (Random Amplified Polymorphic DNA). Intra- and interspecific variation was observed using 20 primers of the 45 primers screened. Some primers showed diagnostic banding patterns and allowed unambiguous differentiation between *B. sexangula* and *B. gymnorhiza* without any indication of hybrid individuals. Species-specific restriction patterns in the genera *Rhizophora* and *Suaeda* were made and also intra-generic variations in three genera *Avicennia*, *Rhizophora* and *Suaeda* were reported (Parani et al. 2002). Analysis of mitochondrial DNA variation in the species of Rhizophoraceae (Lakshmi et al. 2002) was also studied and this again resulted in species specific profiling of different species belonging to the family of Rhizophoraceae. Analysis of expressed sequence tags from the salt-tolerant mangrove species *Avicennia marina* was also reported (Preeti et al. 2005).

The Rhizophoraceae family have only four genera (*Rhizophora*, *Ceriops*, *Kandelia*, *Bruguiera*) including 16 species, live exclusively in mangrove habitats (Tobe and Raven 1988a, b). More recent taxonomic treatments have excluded the Rhizophoraceae from the Myrtales (Johnson and Briggs 1984) and treated them as a separate order Rhizophorales, close to Linales, Malpighiales, and Geraniales (Cronquist 1981; Thorne 1992). To resolve the relationships of the family Rhizophoraceae and Anisophylleaceae, Schwarzbach and Ricklefs (2000) have sequenced the chloroplast gene *rbcL* for members of all tribes of the Rhizophoraceae and two (*Anisophyllea* and *Combretocarpus*) of the four genera of Anisophylleaceae. A large number of putative sister taxa were included to find the closest relatives of the families. This group combined information from three chloroplast markers (*rbcL*, *atpB-rbcL* intergenic spacer, and *trnL-trnF* intergenic spacer), three nuclear ribosomal DNA regions (ITS1, ITS2, and 5.8S), and morphological data to construct a phylogenetic hypothesis for the Rhizophoraceae. In addition, the evolution of morphological characters, especially those traditionally used for delimitation of systematic units in the Rhizophoraceae (e.g. fruit and seed characters) were assessed, and those regarded as adaptations to mangrove habitats (e.g. aerial stilt roots and vivipary). A cladistic analysis of sequences from the chloroplast gene *rbcL* was used to determine the systematic affinities of Rhizophoraceae and Anisophylleaceae. This analysis rejects close relationships of Rhizophoraceae with Celastraceae or Elaeocarpaceae. Tribal and generic relationships within Rhizophoraceae are evaluated with a combination of six molecular data sets (*rbcL*, *atpB-rbcL* intergenic spacer, *trnL-trnF* intergenic spacer, ITS1, ITS2, and 5.8S) and a morphological data set. These relationships are compared with

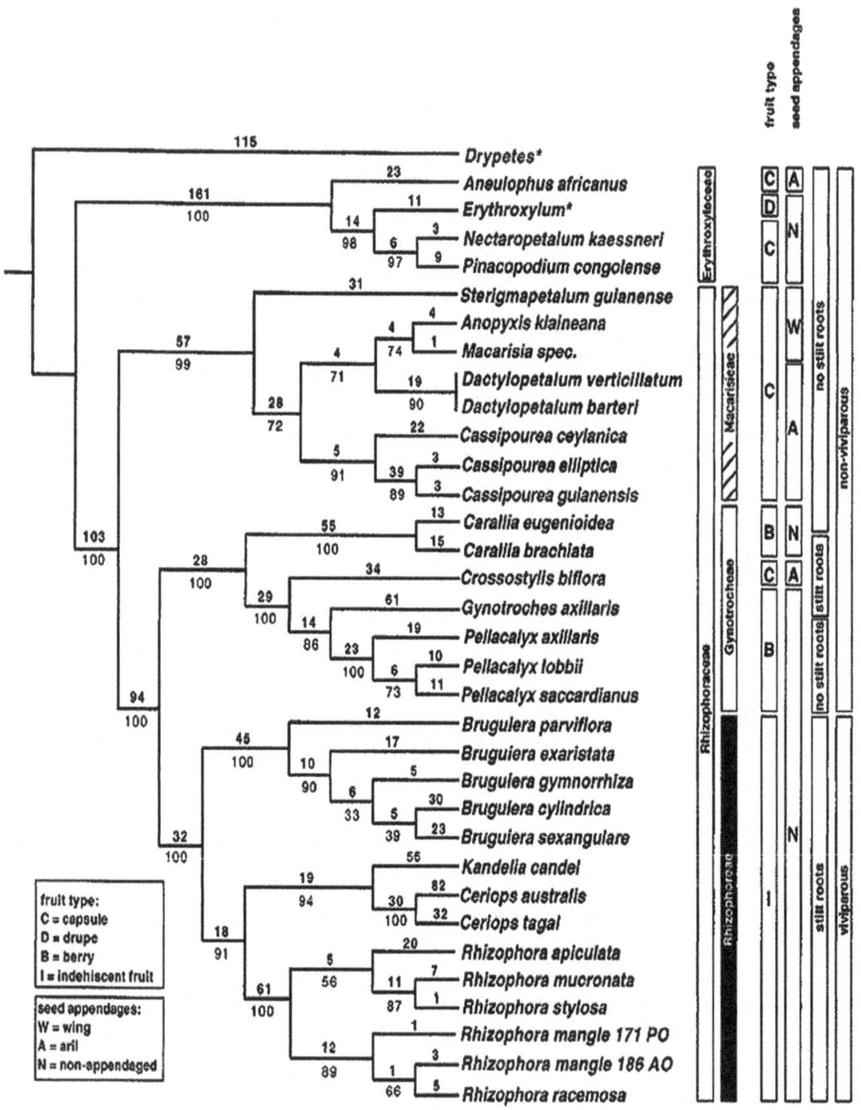

Fig. 16.8 Single most parsimonious tree based on the combined analysis of *rbcL*, *trnL-trnF* spacer, *atpB-rbcL* spacer, ITS1, ITS2, 5.8S, partial 26S sequences, and morphological characters. Figures above branches are number of character changes; figures below branches represent bootstrap values expressed as percentage of 100 bootstrap replications. Asterisks mark taxa where sequences originated from different species of the same genus. (After Schwarzbach and Ricklefs 2000, Am J Bot 87(4):547–564)

results from previous morphological cladistic analyses. Against the background of the molecular results, they briefly discussed the evolution of morphological characters traditionally used for tribal subdivision as well as characters presumably significant for adaptation to mangrove habitats, namely, aerial stilt roots and vivipary (Fig. 16.8).

Maguire et al. (2002) studied genetic diversity of *Avicennia marina* within and among individuals, populations and subspecies of *A. marina* in Australia. At the locus level, SSRs were considerably more variable than AFLPs, with a total of 52 alleles and an average heterozygosity of 0.78. Average heterozygosity for AFLPs was 0.193, but all of the 918 bands scored were polymorphic. Thus, AFLPs were considerably more efficient at revealing polymorphic loci than SSRs despite lower average heterozygosities. Three genetic criteria of importance for defining priorities for ex situ collections or in situ conservation programs (number of alleles, number of locally common alleles and number of private alleles) were correlated between the AFLP and SSR data sets. The congruence between AFLP and SSR data sets suggests that either method, or a combination, is applicable to expanded genetic studies of mangroves. The codominant nature of SSRs makes them ideal for further population-based investigations, such as mating-system analyses, for which the dominant AFLP markers are less well suited. AFLPs may be particularly useful for monitoring propagation programs and identifying duplicates within collections, since a single PCR assay can reveal many loci at once. Maguire et al. (2000a) have also described the characterisation and analysis of 16 SSR loci in *A. marina* and related species. Subsequently, the three most informative SSR loci were used to assess population genetic variation throughout the worldwide range of *A. marina* (Maguire et al. 2000b). Worldwide, a total of 70 alleles were detected at these three loci, but allelic diversity was not distributed equally across the species with populations at the extreme of the species distribution showing reduced heterozygosity and significant inbreeding. Overall, high levels of genetic differentiation were observed among-populations, with some 40 % of the total variation representing among population differences. The finding of strong genetic structure refutes the hypothesis of widespread dispersal, despite the extensive distribution of the species. To complement the SSR study of Maguire et al. (2000b) and given the efficiency promised by the AFLP technique, the aim of this study was to compare the patterns of genetic diversity in the mangrove *A. marina* as detected by AFLPs and SSRs. The specific objectives were to: (1) assess the levels of genetic diversity in *A. marina* within Australia at preselected AFLP and SSR loci, (2) to compare the informativeness of the chosen AFLP and and SSR loci at the level of individuals, populations and subspecies, and (3) to identify any implications of our findings for future mangrove conservation and breeding programs.

Level of genetic variation of mangrove trees, *Avicennia marina*, in three coastlines of Bushehr province (Southwest regions of Iran) was examined by Kahrood et al. (2008) using microsatellite markers. Three microsatellite loci which were applied in the last large-scale study, detected high levels of allelic diversity here (14 alleles in total), essential for an accurate estimation of population genetic parameters. The levels of heterozygosity detected for each population, over all loci, ranged from 0.451 to 0.667 with an average of 0.589, indicating relatively appropriate level of genetic variation. The expected heterozygosity was larger than the observed heterozygosity leading to positive inbreeding coefficients in all three populations. Reduced level of genetic variation was found in the central population indicating strong genetic structure among the other populations with larger area and less exploitation. The RAPD profiles for ten species the family Rhizophoraceae amplified with four random primers. Out of the four genera, *Kandelia* separated from *Ceriops* and *Rhizo-phora*, while *Bruguiera* joined the other genera in the cluster. Similar studies carried out in other mangrove species (Parani et al. 1997, 1998; Parida el al. 1995, 1998a, 1998b; Lakshmi et al. 1997) also suggest that mangrove species show greater variation within their population than across. It is possible that the high levels of polymorphism encountered in mangrove species such as *Avicennia marina* (Parani et al. 1997) and *Excoecaria agallocha* (Parida et al. 1998a). High diversity among individuals in these species may also be indicative of the high micro site variation within a given population. In contrast to the high levels of inter-specific variation, the levels of intra-specific variation were very low. For example, the levels of polymorphism for the two species of *Ceriops* were 29.4 % (in *C. decandra*) and 38.5 % (in *C. tagal*) only. Similarly, the levels of intra-specific polymorphism in *Bruguiera* were 40.6 % (*B. cylindrica*) and 41.17 % (*B. gymnorhiza*). *Kandelia candel* exhibited 36.36 % poly-morphism among its genotypes; the hybrid *R.* × *lamarckii* had an intermediate level of polymorphism (41.1 %) in the amplified loci compared to *R. apiculata* (38.8 %) and *R. mucronata* (44.1 %). It appears that species of genera such as *Bruguiera* and *Ceriops*, which have generally small population sizes, may have limited pollen and seed exchange across populations, thereby leading to narrow variation within but larger variation among populations. Genetic variation of four populations of *Avicennia marina* encompassing the Iranian coastal areas of Persian Gulf were studied using five microsatellite loci by Zolgharnein et al. (2010). The average number of alleles per locus per population ranged from 4 to 4.6, showing no significant difference among the four populations. The observed heterozygosity (H_O), ranging from 0.782 to 0.960 with an average of 0.864, was comparable in the Iranian populations and much higher comparing to the earlier studies on *A. marina* in the worldwide range.

Recently, ISSR (inter simple sequence repeat), has been used in studies of *Sonneratia caseolaris* (Li and Chen 2008), *Excoecaria agallocha* (Zhang et al. 2008), *Ceriops tagal* (Tan et al. 2005), and *Lumnitzera racemosa* (Su et al. 2006). To clarify genetic diversity, genetic structure, the relationship between genetic distance and geographical distance, and to provide basic data and scientific basis for effective protection, ISSR molecular marker technology was used to study the genetic diversity of *K. obovata* in China 7 populations by Chen et al. (2010).Among populations component accounted for 55.48 % of the total variation, whereas the within populations component accounted for 44.52 %, suggesting that genetic differentiation among *K. obovata* populations was relatively high. The results of the genetic diversity and cluster analysis suggest that geographical isolation of *K. obovata* populations mainly results in low gene flow and random genetic drift, leading to genetic differentiation.

16.5.2 Restriction Fragment Length Polymorphism Markers

The RFLP analysis was carried out in the ten species of mangrove family Rhizopho-reaccae, using rDNA probes and genomic clones. A high level of variation was observed for the four genera of mangroves. Nuclear ribosomal RNA genes (rDNA) have been extensively used in taxonomic studies. rDNA sequence was one of the first to be cloned in plants (Gerlach and Bedbrook 1979), and has been reported to occur in multiple copies in the eukaryotic genome. Genes coding for 18S, 5.8S and 26S ribosomal RNA in plants are highly repeated (250–22,000 copies per genome). Heterologous labelled *18S-5.8S-26S* (*pTA* 71) probe showed that the rDNA repeat unit in the Rhizophoraceae species was flanked by the enzyme *Hind III*. The sizes of the rDNA repeat in the three species of *Rhizophora*, *R. apiculata*, *R.* × *lamarckii* and *R. stylosa* were similar (i.e. 7.2 kb), while it was 10.2 kb for *R. mucronata* and for the two species of *Ceriops*, i.e. *C. decandra* and *C. tagal*. There was a high level of conservation in the sizes between the species of genera *Bruguiera* and *Kandelia candel* (about 6.7 kb). Differences among the species using three other probes, *18S*, *26S* and 5.8 were also obtained that indicates that the rDNA is highly variable for most of its length both within and between species. There seems to be a high level of speciation in the genus *Rhizophora*, which is recorded to be the oldest genus among the mangroves, with two different rDNA length. *C. ceriops* shows similarity to *R. mucronata* and could be speculated to have been the subsequent species which evolved in the mangrove habitat, followed by genus *Bruguiera*. Though the genus *Kandelia* shows length similarity to *Bruguiera* in this probe enzyme combination and in all other probe-enzyme combination used in the present study, it stood quite apart from the other three genera. It was observed that *Kandelia* and *Ceriops* shared a number of loci; both in the rDNA analysis and also the genomic clones, thereby indicating a closer affinity of *Kandelia* to *Ceriops*. For most of the species in the three genera, there was a high level of conservation, both for genomic clones and rDNA probes, though there were species-specific patterns obtained for *Bruguiera* and *Ceriops*, whereas the two species C *tagal* and *C. decandra* showed minimal differences in other DNA marker systems. There were no specific differences observed for genus *Rhizophora*. In most probe-enzyme combination *R* × *lamarckii* showed similar patterns as that of *R. apiculata*. It was observed that the hybrid *R.* × *lamarckii* showed an intermediate profile between *R. apiculata* and *R. mucronata*, for the probe *pTA71* in combination with *EcoRI*; while *R. stylosa* remained distinct. This indicates that the putative parents of the hybrid are *R. apiculata* and *R. mucronata* and not *R. stylosa* as recorded earlier. The genomic clones used in the study are useful markers and they revealed genus-specific restriction patterns. High distances observed in the dendrogram for both rDNA and the genomic clones for these genera indicate a high level of their separation from each other over long evolutionary periods.

16.6 Organellar DNA Polymorphism

16.6.1 Mitochondrial DNA

Mitochondrial DNA (mtDNA) evolves at a universally constant rate (Ochman and Wilson 1987) and the gene content is similar although not identical in all eukaryotes. This makes it a useful tool to study evolutionary divergence at all levels. Information on the mitochondrial RFLPs is limited due to the large size of the genome compared to the chloroplast genomes. The mitochondrial genome is known to have evolved at a slower rate and the evolution rates differ between lineages over long evolutionary time periods (Gillespie 1986). Hybridization of total genomic DNA with mitochondrial gene-specific probes revealed that the four genera of tribe Rhizophoreae segregate into three mitotypes; the species of genus *Rhizophora* (*R.* × *lamarckii*, *R. apiculata*, *R. mucronata* and *R. stylosa*) forming one cluster, while the three species of *Bruguiera* (*B. cylindrica*, *B. parviflora* and *B. gymnorhiza*) formed a separate cluster. The two species of *Ceriops* (*C. decandra* and *C. tagel*) formed the third cluster along with *Kandelia candel*. On the whole, the number of loci shared in the species of the genus *Bruguiera* was comparatively more, with *B. cylindrica* showing low level of differences in the restriction patterns while *B. gymnorhiza* and *B. parviflora* showed similar patterns for most probe-enzyme combinations. This therefore, indicates a highly conserved nature of mitochondrial genome between the genera *Ceriops* and *Bruguiera*. Genus *Rhizophora* revealed a higher level of species differentiation than the other two genera. In most probe-enzyme combinations, it was observed that *R. mucronata* and *R. stylosa* gave distinct profiles; while *R.* × *lamarckii* shared number of loci with *R. apiculata*. *R. mucronata* shared a low loci with *R. apiculata* and *R.* × *lamarckii*, while *R. stylosa* was distinct for most probe-enzyme combinations. It is evident from the dendrogram constructed based on mitochondrial data that even though *Kandelia* forms a cluster with *Ceriops*. Three ATP biosynthesis-related proteins are also differentially expressed in wheat under salt stress, and one of these, mitochondrial ATP synthase, is upregulated. Five proteins related to the cytoskeleton, cytokinesis, and the cell cycle are differentially expressed in wheat under salt stress. Wang et al. (2008) suggested that changes in the expression of cytoskeletal and kinesis-associated proteins could help cells adapt to salt stress and secondary osmotic stress.

16.6.2 Chloroplast DNA

The chloroplast genes have a very slow rate of nuclear substitution thereby having a slower rate of evolution than the nuclear genes (Clegg et al. 1994). Chloroplast DNA (cpDNA) is commonly used for establishing relationships from intergeneric to interfamilial levels. The chloroplast region consisting of coding and intergeneric spacer sequences of *trnS* (gene encoding for the tRNA-ser (UGA) and the adjacent *psbC* (PSII *44* kDa) gene regions was amplified with specific primers and restriction digested to reveal differences among the ten species of the family Rhizophoraceae. The PCR–RFLP of two chloroplast genes *rbcL* and *trnS-psbC* revealed that the four genera of Rhizophoreae clustered themselves into three chlorotypes. The genus *Rhizophora* with its four species (*R. apiculata*, *R. mucronata*, *R.* × *lamarckii* and *R. stylosa*) formed one cluster, the two species of *Ceriops* (*C. tagal* and *C. decandra*) along with *K. candel* formed the second cluster and the three species of *Bruguiera* (*B. cylindrica*, *B. gymnorhiza* and *B. parviflora*) formed the third. There was a high level of polymorphism among the genera (59.15 %) in all the enzyme digests, indicating a high degree of separation of the genera over evolutionary time. The amplification of the two genes in the species revealed that the size of amplified fragments is the same as that reported for other species (Ziegenhagen and Fladung 1997). No size differences were observed in any of the analysed species. Therefore the flanking regions of the two different genes (*trnS-psbC*) imply a highly conserved arrangement of the chloroplast gene region. A DNA fragment of 1380 bp representing the entire 1431 bp *rbcL* and the 1.6 kb fragment of the *trnS-psbC* was amplified with the designed primers. These amplified regions on restriction digestion with 19 enzymes revealed that sites were present for only 6 restriction enzymes in *rbcL* region and 7 restriction enzymes in *trnS-psbC* region for these ten species. Based on the restriction pattern obtained it was found that *rbcL* was more conserved in nature than *trnS-psbC*. This gene analysis revealed a very high nature of conservation among the component species of *Ceriops*, *Bruguiera* and *Rhizophora*. However, the number of bands observed in *Rhizophora* was more than the other species, suggesting a relatively recent origin of *Ceriops*, *Bruguiera* and *Kandelia* compared to *Rhizophora*.

Genetic consequence of speciation in plants is not well understood. Speciation in perennial species appears to be due to long-term isolation of large allopatric populations with slowly developing isolation mechanisms (Bousquet et al. 1992). Taxa with small population sizes may more readily result in speciation, and if genetic bottlenecks or genomic reorganization accompanies speciation, it may accelerate the fixation of new variants. This could hold good for the species of *Rhizophora* and *Bruguiera* which showed a very high degree of species separation in all the analysis. Though the genus *Bruguiera* was very variable with most markers, it showed a very highly conserved nature for the cpDNA, and mtDNA and rDNA analysis among its three species *B. cylindrica*, *B. gymnorhiza* and *B. parviflora*. In contrast, genus *Rhizophora* revealed a very high degree of speciation to the extent that the conserved region of cpDNA could reveal species-specific differences.

Genetic diversity of the three species of *Ceriops decandra*, *C. tagal* and *C. australis* was screened (Huang et al. 2008) in 30 populations collected from 23 locations in the Indo West Pacific using Inter-simple sequence repeats (ISSR) and sequences of partial nuclear gene (G3pdh) and chloroplast DNA (trnV-trnM), a cpDNA intergenic spacer as described by Taberlet et al. (1991), whereas the G3pdh is a portion of a single-copy nuclear gene encoding glyceraldehyde 3-phosphate dehydrogenase (Strand et al. 1997). A total of six haplotypes of G3pdh and five haplotypes of trnV-trnM were recognized among the three species. Only *C.*

decandra was detected containing more than one haplotype from each sequence data set (four G3pdh haplotypes and three trnV-trnM haplotypes). At the population level, genetic diversity of *Ceriops* was relativelylow inferred from ISSRs (He = 0.028, 0.023, and 0.053 in *C. decandra, C. tagal*, and *C. australis*, respectively). No haplotype diversity within population was detected from any of the three species. Cluster analysis based on ISSRs, identified three major geographical groups in correspond to the East Indian Ocean, South China Sea, and North Australia in both *C. decandra* and *C. tagal*. A study on *C. tagal* using ISSR from Thailand and South China presented low-genetic variation at the population level; more recently, cpDNA also revealed a high average haplotype diversity (Hd = 0.549) of total populations in *C. tagal* from Malay Peninsula and Borneo (Liao et al. 2006). However, dominant marker methods (RAPD, AFLP and ISSR) were successful to demonstrate differences in amplified DNA products at large-scale geographical distances within *Avicennia* species and to estimate species relationships. Hybrid testing seldom revealed hybridization among tree species. The most promising markers (microsatellites or SSR) were only recently developed and will continue to provide evidence in future studies. SSR loci in *Avicennia* seem to show relatively low levels of polymorphism, though clearly demonstrating that populations located at the edge of the species range can be even more depauperated. Populations located more central in their native range and situated along the same coastline such as reported in *Rhizophora*, are expected to be only weakly differentiated due to increased levels of gene flow (Triest 2008).

Interspecific hybridization has also been frequently reported among mangroves, especially in the four genera *Sonneratia* (Duke 1984, 1994; Tomlinson 1986; Zhou et al. 2005; Qiu et al. 2008), *Rhizophora* (Duke and Bunt 1979; Parani et al. 1997; Duke 2010; Lo 2010), *Bruguiera* (Ge 2001), and *Lumnitzera* (Tomlinson et al. 1978; Tomlinson 1986). Molecular means have been used to confirm the occurrence of hybridization in some cases (Parani et al. 1997; Ge 2001; Zhou et al. 2005, 2008; Qiu et al. 2008; Wu et al. 2009; Lo 2010). *Lumnitzera* genus comprises two species, *Lumnitzera racemosa* and *Lumnitzera littorea*. The two species differ strikingly in petal color, with white petals in *L. racemosa* and red petals in *L. littorea*. *L. racemosa* is widely distributed from East Africa to the West Pacific, including Fiji, Tonga, and northern Australia, while L. littorea largely overlaps with *L. racemosa*, except in East Africa, where *L. littorea* is not found (Tomlinson 1986). Guo et al. (2011) sequenced two low-copy nuclear genes and one chloroplast intergenic spacer (trnS-trnG) in the two *Lumnitzera* species *L. racemosa and L. littorea* species and their putative hybrid. It revealed that there were 9 and 27 nucleotide substitutions at the two nuclear loci, respectively, between one haplotype of *L. racemosa* and *L.littorea*, and that the putative hybrid showed additivity in chromatograms at these sites. Sequencing the chloroplast intergenic region trnS-trnG showed that the two *Lumnitzera* species differed by seven fixed nucleotide substitutions and four fixed insertions/deletions in this region, while the putative hybrid had identical sequences to *L. racemosa*. Molecular data clearly demonstrated that there indeed existed natural hybridization between *L. racemosa* and *L. littorea* and that *L. racemosa* was the maternal parent in this hybridization event.

16.6.3 DNA Bar Coding

Molecular phylogenetic analysis of mangroves for evolution studies of vivipary and salt secretion was studied by Shi et al. (2005). The most remarkable morphological specializations of mangroves are vivipary, salt secretion, and aerial roots. There has been a long debate on whether the complex traits vivipary and secreters have a single origin, the answer to which has profound implications for the mechanism of evolution in mangroves. A large and representative sample across mangroves was analyzed and sequenced the 18S, rRNA, rbcL, and matR genes. Overall evidence suggests the multiple origin for both vivipary and salt secretion in mangroves. Shi et al. (2005) sampled 26 representative genera of major groups (17 families) of mangroves and mangrove associates of which 8 genera (five families) germinate viviparously and 4 genera (four families) control their salt balance by secreting sodium chloride. *Nypa* is the only palm that has a viviparous fruits. Sequence results of all samples of 18S rDNA and matR mtDNA genes, showed about 73 % samples of rbcL cpDNA gene. It was concluded that both vivipary and salt secretion are most likely of multiple origins in mangroves.

16.7 Isolation of Salt Stress Genes

16.7.1 Genes from Avicennia marina

As the mangroves grow in a highly specialized climatic condition, so they are the major source of different stress tolerant genes useful for the development of agricultural crops. Cloning and expression of cycloartenol syntheses from *Rhizophora stylosa* and *Kandelia candel* have been studied by Basyuni et al. (2007, 2009, 2011). Salinity tolerant genes from the mangrove plant *Bruguiera cylindrica* have been isolated using PCR based technique suppression subtractive hybridization (SSH). Several genes in response to NaCI stress have been isolated by Takemura et al. (2002). Genes isolated from mangroves have been utilized for genetic transformation of crop plants (Hiei et al. 1994). Molecular characterization can play a role in uncovering the history, and estimating the diversity, distinctiveness and population structure. Awareness of the level of genetic diversity and the proper management of genetic resources are important issues in the modern setting. New markers deriving from DNA technologies are valuable tools to study genetic variability for conservation purposes. In the near future, the advent of genomics will give an impressive tool for genetic resources evaluation. In mangroves, though of late, efforts have been given to utilize the mangrove diversity but a concerted effort, involving scientist from all groups, is required to utilize it for human welfare. It is also needed to conserve this unique diversity for future use because due to global warming it will be difficult to grow crop plants with normal genetic set up. So, the stress tolerant genes isolated from mangroves will be extremely helpful to fight the global warming through transgenic crops.

Higher plants are generally sensitive to environmental stresses such as high salinity, drought, high temperature, high light intensity, and chilling. These environmental stresses are the main limiting factors for plant growth and distribution, and breeding of stress-tolerant trees and crops is therefore important: (1) to prevent deforestation, which causes elevation of atmospheric CO_2 concentration and global warming, and (2) to expand the cultivation area of crops for the increasing food demand resulting from the rapid growth in world population. The isolation of genes with the capability of improving stress-tolerance (anti-stress genes) is an important factor for the breeding of stress-tolerant plants, and mangrove plants are attractive resources for anti-salt-stress genes, since they grow under a condition of high salinity. In order to isolate anti-stress genes from mangrove plants, a cDNA library of *Avicennia marina* (Mehta et al. 2005) was constructed and screened for anti-stress genes by a functional expression screening with *Escherichia coli* cells. Several stress-related gene homologues, such as chaperonin-60, clpP protease of the clp/Hsp100 family of chaperones, ubiquitin, eEF1A, drought-induced AtDi19 gene of *Arabidopsis thaliana*, and secretory peroxidase, were successfully isolated. The following tables depicts some of the stress related genes in mangroves (Tables 16.2 and 16.3).

16.7.2 Genes from Bruguiera gymnorhiza

About 7029 gene expression patterns in Burmese mangrove under high-salinity stress were reported (Miyama and Hanagata 2007; Ezawa and Tada 2009). To identify genes of potential importance to salt tolerance in Burmese mangrove (*Bruguiera gymnorhiza*), they analyzed the gene expression profiles in salt-stressed mangrove using cDNA microarray containing 7029 clones based on 14,842 expressed sequence tags (ESTs) (Miyama et al. 2006). Combined results from all hybridization, they identified 287 genes with a greater than 5-fold change of at least 1 time point after salt treatment (500 mM) compared with control water-treated plant; 228 genes were up-regulated more than five-fold and 61 genes were suppressed less than one fifth. Venn diagram analysis showed tissue-specific and overlapping of these highly changed genes for up- and down-regulation, respectively. These highly up- and down- regulated genes

Table 16.2 Stress-related genes in *Avicennia marina*

Accession No. (Clone No.)	Identification	% of amino acid (%)	Species
AB049590 (Av117)	Chloroplast chaperon in-60	76	*Brassica napus*
AU108486 (Av129)	Clp protease (clpP)	69	*Arabidopsis thaliana*
AU108479 (Av119)	Ubiquitin	97	*Vicia faba*
AU108477 (Av116)	Translation elongation factor-1 α	87	*Zea mays*
AU108475 (Av113)	Draoght-induced 19 (AtDi19)	41	*Arabidopsis thaliana*
AB049589 (Av114,140)	Secretory peroxidase	84	*Nicotiana tabacum*

Table 16.3 Homologous genes whose anti-stress activity has not been reported in *A. marina*

Accession No (Clone No.)	Description	Accession No.
AU108468 (Av 106)	Protein tyrosine-serine-threonine kinase (APK1)	Dad-AC005825-16
AU108478 (Av118)	Ketol-acid reductoisomerase	Sp-Q05758
AU108481 (Av123)	Photosystem I psaI	Dad-AL049640-15
AU108482 (Av124)	*Arabidopsis thaliana* F10G19.3 protein	Dad-AF000657-3
AU108484 (Av127)	*Arabidopsis thaliana hypothetical F28A23.140 protein*	Pir-T05426
AU108485 (Av128)	*Arabidopsis thaliana* DNA chromosome 4, hypothetical protein	Pir-C71410
AU108491 (Av134)	3-deoxy-D-arabino-heptulosonate 7-phosphate synthase	Dad-Y14797-1
AU108496 (Av139)	*Arabidopsis thaliana* putative protein	Dad-AL021710-18
AU 108502 (Av147)	Endo-beta-1,4-D-glucanase (Cel8)	Pir-T01929
AU108503 (Av148)	Photosystem II 33 kDa protein	Pir-S16586
AU108505 (Av150)	*Arabidopsis thaliana* unknown protein	Dad-Z99707-8
AU108508 (Av153)	*Arabidopsis thaliana* hypothetical F12C20.9 protein	Pir-T02648
AU108512 (Av154	TRAP protein (ER membrane protein)	Dad-AC006264-8
AU108512 (Av157)	Phaseolin G-box binding protein	Dad-U183448-1
AU 108513 (Av158)	ADP-ribosylation factor-like protein	Dad-AC006439-9
AU 108514 (Av159)	Pre-mRNA splicing factor protein	Dad-AF071527-14
AU 108518 (Av164)	12-oxo-phytodienoate reductase	Dad-AF132212-1
AU 108518 (Av165)	Expansion (Exp5)	Dad-AF059489-1
AU 108522 (Av168)	Photosystem II 16 kDa protein	Pir-S00008
AU 108523 (Av169)	Cytosolic phospholopase A2 beta	Dad-AF065215-1
AU 108529 (Av175)	S-adenosyl-L-methionine decarboxylase	Ddbj-U12573
AU 108533 (Av179)	Pyridoxal-phosphate-dependent aminotransferase	Dad-AL021684-15
AU 108535 (Av183)	Phosphoinositide-specific phospholipase C	Ddbj-X93564

were characterized using gene ontology. Based on these genes, it was conducted a hierarchical clustering analysis to reveal the patterns of gene expression. The analysis identified 6 major co-expression clusters that support the interpretation of an adaptive process to high-salinity environment. The genes are as follows in Table (16.4a–f). Miyama and Hanagata (2007) identified many potentially important stress-induced genes in *Bruguiera gymnorhiza* by microarray analysis. The major response observed by microarray analysis was the differential response to salt shock between leaf and roots. cDNA microarray analysis confirmed the stress-responsive expression of a number of previously reported stress-inducible genes, such as CDPK (Berberich and

Table 16.4 List of genes identified from *B. gymnorhiza* through microarray analysis (Adopted from Miyama and Hanagata (2007)

(a) List of genes in the cluster I of *B. gymnorhiza*				(b) List of genes in the cluster II of *B. gymnorhiza*			
EST ID	Leaf [a]	Root [a]	Putative function	EST ID	Leaf [a]	Root [a]	Putative function
Bg04-08_C04	0	1	Unknown protein	Bg02-09_F11	1	0	Hypothetical protein
Bg04-09_G 10	0	1	Hypothetical protein	Bg01-08_G22	2	0	–
Bg04-09_B05	0		Putative protein	Bg01-07_C24	1	0	Leucine-rich repeat family protein
Bg02-03_L07	1		Unknown	Bg04-13_P14	0	7	Beta-glucosidase
Bg04-08_A10	0		–	Bg01-03_O24	2	0	ATPase alpha subunit
Bg04-08_A15	0		Succinate dehydrogenase iron-protein subunit-like	Bg02-08_J10	1	0	Proline-rich protein
Bg04-18_E02	1		Peroxidase	Bg01-03_I05	1	0	Hypothetical protein
Bg04-10_M18	0		CDPK	Bg01-01_E02	32	0	Phtosystem II thylakoid membrane protein
Bg04-08_B20	0		B12D-like protein	Bg01-03_F11	2	0	Hypothetical protein
Bg04-05_J02	0		–	Bg01-03_P07	1	0	Unknown
Bg04-09_E12	0		Hypothetical protein	Bg05-22_K07	1	0	Putative protein
Bg03-01_K04	1		Putative DNA binding protein	Bg04-28_G08	0	3	BURP-domain containing protein
Bg04-07_H12	0		DNAJ-like protein	Bg04-10_L03	0	3	BURP-domain containing protein
Bg04-08_C12	0		Putative GTP binding protein	Bg01-06_P08	6	0	BURP-domain containing protein
Bg04-07_F10	0		Hypothetical protein	Bg04-15_C10	0	1	–
Bg04-07_O14	0		–	Bg04-10_L16	0	1	Hypothetical protein
Bg04-07_N15	0		Polyubiquitin	Bg04-12_G18	0	1	Predicted GI-anchored protein
Bg03-01_D23	1		Alpha-expansin 9 precursor	Bg05-19_D12	1	1	Protein kinase 2

(continued)

Table 16.4 (continued)

(a) List of genes in the cluster I of *B. gymnorhiza*				(b) List of genes in the cluster II of *B. gymnorhiza*			
EST ID	Leaf [a]	Root [a]	Putative function	EST ID	Leaf [a]	Root [a]	Putative function
Bg04-08_C24	2		Actin-depolymerizing factor-2	Bg03-01_O04	1	0	–
Bg03-01_E18	1		Hypothetical protein	Bg03-10_I06	1	0	Methionine aminopeptidase I (MPI)
Bg04-08_M05	0		–	Bg03-06_N14	1	0	Vacuolar sorting protein-like
Bg04-13_G19	10		Thaumatin-like protein	Bg03-07_E12	1	0	Cytoplasmic linker protein-related
Bg04-09_N17	0		1-4, benzoquinone reductase-like	Bg01-01_K16	2	1	At5g08040
Bg04-09_G19	0		–	Bg04-24_L10	0	2	Unknown
Bg04-13_E15	2		Class IV ctitinase	Bg04-09_H15	0	5	Putative COPT5
Bg04-13_E14	0		–	Bg01-04_E19	1	0	–
Bg04-09_D03	0		–	Bg04-29_J09	0	2	Expressed protein
Bg04-08_P08	0		Phosphoserine aminotransferase	Bg03-06_P01	1	0	At4g26490
Bg04-11_J12	0		Vacuolar ATP synthase subunit B isoform 1	Bg05-20_H14	1	0	Hypothetical protein
Bg04-09_B01	0		–	Bg05-19_C15	2	0	–
Bg04-11_F05	0		–	Bg04-28_B13	0	1	–
Bg04-11_B15	0		–	Bg04-17_F19	0	1	Hypothetical protein
Bg04-28_F01	10		Thaumatin-like protein	Bg05-23_J10	1	0	At5g46640
Bg04-11_C24	0		Unknown protein	Bg01-04_C01	2	0	AvrRpt2-induced protein 2-like
Bg04-16_D22	0		Miraculin precursor (MIR)	Bg05-17_J07	1	0	–
				Bg05-19_D13	2	0	–
				Bg05-15_K18	1	0	–
				Bg05-17_B22	1	0	MAP kinase

(continued)

Table 16.4 (continued)

(a) List of genes in the cluster I of *B. gymnorhiza*				(b) List of genes in the cluster II of *B. gymnorhiza*			
EST ID	Leaf [a]	Root [a]	Putative function	EST ID	Leaf [a]	Root [a]	Putative function
				Bg05-25_M07	1	0	–
				Bg05-21_N19	1	0	Lysine decarboxylase-like protein
				Bg03-07_H06	1	0	–
				Bg01-04_B07	1	0	Heat-shock protein
				Bg03-03_M17	2	1	Unknown
				Bg01-04_G07	1	0	–
				Bg01-04_G01	2	0	Glutamine cyclotransferase precursor-like rotein
				Bg03-09_O03	1	0	Unknown
				Bg05-19_B04	1	0	–
				Bg05-22_M09	1	0	Atlg80190
				Bg04-17_P02	0	1	Putative NAC domain protein NAC2
				Bg04-11_H03	1	1	14-3-3-family protein
				Bg05-12_D21	1	0	–

List of genes in the cluster III of *B. gymnorhiza*				List of genes in the cluster VI of *B. gymnorhiza*			
EST ID	Leaf[a]	Root[a]	Putative function	EST ID	Leaf[a]	Root[a]	Putative function
Bg03-03_M09	1	0	Unknown	Bg03-06_O08	10	69	–
Bg03-01_F14	1	0	Putative protein	Bg04-29_D13	34	28	Blight-associated protein p12 precursor
Bg04-07_D24	0	1	F5O11.29	Bg04-02_A15	5	23	Thaumatin-like protein 1 precrsor
Bg04-07_D14	1	0	Bg70	Bg04-11_J14	0	1	Putative TAF5
Bg02-02_112	3	0	3-ketoacyl-CoA thiolase	Bg02-02_N07	1	0	F3M18.8

(continued)

Table 16.4 (continued)

List of genes in the cluster III of *B. gymnorhiza*				List of genes in the cluster IV of *B. gymnorhiza*			
EST ID	Leaf [a]	Root [a]	Putative function	EST ID	Leaf [a]	Root [a]	Putative function
Bg02-01_D12	1	0	Putative Serine/agrinine rich protein	Bg02-02_F05	2	1	Actin
Bg01-05_E11	29	3	Bg70	Bg04-30_J16	2	1	Hypothetical protein
Bg04-12_P20	0	1	Unknown	Bg03-01_I21	2	1	Remorin 1
Bg04-07_C24	0	1	Putative respiratory burst oxidase protein E	Bg04-13_E17	1	1	Unknown protein
Bg04-12_K22	0	1	Hypothetical protein	Bg04-13_N16	1	4	Caffeic acid *O-methyl-transferase*
Bg02-02_N05	3	3	Fiber protein Fb11	Bg04-10_O10	0	26	Bacterial-induced class III peroxidase
Bg03-01_I01	2	0	At4g09730	Bg04-13_B09	0	11	OSJNBb0059K02.17
Bg01-01_B13	85	0	Bg70	Bg04-30_N13	0	11	OSJNBb0059K02.17
Bg04-08_B17	0	1	Unknown	Bg04-10_C24	0	12	–
Bg02-05_D22	89	0	Bg70	Bg04-09_H02	0	1	Alcohol dehydrogenase class iii
Bg02-06_P12	6	4	60S ribosomal protein L10	Bg04-18_K02	0	1	–
Bg02-02_K04	1	1	Serpin-like protein	Bg01-03_I19	12	2	Acidic chitinase iii
Bg01-01_F13	29	8	Bg70	Bg04-01_A10	0	12	Miraculin precursor (MIR)
Bg03-01_F10	2	0	Unknown	Bg04-03_L22	0	1	–
				Bg04-06_G05	0	1	xyloglucan-specific fungal endoglu-canase inhibitor protein precursor

(continued)

Table 16.4 (continued)

List of genes in the cluster III of *B. gymnorhiza*				List of genes in the cluster VI of *B. gymnorhiza*			
EST ID	Leaf [a]	Root [a]	Putative function	EST ID	Leaf [a]	Root [a]	Putative function
List of genes in the cluster IV of *B. gymnorhiza*				Bg13-01_M13	12	9	Endo-1,3-beta-glucanase
EST ID	Leaf[a]	Root[a]	Putative function	Bg04-04_G06	0	4	Bg70
Bg04-11_L10	2	2	60S acidic ribosomalprotein P3	Bg04-03_M21	0	3	–
Bg01-03_H20	20	1	Putative metallothionein like protein	Bg04-17_J19	0	1	–
Bg01-07_O19	1	0	–	Bg05-14_I07	1	0	–
Bg04-14_E13	0	5	–	Bg04-03_B07	0	7	Pathogenesis-related protein PR-6 type
Bg03-03_J10	1	0	–	Bg04-20_L18	0	3	–
Bg02-02_I23	1	0	Similar to WOSE3.3	Bg04-17_B20	0	1	–
Bg04-11_L02	0	1	–	Bg01-01_A10	1	1	Unknown protein
Bg04-30_C22	0	1	Extension-line protein	Bg04-07_L17	4	1	–
Bg04-04_M01	0	2	Putative peroxidase ATP2a	Bg04-11_P18	1	3	Unknown protein
Bg04-11_K09	0	1	Unknown protein	Bg04-16_L08	0	1	At5g26340/f9D12_17
				Bg04-08_B19	2	0	Unknown protein
List of genes in the cluster V of *B. gymnorhiza*				Bg04-16_F18	0	5	Beta-cyanoalanine synthase
EST ID	Leaf[a]	Root[a]	Putative function	Bg02-06_O20	1	4	Caffeic acid O-methyltransferase
Bg03-01_A11	1	0	P0514g12.25	Bg04-07_C15	0	1	Alpha-mannosidase
Bg01-06_I05	19	6	Major latex protein	Bg05-16_O22	2	0	Ankyrin repeat family protein
Bg03-07_E11	1	0	DNAJ like protein	Bg03-01_L04	1	0	Hypothetical protein

(continued)

Table 16.4 (continued)

List of genes in the cluster V of *B. gymnorhiza*				List of genes in the cluster VI of *B. gymnorhiza*			
EST ID	Leaf [a]	Root [a]	Putative function	EST ID	Leaf [a]	Root [a]	Putative function
Bg01-05_N18	2	0	Alpha-expansion 2	Bg04-07_H11	0	1	Unknown protein
Bg01-03_K06	5	1	Bg70	Bg02-02_K14	1	0	Receptor protein kinase
Bg03-02_L03	1	0	–				
Bg01-01_E23	5	4	Aquaporin				
Bg01-06_G23	2	0	Chlorophyllase 1				
Bg03-01_P16	1	0	Expressed protein				
Bg03-01_D19	4	1	At2g47610/ T30B22.8				
Bg03-03_O04	1	0	Oxysterol-binding protein-like				
Bg04-17_P05	74	7	Bg70				
Bg03-02_P11	1	0	Unknown				
Bg02-02_N19	1	0	3′(2′),5′-bisphosphate nucleotidase				
Bg01-03_P12	1	0	CYP82Clp				
Bg01-01_I02	2	3	Light-inducible protein ATLS1				

[a] numbers of ESTs of contig

Kusano 1997), peroxidase, vacuolar ATPase, and several PR-protein family genes (Hoffmann-Sommergruber 2002). These results indicate that there are similar molecular mechanisms of stress tolerance and responses between Burmese mangrove and model plants. Significant up-regulation of unknown genes or mangrove specific genes (Bg70) was also observed. Furthermore, several genes including BURP-domain containing protein (RD-22 homologue) showed completely opposite results (Banzai et al. 2002a). Although these results identify hundreds of potentially important transcriptome changes, the biochemical functions of many stress-regulated genes remain unknown. Computational analysis of these clones allows annotation of putative gene functions through similarity searches in nucleic acid and protein databases, over one-third (34.5 %) of the possible coding sequences have no matches in current public databases and, hence, remain novel sequences with unknown functions. Determination of the biological functions of these genes is among the greatest challenge for post-genomic research. Further analysis using transgenic plants of these stress inducible genes will identify salt tolerant determinants from these results set and provide more information about the functions of the stress-inducible genes involved in stress tolerance. cDNA expression libraries were constructed from salt-treated roots and leaves using the host organism *Agrobacterium tumefaciens* to identify key genes in the regulation of salt tolerance in the mangrove plant *Bruguiera gymnorhiza* by Ezawa and Tada (2009).

Functional screening of the *Agrobacterium* libraries identified 44 putative salt tolerance genes in *B. gymnorhiza* (Table 16.5). A cDNA clone which is homologous to an unknown cDNA from the mangrove plant *K. candel* and the *cyc02* gene from *Catharanthus roseus* conferred the highest level of salt tolerance at 450 mM NaCl to *A. tumefaciens* indicating that it plays a major role in the regulation of salt tolerance in mangrove plants. Transgenic *Arabidopsis* plants expressing *Bg70* and *cyc02* homologue exhibited increased tolerance to NaCl. Although the generation and analysis of transgenic plants expressing each identified gene need to properly evaluate the gene function, at least two genes, *Bg70* and cyc02 homologue, successfully conferred increased salt tolerance to *Arabidopsis* plants.

Table 16.5 cDNA sequences from B. gymnorhiza that confer salt tolerance to *Agrobacterium*

Putative identity and accession number of matching sequence	Full length[a]	RNA origin[b]
Reported genes from *B. gymnorhiza*		
B. gymnorhiza Bg70 mRNA. complete cds	Y	L
BURP domain-containing protein (*B. gymnorhiza*)	N	L
BURP domain-containing protein (*B. gymnorhiza*)	N	L
B. gymnorhiza psoA mRNA for photosystem II Qb protein	N	L
B. gymnorhiza 1AS mRNA for lipoic acid synthase	N,N,N,	L,U
Known functions (except for genes from *B. gymnorhiza*)		
Vegetative storage protein (*Populus tmlsamifera subsp. trichocarpax Populus deltoides*)	N,Y	L.L
Major storage protein (*Populus x canadensis*)	Y,N	L.L
Ribulose-1,5-bisphophate carboxylase/oxygenase small subunit (*Vinspse udoreticuiata*)	Y	L
Ribulose 1,5-folsphosphate carboxylase small subunit	Y	L

(continued)

Table 16.5 (continued)

Putative identity and accession number of matching sequence	Full length[a]	RNA origin[b]
CYP92B3 (*Affcoffana tatacum*, (P450)	Y	L
CYP82C1p (*Glycin max*).(P450)	N	L
Citrus sinensis allene oxide synthase (AOS) mRNA (P450)	N	L
Putative metallothionein-like protein (*Vitis vinifera*).	Y	L
Non-specific lipid transfer protein precursor (*Fragaria xananassa*).	N	L
Caucus carota glyceraldehyde 3-phosphate dehydrogenase (C3PDH)	Y	R
Abscisic stress ripening protein (*Vitis pseudareticutota*)	N	L
Pranus aw'um putative glycine-rich RNA-binding protein mRNA	N	L
Zinc finger (C3HC4-type RING finger) family protein (*A thaliana*)	N	L
Citrus medico partial mRNA for putative extensih-like protein	N	L
A thaliana H * -transporting ATP synthase-like protein	N	L
Fyrus communis PVP3 mRNA for vacuolar proton-inorganic pyrophosphatase	N	R
A thaliana ENTH domain-containing protein (AT5C35200) mRNA	N	L
A t/iafimra 0 J-1 family protein/protease-related (AT3G02720) mRNA	N	L
A, thaliana DNAJ heat shock N-terminal domain-containing protein (AT4G09350) mRNA	Y	L
V. vinifira cultivar Maxxa chloroplast	N	L
Cicer arietinum clone CaHal 1 CA-like protein mRNA, partial cds	N	L
A. thaliana 60S ribosornal protein L17 (T27C4.4)mRNA	N	R
A thaliana 60S ribosomal protein L7A (RPL7aB)	Y	L
Unknown function		
K. candel unknown mRNA	Y,Y	U
Solanum lycopersicum cDNA. clone: LEFL1020AC12. HTC in leaf.	N	L
Oryza sativa (indica cultivar-group) cDNA clone:OSIGCFA236El 1	Y	L
A. thaliana binding (AT5C54440) mRNA	Y,Y	U
P. trichocarpa x P. deltoides clone WS0138_O15 unknown mRNA	N	L
P. trichocarpa clone WSO 1 1 5_F1 4 unknown mRNA	Y	L
P. trichocarpa clone WS0124_K1 6 unknown mRNA	Y	L
P. trichocarpa clone WS01 120J23 unknown mRNA	Y	L
P. tricnocarpa done WS01228JM8 unknown mRNA.	Y.Y	R.L
P. trichocarpa clone WS0126J04 unknown mRNA	N	L
Populus EST from severe drought-stressed opposite wood	N	R
Populus EST from mild drought-stressed leaves	N	L
No significant similarity found.	–	All

[a] Putative full length or partial cDNA are represented as Y and N, respectively
[b] Isolation from the library constructed with RNA from root (R) or leaf (L)

16.7.3 Genes from *Bruguiera sexangula* and *B. cylindrica*

In the recent years, many studies (Takemura et al. 2002) have been performed on *Bruguiera* species to explore the potential of genes belonging to trees in this family in salinity tolerance. A Na +/H + antiporter, catalyzing the exchange of Na+ ion for H+ ion across the vacuolar membrane, and may be responsible for the ultrafiltration in roots, was reported from *Bruguiera sexangula* (Tanaka et al. 2000). Through the isolation and characterization of genes involved in salinity stress from *Bruguiera* species, detoxification of reactive oxygen species (ROS) by superoxide dismutase (Takemura et al. 2000), osmotic adjustments via sucrose biosynthesis (Banzai et al. 2002a), increase in total amino acid pool especially proline (Parida et al. 2002) and polyphenol, were suggested as some of the mechanisms that also contribute to salinity tolerance of these mangrove trees. Among the genes that were isolated were also genes involved in other salinity tolerance adjustments, such as the BURP domain-containing proteins (Banzai et al. 2002a), allene oxide cyclase (AOC) or 'mangrin' (Yamada et al. 2002a, b) and cytosolic chaperonin containing TCP-1a (CCTa) homologue (Yamada et al. 2002c). Isolation of salinity tolerant genes from the mangrove plant, *Bruguiera cylindrica* by using suppression subtractive hybridization (SSH) and bacterial functional screening was reported by Wong et al. (2005). In this study, they have identified and isolated 126 salinity tolerant cDNAs from the root of a mangrove plant, *Bruguiera cylindrica* by using suppression subtractive hybridization (SSH) and bacterial functional screening. Sequencing of 51 subtracted cDNA clones that were differentially expressed in the root of *B. cylindrica* exposed to 20 parts per thousand (ppt) NaCl water revealed 10 tentative unique genes (TUGs) with putative functions in protein synthesis, storage and destination, metabolism, intracellular trafficking and other functions; and 9 unknown proteins. Meanwhile, the 75 cDNA sequences of *B. cylindrica* that conferred salinity tolerance to *Escherichia coli* consisted of 29 TUGs with putative functions in transportation, metabolism and other functions; and 33 with unknown functions. Both approaches yielded 42 unique sequences that have not been reported else where to be stress related and might provide further understanding of adaptations of this plant to salinity stress. Tables 16.6 and 16.7 will provide some insights into the isolated salt stress genes in *B. cylindrica*.

16.7.4 Transcriptome and Transcriptional Analysis in Salt Stressed Mangroves

Transcriptional response of *Bruguiera gymnorhiza* to high salinity (salt stress; 500 mM NaCl) and hyperosmotic stress (osmotic stress; 1 M sorbitol) by microarray analysis was done by Miyama and Tada (2008). It revealed that 865 of 11,997 genes showed significant differential expression under salt and osmotic stress. Hierarchical clustering of the 865 genes showed that expression profiles under salt stress were distinctly different from those under osmotic stress.

Table 16.6 Subtracted cDNA sequences from *B. cylindrica* that were isolated by suppression subtractive hybridization (SSH)

Putative identity	Number of cDNAs	Origin of matching sequence	Source	Score
Ubiquitin	1	AAC49014	*Zea mays*	111
NTCP23-like cysteine proteinase	1	aak07729	*Nicotiana tabacum*	57
Aspartic protease	1	aak55849	*Manihot esculenta*	207
Putative translation factor	1	cac84489	*Pinus pinaster*	138
Ubiquinol-cytochrome C reductase complex 7.8 kDa protein	6	np_172964	*Arabidopsis thaliana*	137
Putative glutathion S-transferase	1	aag16758	*Lycopersicon esculentum*	64
Small-Ras-like GTP-binding protein	5	aaq54569	*Malus x domestica*	101
GIGANTEA	6	aaf00023	*Arabidopsis thaliana*	125
Probable membrane protein ylaB	1	h64775	*Escherichia coli*	292
Putative reverse transcriptase	1	cad59768	*Cicer arietinum*	58

Comparison of gene ontology (GO) categories of differentially expressed genes under the stress conditions revealed that the adaptation of Burmese mangrove to salt stress was accompanied by the upregulation of genes categorized for "cell communication," "signal transduction," "lipid metabolic process," "photosynthesis," "multicellular organismal development," and "transport," and by down-regulation of genes categorized for "catabolic process." Burmese mangrove maintained its leaf water potential and recovered from its photosynthesis rate that declined temporarily under salt stress, but not under osmotic stress. It was suggested that salt tolerance of *B. gymnorhiza* might be attributed to their ability to accumulate high concentrations of Na^+ and Cl^-, even under non-stressed conditions; to uptake additional Na^+ and Cl^- for use as osmolytes; and to maintain K^+ homeostasis under salt stress. The transcriptome of a highly salt tolerant mangrove species, *Sonneratia alba*, was sequenced by Chen et al. (2011). Over 15 million 75-bp paired-end reads were assembled into 30,628 unique sequences with an average length of 581 bp. Of them 2358 SSRs were detected, with di-nucleotide repeats (59.2 %) and tri-nucleotide repeats (37.7 %) being the most common. Analysis of codon usage bias based on 20,945 coding sequences indicated that genes of *S. alba* were less biased than those of some microorganisms and *Drosophila* and that codon usage variation in *S. alba*. Genome-wide gene ontology (GO) assignments showed that *S. alba* shared a similar GO slim classification with *Arabidopsis thaliana*. High percentages of sequences assigned to GO slim category 'mitochondrion' and four KEGG pathways, such as carbohydrates and secondary metabolites metabolism, may contribute to salt adaptation of *S. alba*. In

Table 16.7 cDNA sequences from *B. cylindrica* that confer salinity tolerance to *E. coli*

Putative identity	Number of cDNAs	Source	Score
COG1226: Kef-type K+ transport systems predicted NAD-binding protein	2	*Desulfovibrio desulfuricans* G20	230
Putative potassium transporter HAK1p	1	*Mesembranthemum crystallinum*	89.4
COG1629: outer membrane receptor proteins mostly Fe transport	1	*Burkholderia fungoram* LB400	151
GNS1/SUR4 membrane family protein	1	*Arabidopsis thaliana*	150
Bifunctional carbohydrate binding and transport protein	1	*Rhodopirellula baltica* SH 1	111
Oligopeptide ABC transporter, permease protein	1	*Thermotoga maritima* MSB8	120
Cytochrome B6-F complex iron-sulfur subunit 1, chloroplast	1	*Nicotiana tabacum*	357
COG1845: heme/copper-type cytochrome/quinol oxidase, subunit 3	1	*Bnovosphingobium aromaticivorans* DSM 12444	108
Isocitrate dehydrogenase	1	*Vibrio vulnificus* YJ016	153
Probable cytochrome P450 monooxygenase (NDP forming)	1	*Zea mays*	109
COG1042: acyl-CoA synthetase (NDP forming)	1	*Ralstonia metallidurans* CH34	105w
FrdA	1	*Mycobacterium avium subsp. paratuberculos* str. K10	213
COG0774:UDP-3-O-acetylglucosamine deacetylase	1	*Pseudomonous fluorescens* PfO-1	99.8
Cytidine deaminase (CDD) cytidine aminohydrolase	1	*Arabodopsis thaliana*	282
COG1024: enoyl-CoA hydratase/camithin racemase	1	*Ralstonia eutropha* JMP 134	59.3
Probable NADH2 dehydrogenase (ubiquinone)	1	*Citrullus lanatus*	61.6
COG1529: aerobic-type carbon monoxide dehydrogenase, large sub unit CoxL/CutL homologs	2	*Rhodospirillum rubrum*	108
COG1974: SOS-response transcriptional repressors	4	*Rubrivivax gelatinosus* PM1	233
Putative zinc finger protein	1	*Oryza sativa (japonica* cultivar group)	127
LexA repressor	2	*Salmonella enterica* subsp. *enterica serovar* Typhi str. CT18	371
Ribosomal protein L36	1	*Arabidopsis thaliana*	74.3
Probable tRNA-guaninine transglycosylase	1	*Arabidopsis thaliana*	164
COG1706:flagellar basal-body P-ring protein	1	*Magnetococcus* sp MC-1	102
Ethylene responsive family protein	1	*Arabidopsis thaliana*	200

(continued)

Table 16.7 (continued)

Putative identity	Number of cDNAs	Source	Score
Hemolysin		*Acanthamoeba polyphyga*	53.1
Putative senescence-associated protein	2	*Pisum sativum*	254
Putative senescence-associated protein	1	*Pisum sativum*	85.9
Tie20 family protein (Arabidopsis thaliana)	1	*Arabidopsis thaliana*	57.8
Probable disease resistance protein At4g33300	2	*Arabidopsis thaliana*	244

addition, 1266 unique sequences matched to 273 known salt responsive genes (gene families) in other species were screened as candidates for salt tolerance of *S. alba*, and some of these genes showed fairly high coverage depth. They identified four genes (AtRUB1 (related to uniquitin), cyclin D3-2, LAG1 homolog 2, Metallothionein-like protein 2) with signals of strong diversifying selection by comparing the transcriptome sequences of *S. alba* with 249 known ESTs from its congener *S. caseolaris*. Abundant SSR markers, salt responsive genes and four genes with signature of natural selection obtained from *S. alba* provide abundant sequence sources for future genetic diversity, salt adaptation and speciation studies.

Key genes in the regulation of salt tolerance in the mangrove plant *Bruguiera gymnorhiza*, transcriptome profiling in the lateral and main roots under conditions of salt stress was performed by Yamanaka et al. (2009). It revealed that 175 and 403 of 11,997 genes showed significantly increased high expression in the lateral and main roots respectively. One hundred and sixty genes were up-regulated in both types of roots in the early time period, 1 to 12 h after salt treatment. Expression vectors for 28 selected salt responsive genes were constructed and transformed in *Agrobacterium tumefaciens*, and then screened for salt tolerance. *A. tumefaciens* transformed with genes for lipid transfer, zinc finger, and ankyrin repeat proteins showed enhanced salt tolerance. Transgenic *Arabidopsis* plants expressing these three genes also exhibited increased tolerance to NaCl. The primer sequences of the RT PCR and coding region of each cDNA used by the Yamanaka et al. (2009) are given below (Tables 16.8, 16.9). The use of plants as host organisms in overexpression screening should result in the identification of different sets of salinity tolerance genes. Analysis of transgenic plants expressing the genes identified in these type of study, and functional screening using plants as hosts, which should identify additional genes involved in salt tolerance in the mangrove plant.

Yang et al. (2011) explored the expression profiles of *Hibiscus tiliaceus* under salt stress using a full-length cDNA microarray. Four hundred and eighty-six salt-responsive unigenes were identified in *H. tiliaceus*; 224 of which had high sequence similarity to *Arabidopsis*. Many genes identified are known to be salt-stress responsive. The-physiological performance of *H. tiliaceus* under salt stress suggests decrease in ratio of K^+/Na^+ and negative influence on photosynthesis of *H. tiliaceus*. It was evident that, *H. tiliaceus* evolved its own mechanisms to regain both ionic and osmotic homeostasis through coordinated engagement of genes associated

Table 16.8 List of primer sequences for real-time PCR of *B. gymnorhiza*

Gene ID	Annotation	Size of RT-PCR products (bp)	Forward primer	Reverse primers
LC354contig I	Unknown protein	120	AGTGGCCATGGATATGAAGG	TTTGCTGGTCCTACGGAAAC
Bg04_13_101	Zinc-finger protein	103	GTCGTACCAAGCCCTAGGTG	GACCGTGGGAACAGAGTTGG
CL1291contig 1	Ankyrin repeat protein	104	GCAGTTGAACTGGGAGATGTG	GCAACGTGGAAAGCAGTGTC
CL1184Ccntig 1	ACC oxydase	115	CTTCCGTGGGGACAAAAGTGG	CGATCGTCCTGCAGCAAGAG
CL1228Ccntig 1	DnaJ-like protein	115	CCGAAGATTAGCTAGGACTTG	CTCGGGATCGGATAGAGTAG
Bg04-02_P08	BURP domain containing protein	100	CGCAGCAACAAACCGTAACC	AACGGCGGTTGTACCTCAAG
Bg05-07C15	Glyceraldehyde 3-phosphate dehydrogenase (GAPD)	91	TCCTTCCTCTCACTGCTCAAATC	CACCAGCCTACCGATCCTTC
	A. tumefaciens translation initiation factor IF-2	100	AATGCCGAGAGCACCGATCAG	TGGCCTTCTGACGCTCGTTTC
	A. tumefaciens 30S ribosomal protein S15	110	GGGCGTGTCGATGAAGGTGA	CCATCCGTCAGGCAAATGTGG

Table 16.9 List of primer sequences for cDNA amplification in *B. gymnorhiza*

Gene ID	Annotation	Forwarded primer for cDNA amplification	Reverse primers for cDNA amplification
CL894Contig 1	Lipid transfer protein	CACCATGGCTCATTACATGAGAGCTTGC	TGGCCTTGCCATATTATGGAAC
Bg01-06_E13	Vegetative storage protein	CACCATGGCTGTGGCTGGCCGCATTC	GACTTCACAGTACTTGCATACAGC
CL793Contig1	Harpin-induced protein	CACCATGGCCGAGAAACAACCCC	AAGGAACCTTCAAGTCGGACTCAG
Bg05-10D05	CCR protein	CACCATGACTATAATCCCACCAAAG	CTCTTAATAGACTCCCCTCTG
Bg04-23_N13	Hypothetical protein	CACCATGGATCTCCGACCGCTACTTC	AGAAGACTGTCAATGGGGACC
CL354Contig 1	Unknown protein	CACCATGAAGAAAGTCATATTGAAATTGG	TTAACAAATGACACAAGCATTAGGG
Bg04-04_P02	Transcription factor	CACCATGTTGGCGAGGAAGCCACCG	CGGTCATCAAGCTGTTTTCACC
Bg02-08_D17	Aquaporin	CACCATGCCGATCACTAGAATTGCAT	GACTTCAATCTTTCAAGTAATCTGC
Bg04-02_K01	Late embryogenesis abundant protein	CACCATGGCGGGCTTGCTGGACAAG	GGCTTGAGATTAGCCCAAGTG
Bg04-13_I01	Zinc-finger protein	CACCATGGCACTAGAAGCTTTGAATTCT	GCGGAGAAATTAATCTTGAGATGC
Cl528Contig 1	Hypothetical protein	CACCATGGCAACGGTGTTGGCTG	CAATTCGGTTCATGACAAAACC
Bg02-09_M07	TPA	CACCATGCCTTACTACCTCGCACAA	TACTCAATCGCGGATAGCGTG
Bg02-08_H08	Hypothetical protein	CACCATGGCGTCTCCTTGTTTAACT	CATCTCACAAATTCATATTGACC
Bg04-28_G06	Transcription factor	CACCATGTACGGACAGAGCATGGCT	CTTGGCGGACTCTATTTACCA
Bg04-25_H02	Hypothetical protein	CACCATGGCACGGCTGCAGCAATCC	ACTTAATCATTTGGAGATCTC
CL1291Contig 1	Ankyrin repeat protein	CACCATGAATCAGGATTTAAAGGAGGC	ATGTGCTCTCTAGTCTCCCTG
Bg04-16_K15	Zinc-finger protein	CACCATGGCACTGGAAGCTTTGAATTCTCC	TCCTGCTCAGAAGATGAGTCG
Bg05-20_B04	Hypothetical protein	CACCATGCAGGATTACAGCAATATG	CATCCGTCAAGCGATAGCC
CL1132Contig 1	Myb	CACCATGGGGAGGGCTCCTTGCTG	AATTCAGGCTGCTTGGATGGG
CL1184Contig 1	ACC oxydase WRKY	CACCATGGAGATTCCAGTGATTAG	TGTGGTAAATTAGGCCTGGAG
CL764RACE	WRKY	CACCATGGAATCAAAACGGGTGAAC	ACCTCAAAATGTGGTTCCTGAGAGGAC
Bg04-13_B17RACE	Heavy metal transporter	CACCATGAAGCTAGCAAAACAGTTGCG	GTGAAATTAAGAGAACCCTGGGGTATC
CL1467Contig 1	Hypothetical protein	CACCATGGTCATCAAGTTATCATGGATG	AGCATATGGTCACACGATGGA
Bg05-25_I07	Hypothetical protein	CACCATGGAGTTTCATCAACAAAAG	CCCTATTCATCTTGATTCCTG
Bg05-15_I02	Nam-like	CACCATGAGCCCTGGTGGGTCATA	GCTAGTAACTTATAGCTCTG
CL311Contig 1	Transcription factor	CACCATGAAAGGAGCAGAGATAGAG	GAGTTATTACAATGGCTTCTGCC
CL666Contig 1	Hypothetical protein	CACCATGGAAGACGAGTCTCAGAGG	TGCTGCTTACGCCACCATCTG
Bg04-24_K17		CACCATGCAGAACACTGAGAATAAAGAG	AATCTAAGGTCTTTGGCC

Table 16.10 Primer sequences for quantitative RT-PCR used in *H. tiliceus*

Gene	Forward primer sequences (5′-3′)	Reverse primer sequences (5′-3′)
18S rRNA	GGCTGAGGTCTCGCTCGTTA	ATTGACGGAAGGGCACCAC
INPS1	AACACATGCGAGGACTCCCT	GGGCCTTGGTGAGGTAACTG
NHX1	TGAGTGCCTGCCAAAGTACA	TGCTTTTCAATGCAATCCAG
ERD3	TGTTATCGACGATCCCCTATGG	TGGCTCTCGGCAGTGAAGA
GMP1	GGTCAGCCGAGGGACTACATT	AACGCAGCCTGGACCAATT
No-hit transcript	GGAAAATCGACGTGGCAGTC	CAAAGGAACCGTGTAAGCCC

with gene transcription, signaling, and downstream cell transport and detoxification pathways. The primer sequences used in RT-PCR for *H. tiliceus* are given at Table 16.10. Salt stress, one of the primary abiotic stresses, greatly inhibits plant growth and development and affects crop production. More than 800 million hectares of land are salt affected (FAO 2008), accounting for more than 6 % of the total land area in the world. Increased salinization of arable land is expected to have devastating global effects: 30 % land loss is predicted within the next 25 years, and up to 50 % by 2,050 (Wang et al. 2003). Salinity stress response is multigenic, since many processes are involved in modulation of biochemical activities and development based on stress sensing and salt-stress responsive signal transduction, such as various compatible solute/osmolytes, polyamines, reactive oxygen species (ROS) and antioxidant defense mechanism, ion transport, and compartmentalization of injurious ions (Hasegawa et al. 2000a, c; Zhu 2001, 2002). Many non-model plants (e.g. mangroves) highly adapt to local saline conditions and have attracted more and more attention in salt-tolerance studies (Fu et al. 2005; Miyama and Hanagata 2007). Expression profiles of *Hibiscus tiliaceus*, a mangrove associate, under salt stress using a full-length cDNA microarray was studied by Yang et al. (2011). Four hundred eighty-six salt-responsive unigenes were identified in *H. tiliaceus*; 224 of which had high sequence similarity to *Arabidopsis*. Many genes identified are known to be salt stress responsive. Physiological analysis displayed decrease in ratio of K^+/Na^+ and negative influence on photosynthesis of *H.tiliaceus*. It reveled that to survive under high salinine intertidal environments, *H. tiliaceus* evolved its own mechanisms to regain both ionic and osmotic homeostasis through coordinated engagement of genes associated with gene transcription, signaling, and downstream cell transport and detoxification pathways.

The transcriptome of a highly salt tolerant mangrove species, *Sonneratia alba*, was sequenced using the Illumina Genome Analyzer (Chen et al. 2011). Over 15 million 75-bp paired-end reads were assembled into 30,628 unique sequences with an average length of 581 bp. Of them, 2358 SSRs were detected, with di-nucleotide repeats (59.2 %) and tri-nucleotide repeats (37.7 %) being the most common. Genome-wide gene ontology (GO) assignments showed that *S. alba* shared a similar GO slim classification with *Arabidopsis thaliana*. High percentages of sequences assigned to GO slim category 'mitochondrion' and four KEGG pathways, such as carbohydrates and secondary metabolites metabolism, may contribute to salt adaptation of *S. alba*. In addition, 1266 unique sequences matched to 273 known salt responsive genes (gene families) in other species were screened as candidates for salt tolerance of *S. alba*, and some of these genes showed fairly high coverage depth. This group identified four genes with signals of strong diversifying selection (Ka/KsN1) by comparing the transcriptome sequences of *S. alba* with 249 known ESTs from its congener *S. caseolaris*. Recently, the development of novel high-throughput DNA sequencing methods has provided an opportunity to address this question by de novo assembly or mapping and quantification of transcriptomes (Wang et al. 2009). As an early step in sequencing the transcriptomes of mangrove species, these high-throughput sequencing technologies were used to sequence the transcriptome of two mangrove species, Rhizophora mangle and *Heritiera littoralis* (Dassanayake et al. 2009). In this study, they sequenced the transcriptome of *S. alba* using the Illumina platform. After de novo assembly of the transcriptome, simple sequence repeats (SSRs) were identified and codon usage bias was analyzed. Based on known salt responsive genes, we searched their homologous sequences in *S. alba* and calculated their coverage depth. By comparing the transcriptome sequences of *S. alba* with the ESTs from *S. caseolaris*, genes under positive selection were identified and we wish to gain insights into the role of natural selection in adaptation of mangroves to their environments.

16.8 Biotic and Abiotic Stress Tolerance Genes from Mangrove Trees

Generation and analysis of expressed sequence tags from the salt-tolerant mangrove species *Avicennia marina* was reported by Jithesh et al. (2006). Construction of a cDNA library of *A. marina* genes is reported and random expressed sequence tag (EST) sequencing of 1,841 clones produced 1,602 quality reads. The abundance of dehydrin clones in the *A. marina* ESTs and up-regulation of BM173212in response to salt stress correlates with their role in dessication tolerance. The BADH gene isolated showed 97 % identity (amino acid) with the clone reported by Hibino et al. (2001; accession no. AB043539). The EST for the proline/betaine transporter (BM173094/partial clone) shows maximum identity with AmTl, the betaine/pro-line transporter reported by Waditee et al. (2002). The unknown category of genes forms 30 % of the ESTs analyzed and is a useful starting point from which to isolate new genes that govern salt tolerance. Thus, 52 genes from this pool were selected for further analysis. Existence of stress-responsive genes using Expressed Sequence Tags (ESTs) was also reported by (Mehta et al. 2005). For studying the role of individual genes in *A. marina*, antioxidant response to salt, iron, and light and hyperosmotic stress by monitoring the mRNA levels, were reported particularly on three genes; Cu–Zn SOD, catalase and ferritin. Mangrove plants like *A. marina* grow in anoxic soils in coastal areas with high salinity, often under conditions of high temperature and light that could potentially lead to ROS production (Mehta et al. 2005). *A. marina* may serve as a good model to study antioxidant response in mangroves, because mangroves have developed effective protection from photosynthetic damage by the active involvement of ROS scavengers (Cheeseman et al. 1997). A full-length cDNA inserts encoding *SodI*, *CatI* and *FerI* were isolated to study the response to abiotic stress treatments. Salt stress elicits short-term response of *CatI* and *FerI* mRNA while maintaining constant mRNA levels for *SodI*. The accumulation of ROS during salt stress is mainly attributed to the inhibition of photosynthesis and a decline in CO_2 fixation. In agreement with this, concentrations of H_2O_2 in shoot tissues of rice have been shown to increase upon salt stress. Increased time of exposure to NaCl stress caused a decline in *CatI* and *FerI* mRNA levels. There are few reports on the effect of NaCl stress on antioxidant mRNA levels in mangroves though enzyme activity studies have been reported. Salt stress inhibition of antioxidant enzymes like SOD and CAT were observed at high NaCl concentrations in *A. marina* (Cherian et al. 1999); *Suaeda nudiflora* Moq (Cherian and Reddy 2003) and in *Bruguiera parviflora* (Parida et al. 2004b). Increased mRNA levels of *CatI* and *FerI* in leaf tissues with 12 h of NaCl stress treatment were observed. It was evidient that in *A. marina*, *CatI* and *FerI* mRNA levels increase in response to short-term salt stress. On the contrary, *SodI* mRNA levels were unaltered during NaCl stress treatment. Cyt Cu–Zn SOD used in this particular study, showed no distinct dissimilarities at the amino acid level in comparison to the salt stress inducible (82 % with *O. sativa SODCc2*) and non-inducible forms of SODs in maize and rice (81 % with *O. sativa SODCc1*). Therefore, the differences between different Cu–Zn SOD isoforms might reside in functional responses (Fink and Scandalios 2002; Benavente et al. 2004). *A. marina* plants showed physiological signs of wilting in stress treatments, indicative of H_2O_2 stress (Guan et al. 2000). Additionally, *SodI* and *FerI* mRNA levels were also up regulated. Hence H_2O_2, which acts as an intermediate molecule mediating antioxidant responses to different stresses, activates all the three genes directly and confirms the participation *of Sodl*, *Catl* and *Ferl* in the antioxidative pathway. The observed increase in abundance for *SodI*, *CatI* and *FerI* mRNA levels could reflect the regulation of these genes upon light-mediated oxidative stress. In order to test this, analysis of light responsiveness of ubquitin (*Ubc*) were made. *Ubc* mRNA levels were reported to be unaffected by salt stress, which is a composite stress including oxidative, osmotic and ionic stress (Parani et al. 2002). It was confirmed that oxidative stress positively regulates the functioning of *SodI*, *CatI* and *FerI*. These genes are upregulated upon oxidative stress to counter the production of increased amounts of accumulating H_2O_2. Osmotic stress, on the other hand, either down-regulates (*SodI* and *CatI*) or causes no alteration (*FerI*) in mRNA levels at the time-intervals studied. Mangroves have to deal with many environmental stress factors, especially, salt stress and light. Taken together, the time response of these genes with salt stress reveals an induction with 12 h of NaCl beyond which osmotic and ionic stress may take effect. NaCl stress in plants, apart from its increased osmolarity, is ionic in its toxic effects (Streb et al. 1993; Munns 2002). It is also possible that after 12 h, NaCl specific effects could be severe in hydroponically grown *A. marina* plants. A study of ionic accumulation in *A. marina* leaves revealed that Na^+ and Cl^- ions reached saturation levels by 18 h of 500 mM NaCl stress treatment (Ashihara et al.

1997). The response of *CatI* and *FerI* can be categorized as short-term response while *SodI* performs a constitutive role in NaCl stress. This particular isoform, *SodI* may not have a contributory role in salt stress response of *A. marina*.

Comparative studies on the photosynthetic performance with high light intensity revealed that *A. marina* was more sensitive to changes in light intensity, has a higher light requirement than *Bruguiera gymnorhiza*, and shows photoinhibition. Oxidative stress-induced increase in transcripts like *SodI*, *CatI* and *FerI* might have a role in mitigating the deleterious effects of oxidative stress with increased incident light in *A. marina*. Antioxidant gene expression in a mangrove, *A. marina* confirmed that observed differences in mRNA levels with either an induction or decrease or a constitutive expression of each of these stresses could be easily differentiated. Additionally, an early or delayed response was also shown with these genes in different stresses, confirming the specificity of individual stress components on differential gene regulation. This study and the earlier study (Willekens et al. 1994b) confirm that relative mRNA expression levels could be used as indicators to study the role of individual genes of multigene family in each of these stresses.

16.9 Transfer of Salinity Tolerance Gene from Mangroves

The Mangrove genes isolated from the mangroves were transferred into crop plants using *Agrobacterium tumefaciens* mediated transformation (Hiei et al. 1994). Specific genes isolated from *A. marina* cDNA library were cloned in binary vectors. These genes were expressed under the control of constitutive promoters. These gene constructs were then transformed into *Agrobacterium tumefaciens* and used for co-cultivation with rice calli and tobacco leaf explants. The integration and expression of the transgenes were confirmed in these plants using various molecular analyses. The homozygous lines from these transgenic plants were raised and were tested with various abiotic stresses such as salt stress and drought stress. Initial analyses in the laboratory have been promising. However, further analyses would need to be carried out to evaluate the performance of these transgenics as reported by Prasant et al. (2008). A number of salt related genes were reported from a number of tree mangrove (Tables 16.11 and 16.12).

16.9.1 Mangrove Genes for Heavy Metal Detoxification

Metallothioneins are involved in detoxification of heavy metals. Metallothioneins (MTs) are a family of low-molecular-weight Cys-rich proteins and are thought to play possible role in metal metabolism or detoxification. A cDNA encoding type 3 metallothionein (*PcMTS*) was isolated from the salt stressed leaf cDNA library of *Porteresia coarctata cv.* Tateoka (wild rice) that grows well in the heavy metal laden estuarine soils by Usha, et al. (2011). The *PcMTS* cDNA (581 bp)

Table 16.11 Salt-related genes reported in mangroves

Mangrove species	Genes	Description	Function	References
Aegiceras corniculatum	P5CS	Delta 1-pyrroline 5-carboxylate synthetase	A key enzyme of proline synthesis pathway- accumulation of transcript of this gene under salinity tended to accompany recruitment of proline in *A. corniculatum*	Fu et al. (2005)
	PIP1	PIP1 aquaporin	This gene was upregulated by salt stress	Fu et al. (2005)
	PIP2	PIP2 aqaporin	This gene was upregulated by salt stress	Fu et al. (2005)
	NHA	Na+/H+ antiporter	This gene was upregulated by salt stress	Fu et al. (2005)
	CP1	Cysteine proteinase inhibitor	Of this gene in transgenic Arabidopsis enhanced tolerance capacity of high saline medium	Fu (2006)
	BADH	Betaine-2 aldehyde dehydrogenase	High salinity induced increase of transcript level and such an increase was accompanied by accumulation of betaine. Although activity of this enzyme decreases with an increase in salinity, the extent of decrease is less than in homologs in *E. coli and spinage*	Hibino et al. (2001)
Avicennia marina	SodI	CU/Zn superoxide dismutage	High salinity did not lead to transcriptional change but osmotic stress decreased transcript level of this gene. Under oxidative stress, its transcription was transiently upregulated	Prashanth et al. (2008a, b)
	CatI	Catalase	It was upregulated by saline or oxidative stress but downregulated by osmotic stress	Jithesh et al. (2006)
	FerI	Ferritin 1	It was transcriptionally upregulated by saline or oxidative stress but did not change under osmotic stress	Jithesh et al. (2006)
	AmT1; AmT2	Betaine/Proline transporter	Transgenic E. coil with such gene could accumulate betaine under salt stress; in *A. marina, salt stress induced transcription of such gene in root and leaf*	Wadit et al. (2002)
	AmT3 (partial)	Do	Do	(do)
Bruguiera gymnorhiza	OEE1	OEE1 is one component of PSII	High salinity induced accumulation of its transcript and protein	Sugihara et al. (2000)

(continued)

Table 16.11 (continued)

Mangrove species	Genes	Description	Function	References
	DLDH	Dihydrolipoamide dehydrogenase	Upregulated when treated with 500 mmol/L NaCl for 1 d	Banzai et al. (2002a)
	LAS	Lipoic acid synthase	Being upregulated when treated with 500 mmol/L NaCl for 1d	Banzai et al. (2002a)
	Unnamed gene	Fructose-6 phosphate, 2-kinase/fructose-2, 6-biphosphatase	Transcription of this gene increased after 6 h of salt stress. It was supposed to act in osmotic regulation process by controlling the content of Fru-2,6-P2	Banzai et al. (2002a), (2003)
	Cytosolic Cu/Zn SOD	Cytostolic Cu/Zn superoxide dismutase	High salinity, mannitol and ABA induced accumulation of its transcripts in leaves; Transcript was induced by high salinity in young and mature leaves rather than in old leaves	Takemura et al. (2002)
	Cytosolic CAT (partial)	Catalse	No significant change occurred in the expression of this gene during the treatment with NaCl, mannitol and ABA, but CEPA (2-chloroethylphosphonic acid) can increase its transcript level	Takemura et al. (2002)
Bruguiera sexangula	CCTα	α subunit of CCT complex	Transgenic E. coli with one domain of this subunit displayed enhanced tolerance to high salinity	Yamada et al. (2002a)
	Mangrin	Partially homologous to gene encoding Allene Oxidase Cyclase (AOC)	It was upregulated by high salinity and its overexpression enhanced salt tolerance of transgenic yeast and tobacco cells.	Yamada et al. (2002a)
Kandelia candel	SIGKC1 & 2	Cytosolic low molecular mass heat shok protein (sHSPs)	It act as molecular chaperones to prevent thermal aggregation of protein by binding non-native intermediates	Huang et al. (2003)
	SIGKC3	AFP-ribosylation factor (AFR)	A ubiquitous, high conserved 21 kDa GTP-binding protein. The ARF proteins are through to function as regulators of membrane traffic	Huang et al. (2003)
	SIGKC4 & 5	Unknown	–	Huang et al. (2003)

Table 16.12 Genome scale studies in mangroves

Mangrove species	Description	Research topic covered	References
Aegiceras corniculatum	Constructing a leaf SSH library		Fu et al. (2005)
Avicennia marina	Identifying differentially expressed genes in response to 50 % sea water in leaves of 50 d old seedlings by differentially display technique. Constructing a leaf cDNA library from 500 mM NaCl-treated seedling	Random sequencing generated 1602 ESTs which were grouped into 13 categories, among those, 7 % were homologous with stress-responsive genes	Mehta et al. (2005)
Bruguiera gymnorhiza	Large-scale sequencing of ESTs collected from high salinity or hormone treated leaves and roots. Monitoring salt-responsive transcript profiling of 7029 unique genes in leaf and root tissues using microarray techniques	Statistically-confident genes were grouped into 4 clusters depending on their EST frequency and each group has specific pattern of transcript profiling under high salinity	Banzai et al. (2002a)
Ceriops tagal	Constructing a leaf SSH library; and a root cDNA library; monitoring time-course transcripts profiling through microarray	Totally 98 differentially expressed EST's induced by 500 mM NaCl were identified and some of them were cooperatively regulated by salt to acting on the process. When compared with glycophytes global transcription of this species is stable in saline environment, including homeostasis mechanism in salt-related adaptation	Liang (2007)
Hibiscus tiliaceus	Constracting a leaf cDNA library; monitoring time-course transcrip profilings through microarray	Totally 1,220 differentially expressed ESTs induced by high salinity were identified. Among these 434 ESTs were assigned to function-known genes and some responded to high salinity in ecotype-specific manners. Genes of transcription factors responded to salt stress more rapidly, and changed much more in littoral ecotypes than the terrestrial one	Yang (2007)

(continued)

Table 16.12 (continued)

Mangrove species	Description	Research topic covered	References
Acanthus ebracteatus	Constructing a leaf cDNA library from seawater growing seedlings	Random sequencing generated 521 readable sequences and 67 % of them matched function-known genes by homolog searching among which 18 % were predicted to function in stress response 23.9 % in metabolism, 7.3 % in regulation of transcription and 2.7 % in others	Nguyen et al. (2006)
Sesuvium protulacastrum	A SSH cDNA library from mangrove associate in salt treatment	P66, P175 and P233 are novel clones of salt stress were obtained upon alignment with GeneBank database. Clone P89 demonstrated high homology with NADPH of *Arabidopsis thaliana*, whereas clone P152 was highly homologous with the gene encoding late embryogenesis abundant (LEA) protein of *A. thaliana*	Zeng et al. (2006)

encodes a protein of 64 amino acids. PcMTS is highly homologous (82 %) to OsMT-I-3a of rice, but is unique from other type 3 plant MTs due to the presence of an additional glycine residue in the C-terminal domain. Analysis of the 5′ upstream region *of PcMTS* showed the presence of cw-acting elements like the CG box and STRE previously reported to be involved in gene expression under heavy metal stress. They reported more than one copy of *PcMTS-like* sequences in the *P. coarctata* genome. Analysis of genomic clone of *PcMT3* revealed the presence of two introns. A comparison of the genomic sequence of *PcMT3* with closely similar type 3 MTs from rice and mangrove species revealed conservation in the number and position of introns. Transcript profiling for *PcMTS* in *P. coarctata* leaves in the presence of Cd, Cu and Zn showed an increase in transcript accumulation. In future, these genes are of high importance for developing heavy metal tolerance rice.

To shed light on the role of type 2 MT in *Bruguiera gymnorhiza* (*BgMT2*) under heavy metal stress, *B. gymnorhiza* seedlings were exposed to different concentrations of $CdCl_2$ (2 μM, 10 μM, 20 μM or 40 μM) for 3,7 and 11 d. Real-time quantitative PCR analysis demonstrated that *BgMT2* gene transcripts in leaves of *B.gymnorhiza* increased in all Cd concentrations and exposure durations (Huang et al. 2011). *BgMT2* was overexpressed in *Escherichia coli* BL21 (DE3) as a fusion protein (GST-BgMT2), and bacteria expressing the fusion protein had higher tolerance to Zn, Cu, Pb and Cd than control cells. Moreover, by analysis of metal-binding properties of GST-BgMT2 fusion protein, the expression of *BgMT2* endowed resistance in *E. coli* to Zn, Cu, Pb and Cd by *BgMT2* protein sequestering the four metals. *BgMT2* gene transcripts in *B. gymnorhiza* were induced by Cd and other heavy metals (Huang and Wang 2009, 2010). Besides, *BgMT2* protein could contribute to heavy metal tolerance by binding to heavy metals when it was overexpressed in *E coli*. It suggested that, *BgMT2* may be involved in the homeostasis or detoxification of heavy metals in *B. gymnorhiza*. It is needed to isolate the promoter region of *BgMT2*.

16.9.2 Genetic Engineering of Mangrove Genes to Crop Plants with High Salinity Tolerance

As a first step towards characterizing genes that contribute to combating salinity stress, constructed a cDNA library from a mangrove species *Avicennia marina*. The availability of the full genome sequences of *Arabidopsis thaliana* and *Oryza sativa* allows comparison with other plant databases. An analysis revealed that 88 % of *Avicennia marina* sequences in the unknown category matched the *A. thaliana* genome. In the other categories 25–65 % of the ESTs showed highest homology to *A. thaliana* genes. A very small percentage, less than 10 % in most classes, matched the *O, sativa* database. All 1841 ESTs were compared against sequences in the non-redundant database (nr) at the NCBL using the program BLASTX. The result of each comparison was screened manually and 1602 readable' ESTs (87 %) retained after screening. Assignment of putative functions to the ESTs was based on the numerical cutoff values obtained in the BLAST comparisons and supplemented with information from PubMed. Of 1602 ESTs, 1155 had significant homology to previously identified genes and these were grouped into 12 functional categories. The remaining 'unknown' genes (447) refer to that subset of the ESTs that show a significant similarity (high e-value) to genes reported in the public database but do not have an assigned function to date. Unknown genes form the largest category at 30 % and followed by genes required for primary metabolism (13 %). Genes involved in transcription and chromatin organization, protein synthesis and processing each represent 10 % of the sequenced ESTs while those involved in membrane transport and intracellular trafficking represent 9 % of the ESTs. 8 % of the ESTs relate to signal transduction while 7 % are similar to previously reported stress induced genes.

Over expression of cytosolic copper/zinc superoxide dismutase from a mangrove plant *Avicennia marina* in *indica* Rice var Pusa Basmati-1 confers abiotic stress tolerance was reported. They have previously reported the isolation of *Sodl*, a cDNA encoding a cytosolic copper zinc superoxide dismutase from the mangrove plant *Avicennia marina* and its mRNA expression pattern during various oxidative and abiotic stresses. Southern hybridization of *A. marina* genomic DNA using *Sodl*, revealed that this gene in *A. marina* genome is present as a single copy. The cDNA was cloned into a binary vector (pCAMBIA 1300) and transformed into *indica* rice var Pusa Basmati-1. Southern hybridization analysis of transgenic rice plants revealed stable integration of the *Sodl* transgene in the rice genome. Cheeseman et al. (1997) reported that Superoxide dismutase (SOD) is one of the important antioxidative enzymes implicated in protection of the mangroves against active oxygen species. SOD forms the first line of defense in the enzymatic pathway of defense against free oxygen radicals. This enzyme dismutates super-oxide radical into hydrogen peroxide. The SOD activities of field grown mangroves are almost 40 times higher than the SOD activity found in common crop plants like pea (Cheeseman et al. 1997). Jithesh et al. (2006) isolated a cDNA (*Sodl*) encoding a cytosolic Cu/ZnSOD from the mangrove species A. *marina* and studied the effect of salinity and oxidative stress on the mRNA transcript accumulation of this gene. Over expression of *Sodl* encoding a cytosolic Cu/ZnSOD isolated from the mangrove species *A. marina* Forsk. in an *indica* rice variety Pusa Basmati-1 was also reported by Jitesh et al. (2006). Overexpressed a cytosolic Cu/ZnSOD without targeting it into any particular organelle and could observe abiotic stress tolerance in the transgenic plants. However it has been suggested that overexpressing SOD in the chloroplasts is more effective than overexpressing it in the cytosol in transgenic plants in terms of providing tolerance to methylviologen mediated oxidative stress, in transgenic plants. Parida et al. (2004a) and Wang et al. (2004c) reconfirms the importance of cytosolic Cu/ZnSOD isoform in conferring abiotic stress tolerance in plants SOD genes exist as multiple gene families in mangroves. Isolation of genes coding for different isoforms of SOD from *A. marina* and other mangroves and studying the gene regulation of these isoforms would provide more input on the role of SOD enzyme in providing protection to the mangrove plants against abiotic stresses. These studies would in turn throw light on how these genes can be utilized for transforming to crop plants for enhancing their abiotic stress tolerance. Hence, mangroves could be potential sources for isolating novel genes for abiotic stress tolerance given the fact that they are halophytes and are capable of tolerating a high degree of salinity and abiotic stress.

Vacuolar V-ATPase plays a major role in the maintenance of cellular pH through the ion pump and influences the transport of cations into the vacuoles of plant cells. A cDNA clone (*PcVHA-cl*) encoding the subunit of V-ATPase was isolated from the salt-tolerant wild rice, *Porteresia coarctata.* The DNA sequence of *PcVHA-cl* showed significant homology with V-ATPase subunit of rice. Southern analysis suggested the presence of multiple coding regions for subunit c in *P. coarctata.* Northern and Western analyses of salt-treated *P. coarctata* plants revealed that subunit c of V-ATPase is up regulated by NaCl treatment at both transcriptions and translations level. Total RNA from leaf and root tissues of *P. coarctata* was isolated Northern analysis was initially carried out with the full-length *PcVHA-cl* probe. The leaf RNA blot was then stripped and re-probed with *PcVHA-cl* UTR-specific fragment. The primers used for PCR amplification of the UTR region of V-ATPase are: V-ATPase forward- 5′-TTG CAG TAG CAA TCC GCA GTT-3′; V-ATPase reverse-5′-GGA GCA GCA AAT CCA GCC

TA-3′. Total RNA was isolated from the leaf tissue of P. *coarctata* plants treated with 0.5 M NaCl for 48 h and mRNA was purified by oligo-dT cellulose column chromatography. Double-stranded cDNA was synthesized from purified mRNA using Superscript II reverse transcriptase and ligated to pSPORTl plasmid vector in the *Noil* and *Sail* restriction sites. The ligation product was then transformed into *Escherichia coli DHA-α5* and maintained as plasmid DNA lots. A full-length cDNA clone designated *PcVHA-cl* (*P. coarctata* Vacuolar ATPase c-subunit gene number I), coding for the V-ATPase subunit c was obtained from *P. coarctata* cDNA library (Senthilkumar et al. 2005) suggest that there is a contributory role for the V-ATPase subunits in the salt stress response in *P. coarcata.*

A fructose-1,6-bisphosphate aldolase gene, designated *SpFBA*, was isolated and characterized from *Sesuvium portulacastrum*, a halophytic plant from the seashore, roots in response to seawater by Fan and Zhang (2009). The *SpFBA* cDNA has a total length of 1452 bp with an open reading frame of 1071 bp, and is predicted to encode a precursor protein of 357 amino acid residues sharing high degree of homology with class I FBAs from other plants. Semi-quantitative RT-PCR analysis indicated that the *SpFBA* was more strongly expressed in roots than in leaves and stems, and the abiotic stimuli such as Seawater, NaCl, ABA, and PEG could trigger a significant induction of *SpFBA* in *S. portulacastrum* roots within 2–12 h. Overproduction of Recombinant SpFBA resulted in an increased tolerance to salinity in transgenic *Escherichia coli.* Some halophytes such as the *M. crystallinum*, operate in the C_3 mode of photosynthesis under water abundance, and can be converted to the water-conserving crassulacean acid metabolism pathway (CAM) mode while subjected to water deficits. This is an adaptive and protective mechanism for some halophytes under saline conditions (Winter and Willert 1972). This kind of photosynthesis-mode change will consequentially lead to increase of expression and activity of carbohydrate metabolism-related enzymes. Cofactor-independent phos-phoglyceromutasel (PGM1) is an enzyme involved in glycolysis and gluconeogenesis in *M. crystallinum* and salinity could result in an accumulation of the mRNA and protein of PGM1 in *M. crystallinum*, proposing PGM1 contributes to the maintenance of efficient carbon flux through glycolysis/gluconeogenesis in conjunction with the stress-induced shift to CAM photosynthesis. The expression *of SpFBA* in *S. portulacastrum* roots was rapidly up-regulated by high salinity, ABA and dehydration, but it is also not clear whether the *SpFBA* is involved in responding to the change of photosynthesis mode. Investigation of carbohydrate metabolism and sucrose biosynthesis might be helpful in the *SpFBA* transgenic plants under salt stress.

A crucial strategy of producing stress-tolerant crops has involved the use of 'extremophiles' as source material for stress tolerance genes because of the capacity to survive and flourish under extreme or fluctuating environment regimes (Holmberg and Bulow 1998; Banzai et al. 2002a). Mangroves are halophytes that grow in the tropical and subtropical regions of the world. Mangroves are defined as woody trees and shrubs which flourish in mangrove habitats. Mangroves could be classified into (1) true or exclusive mangroves and (2) nonexclusive mangrove species otherwise termed 'also mangrove associates'. True or exclusive mangroves are those that occur only in mangrove habitats, or rarely elsewhere. Mangrove associates include species of plants typically occurring on the landward margin of mangrove habitats, and often in non-mangrove habitats such as salt marshes. These include (1) halophytes such as *Salicornia* spp., *Sesuvium* spp., *Suaeda* spp. and *Porteresia coarctata* that are able to grow along with mangroves, and (2) terrestrial species like some ferns that are often found associated with mangroves and are unable to tolerate high salinity and therefore do not penetrate deep into the mangrove wetlands (Tomlinson 1986; Hogarth 1999). One of the characteristic features of halophytes is the presence of physical adaptations that help confer salinity tolerance (Tomlinson 1986), and these are highly developed in mangroves. These physical adaptations play an important role in protection of these halophytes in their saline environment. Most of the early studies *in* mangroves have dealt with the effects of salinity on photosynthesis (Ball and Farquhar 1984) and respiration (Burchett et al. 1989; Fukushima et al. 1997). However, recently, there has been a growing interest in the relation of salinity to antioxidant enzyme status in mangroves (Takemura et al. 2000; Cherian and Reddy 2003; Parida et al. 2004a, b, c; Jithesh et al. 2006).

16.9.3 Epigenetics: What is Next in Terms of Biotechnological Application?

Next to transcriptional regulations of abiotic stress responses, epigenetic processes are becoming a new and current chapter in plant environmental adaptation. Efficiency of gene expression is highly influenced by chromatin structure that might be modulated epigenetically by processes such as DNA methylation and posttranslational modifications of histones. The histone-mediated structure of nucleosomes in the chromatin might be posttranslationally modified at the N-terminal tails of the core histone complexes (H2A, H2B, H3, H4) and thus influence nucleosome density, binding efficiency of TFs, and transcriptional activity (Chinnusamy and Zhu 2009; Kim et al. 2010). In addition to methylations of histones, also acetylations and phosphorylations as well as other posttranslational modifications of histones as ubiquitination, biotinylation, and sumoylation might have a modulating impact on the regulation of stress-specific gene expression (Chinnusamy et al. 2008). Meanwhile, it is accepted knowledge that phenotypes within one species may transmit different epigenetic information based on covalent modifications of DNA or histones (Fazzari and Greally 2004). Thus, plant populations from stress exposed habitats may carry inherited memories of stress adaptation and transfer this epigenetically to next generations. As an example, the desert shrub *Zygophyllum dumosum* was posttranslationally methylated at histone H3 under wet but less under dry growth conditions indicating posttranslational regulation of gene expression activity (Granot et al. 2009). As it was also reported recently, natural populations of mangroves were DNA hypomethylated when grown under saline conditions in contrast to populations non-saline sites

(Lira-Medeiros et al. 2010). Based on these results, it seems obvious to think on simulation of inherited memories of stress adaptation in biotechnological applications to confer increased drought and salt tolerance to naturally sensitive species. However, in contrast to the detailed knowledge on influences of epigenetic mechanisms on developmental processes, information on epigenetic regulation of abiotic stress resistance is still rare. As a few examples, salinity-induced phosphorylation of histone H3 and acetylation of histone H4 in *A. thaliana* and tobacco have been reported (Sokol et al. 2007). In addition, altered acetylation as well as trimethylation of histone H3 under drought stress in drought-responsive genes of *A. thaliana* has been observed (Kim et al. 2008). In rice, expression of cytosine DNA methyltransferases was modified by salt stress indicating functional importance of epigenetic modulation of genome activity also in monocot species (Sharma et al. 2009). Detailed knowledge on the specific mechanisms that underlay epigenetic regulation under environmental exposure is, however, only slowly emerging. Thus, trans-generational modifications of stress adaptations as salt stress include altered genomic DNA methylation as well as function of Dicer-like proteins suggesting involvement of small RNA pathways in epigenetic regulations (Boyko et al. 2010). Interestingly, in barley expression of Polycomb proteins with function in histone methylation was influenced by abscisic acid (ABA) suggesting involvement of ABA-mediated pathways in epigenetic modifications (Kapazoglou et al. 2010).

Thus, according to the current knowledge, an application of epigenetic processes to improve the stress-regulating function of TFs will be a challenging and novel biotechnological approach for the engineering of plant tolerance to drought and salinity, however, much detailed information is still missing. Particularly, despite the importance of elucidating epigenetic mechanisms in model plants, it will be obligatory to extend investigations to systematic and comprehensive comparisons of stress relevant epigenetics in sensitive- and naturally tolerant species. Linking epigenetic processes to the key regulatory components of the general stress adaptive frameworks will be essential to further support the feasibility of epigenetics in the customized engineering of stress adaptation.

16.10 Conclusion and Future Perspectives

Cellular effects of environmental stresses such as drought and salinity are not only imbalances of ionic and osmotic homeostasis but also impaired photosynthesis, cellular energy depletion, and redox imbalances. Regulatory systems inclusive TFs that link sensing and signaling of the environmental conditions and the cellular adaptive responses are emerging but are not well understood yet. As a next step, it will be important to identify master regulators and master pathways of stress adaptation in naturally stress-tolerant species as well as integration of the diverse regulatory factors in the network of intracellular stress adaptation pathways. More detailed understanding of shared and competing transcriptional regulation as well as modulated intramolecular interactions of different factors and epigenetic processes will be essential for targeted and efficient genetic engineering of improved drought and salt tolerance in plants. Transgenic approaches for increasing plant salt tolerance are feasible. So far, results obtained with many genes are encouraging, and recent results obtained from in transgenic plants harbouring genes encoding an Na/H antiporter or a transcription factor show the possibility of increasing the salt tolerance. Salt tolerance responses are being investigated in several species through large scale gene expression analyses. But, it should be noted that this is only the beginning. Transformation of agronomically important crops and the identification of uncovered tolerance determinants and stress inducible promoters must be further explored to obtain plants with increased tolerance to salt stress. The use of promoters that direct the expression at the proper time and place will maximize salt tolerance. In addition, it will avoid potential un desiderable pleiotropic effects resulting from the ectopic expression of foreign genes.

Salt tolerance is a complex trait in plants, but molecular and genetic approaches are beginning to characterize the diverse biochemical events that occur in response to salt stress. In the short term, it will remain a challenge to manipulate the essential protective mechanisms in plants and to utilize our biochemical knowledge for optimal molecular engineering of salt tolerance in plants. A major unresolved question is the extent and importance of both short-term and long-term stress responses for sustained tolerance and their effects on agriculturally desirable traits in crop plants. With the recognition that the enhanced expression of a number of functionally related genes may be required for optimal improvements in salt tolerance, molecular engineering has been expanded to include proposals for multiple gene transfers to enhance salt tolerance (Bohnert and Jensen 1996). An equally promising approach to manipulating many genes may emerge as we learn more about the specificity of signalling pathways that turn on transcription of related genes that counteract salt stress at the cellular level. Redundancy of the intersecting signalling pathways and communication between the different pathways, however, is likely to create difficulties in using this information in a directed approach at improving salinity tolerance in the near future. Transcriptional regulation is another new area with potential to coordinate regulation of genes relevant to tolerance, but will require identification of factors limiting the sustained response so that their expression may be manipulated in a tissue targeted manner. Overall, the continuing significant progress in our understanding and ability to modify salt tolerance by molecular engineering using both model and crop plants based on knowledge of how salinity affects plant biochemistry and physiology through gene expression will help to develop salt tolerant crops in near future.

References

Abeysinghe PD, Triest L, Greef BD, Koedam N, Hettiarachi S (2000) Genetic and geographic variation of the mangrove tree *Bruguiera* in Sri Lanka. Aquat Bot 67:131–141

Alscher RA, Erturk N, Heath LS (2002) Role of superoxide dismutases (SODs) in controlling oxidative stress in plants. J Exp Bot 53:1331–1341

Arora A, Sairam RK, Srivastava GC (2002) Oxidative stress and antioxidative systems in plants. Curr Set 82:1227–1238

Ashihara H, Adachi K, Otawa M, Yasumoto E, Fukushima Y, Kato M, Sano H, Sasamoto H, Baba S (1997) Compatible solutes and inorganic ions in the mangrove plant *Avicennia marina* and their effects on the activities of enzymes. Z Naturforsh c 52:433–440

Askari H, Edqvist J, Hajheidari M, Kafi M, Salekdeh GH (2006) Effects of salinity levels on proteome of *Suaeda aegyptica* leaves. Proteomics 6:2542–2554

Baba S, Ogino K, Naksuga T (1989) Isozyme patterns of six mangrove tree species in Okinawa. Galaxea 8:79–83

Ball MC, Farquhar GD (1984) Photosynthetic and stomatal responses of two mangrove species, *Aegiceras corniculatum* and *Avicennia marina* to long term salinity and humidity conditions. Plant Physiol 74:1–6

Ballment ER, Smith TJ, Stoddaart JA (1988) Sibling species in the mangrove genus *Ceriops* (Rhizophoraceae), detected using biochemical genetics. Aust Syst Bot 1:39–97

Banerjee LK, Sastry ARK, Nayar MP (1989) Mangroves in India: identification manual. Botanical Survey of India, Calcutta

Banzai T, Hershkovits G, Katcoff DJ, Hanagata N, Dubinsky Z, Karube I (2002a) Identification and characterization of mRNA transcripts differentially expressed in response to high salinity by means of differential display in the mangrove, *Bruguiera gymnorrhiza*. Plant Sci 162:499–505

Banzai T, Snmiya K, Hanagata N, Dubtnsky Z, Karube I (2002b) Molecular cloning and characterization of genes encoding BURP domain-containing protein in the mangrove, *Bruguiera gymnorrhiza*. Trees 16:87–93

Basyuni M, Baba S, Inafuku M, Iwasaki H, Kinjo K, Oku H (2009) Expression of terpenoid synthase mRNA terpenoid content in salt stressed mangrove. J Plant Physiol 166:1786–1800

Basyuni M, Kinjo Y, Baba S, hinzato N, Iwasaki H, Siregar Edy BM, Oku H (2011) Isolation of salt stress tolerance genes from roots of mangrove plant, *Rhizophora stylosa* Griff., using PCR-based suppression subtractive hybridization. Plant Mol Biol Rep 29:533–543

Basyuni M, Oku H, Tsujimoto E, Kinjo K, Baba S, Takara T (2007) Triterpene synthases from the Okinawan mangrove tribe, Rhizophoraceae. FEBS J 274:5028–5042

Beauchamp CO, Fridovich I (1971) Superoxide dismutase: improved assay and an assay applicable to acrylamide gels. Anal Biochem 44:276–287

Behera B, Das AB, Mittra B (2009) Changes in proteins and antioxidative enzymes in tree mangroves *Bruguiera parviflora* and *Bruguiera gymnorrhiza* under high NaCl stress. Biol Divers Conserv 2:71–77

Ben Amor N, Ben Hamed K, Debez A, Grignon C, Abdelly C (2005) Physiological and antioxidant responses of the perennial halophyte *Crithmum maritimum* to salinity. Plant Sci 68:889–899

Benavente ML, Teixeira FK, Kamei CLA, Pinheiro MM (2004) Salt stress induces altered expression of genes encoding antioxidant enzymes in seedlings of a Brazilian *indica* rice (*Oryza saliva* L.). Plant Sci 166:323–331

Berberich T, Kusano T (1997) Cycloheximide induces a subset of low temperature-inducible genes in maize. Mol Gen Genet 254:275–283

Bohnert HJ, Jensen RG (1996) Strategies for engineering water-stress tolerance in plants. Trends Biotechnol 14:89–97

Bousquet I, Strauss SH, Doerksen AH, Price RA (1992) Extensive variation in evolutionary rate of *rbcl.* gene sequences among seed plants. Proc Natl Acad Sci USA 89:7844–7848

Bowler C, Slooten L, Vandenbranden S, De Rycke R, Botterman J, Sybesma C, Van Montagu M, Inze D (1991) Manganese super-oxide dismutase can reduce cellular damage mediated by oxygen radicals in transgenic plants. EMBO J 10(7):1723–1732

Bowler C, Van Montagu M, Inze D (1992) Superoxide dismutase and stress tolerance. Anna Rev Plant Physiol Plant Mol Biol 43:83–116

Boyko A, Blevins T, Yao Y, Golubov A, Bilichak A, Ilnytskyy Y, Hollander J, Meins F Jr, Kovalchuk I (2010) Transgenerational adaptation of *Arabidopsis* to stress requires DNA methylation and the function of Dicer-like proteins. PLoS ONE 5:e9514

Burchett MD, Clark CJ, Field DC, Pulkownik A (1989) Growth and respiration in two mangrove species at a range of salinities. Physiol Plant 75:299–303

Canvin DT (1990) Photorespiration and CO_2 concentrating mechanisms. In: Dennis DT, Turpin DH (eds) Plant physiology, biochemistry and molecular biology. Longman Scientific and Technical, Harlow, pp 253–273

Cheeseman JM, Herendeen LB, Cheeseman AT, Clough BF (1997) Photosynthesis and photoprotection in mangroves under field conditions. Plant Cell Environ 20:579–588

Chen S, Ding W, Qiu J, Wang G, Zhou Z, Chen J, Ai W, Wang C, Xie Q (2010) The genetic diversity of the mangrove *Kandelia obovata* in China revealed by ISSR analysis. Pak J Bot 42(6):3755–3764

Chen S, Zhou R, Huang Y, Zhang M, Yang G, Zhong C, Shi S (2011) Transcriptome sequencing of a highly salt tolerant mangrove species *Sonneratia Alba* using illumine platform. Mar Genomics 4:129–136

Cherian S, Reddy MP (2003) Evaluation of NaCl tolerance in the callus cultures of *Suaeda nudiflora* Moq. Biol Plant 46(2):193–198

Cherian S, Reddy MP, Pandya JB (1999) Studies on salt tolerance in *Avicennia marina* (Forsk.) vierh. Effect of NaCl on growth, ion accumulation and enzyme activity. Indian J Plant Physiol 4(4):266–270

Chinnusamy V, Gong Z, Zhu JK (2008) Abscisic acid-mediated epigenetic processes in plant development and stress responses. J Integr Plant Biol 50:1187–1195

Chinnusamy V, Zhu JK (2009) Epigenetic regulation of stress responses in plants. Curr Opin Plant Biol 12:133–139

Clegg MT, Brandon SG, Gerald HL Jr, Morton BR (1994) Rates and patterns of chloroplast rDNA evolution. Prof Nad Acad Sci USA 91:6795–6801

Cronquist A (1981) An integrated system of classification of flowering plants. Columbia University Press, New York

Das AB, Mukherjee AK, Das P (2002) Molecular phylogeny of *Heritiera* Aiton (Streculiaceae), a tree mangrove: variations in RAPD markers and nuclear DNA content. Bot J Linn Soc 136(2):221–229

Das AB, Jena S (2008) Chromosome stability and interpopulation genetic variability in a tree mangrove *Xylocarpus granatum* Koen. (Meliaceae) as revealed by RAPD markers. Cytologia 73:105–113

Das AB, Jena S, Sahoo P, Mohanty S (2004) Karyotype, genome size and RAPD markers derived molecular phylogenetic relationship among five Indian mangroves associate legumes. Genetica 122:217–226

Das AB, Mishra S, Mohanty P (2006) Physiology of mangroves: strategies on stress adaptation with special emphasis on tolerance to high salinity. In: Hamantaranjan A (ed) Advances of plant sciences vol 9(6). Scientific Publishers India, Jodhpur, pp 101–154

Dassanayake M, Haas JS, Bohnert HJ, Cheeseman JM (2009) Shedding light on an extremophile lifestyle through transcriptomics. New Phytol 183:764–775

Dey SK, Kar M (1995) Antioxidant efficiency during callus initiation from mature rice embryo. Plant Cell Physiol 36:543–549

Duke NC (1984) A mangrove hybrid, *Sonneratia* × *gulngai* (Sonneratiaceae) from north-eastern Australia. Austrobaileya 2:103–105

Duke NC (1994) A mangrove hybrid, *Sonneratia* × *urama* (Sonneratiaceae) from northern Australia and southern New Guinea. Aust Syst Bot 7:521–526

Duke NC (2010) Overlap of eastern and western mangroves in the South-western Pacific: hybridization of all three Rhizophora (Rhizophoraceae) combinations in New Caledonia. Blumea 55:171–188

Duke NC, Benzie JAH, Goodall JA, Ballment E (1998) Genetic structure and evolution of species in the mangrove genus *Avicennia* (Avicenniaceae) in the Indo-West Pacific. Evolution 52:1612–1626

Duke NC, Bunt JS (1979) The genus Rhizophora (Rhizophoraceae) in north-eastern Australia. Aust J Bot 27:657–658

Eising R, Trefease RN, Ni WT (1990) Biogenesis of catalase in glyoxysomes and leaf-type peroxisomes of sunflower cotyledons. Arch Biochem Biophys 278:258–264

Eltsner FF, Formmeyer D (1979) Analysis of different mechanisms of photosyuthetic oxygen reduction. Biochim Biophys Acta 325:162–188

Ezawa S, Tada Y (2009) Identification of salt tolerance genes from the mangrove plant *Bruguiera gymnorrhiza* using *Agrobacterium* functional screening. Plant Sci 176:272–278

Fan W, Zhang Z (2009) Cloning and molecular characterization of fructose-1,6-bisphosphate aldolase gene regulated by high-salinity and drought in *Sesuvium portulacastrum*. Plant Cell Rep 28:975–984

Fang ZQ, Yuan LY, Hong PC, Ming LC, Shan WB (2005) NaCl enhances thylakoid-bound SOD activity in the leaves of C3 halophyte *Suaeda salsa* L. Plant Sci 168:423–430

FAO (2008) FAO land and plant nutrition management service. Available at: http://www.fao.org/ag/agl/agll/spush. Accessed 25 Apr 2008

Farnsworth EJ, Ellison AM (1997) The global conservation status of mangroves. Ambio 26:328–334

Fazzari M, Greally J (2004) Epigenomics: beyond CpG islands. Nat Rev Genet 5:446–455

Fink RC, Scandalios JG (2002) Molecular evolution and stnicturefunction relationships of the superoxide dismutase gene families in Angiosperms and their relationship to other eukaryotic and prokaryotic superoxide dismutases. Arch Biochem Biophys 399:19–36

Frugoli JA, Zhong HH, Nuccio ML, McCourt R, McPeek MA, Thomas TL, McClung CR (1996) Catalase is encoded by a multigene family in *Arabidopsis thaliana* (L.) Heyuh. Plant Physiol 112:327–336

Fu XH (2006) Cloning and functional expression of salt tolerance related genes in *Aegiceras corniculatum* and their adaptive evolution analysis (in Chinese). Ph. D. Thesis. Sun Yat-sen University, Guangzhou

Fu X, Huang Y, Deng S, Zhou R, Yang G, Ni X, Li W, Shi S (2005) Construction of a SSH library of *Aegiceras corniculatum* under salt stress and expression analysis of four transcripts. Plant Sci 169:147–154

Fukushima Y, Sasamoto H, Baba S, Ashihara H (1997) The effect of salt stress on the catabolism of sugars in leaves and roots of the mangrove plant, *Avicennia marina*. Z. Naturforsch 52:187–192

Gaymard F, Boucherez J, Briat JF (1996) Characterization of a ferritin mRNA from *Arabidopsis thaliana* accumulated in response to iron through an oxidative pathway independent of abscisic acid. Biochem J 318:67–73

Ge XJ, Sun M (1999) Reproductive biology and genetic diversity of a cryptoviviparous mangrove *Aegiceras corniculatum* (Myrsinaceae) using allozyme and intersimple sequence repeat (ISSR) analysis. Mol Ecol 8:2061–2069

Ge X (2001) Genetic diversity and conservation genetics of mangrove species in South China and Hong Kong. University of Hong Kong, Hong Kong

Gerlach WL, Bedbrook JR (1979) Cloning and characterization of ribosomal RNA genes from wheat and barley. Nucl Acids Res 7(7):1869–1885

Gillespie JH (1986) Variation of evolutionary rates of DNA. Genetics 112:1077–1091

Gong Q, Li P, Ma S, Rupassara SI, Bohnert HJ (2005) Salinity stress adaptation competence in the extremophile *Thelungiella halophila* in comparison with its relative *Arabidopsis thaliana*. Plant J 44:826–839

Granot G, Sikron-Persi N, Gaspan O, Florentin A, Talwara S, Paul LK, Morgenstern Y, Granot Y, Grafi G (2009) Histone modifications associated with drought tolerance in the desert plant *Zygophyllum dumosum* Boiss. Planta 23:27–34

Guan LM, Zhao J, Scandalios JG (2000) Cis-elements and trans-factors that regulate expression of the maize Call antioxidant gene in response to ABA and osmotic stress: H_2O_2 is the likely intermediary signaling molecule for the response. Plant J 22(2):87–95

Guan L, Scandalios JG (1993) Characterization of the catalase antioxidant defense gene *Cat1* of maize, and its developmentally regulated expression in transgenic tobacco. Plant J 3:527–536

Guo M, Zhou R, Huang Y, Ouyang J, Shi S (2011) Molecular confirmation of natural hybridization between *Lumnitzera racemosa* and *Lumnitzera littorea*. Aquat Bot 95:59–64

Hasegawa M, Bressan R, Pardo JM (2000a) The dawn of plant salt tolerant genetics. Trends Plant Sci 5:317–319

Hasegawa PM, Bressan RA, Zhu JK, Bohnert HJ (2000b) Plant cellular and molecular responses to high salinity. Annu Rev Plant Physiol Plant Mol Biol 51:463–499

Hasegawa PM, Bressan RA, Zhu JK, Bohnert HJ (2000c) Plant cellular and molecular responses to high salinity. Annu Rev Plant Physiol Plant Mol Bio 51:463–499

Havir EA, McHale NA (1987) Biochemical and developmental characterization of multiple forms of catalase in tobacco leaves. Plant Physiol 84:450–455

Hibino T, Meng YL, Kawamitsu Y, Uehara N, Matsuda N, Taiiaka V (2001) Molecular cloning and functional characterization of two kinds of betaine-aldehyde dehydrogenase in betaine-accumulating mangrove *Avicennia marina* (Forsk.) Vierh. Plant Mol Biol 45:353–363

Hiei Y, Ohta S, Komari T, Kumashiro T (1994) Efficient transformation of rice (*Oryza sativa* L.) mediated by *Agrobacterium* and sequence analysis of the boundaries of the T-DNA. Plant J 6:271–282

Hoffmann-Sommergruber K (2002) Pathogenesis-related (PR)-proteins identified as allergens. Biochem Soc Trans 30:930–935

Hogarth PJ (1999) The biology of mangroves, 1st edn. Oxford University Press, Oxford

Holmberg N, Bulow L (1998) Improving stress tolerance in plants by gene transfer. Trends Plant Sci 3:61–66

Huang G, Wang Y, Ying G (2011) Cadmium-inducible *BgMT2, a* type 2 metallothionein gene from mangrove species (*Bruguiera gymnorrhiza*), its encoding protein shows metal-binding ability. J Exp Marine Biol Ecol 405:128–132

Huang GY, Wang YS (2009) Expression analysis of type 2 metallothionein gene in mangrove species (*Bruguiera gymnorrhiza*) under heavy metal stress. Chemosphere 77:1026–1029

Huang GY, Wang YS (2010) Expression and characterization analysis of type 2 metallothionein from grey mangrove species (*Avicennia marina*) in response to heavy metal stress. Aquat Toxicol 99:86–92

Huang W, Fang XD, Li GY, Lin QF, Zhao WM (2003) Cloning and expression analysis of salt responsive gene from *Kandelia candel*. Biol Plant 47:501–507

Huang Y, Tan F, Su G, Deng S, He H, Shi S (2008) Population genetic structure of three tree species in the mangrove genus *Ceriops* (Rhizophoraceae) from the Indo West Pacific. Genetica 133:47–56

Hurst AC, Grams TEE, Ratajczak R (2004) Effects of salinity, high irradiance, ozone, and ethylene on mode of photosynthesis, oxidative stress and oxidative damage in the C3/CAM intermediate plant *Mesembryanthemum crystallinum* L. Plant Cell Environ 27:187–197

Jena S, Das AB (2006) Intra-population variation of chromosome and RAPD markers of *Suaeda nudiflora* (Willd.) Moq. a mangrove species in India. African J Agri Res 1(4):137–142

Jithesh MN, Prashanth SR, Sivaprakash KR, Panda AK (2006) Monitoring expression profiles of antioxidant genes to salinity, iron, oxidative, light and hyperosmotic stresses in the highly salt tolerant grey mangrove, *Avicennia marina* (Forsk.) Vierh. by mRNA analysis. Plant Cell Rep 25:865–876

Johnson LAS, Briggs BG (1984) Myrtales and Myrtaceae—a phylogenetic analysis. Ann Mo Bot Gard 71:700–756

Kahrood HV, Korori SAA, Pirseyedi M, Shirvany A, Danehkar A (2008) Genetic variation of mangrove species *Avicennia marina* in Iran revealed by microsatellite markers. Afr J Biotech 7(17):3017–3021

Kant S, Kant P, Raveh E, Barak S (2006) Evidence that differential gene expression between the balophyte *Thellungiella halophila* and *Arabidopsis thaliana* is responsible for higher levels of the compatible osmolyte proline and tight control of Na+ uptake in *T. halophila*. Float. Cell Environ 29(7):1220–1234

Kapazoglou A, Tondelli A, Papaefthimiou D, Ampatzidou H, Francia E, Stanca MA, Bladenopoulos K, Tsaftaris AS (2010) Epigenetic chromatin modifiers in barley: IV. The

study of barley polycomb group (PcG) genes during seed development and in response to external ABA. BMC Plant Biol 10:73–78
Kathiresan K, Qasim (2005) Biodiversity of mangrove ecosystems. Hindustan Publishing Corporation, New Delhi
Kawano T (2003) Roles of the reactive oxygen species-generating peroxidase reactions in plant defense and growth induction. Plant Cell Rep 21:829–837
Kennedy CH, Maples KR, Mason RP (1990) In vivo detection of free radical metabolites. Pure Appl Chem 62:295–299
Kim JM, To TK, Ishida J, Morosawa T, Kawashima M, Matsui A, Toyoda T, Kimura H, Shinozaki K, Seki M (2008) Alterations of lysine modifications on the histone H3N-tail under drought stress conditions in *Arabidopsis thaliana*. Plant Cell Physiol 49:1580–1588
Kim JM, To TK, Nishioka T, Seki M (2010) Chromatin regulation functions in plant abiotic stress responses. Plant Cell Environ 33:604–611
Kliebenstein DJ, Monde RA, Last RL (1998) Superoxide dismutase in *Arabidopsis*: an eclectic enzyme family with disparate regulation and protein localization. Plant Physiol 118:637–650
Klotz MG, Klassen GR, Loewen PC (1997) Phylogenetic relationships among prokaryotic and eukaryotic catalases phytogeny. Mol Biol Evol 14:951–958
Koreeda S, Cushman MA, Akselrod I, Bufford D, Fredrickson M, Clark E, Cushman JC (2004) Transcript profiling of salinity stress responses by large-scale expressed sequence tag analysis in *Mesembryanthemum crystallinum*. Gene 341:83–92
Lakshmi M, Parani M, Nivedita R, Parida A (2000) Molecular phylogeny of mangroves. VI. Intraspecific genetic variation in mangrove species *Excoecaria agallocha* L. (Euphorbiaceae). Genome 43:110–115
Lakshmi M, Parani M, Senthilkumar P, Parida A (2002) Molecular phylogeny of mangroves VIII: analysis of mitochondrial DNA variation for species identification and relationships in Indian mangrove *Rhizophoraceae*. Wetlands Ecol Manag 10:355–362
Lakshmi M, Rajalakshmi S, Parani M, Anuratha CS, Parida A (1997) Molecular phylogeny of mangroves I. Use of molecular markers in assessing the intraspecific genetic variability in the mangrove species. *Acanthus ilicifolius* Linn. (Acanthaceae). Theor Appl Genet 94:1121–1127
Li H, Chen G (2004) Genetic diversity of *Sonneratia alba* in China detected by Inter-simple Sequence Repeats (ISSR) analysis. Acta Bot Sinica 46(5):515–521
Li HS, Chen GZ (2008) Genetic relationship among species in the genus *Sonneratia* in China as revealed by inter-simple sequence repeat (ISSR) markers. Biochem Syst Ecol 36:392–398
Liang S (2007) Transcript profile of *Ceriops tagal* in response to salinity and its implication for adaptive evolution. Ph. D Thesis. Sun Yat-sen University, Guangzhou (in Chinese)
Liangs S, Zhou R, Dong S, Shi S (2008) Adaptation to salinity in mangroves: mplication on the evolution of salt-tolerance. Chin Sci Bull 53(11):1708–1715
Liao PC, Havanond S, Huang S (2006) Phylogeography of *Ceriops tagal* Rhizophoraceae) in Southeast Asia: the land barrier of the Malay Peninsula has caused population differentiation between the Indian Ocean and South China Sea. Conserv Genet 8:89–98
Lira-Medeiros CF, Parisod C, Fernandes RA, Mata CS, Cardoso MA, Ferreira PCG (2010) Epigenetic variation in mangrove plants occurring in contrasting natural environment. PLoS ONE 5(4):e10326
Lo EY (2010) Testing hybridization hypotheses and evaluating the evolutionary potential of hybrids in mangrove plant species. J Evol Biol 23:2249–2261
Maguire TL, Peakall R, Saenger P (2002) Comparative analysis of genetic diversity in the mangrove species *Avicennia marina* (Forsk.) Vierh. (Avicenniaceae) detected by AFLPs and SSRs. Theor Appl Genet 104:388–398
Maguire TL, Edwards KJ, Saenger P, Henry R (2000a) Characterization and analysis of microsatellite loci in a mangrove species *Avicennia marina* (Forsk.) Vierh. (Avicenniaceae). Theor Appl Genet 101:279–285
Maguire TL, Saenger P, Baverstock P, Henry R (2000b) Microsatellite analysis of genetic structure in the mangrove species *Avicennia marina* (Forsk.) Vierh. (Avicenniaceae). Mol Ecol 9:1853–1862

Markley JL, McMillan C, Thompson GA Jr (1982) Latitudinal differentiation in response to chilling temperatures among populations of three mangroves, *Avicennia germinans*, *Laguncularia racemosa*, and *Rhizophora mangle* from the western tropical Atlantic and Pacific Panama. Can J Bot 60:2704–2715

McMillan C (1986) Isozyme patterns among populations of black mangrove, *Avicennia germinans* from the gulf of Mexico-Caribbean and Pacific Panama Contrib. Mar Sci 29:17–25

Meher-Homji VM (1988) The Pichavaram mangroves. Blackbuck 4:1–12

Mehta PA, Sivaprakash K, Parani M, Venkataraman G, Parida AK (2005) Generation and analysis of expressed sequence tags from the salt-tolerant mangrove species *Avicennia marina* (Forsk) Vierh. Theor Appl Genet 110:416–424

Mishra S, Das AB (2004) Effect of short-term exposure to NaCl on photochemical activity and antioxident enzymes in Bruguiera parviflora, a non-secretor mangrove. Acta Physiol Plant 26(3):317–326

Mishra S, Das AB (2003) Effect of NaCl on leaf secretion and antioxidative enzyme level in roots of a mangrove, *Aegiceras corniculatum*. Indian J Exp Biology 41:160–166

Mittler R (2002) Oxidative stress, antioxidants and stress tolerance. Trends Plant Sci 7(9):405–410

Miyama M, Hanagata N (2007) Microarray analysis of 7029 gene expression patterns in Burma mangrove under high-salinity stress. Plant Sci 172:948–957

Miyama M, Simizu H, Sugiyama M, Hanagata N (2006) Sequencing and analysis of 14,842 expressed sequence tags of Burma mangrove, *Bruguiera gymnorrhiia*. Plant Sci 171:234–241

Miyama M, Tada Y (2008) Transcriptional and physiological study of the response of Burma mangrove (*Bruguiera gymnorhiza*) to salt and osmotic stress. Plant Mol Biol 68:119–129

Mukherjee AK, Acharya LK, Mattagajasingh I, Panda PC, Mohapatra T, Das P (2003) Molecular characterization of three Heritiera species using AFLP markers. Biol Plant 47(3):445–448

Mukherjee AK, Acharya LK, Panda PC, Mohapatra T, Das P (2004) Genomic relations among two non- mangroves and nine mangroves species of Indian Rhizophoraceae. Z. fur Naturforsch C 59:572–578

Munns R (2002) Comparative physiology of salt and water stress. Plant Cell Environ 25:239–250

Nguyen PD, Ho CL, Harikrishna JA, Wong MCVL, Rahim RA (2006) Generation and analysis of expressed sequence tags from the mangrove plant, *Acanthus ebracteatus* Vahl. Tree Genet Genomes 2:196–201

Ochman H, Wilson AC (1987) Evolution on bacteria: evidence for a universal substitution rate in celluar genomes. Mol Evol 26:74–86

Ohlrogge J, Benning C (2000) Unraveling plant metabolism by EST analysis. Curr Opin Plant Biol 3:224–228

Ota Y, Ario T, Hayashi K, Nakagawa T, Hattori T, Maeshima M, Asahi T (1992) Tissue-specific isoforms of catalase subunits in castor bean seedlings. Plant Cell Physiol 33:225–232

Paramonova NV, Shevyakova NI, Kuznetsov VV (2004) Ultrastructure of chloroplasts and their storage inclusions in the primary leaves of *Mesembryanthemum crystallinum* affected by putrescine and NaCl. Russ J Plant Physiol 1:86–96

Parani M, Jithesh MN, Lakshmi M, Parida A (2002) Cloning and characterization of a gene encoding ubiquitin conjugating enzyme from the mangrove species, *Avicennia marina* (Forsk.) Vierh. Ind J Biotech 1:164–169

Parani M, Lakshmi M, Senthilkumar P, Nivedita R, Parida A (1998) Molecular phylogeny of Mangroves VI: analysis of genome relationships in mangrove species using RAPD and RFLP markers. Theor Appl Genet 97:617–625

Parani M, Lakshmi M, Zeigenhagen B, Faldung M, Senthilkumar P, Parida A (2000) Molecular phylogeny of mangroves VII. PCR-RFLP of trnS-psbC and rbcL gene regions in 24 mangrove and mangrove-associate species. Theor Appl Genet 100:454–460

Parani M, Rao CS, Mathan N, Anuratha CS, Narayanan KK, Parida A (1997) Molecular phylogeny of mangroves III parentage analysis of a Rhizophora hybrid using random amplified polymorphic DNA and restriction fragment length polymorphism markers. Aquat Bot 58:165–172

Parida AK, Das AB (2005) Salt tolerance and salinity effects on plants: a review. Ecotox Environ Safe 60(3):324–349
Parida AK, Das AB, Mohanty P (2004a) Defense potentials to NaCl in a mangrove, Bruguiera parviflora: differential changes of iso-forms of some antioxidative enzymes. J Plant Physiol 161:531–542
Parida AK, Das AB, Das P (2003a) Salt induced regulations of isoforms of some antioxidative enzymes. Ind J Plant Physiol (Special issue):288–297
Parida AK, Das AB (2004) Effects of NaCl stress on nitrogen and phosphorous metabolism in a true mangrove Bruguiera parviflora under hydroponic culture. J Plant Physiol 161:921–928
Parida AK, Das AB, Mitra B, Mohnaty P (2004b) Salt stress induced alterations in protein profile and protease activity in the mangrove *Bruguiera parviflora*. Z. Naturforsch c 59:408–414
Parida AK, Das AB, Mittra B (2003b) Effects of NaCl stress on the structure, pigment complex composition, and photosynthetic activity of mangrove *Bruguiera parviflora* chloroplasts. Photosynthetica 41(2):191–200
Parida AK, Das AB, Mittra B (2004c) Effects of salt on groath, ion accumulations, photosynthesis and leaf anatomy in the mangrove *Bruguiera parviflora*. Trees-Struct Funct 18:167–174
Parida AK, Das AB, Mohanty P (2004d) Defense potentials to NaCl in a mangrove, *Bruguiera parviflora*: differential changes of isoforms of some antioxidative enzymes. J Plant Physiol 161(91):10450–10454
Parida AK, Das AB, Mohanty P (2004e) Defense potential to NaCl in a mangrove, *Bruguiera parviflora*: differential changes of isoforms of some antioxidative enzymes. J Plant Physiol 161(5):531–542
Parida AK, Das AB, Sanada Y, Mohnaty P (2004f) Effects of salinity on biochemical components of the mangrove, *Aegiceras corniculatum*. Aquat Bot 80(2):77–87
Parida AK, Das AB, Mohnaty P (2004g) Investigations on the antioxidative defense responses to NaCl stress in a mangrove, *Bruguiera parviflora*: differential regulations of isoforms of some antioxidative enzymes. Plant Growth Regul 42(3):213–226
Parida AK, Jha B (2010a) Salt tolerance mechanisms in mangroves: a review. Trees 24:199–217
Parida AK, Jha B (2010b) Antioxidative defense potential to salinity in the euhalophyte *Salicornia brachiata*. J Plant Growth Regul 29:137–147
Parida AK, Mittra B, Das AB, Das TK, Mohnaty P (2005) High salinity reduces the content of a highly abumdant 23-kDa protein of the mangrove *Bruguiera parviflora*. Planta 221:135–140
Parida A, Anuratha CS, Lakshmi M, Parani M, Kurjen J (1995) Application of molecular markers in assessing genetic diversity in Indian mangroves. In: Use of induced mutations and molecular techniques in crop improvement, FAO/IAEA, Vienna, pp 595–600
Parida A, Das AB, Das P (2002) NaCl stress on photosynthetic pigments, proteins and other metabolic changes in the leaves of *Bruguiera parviflora*: studies under hydroponic culture conditions. J Plant Biol 45:28–36
Parida A, Parani M, Lakshmi M, Nivedita R, Elango S, Anuratha CS (1998a) Nature and extent of genetic variation and species diversity in Indian mangroves. IAEA Techdoc 1047:95–105
Parida A, Parani M, Laxmi M, Nivedita R, Elango S, Anuratha CS (1998b) Molecular phylogeny of mangrove IV. Nature and extend of genetic variation and species diversity in Indian mangroves. In: Molecular markers in crop improvement, FAO/IAEA, Vienna
Petit JM, Briat JF, Lobreaux S (2001) Structure and differential expression of the four members of the *Arabidopsis thaliana* ferritin gene family. Biochem J 359:575–582
Prashanth SR, Sadhasivam V, Panda A (2008a) Over expression of cytosolic copper/zinc superoxide dismutase from a mangrove plant *Avicennia marina* in indica Rice var Pusa Basmati-1 confers abiotic stress tolerance. Transgenic Res 17:281–291
Prashanth SR, Sadhasivam V, Parida A (2008b) Over expression of cytosolic copper/zinc superoxide dismutase from a mangrove plant *Avicennia marina* in indica Rice var Pusa Basmati-1 confers abiotic stress tolerance. Transgenic Res 17:281–291
Preeti AM, Sivaprakash KR, Parani M, Venkataraman Gayatri, Parida A (2005) Generation and analysis of expressed sequence tags from the salt-tolerant mangrove species *Avicennia marina* (Forsk) Vierh. Theor Appl Genet 110:416–424

Qiu S, Zhou RC, Li YQ, Havanond S, Jaengjai C, Shi SH (2008) Molecular evidence for natural hybridization between *Sonneratia alba* and *S. griffithii*. J Syst Evol 46:391–395

Rabbani MA, Maruyama K, Abe H, Khan MA, Katsura K, Ho Y, Yoshiwara K, Seki M, Shinozaki K, Yamaguchi-Shinozaki K (2003) Monitoring expression profiles of rice genes under cold, drought, and high-salinity stresses and abscisic acid application using cDNA microarray and RNA gel-blot analyses. Plant Physiol 133:1755–1767

Sahoo P, Jena S, Mohanty S, Das AB (2007) Molecular phylogenetic relationships among the four species of *Bruguiera* L. of the family Rhizophoraceae as revealed by RAPD, genome size, karyotype, Isoenzyme and Protein polymorphism. Rev Biol Trop 55:437–448

Salin ML (1987) Toxic oxygen species and protective systems of the chloroplast. Physiol Plant 72:681–689

Scandalios J, Tong WJ, Roupaklas DG (1980) Cat3, a third gene locus coding for a tissue-specific catalase in maize: genetics, intercellular location, and some biochemical properties. Mol Gen Genet 179:33–41

Schwarzbach AE, Ricklefs RF (2000) Systematic affinities of Rhizophoraceae and Anisophylleaceae and Intergeneric relationships within Rhizophoraceae, based on chloroplast DNA, nuclear Ribosomal DNA and morphology. Am J Bot 87(4):547–564

Senthilkumar P, Jithesh MN, Parani M, Rajalakshmi S, Praseetha K, Parida AK (2005) Salt stress effects on the accumulation of vacuolar H+-ATPase subunit c transcripts in wild rice, *Porteresia coarctata* (Roxb.) Tateoka. Curr Sci 89:1386–1394

Sharma R, Mohan Singh RK, Malik G, Deveshwar P, Tyagi AK, Kapoor S, Kapoor M (2009) Rice cytosine DNA methyltransferases—gene expression profiling during reproductive development and abiotic stress. FEBS J 276:6301–6311

Shi S, Huang Y, Zeng K, Tan F, He H, Huang J, Fu Y (2005) Molecular phylogenetic analysis of mangroves: independent evolutionary origins of vivipary and salt secretion. Mol Phylogenet Evol 34:159–166

Slesak I, Miszalski Z, Karpinska B, Niewiadomska E, Ratajczak R, Karpinski S (2002) Redox control of oxidative stress responses in the C3-CAM intermediate plant *Mesembryanthemum crystallinum*. Plant Physiol Biochem 40:669–677

Sokol A, Kwiatkowska A, Jerzmanowski A, Prymakowska-Bosak M (2007) Up-regulation of stress-inducible genes in tobacco and Arabidopsis cells in response to abiotic stresses and ABA treatment correlates with dynamic changes in histone H3 and H4 modifications. Planta 227:245–254

Strand AE, Leebens-Mack J, Milligan BG (1997) Nuclear DNA based markers for plant evolutionary biology. Mol Ecol 6:113–118

Streb P, Knauf AM, Feierabend J (1993) Preferential photoin photoinactivation of catalase and photoinhibition of photosystem II are common early symptoms under various osmotic and chemical stress conditions. Physiol Plant 88:590–598

Su GH, Huang YL, Tan FX, Ni XW, Tang T, Shi SH (2006) Genetic variation in *Lumnitzera racemosa*, a mangrove species from the Indo-West Pacific. Aquat Bot 84:341–346

Sugihara K, Hanagata N, Dubinsky Z, Baba S, Karube I (2000) Molecular characterization of cDNA encoding oxygen evolving enhancer protein 1 increased by salt treatment in the mangrove *Bruguiera gymnorrhiza*. Plant Cell Physiol 41:1279–1285

Taberlet P, Gielly L, Pautou G, Bouvet J (1991) Universal primers for amplification of three non-coding regions of chloroplast DNA. Plant Mol Biol 17:1105–1109

Taji T, Seki M, Satou M, Sakurai T, Kobayashi M, Ishiyama K, Narusaka Y, Narusaka M, Zhu JK, Shinozaki K (2004) Camparative genomics in salt tolerance between Arabidopsis and Arabidopsis-related halophyte salt cress using Arabidopsis microarray. Plant Physiol 135:1697–1709

Takemura T, Hanagata N, Dubinsky Z, Karube I (2002) Molecular characterization and response to salt stress of mRNAs encoding cytosolic Cu/Zn superoxide dismutase and catalase from *Bruguiera gymnorrhiza*. Trees 16:94–99

Takemura T, Hanagata N, Sugihara K, Baba S, Karube I, Dubinsky Z (2000) Physiological and biochemical responses to salt stress in the mangrove, *Bruguiera gymnorrhiza*. Aquat Bot 68:15–28

Tan FX, Huang YL, Ge XJ, Su GH, Ni XW, Shi SH (2005) Population genetic structure and conservation implications of *Ceriops decandra* in Malay Peninsula and North Australia. Aquat Bot 81:175–188

Tanaka Y, Fukuda A, Nakamura A, Yamada A, Saito T, Ozeki Y, Mimura T (2000) Molecular cloning and characterization of mangrove Na/H antiporter cDNA. Plant Cell Physiol 41:27–32

Tanaka Y, Hibino T, Hayashi Y, Tanaka A, Kishitani S, Takabe T, Yokota S, Takabe T (1999) Salt tolerance of transgenic rice overexpressing yeast mitochondrial Mn-SOD in chloroplasts. Plant Sci 148:131–138

Thorne RF (1992) An updated phylogenetic classification of the flowering plants. Aliso 13:365–389

Tobe H, Raven PH (1988a) Floral morphology and evolution in Anisophylleaceae. Bot J Linnean Soc 98:1–25

Tobe H, Raven PH (1988b) Seed morphology and anatomy of Rhizophoraceae, inter- and intrafamilial relationships. Ann Mo Bot Gard 75:1319–1342

Tomlinson PB (1986) The botany of mangroves. Cambridge University Press, Cambridge

Tomlinson PB, Bunt JS, Primack RB, Duke NC (1978) *Lumnitzera* × *rosea* (Combretaceae)—its status and floral morphology. J Arnold Arboretum 59:342–351

Triest L (2008) Molecular ecology and biogeography of mangrove trees towards conceptual insights on gene flow and barriers: a review. Aquat Bot 89(2):138–154

Usha B, Keeran NS, Harikrishnan M, Kavitha K, Parida A (2011) Characterization of a type 3 metallothionein isolated from *Porteresia coarctata*. Biol Plant 55(1):119–124

Van Breusegem F, Vranova E, Dat JF, Inz′e D (2001) The role of active oxygen species in plant signal transduction. Plant Sci 161:405–414

Van-Camp W, Capiau K, Van-Montagu M, Inze D, Slooten L (1996) Enhancement of oxidative stress tolerance in transgenic tobacco plants overproducing Fe-superoxide dismutase in chloroplasts. Plant Physiol 112:1703–1714

Waditee R, Hibino T, Tanaka Y, Nakamura T, Incharoensakdi A, Hayakawa S, Suzuki S, Futsuhara Y, Kawamitsu Y, Takabe T, Takabe T (2002) Functional characterization of betaine/proline transporters in betaine-accumulating mangrove. J Biol Chem 277:18373–18382

Wang B, Davenport RJ, Volkov V, Amtmann A (2006) Low unidirectional sodium influx into root cells restricts net sodium accumulation in *Thellungiella halophila*, a salt-tolerant relative of Arabidopsis thaliana. J Exp Bot 57:1161–1170

Wang B, Luttge U, Ratajczak R (2004a) Specific regulation of SOD isoforms by NaCl and osmotic stress in leaves of the C3 halophyte *Suaeda salsa* L. J Plant Physiol 161:285–293

Wang MC, Peng ZY, Li CL, Li F, Liu C, Xia GM (2008) Proteomic analysis on a high salt tolerance introgression strain of *Triticum aestivum/Thinopyrum ponticum*. Proteomics 8:1470–1489

Wang WX, Vinocur B, Altman A (2003) Plant responses to drought, salinity and extreme temperatures: towards genetic engineering for stress tolerance. Planta 218:1–14

Wang Y, Ying Y, Chen J, Wang XC (2004b) Transgenic Arabidopsis overexpressing Mn-SOD enhanced salt-tolerance. Plant Sci 167:671–677

Wang ZL, Li PH, Fredricksen M, Gong ZZ, Kim CS, Zhang C, Bohnert HJ, Zhu JK, Bressan RA, Hasegawa PM, Zhao YX, Zhang H (2004c) Expressed sequence tags from *Thellungiella halophila*, a new model to study plant salt-tolerance. Plant Sci 166:609–616

Wang Z, Gerstein M, Snyder M (2009) RNA-Seq: a revolutionary tool for transcriptomics. Nat Rev Genet 10:57–63

Willekens H, Chamnongpol S, Davey M, Schraudner M, Langebartels C, Van Montagu M (1997) Catalase is a sink for H_2O_2 and is indispensable for stress defense in C3 plants. EMBO J 16:4806–4816

Willekens H, Inze D, Van Montagu M, Van Camp W (1995) Catalase in plants. Mol Breed 1:207–228

Willekens H, Langebartels C, Tire C, Van Montagu M, Inze D, Van Camp W (1994a) Differential expression of catalase genes in *Nicotiana plumbaginifolia* (L.). Proc Natl Acad Sci USA 91:10450–10454

Willekens H, Villarroel R, Van Montagu M, Inze D, Van Camp W (1994a) Molecular identification of catalases from *Nicotiana plumbaginifolia* (L.). FEES Lett 352(1):79–83

Winter K, Willert DJ (1972) NaCl induzierter Crassulaceen Sa urestoffwechsel bei *Mesembryanthemum crystallinum*. Z Pflanzenphysiol 67:166–170

Wong CE, Li Y, Whitty BR, Diaz-Camino C, Akhter SR, Brandle JE (2005) Expressed sequence tags from the Yukon ecotype of Thellungiella reveal that gene expression in response to cold, drought and salinity shows little overlap. Plant Mol Biol 58:561–574

Wu W, Dai S, Bao H, Zhou RC (2009) Hybrid status of *Tibouchina urvilleana* Cogn. revealed by ITS sequences of nuclear ribosomal DNA. Biochem Syst Ecol 37:640–644

Wydrzynski T, Angstrom J, Vangaurd T (1989) H_2O_2 formation by photosystem II. Biochim Biophys Acta 1973:23–28

Xiong L, Zhu JK (2002) Molecular and genetic aspects of plant responses to osmotic stress. Plant Cell Environ 25:131–139

Yamada A, Saitoh T, Mimura T, Ozeki Y (2002a) Detection of differences in mRNA expression regulated by salt-stress in mangrove cultured cells. Plant Biotech 19:145–148

Yamada A, Saitoh T, Mimura T, Ozeki Y (2002b) Expression of mangrove allene cyclase enhances salt tolerance in *Escherichia coli*, yeast and tobacco cells. Plant Cell Physiol 43:903–910

Yamada A, Sekiguchi M, Mimura T, Ozeki Y (2002c) The role of plant CCTa in salt- and osmotic-stress tolerance. Plant Cell Physiol 43:1043–1048

Yamanaka T, Miyama M, Tada Y (2009) Transcriptome profiing of the mangrove plant *Bruguiera gymnorhiza* and identification of salt stress genes by *Agrobacterium* functional screening. Biosci Biotechnol Biochem 73(2):304–310

Yan L, Guizhu C (2007) Physiological adaptability of three mangrove species to salt stress. Acta Ecologica Sinica 27(6):2208–2214

Yang GL (2007) Comparative genomics in salt tolerance between two ecotypes of *Hibiscus tiliaceus* using cDNA microarray. PhD Thesis. Sun Yat-sen University, Guangzhou

Yang G, Zhou R, Tang T, Chen X, Ouyang J, He L, Li W, Chen S, Guo Li M, Zhong C, Shi S (2011) Gene expression profiles in response to salt stress in hibiscus tiliaceus. Plant Mol Biol Rep 29:609–617

Yenney-Esinsine (1980) Elements of tropical ecology. Heine-mann Educational Books, London

Zeng HC, Deng LH, Zhang CF (2006) Cloning of salt tolerance related cDNAs from the Mangrove plant *Sesuvium portulacastrum* L. J Integr Plant Biol 48:952–957

Zhang L, Ma XL, Zhang Q, Ma CL, Wang PP, Sun YF, Zhao YX (2001) Expressed sequence tags from a NaCl-treated Suaeda salsa cDNA library. Gene 267:193–200

Zhang ZH, Zhou RC, Tang T, Huang YL, Zhong Y, Shi SH (2008) Genetic variation in central and peripheral populations of *Excoecaria agallocha* from Indo-West Pacific. Aquat Bot 89:57–62

Zhou R, Gong X, Boufford D, Wu CI, Shi S (2008) Testing a hypothesis of unidirectional hybridization in plants: observations on *Sonneratia, Bruguiera* and *Ligularia*. BMC Evol Biol 8:149–153

Zhou R, Shi S, Wu CI (2005) Molecular criteria for determining new hybrid species-an application to the *Sonneratia* hybrids. Mol Phylogenet Evol 35:595–601

Zhu JK (2001) Plant salt tolerance. Trends Plant Sci 16:66–67

Zhu JK (2002) Salt and drought stress signal transduction in plants. Annu Rev Plant Biol 53:247–273

Zhu JK, Hasegawa PM, Bressan RA (1997) Molecular aspects of osmotic stress. CRC Crit Rev Plant Sci 16:253–277

Zhu JK, Liu J, Xiong L (1998) Genetic analysis of salt tolerance in Arabidapsis. Evidence for a critical role of potassium nutrition. Plant Cell 10:1181–1191

Ziegenhagen B, Fladung M (1997) Variation in the psbC gene region of gymnosperms and angiosperms as detected by a single restriction site poly-morphism. Theor Appl Genet 94:1065–1071

Zimmermann P, Heinlein C, Orendi G, Zentgraf U (2006) Senescence-specific regulation of catalases in *Arabidopsis thaliana* (L.) Heynh. Plant Cell Environ 29:1049–1060

Zolgharnein H, Kamyab M, Keyvanshokooh S, Ghasemi A, Nabavi SMB (2010) Genetic diversity of *Avicennia marina* (Forsk.) Vierh. populations in the persian gulf by microsatellite markers. J Fish Aquat Sci 5:223–229

Index

S. M. Jain and S. Dutta Gupta (eds.), *Biotechnology of Neglected and Underutilized Crops*, DOI: 10.1007/978-94-007-5500-0,

The manufacturer's authorised representative in the EU is Springer Nature Customer Service Centre GmbH, Europaplatz 3, 69115 Heidelberg, Germany. If you have any concerns regarding our products, please contact ProductSafety@springernature.com

Printed and bound by CPI Group (UK) Ltd, Croydon, CR0 4YY
22/09/2026
02220979-0002